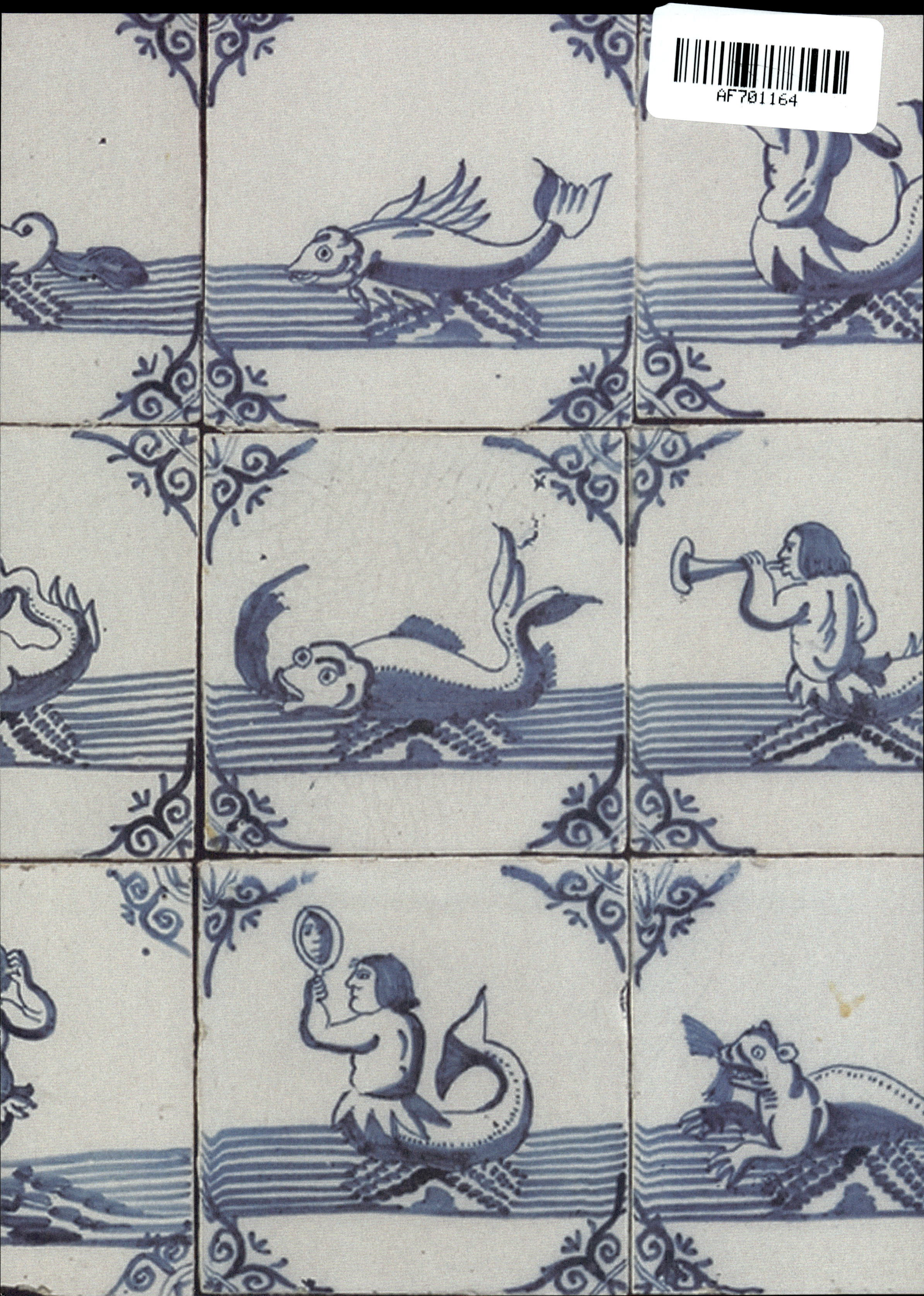

Ceramics in America

CERAMICS IN AMERICA 2006

Edited by Robert Hunter

Published by the CHIPSTONE FOUNDATION

Distributed by University Press of New England

Hanover and London

Cover Illustration: Face jug fragments, Miles Mill, Edgefield, South Carolina, ca. 1867–1872. (Private collection; photo, Gavin Ashworth.)

Design: Wynne Patterson, VT
Copyediting: Mary Gladue, CT
Proofreading: Fronia W. Simpson, VT
Printing: Balding + Mansell, Norwich, UK
Type: Aardvark Type, CT

Published by the Chipstone Foundation
7820 North Club Circle, Milwaukee, WI 53217

Printed in England 5 4 3 2 1
ISSN 1533-7154
ISBN 0-9724353-7-9

Contents

Editorial Statement

Ceramics in America is an interdisciplinary journal intended for collectors, historical archaeologists, curators, decorative arts students, social historians, and contemporary studio potters. Authors are encouraged to submit articles on the broad role of historical ceramics in the American context including essays on ceramic history, archaeological research, ceramic technology, social history, studio pottery, and ceramic collecting. Short, illustrated notes on new ceramic discoveries are particularly wanted. References to be compiled in an annual bibliography, including electronic sources, may also be submitted. Manuscripts must be typed, double-spaced, illustrated with high-quality color transparencies, and prepared in accordance with the *Chicago Manual of Style*. Computer disk copy is requested but not required. The Chipstone Foundation will offer significant honoraria for manuscripts accepted for publication and reimburse authors for all photography approved in writing by the editor.

Low-resolution digital images or photographs of poor quality will not be accepted for publication.

A style guide is available for prospective authors by request and can be transmitted via email.

Robert Hunter

Robert Hunter

Introduction

▼ FOR SEVERAL MILLENNIA ceramics have helped record the history of human culture. The shattered remains of clay pots have outlasted most other types of historical artifacts that embody information about general trends or specific people and places. As well, the heroes and myths of long-forgotten times often have been immortalized on ceramics, from the carved Moche pots of the Peruvian coast to the brightly painted wares of the Indus River Valley.

Yet the function of ceramics as documents of economic and political history or social and ideological rituals is not restricted to the distant past. During the formative years of American colonization, new ceramic forms were invented for imbibing tea, punch, and tobacco, and they were integrated into ritualized behaviors, becoming part of the everyday cultural landscape. With the onset of the Industrial Revolution, mass-produced commemorative ceramics now read like a concise account of the events in the Western world. Ceramics made and used in America, both new and old, continue to be a rich source for understanding the myths, mythology, and rituals of our own heritage.

It is perhaps fitting that the first article of the 2006 issue of *Ceramics in America* is an examination by Barbara Magid of ceramics made to honor the accomplishments and memory of George Washington, the most readily acknowledged hero in American mythology. Magid's viewpoint derives from pieces of broken commemorative objects that have been recovered in archaeological excavations in Washington's hometown of Alexandria, Virginia. From these sherds Magid demonstrates little-known connections to the Washington mythology that still thrives in Alexandria and elsewhere.

Besides glorifying political leaders, ceramics have also documented the most despicable of America's social ills. With the advent of transfer printing in the nineteenth century, many social, moral, and religious messages could quickly be disseminated among the middle classes. Many of these messages derived from literary works of the day. Jill Fenichell describes the outpouring of ceramics related to Harriet Beecher Stowe's *Uncle Tom's Cabin,* arguably the most important novel of the nineteenth century. The visual imagery taken from Stowe's book translated into a huge ceramic repertoire used in the cause for the abolishment of slavery.

A quite different connection between the written word and ceramics is examined by Arthur Goldberg and James Witkowski in their comprehensive analysis of the life and times of Southern potter David Drake, one of many slaves making large alkaline-glazed stoneware vessels in South

Carolina's Edgefield District in the nineteenth century. While most of these potters have remained anonymous, Dave is known from his signed pots, many of which he inscribed with personal, poetic expressions. As the authors poignantly explain, Dave's work is a contradictory legacy, for his inscriptions were written in a time when slave literacy was not permitted in South Carolina. Dave's pots show him to have been a master potter by any standard, but one who, during his prolific life, was bought and sold by several masters. Goldberg and Witkowski acknowledge the liberal stance of one of them, Lewis Miles, who at the risk of sanction permitted Dave to inscribe his name and poems for posterity.

Beyond the tales of individual potters, the history of specific forms tells a cultural story. John Burrison traces the jug from its beginnings in the Near East through Europe and the evolution of its form in America. In telling the story of this utilitarian vessel, Burrison makes connections with people and things and behaviors often overlooked. In particular he reviews the mystery behind the African-American face jug phenomenon, a distinctive horizon in American ceramic history and an almost ubiquitous emblem of the twentieth-century Southern "folk" potter. These African-American–made face jugs continue to tantalize collectors and historians as to their original purpose. Although the heyday of the ceramic jug has passed, Burrison's passionate participation with the traditional stoneware potters working in the rural South will help keep the jug's memory alive for future generations of material culture students.

For students of the jug nothing could be more exciting than Mark Newell's discovery of an archaeological site, located in the Edgefield District of South Carolina, where African-American face vessels were made. The first archaeologically recovered evidence of such activity, the face jug fragments Newell presents are attributed to a previously undiscovered John Miles pottery. In an attempt to understand the construction techniques of these fascinating objects, Newell enlisted contemporary potter Peter Lenzo to study and replicate the process. The several facsimile vessels that he created are illustrated here in marvelous detail and should facilitate assigning attributions of other extant face jugs to specific makers or places.

In a stark leap from the slave potteries of South Carolina to the opulent mansions of Maryland, Diana Edwards and Lynne Hastings take a meticulous look at the household ceramic acquisitions of the most prominent family in Baltimore during the eighteenth and nineteenth centuries. Using estate inventories and the surviving ceramics still preserved at Hampton Mansion, home of the Ridgely family, the authors examine what ceramics were chosen by succeeding Ridgely generations. Their analysis weaves together a story of changing styles, tastes, household economy, and familial connections. Such evidence is generally extremely rare and provides important insights into the specific acquisition behaviors of an American upper-class household.

As the above article illustrates, ceramics play many roles within a household and are integral to many daily rituals that we take for granted. In the seventeenth century, no single ceramic object had more cultural and ritual significance than the clay tobacco pipe. Al Luckenbach and Taft Kiser contribute an important new dimension to understanding the role of this implement in the Anglo-American Chesapeake. The locally made terracotta tobacco pipe, long a hallmark of seventeenth-century life in the Chesapeake, has been poorly understood. As many of these pipes are decorated with seemingly strange geometric and figurative designs, they have been the subject of much speculation. Through a careful analysis of archaeological evidence and corresponding information from sites in England, Luckenbach and Kiser provide a new framework for understanding the manufacture and distribution of these items. By rightfully elevating these decorative objects to "folk art," they hold great promise for attracting scholars from sister disciplines to participate in an ongoing dialogue.

In Part 1 of John Austin's look at ceramicist Palin Thorley's career, presented in the 2005 volume of *Ceramics in America,* we saw Thorley begin his professional life and work as an apprentice in 1906 with the Wedgwood factory and later travel to the United States to work as a designer for a number of American ceramic firms. Part 2 picks up the story with Thorley's work for Colonial Williamsburg's Craft House, for which he first designed and later produced ceramics as one of its licensed manufacturers. Thorley moved to Williamsburg in 1949, where he set up his studio to produce many of his licensed products. His creations from this period are beyond mere reproductions; they represent an important contribution in the colonial revival movement. Thorley's other works reflect a broad spectrum of artistic and technological accomplishment. Ceramics historians will forever owe John Austin a debt of gratitude for his foresight and perseverance in his effort to preserve and document Thorley's legacy.

Another ceramic type that exudes "folk art" but is little acknowledged for its social and historical significance is the delft wall or fireplace tile. Drawing from his diverse collecting experience, Ivor Noël Hume combines his interest in sea mythology and ceramics with an essay on collecting Dutch delft maritime tiles. In the course of his essay we are taken on a journey that travels from the primordial mermaid archetype to delft tile examples influenced by Greek and Roman mythology to the fireplace surrounds of colonial American homes and ultimately to Noël Hume's own delft-tiled bathroom—proof positive that Noël Hume is one collector and scholar who practices what he teaches.

The articles in this year's journal should make it clear that virtually any ceramic object can contain important messages about a particular time and place. The continued interest in deciphering some of these messages is revealed by this year's diverse New Discoveries, edited by Merry Outlaw. Further evidence of this interest is confirmed by the twelve newly released books on ceramics, reviews of which have been edited by Amy Earls.

Ceramics in America

Figure 1 Detail of the pitcher illustrated in fig. 9. (Unless otherwise noted, all objects are from the collection of the Alexandria Archaeology Museum; unless otherwise noted, all photos by Gavin Ashworth.) This George Washington cartouche adorns a creamware mourning pitcher that was recovered from an archaeological site in Alexandria.

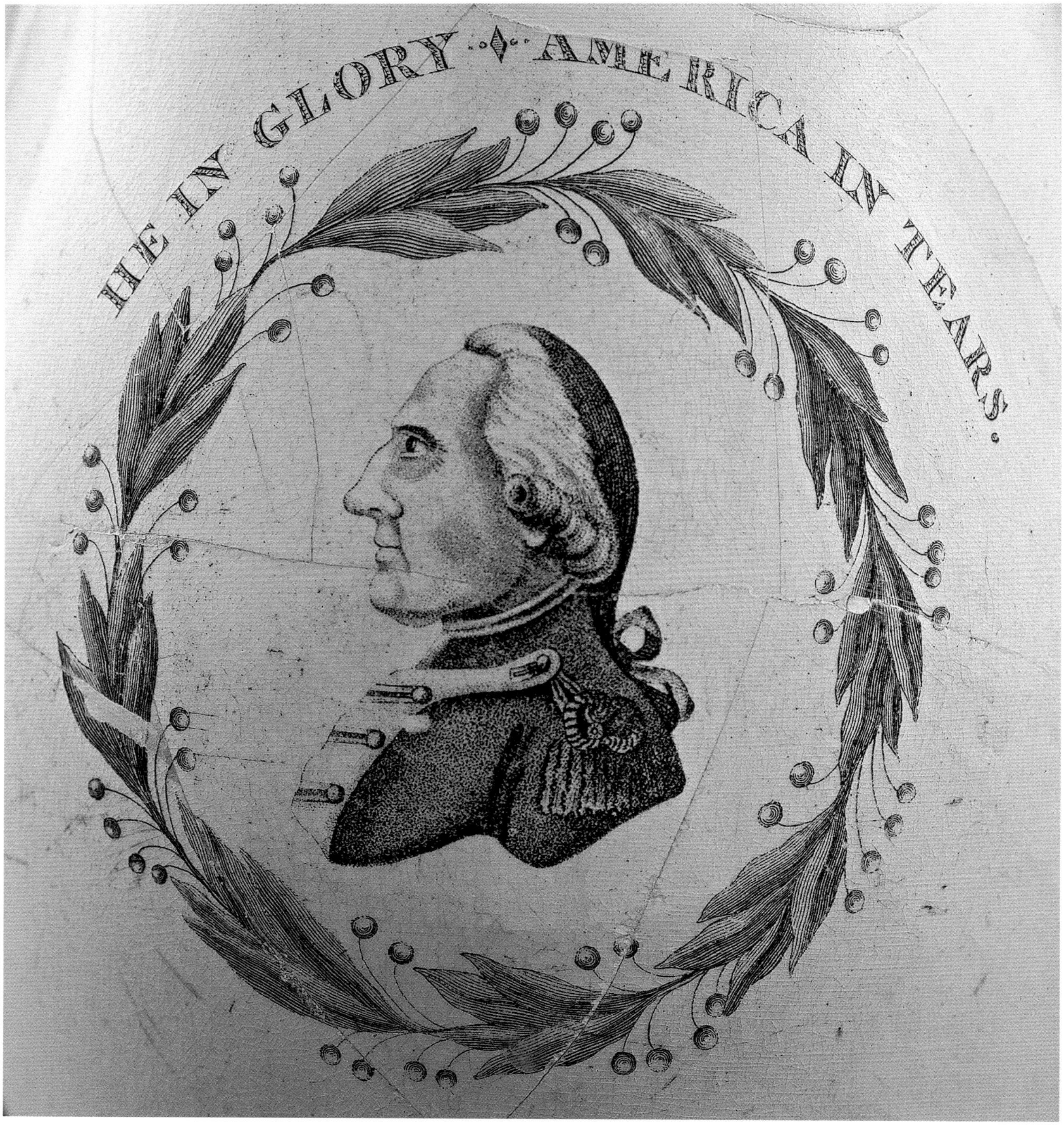

Barbara H. Magid

Commemorative Wares in George Washington's Hometown

▼ "THE PEOPLE OF ALEXANDRIA were always patriotic and public spirited," wrote an Alexandrian on December 21, 1861, reminiscing about the city in the early nineteenth century.[1] Founded in 1749, Alexandria, Virginia, is certainly not among America's oldest cities, but justifiably prides itself as being the hometown of George Washington (fig. 1) and Robert E. Lee. In the 1980s a billboard at the entrance to town declared it to be "America's Most Historic City." Events such as George Washington's Birthnight Ball and Lafayette's 1824 visit to Alexandria, both causes for great celebration, are re-created annually. These and other occasions of local and national significance were commemorated in ceramics that have been recovered from archaeological excavations at many of Alexandria's private homes and taverns. Several of the city's historic sites are also commemorated in ceramics, some of which are preserved in the collection of The Lyceum: Alexandria's History Museum.

Commemorative wares first reached the American market during the colonial period and the early years of transfer printing. Only a few commemorative ceramics from the colonial period have been found in Alexandria, however, and these depict English heroes—Frederick the Great, Major-General James Wolfe, and Lord Rodney—and celebrate British alliances and victories in the Seven Years' War, known in the United States as the French and Indian War.

After the Revolution, English potters recognized a fertile market in America and began to sell patriotic creamware there starting about 1800. Popular patterns included depictions of George Washington and other notable American personages. A pattern known as Britannia, or Pax, made to commemorate the Treaty of Amiens in 1802, was also popular. These wares celebrated not only the British victory against Napoleon but the end of America's Quasi-War with France.

Fewer English wares reached American shores after the Embargo Act of 1807 and before the Treaty of Ghent, which ended the War of 1812 on Christmas Day 1814. After the resumption of trade, English-made American historical wares reached Alexandria and other ports in great quantities. Shell-edged plates depicted the American eagle (adopted as the Great Seal of the United States in 1782), and various commemorative wares celebrated peace and such American heroes of the War of 1812 as Zebulon Pike and Jacob Jones.[2] Lafayette and Cornwallis were honored after Lafayette's celebrated visit to America, and Presidents Andrew Jackson and William Henry Harrison were portrayed in ceramics in 1829 and 1841, respectively.

In the nineteenth century ceramics celebrated America itself, portraying views of historic and scenic sites. Transfer-printed ceramics, imported from England, depicting scenes from Washington, D.C., to Boston have been found in excavations in Alexandria. Although not found archaeologically, scenes of Alexandria were depicted on ceramics at the turn of the twentieth century; two of these sites, Christ Church and Braddock House (known today as the Carlyle House), have national significance. Later in the twentieth century, popular local properties, such as the Washington Street Methodist Church and the George Washington Masonic National Memorial, commissioned their own commemorative wares.

Alexandria and the French and Indian War

The French and Indian War (1754–1763) resulted in France losing most of its North American territories to England. British Major General Edward Braddock, commander-in-chief of all British forces in North America, arrived in Alexandria with about 1,600 British troops. For three weeks in 1755, he was headquartered at the house of John Carlyle (fig. 2), a prominent merchant and one of the founding fathers of Alexandria. Braddock

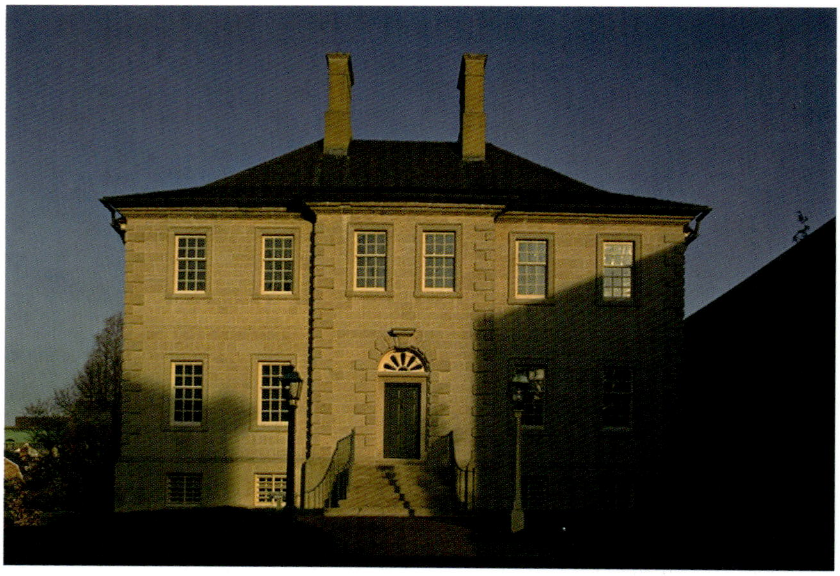

Figure 2 The Carlyle House, 121 North Fairfax Street, Alexandria, Virginia. (Courtesy, Office of Historic Alexandria; photo, Eric Kvalsvik.) Now a museum operated by the Northern Virginia Park Authority, Braddock headquartered here in 1755.

appointed Carlyle, who had just resigned as overseer of the town, to provision the British forces.[3]

During Braddock's visit, a congress of colonial governors met with military leaders at Carlyle's house to plan their strategy against the French. In a letter to William Fairfax dated April 23, 1755, George Washington wrote: "Alexandria has been honoured with 5 Governours in Consultation—a favorable presage I hope, not only of the success of this Expedition, but of the future greatness of this town for surely, such a meeting must have been occasioned by the Commodious, and pleasant situation of the place."[4] John Carlyle, however, was less than pleased with his illustrious visitors. On August 15, 1755, he wrote to his brother George:

> They differed, us & them, and . . . came In so prejudiced against us, our Country & etc. that they used us Like an Enemy Country: took everything they wanted & paid Nothing, or Very little for it. . . . The General & his Aid de Camps, Secretary, & Servants Lodged with Me. He took everything he wanted, abused my home & furniture, and made me little or no satisfaction—tho Expressed a Great deal of Friendship for me & Gave me a Commission as Keeper of the Kings Storehouses, which he assured Me Should be worth £100 per Anno to me, and paid me £30 for the use of my house for a Month. . . .

Carlyle characterized Braddock as "a man (if I am a judge) of Week [*sic*] understanding, Positive, & Very Indolent; Slave to his Passions, Women & Wine . . . Tho' a brave man." Nevertheless, he wrote of the gathering as "the Greatest Congress held at My home ever known on This Continent. . . ."[5]

The war spread from America to Europe in 1756. Frederick the Great of Prussia allied with Britain against Maria Theresa of Austria, who was allied with the French. The British gave active aid to Prussia beginning in 1758, and their battles against French forces were commemorated with white salt-glazed stoneware plates marked "Success to the King of Prussia" (fig. 3).[6] In America, Braddock led a band of British and colonial soldiers,

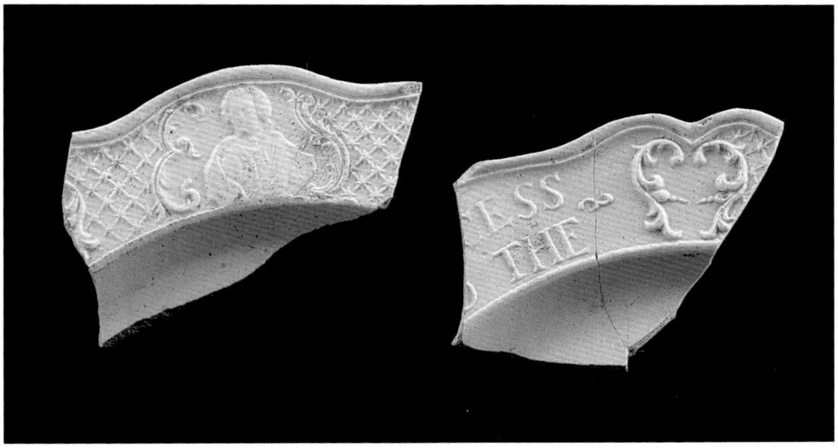

Figure 3 Plate fragments, England, ca. 1758. White salt-glazed stoneware. D. 9". The plate bore a bust of Frederick the Great and the slogan "SUCCESS TO THE KING OF PRUSSIA." Beginning in 1758 the British supported Prussia against the Austrians and their French allies. This plate was owned by a family residing at 418 King Street, one block from the Carlyle House, and its fragments were discarded in a privy.

among them George Washington, against the French at Fort Duquesne (located in what is now Pittsburgh, Pennsylvania). He died of battle injuries, and Washington led the surviving troops to safety.

Braddock's visit, the involvement of George Washington, and the participation of local troops gave immediacy to the French and Indian War for Alexandrians, and this was reflected in everything from wares to street names. In addition to a King of Prussia plate, archaeologists found later wares, one of which depicts General Wolfe, who led British forces in the capture of the city of Quebec from the French. The battle on the Plains of Abraham, on September 13, 1759, lasted less than an hour but is considered the decisive battle of the French and Indian War. Both Wolfe and his rival, the French marquis de Montcalm, died from battle wounds. A letter written five days later by Rear Admiral Charles Holmes described the battle:

> About Ten, [Montcalm] marched in order of Battle & attacked us on the Plains of Abraham. His Disposition was excellent, & his Army advanced

in very good order. Wolfe was entirely prepared; & overjoyed at the unexpected Resolution of the Enemy.

They fired twice before he made any return; Then commenced the Battle which hardly lasted a quarter of an Hour . . . Gen'r Wolfe fell early in the action, & only lived long enough to know that his Troops were victorious; On learning of which, he said—"Since I have conquered I dye satisfied."[7]

The Death of Wolfe, a well-known neoclassical painting of 1770 by Benjamin West, is the source of an engraving on a English creamware pitcher fragment found in Alexandria (fig. 4) as well as on similar wares made by other

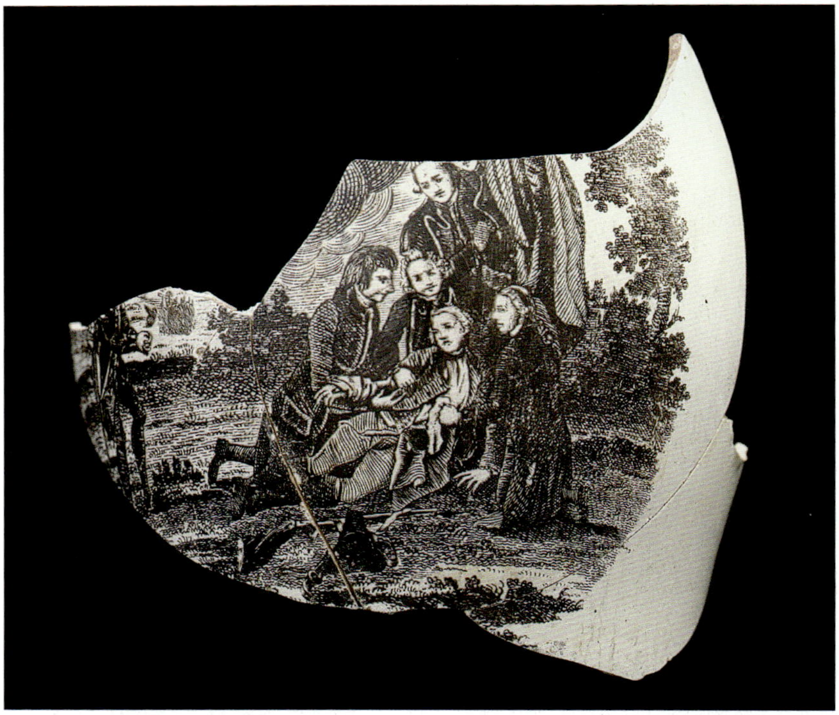

Figure 4 Pitcher fragment, England, ca. 1770s. Creamware. The scene depicts the death of General Wolfe, who successfully led British forces against the city of Quebec in 1759, during the French and Indian War.

English potteries. The Alexandria pitcher differs from a Wedgwood pitcher illustrated by Ellouise Baker Larsen in several ways.[8] The most significant involves a standing figure to the left of the group surrounding the dying general. This figure is a reverse image of an officer with clasped hands who stands to the right in the original painting.[9]

A later ware found by archaeologists in Alexandria depicts Lord Rodney, who by capturing Martinique from the French became another hero of the French and Indian War. However, in the American Revolution Rodney became an enemy of the colonists, defeating the Spanish fleet off Cape St. Vincent and then the French fleet (both countries were American allies) at the Battle of the Saints. After his successes for the British, Rodney was made a baron in 1782, and his likeness appeared on pub signs as well as commemorative pottery.[10]

Rodney's exploits during the Revolution were certainly known to Alexandrians. An extract of a letter from Jamaica, dated March 17, 1784, was reprinted in the *Alexandria Gazette*: "A military gentleman, sailing here from Antigua, called at St. Eustatius. There he found the inhabitants

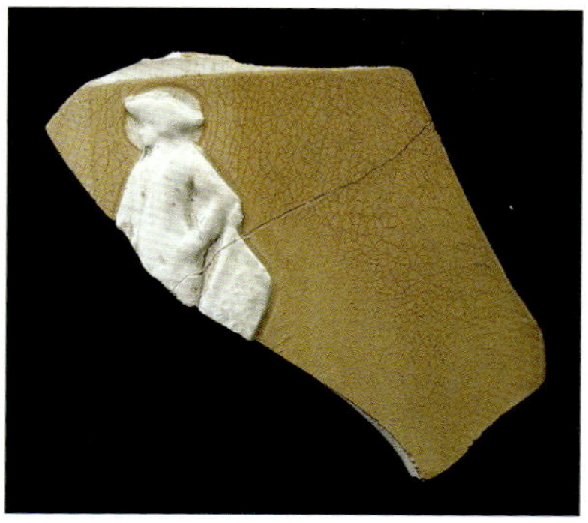

Figure 5 Tankard fragment, England, ca. 1782. Banded creamware with relief-molded appliqué of Lord Rodney. This fragment and others, discarded about 1790, were found at the bottom of a brick shaft at 109 South Fairfax Street, Alexandria, Virginia.

amusing themselves by burning Lord Rodney and General Vaughan in effigy, the island having been returned to the Dutch the day before."[11] A fragment of a tankard depicting Lord Rodney (fig. 5) that was found on an archaeological site in Alexandria was made around the time that this letter was published. Its presence may indicate loyalist sentiments lurking in an Alexandria household, or it may simply commemorate a well-known figure of the day.

Alexandria Mourns George Washington

Alexandria has always taken great pride in being the hometown of George Washington. A replica of his town house, which was demolished in 1857, stands at 508 Cameron Street (fig. 6), and his family pew at Christ Church (fig. 7) still bears his name.[12] Washington frequented many of Alexandria's taverns for both business and pleasure. He dined at Lomax's Tavern with Lafayette in 1784 and celebrated the ratification of the Constitution at Wise's Tavern in 1788. At Gadsby's Tavern, also known as the City Hotel, he danced at his own Birthnight Balls, which are still celebrated today (fig. 8).

The importance of Washington's birthday celebrations in the 1830s was recounted in the reminiscences of an "Old Bachelor" on his return to Alexandria thirty years later.

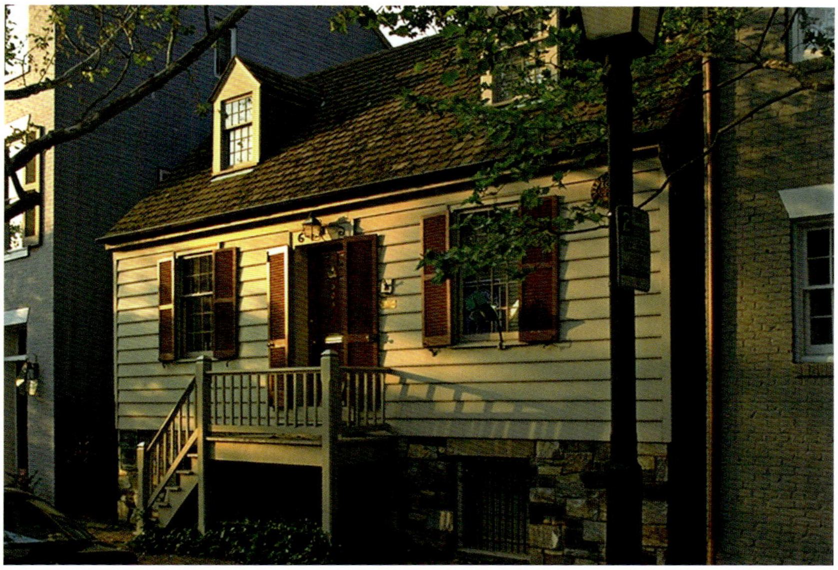

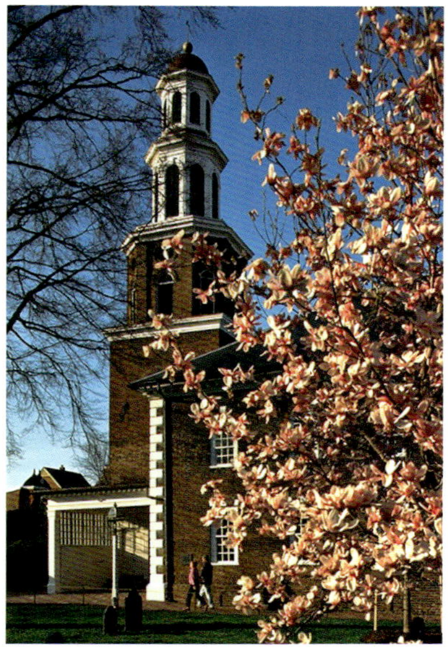

Figure 6 Replica of George Washington's town house. (Courtesy, Office of Historic Alexandria; photo, Eric Kvalsvik.) Washington's town house at 508 Cameron Street was built by 1769 and demolished in 1857. This replica was constructed on the same site in 1960. The *Alexandria Gazette* (June 7, 1851) described the original house as "a small frame house, the ascent to which is by a long narrow flight of steps. How simple the habits, how unostentatious the movements and the mansion of the deliverer of his country. . . . A stranger might pass and re-pass this house many times without taking any notice of it." Although the house was often rented or lent to friends and family, Washington is thought to have spent some nights there in the early 1770s, and his letters give detailed orders for fencing the property, painting, and other repairs.

Figure 7 Christ Church, 118 North Washington Street, Alexandria, Virginia. (Courtesy, Office of Historic Alexandria; photo, Eric Kvalsvik.) George Washington's family pew is preserved here. The church was constructed in 1773; the bell tower was added in 1818.

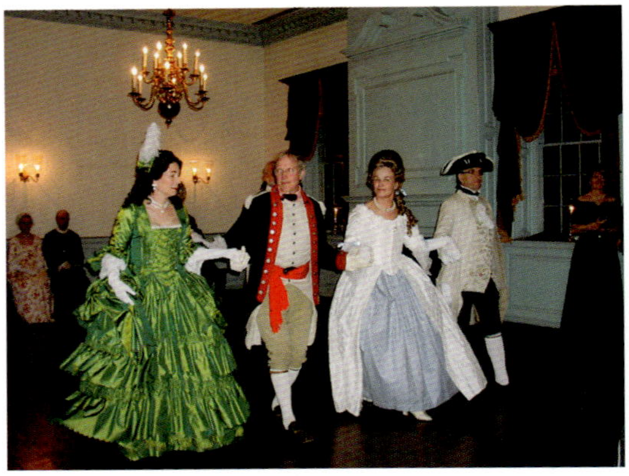

Figure 8 The ballroom at Gadsby's Tavern Museum, 134 North Royal Street, Alexandria, Virginia. (Photo, Louise Krafft.) George and Martha Washington danced here at his Birthnight Balls in 1797 and 1798. The celebratory balls continued to be held long after his death, until the Civil War. Reenactments of the ball were held in 1932, on the bicentennial of Washington's birth, and sporadically until the bicentennial of the nation, when the ball once again became an annual event.

> The great day with us, in old Alexandria, was Washington's Birth Day—the 22nd of February—the anniversary of the birth of our fathers' friend and neighbor, as well as the Father of his Country. . . . There was a more general public, united celebration of that day, than there was even of the 4th of July . . . . Then the military paraded and kept together. A long procession was formed, consisting of the Town Authorities, the reverend clergy, the civic associations—and there were several, then of the highest respectability as to character and numbers—the Fire Companies—the Free School children (the latter all neatly dressed, headed by their teachers—the most interesting part of the pageant)—the volunteer companies with their bands of music—followed by a long line of citizens who made it a rule always to unite in the ceremonies of the occasion—and escorted the Reader of the Farewell Address and the Orator of the Day, to some one of the Churches, where the exercises took place. There was always a fine Ode sung—an appropriate prayer made—and then succeeded the reading, and the speech. . . . Flags were hoisted, cannon fired, and huzzas shouted in every direction, throughout the day. . . .
>
> On the night of the 22nd, there was the "Birth Night Ball." . . . Washington, himself, more than once attended the balls—and there is, or was, in the Museum, an autograph letter of his, framed, sending his apology, on one occasion, for not being able to come into town to visit his friends, at this, the most delightful of their social meetings. . . . At 12 o'clock precisely, supper was announced—and these suppers were equal to any of the feasts ever prepared at the old City Hotel, then and since famous for the excellence and taste of its entertainments.[13]

Drs. James Craik and Elisha Cullen Dick, who attended Washington on his deathbed, were both Alexandrians. Their account of Washington's final illness, which was published in the *Virginia Herald,* ended with these words: "During the short period of his illness, he economized his time, in the arrangement of such few concerns as required his attention, with the utmost serenity; and anticipated his approaching dissolution with every demonstration of that equanimity for which his whole life has been so uniformly and singularly conspicuous."[14]

The bells of the Old Presbyterian Meeting House tolled for four days and nights after Washington's death, and a memorial service was held there on December 29, 1799.[15] The news of Washington's death and its effect on Alexandrians appeared in the *Alexandria Times and District of Columbia Advertiser*:

> Washington in Glory—America in Tears
>
> The effect of the sudden news of his death upon the inhabitants of

> Alexandria can better be conceived than expressed. At first a general disorder, wildness, and consternation pervaded the town. The tale appeared as an illusory dream, as the raving of a sickly imagination. But these impressions soon gave place to sensations of the most poignant sorrow and extreme regret. On Monday and Wednesday the stores were all closed and all business suspended, as if each family had lost its father. From the time of his death to the time of his interment the bells continued to toll, the shipping in the harbor wore their colors half mast high, and every public expression of grief was observed. On Wednesday, the inhabitants of the town, the county, and the adjacent parts of Maryland proceeded to Mount Vernon to perform the last offices to the body of their illustrious neighbor. All the military within a considerable distance and three Masonic lodges were present. The concourse of people was immense. Till the time of interment the corpse was placed on the portico fronting the river, that every citizen might have an opportunity of taking a last farewell of the departed benefactor.[16]

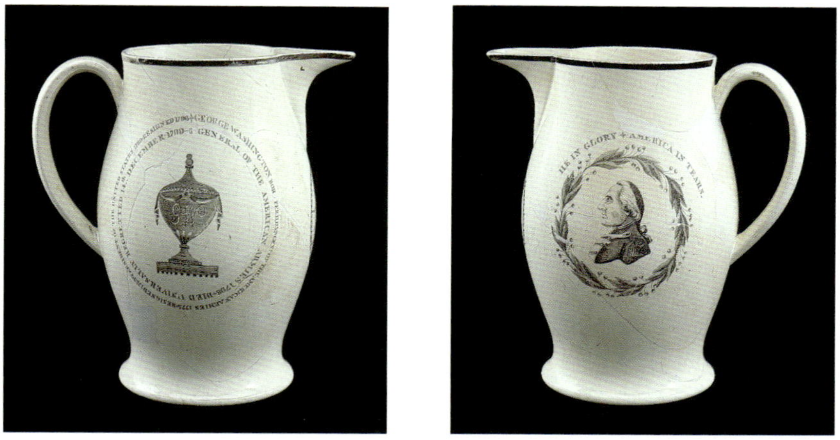

Figure 9 Pitcher, attributed to the Herculaneum Pottery, Liverpool, ca. 1800. Creamware. H. 8½". This black-rimmed mourning pitcher depicts Washington in profile surrounded by a wreath, with the words "HE IN GLORY · AMERICA IN TEARS." On the reverse is an urn with the initials GW and the words "GEORGE WASHINGTON BORN FEB. 11, 1732 / GEN'L OF THE AMERICAN ARMIES 1775 / RESIGNED 1785 / PRESIDENT OF THE UNITED STATES 1789 / RESIGNED 1796 / GENERAL OF THE AMERICAN ARMIES 1798 / DIED UNIVERSALLY REGRETTED 14TH DECEMBER 1799." Underneath the spout are the words "A MAN WITHOUT EXAMPLE / A PATRIOT WITHOUT REPROACH."

Similar to the title of that article, the words "HE IN GLORY · AMERICA IN TEARS" were printed on English creamware pitchers commemorating Washington's death. One of them, a mourning pitcher recovered from excavations in a well behind 416–418 King Street, Alexandria (fig. 9), has been attributed to the Herculaneum Pottery in Liverpool, England.[17] A tanner, Robert William Kirk, owned a store and dwelling on this property from the time of Washington's death until 1804, when it was sold to Jacob Fortney, who owned a stable and blacksmith shop.

Commemorative wares were again issued at special anniversaries of Washington and the nation. A surviving fragment of a Rockingham teapot or pitcher depicting Washington's bust was discovered (fig. 10), along with other artifacts from the 1840s and 1850s, during excavations in Alexandria. The teapot or pitcher probably was made in 1849 to commemorate the fiftieth anniversary of Washington's death, reflecting his

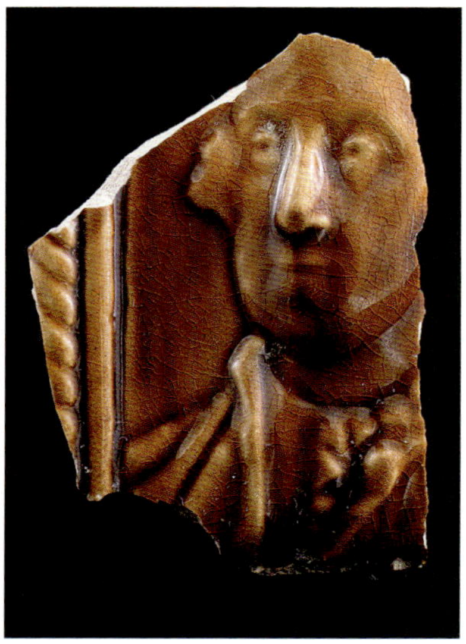

Figure 10 Teapot or pitcher fragment, United States, after 1845. Rockingham pottery. This sherd depicting George Washington was found during excavations at a home at 809 Duke Street, Alexandria, Virginia.

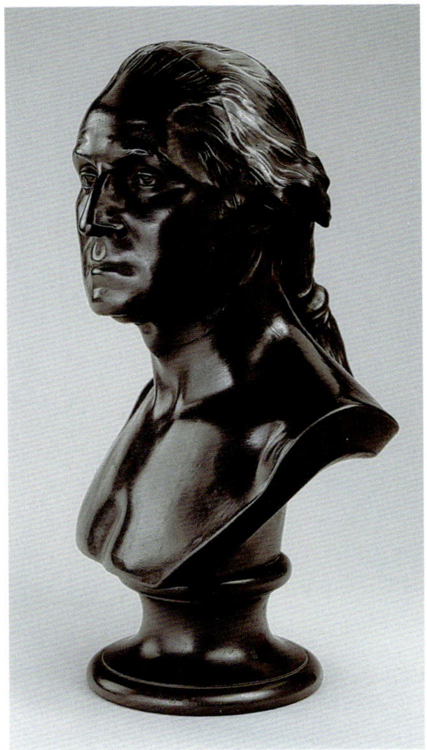

Figure 11 Bust of George Washington, Wedgwood, Staffordshire, 1875. Basalt stoneware. H. 18¼". (Courtesy, The Lyceum: Alexandria's History Museum.) Inside the base are the letters XCD. This large bust was made in time to celebrate the nation's centennial.

continued importance to Americans (and Alexandrians in particular, who hold the largest parade in the country in honor of Washington's birthday). In 1875, in time for America's centennial and just after the seventy-fifth anniversary of Washington's death, Wedgwood made a bust of Washington based on the famous sculpture by the French sculptor Jean-Antoine Houdon (fig. 11).[18]

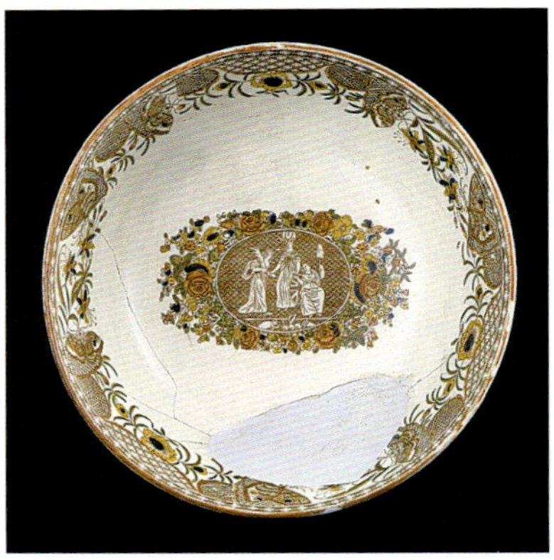

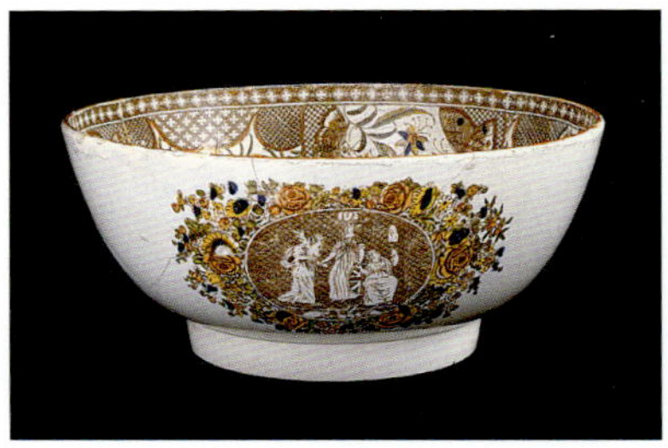

Figure 12 Punch bowl, England, ca. 1802. Pearlware. D. 10¼". Britannia is pictured with her shield. To her left is Victory, holding a palm frond, and seated to her right is France holding the Liberty Cap. The border includes a lovely butterfly motif. This punch bowl, made in celebration of the Treaty of Amiens, is from the household of Samuel Snowden, editor of the *Alexandria Gazette*. It was discarded in the 1820s, in a privy behind his home and office at 105 South Royal Street.

Alexandria and America's Quasi-War with France

The household of Samuel Snowden, editor of the *Alexandria Gazette,* discarded a large punch bowl with a transfer-printed pattern known as Britannia or Pax (fig. 12). Archaeological excavations also show that a neighboring house around the corner on the 400 block of King Street had a set of teacups and saucers in the same pattern (fig. 13), as did McKnight's Tavern across the street (fig. 14). The pattern celebrates the Treaty of Amiens

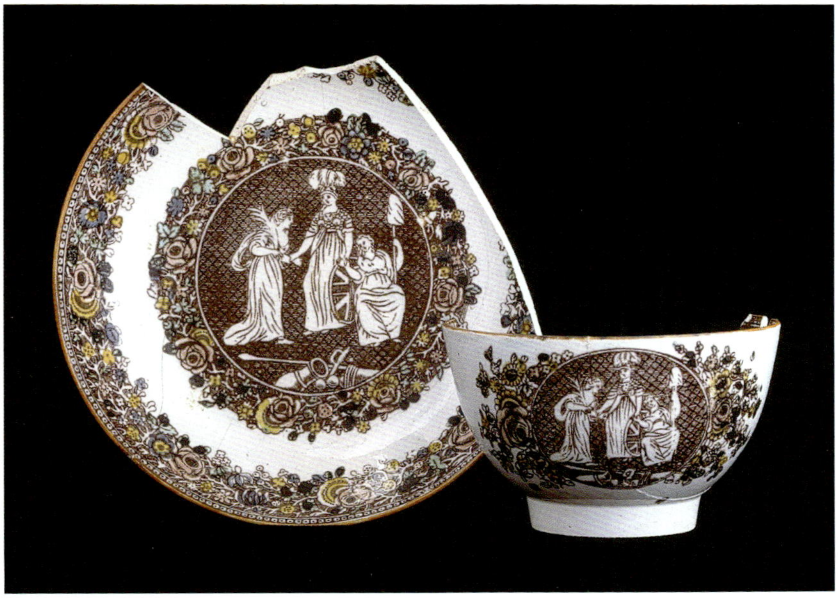

Figure 13 Cup and saucer, England, ca. 1802. Pearlware. D. of cup 3⅝"; D. of saucer 5½". Cups and saucers depicting Britannia came from the household of one of Snowden's neighbors, on the 400 block of King Street.

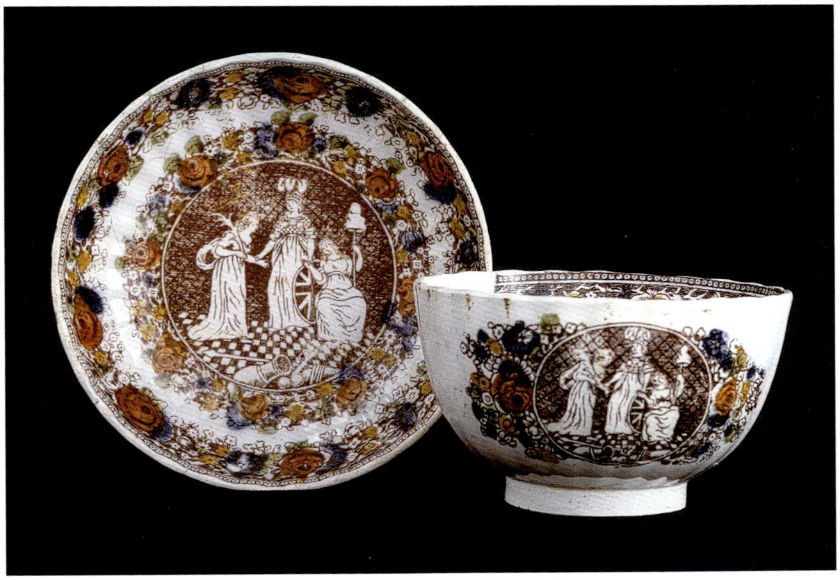

Figure 14 Cup and saucer, England, ca. 1802. Pearlware. D. of cup 4"; D. of saucer 5". Cups, saucers, a cream pitcher, and a slop bowl from this tea set were found at the site of McKnight's Tavern, at the corner of King and Royal Streets.

in 1802 and symbolizes Britain's victory over France, although Americans may have purchased these wares to celebrate their country's own victories over the French and Napoleon.[19]

The Britannia wares were first manufactured in 1802 and may have reached America any time before the Embargo of 1807, which was instituted by President Thomas Jefferson against the French and English for interfering with American shipping. Pearlware vessels such as these, in which the outline of the design was transfer printed in brown and then painted in polychrome colors, are sometimes known as Salopian ware (i.e., from Shropshire), although the wares were made by a number of English factories in the early part of the nineteenth century. They were commonly made for the American market and are found more often in this country than in England.[20]

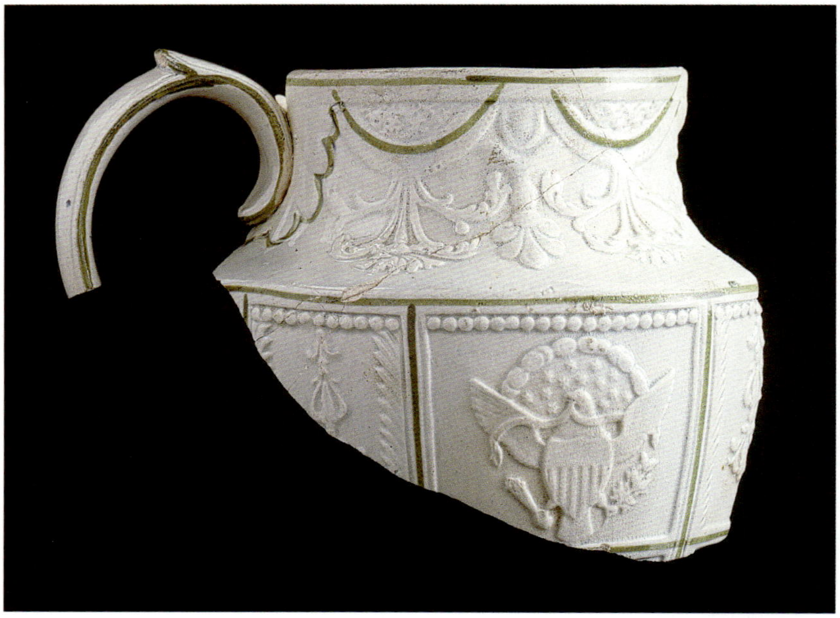

Figure 15 Pitcher fragment, attributed to Sowter and Company, Mexborough, Yorkshire, ca. 1800–1807. Castleford-style white feldspathic stoneware. This pitcher fragment, which bears the Great Seal of the United States, was found behind a house at 113 South Royal Street, Alexandria. It may have been discarded by the owner of McKnight's Tavern, William H. McKnight, who moved there in 1816. The pitcher has the same molded relief pattern as a black stoneware teapot attributed to Sowter and Company.

A variety of wares depicting the American eagle and patriotic themes also date from these years, before the 1807 embargo. The eagle from the Great Seal of the United States is featured on two pitchers made in the style of the Castleford Pottery, with molded borders separating panels with molded relief patterns. The Castleford Pottery is not known to have made wares with American symbols, but their style of pottery was copied by several other manufacturers, and attributions can sometimes be made based on the molded relief patterns used as borders. A Castleford-style white feldspathic stoneware pitcher (fig. 15) has been attributed to Sowter and Company,[21] and a pearlware pitcher (fig. 16) has been attributed to the Leeds Pottery based on its similarity to a feldspathic stoneware sucrier commemorating Admiral Lord Nelson and General Howe.[22] The eagle emblem has also been found on a soft-paste porcelain saucer (fig. 17) and a bone china coffee can (fig. 18), both manufactured in England.[23] A patriotic pitcher attrib-

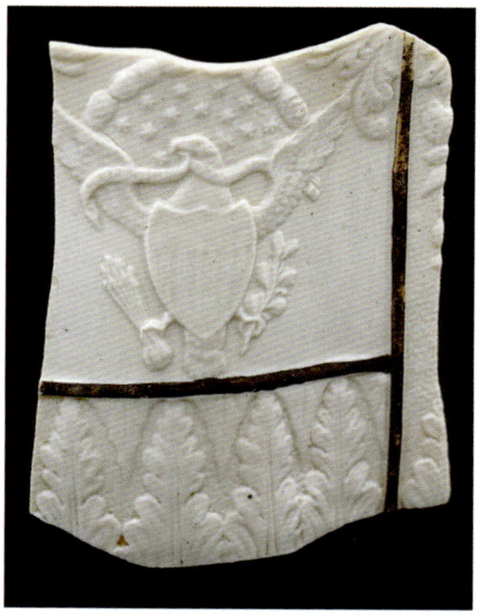

Figure 16 Pitcher fragment, attributed to the Leeds Pottery, ca. 1805–1807. Pearlware. The enameled lines that separate panels with molded relief decoration of the Great Seal are in the style of the Castleford Pottery, but the molded border on the panels is identical to a Leeds sucrier in feldspathic stoneware. This fragment was found in a well behind the workshop of James Green, cabinetmaker, which stood in the interior of the square bounded by Fairfax, Prince, Royal, and King Streets and near the intersection of Royal and King Streets. The sherd was in a deposit relating to a fire in 1823 but probably had reached Alexandria before the Embargo of 1807.

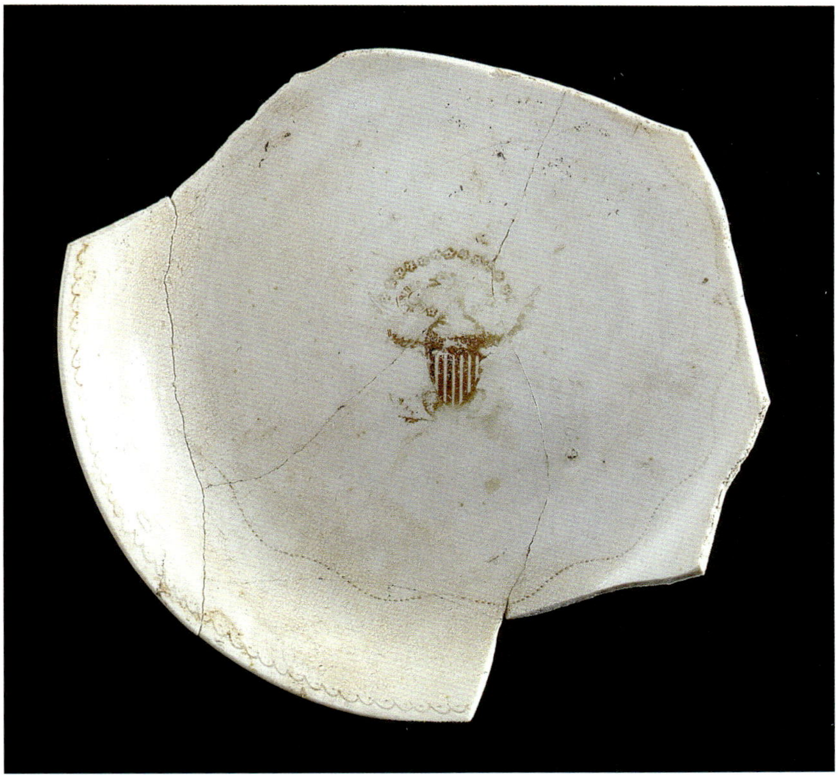

Figure 17 Saucer, England, possibly Enoch Wood and Sons, ca. 1784–1792. Soft-paste porcelain. D. 6". The Great Seal, with the motto "E PLURIBUS UNUM," appears on this saucer, which was found in a well associated with Gadsby's Tavern at 134 North Royal Street.

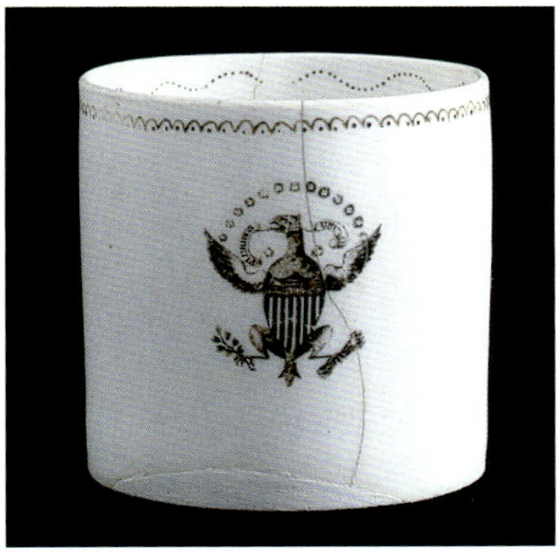

Figure 18 Coffee can, England, ca. 1795–1807. Bone china. H. 2½". This cup, depicting the Great Seal, was found behind the James Green cabinet shop along with the Castleford-style pearlware sherd illustrated in fig. 16.

uted to the Herculaneum Pottery shows a reproduction of the print *An Emblem of America* (fig. 19), first published in 1799. It depicts a central female figure, draped in classical garb and holding an American flag, flanked by Indians and portrait medallions of some of America's best-known figures, including Christopher Columbus, Americus, Sir Walter Raleigh, Benjamin Franklin, George Washington, and John Adams.[24]

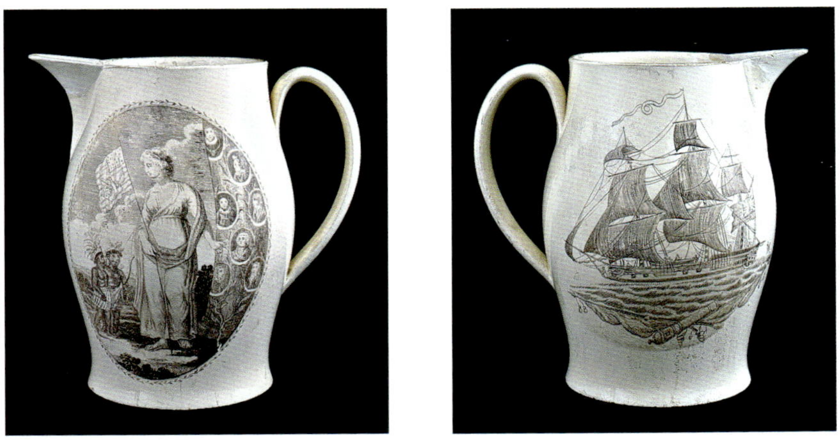

Figure 19 Pitcher, attributed to the Herculaneum Pottery, Liverpool, ca. 1800. Creamware. H. 8½". In this reproduction of *An Emblem of America,* first published in 1799, a central figure holds an American flag. At left are two Indians and at right are portrait medallions of Christopher Columbus, Americus, Sir Walter Raleigh, Benjamin Franklin, George Washington, John Adams, and one unidentified person. On the reverse is a ship and cannon. Samuel Snowden, editor of the *Alexandria Gazette,* owned this patriotic piece.

In 1781 America's French allies under Jean-Baptiste-Donatien de Vimeur, comte de Rochambeau, encamped in Alexandria, near Four Mile Run, on their way to Yorktown and the defeat of the British. "The most elegant and handsome young ladies of the neighborhood danced with the (French) officers on the turf, in the middle of the camp, to the sound of military music."[25] America's alliance with the French had been instrumental in winning American independence, and Americans were supportive of the goals and ideals of the French Revolution through the storming of the Bastille in 1789 and the beginnings of the French Republic in 1792. A few months after Louis XVI was deposed, however, the Reign of Terror began, and thousands were executed within the year, events that horrified Americans. As French power spread through Europe, Americans became mistrustful of France's growing empire and concerned with French expansion in Louisiana. In 1794 Congress declared neutrality in the Anglo-French conflict, a move the French believed violated a 1778 treaty that required America to help protect French colonies in the West Indies against the British. In retaliation, French privateers attacked American merchant ships, leading to an undeclared naval war that became known as the Quasi-War.[26]

The port of Alexandria played a role in the attack of the fledgling U.S. Navy against privateers in the West Indies. In May 1798 Lieutenant James Triplett was sent from Baltimore to Alexandria to "engage as many Seamen and Marines as he can find disposed to enter on board the Frigate Constellation." In the next year at least three privateers were built at Alexandria shipyards, each carrying twenty-two guns and fifty-four crewmen.[27] By 1800 many of Alexandria's wharves had been converted for construction of ships for the Quasi-War. Even the County Wharf, which had been in disrepair, was refurbished for shipbuilding. The French seized at least twenty of the Alexandria-built ships in the West Indies.[28] In April 1799 the secretary of the navy authorized a brig with eighteen guns to be built in Alexandria. However, the order was rescinded three weeks later, amid concerns about how quickly the ship could be built. The navy instead ordered 40,000 pounds of ships' bread from the Jamieson bakery in Alexandria.

Trade with Europe and the West Indies was immensely important for Alexandria, one of the largest ports in the United States at the beginning

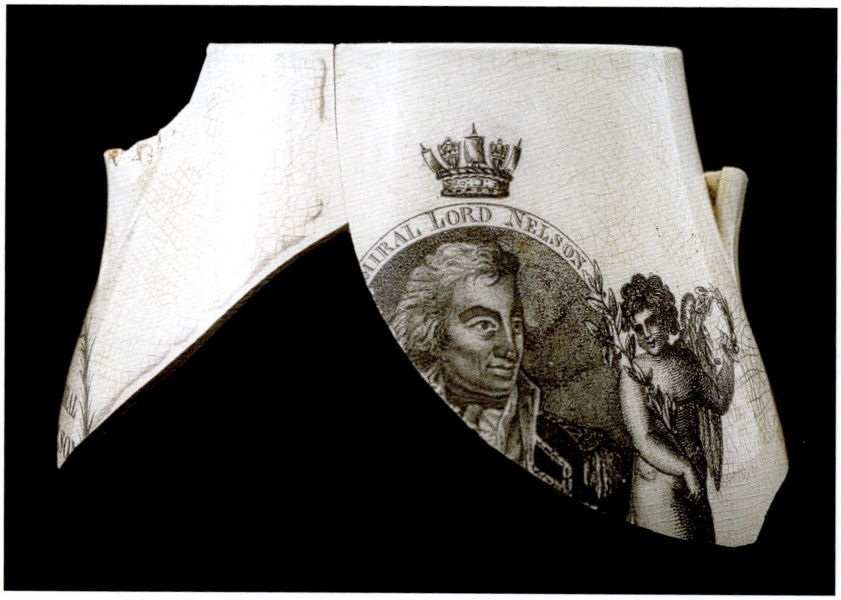

Figure 20 Mourning pitcher fragment, England, ca. 1805. Creamware. Found at the site of Arell's Tavern, this pitcher has the words "ADMIRAL LORD NELSON" on a wreath cartouche below the spout. On one side is a portrait of Nelson flanked by cherubs. The missing portion shows a plan of the Battle of the Nile beneath the portrait. On the reverse are more cherubs within a wreath border. Nelson died tragically at Trafalgar as his troops defeated Napoleon.

of the nineteenth century, and the town had suffered as a result of the Anglo-French and American-French hostilities. The end of the Quasi-War with France aroused patriotic fervor in Alexandria—and, indeed, the nation—and these commemorative wares were one manifestation of the sentiment. Alexandrians were celebrating not only the Treaty of Amiens and the new, albeit short-lived, peace in Europe but also the end of American-French hostilities in the West Indies and the Louisiana Purchase in 1803, which ended French occupation in North America and nearly doubled the size of the United States.

Two great English heroes of the long Napoleonic Wars are portrayed on ceramics found in Alexandria: Vice Admiral Horatio Nelson on a pitcher from Arell's Tavern (fig. 20), and the duke of Wellington on a basalt stoneware creamer and a creamware pitcher from homes nearby (figs. 21, 22). Although Nelson died tragically at Trafalgar on October 21, 1805, his

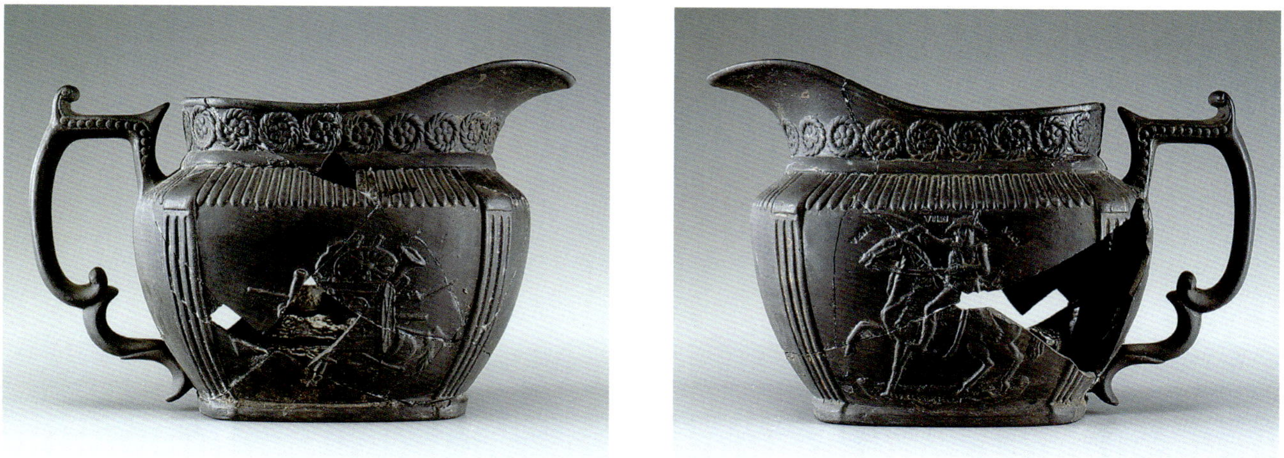

Figure 21 Creamer, England, ca. 1815. Basalt stoneware with glazed interior. L. 5¾". Tools of war appear on one side of this creamer; on the other side, inscribed above the figure of Wellington on horseback, are the words "VENI VIDI VICI." This famous Latin phrase complements the creamer's neoclassical form.

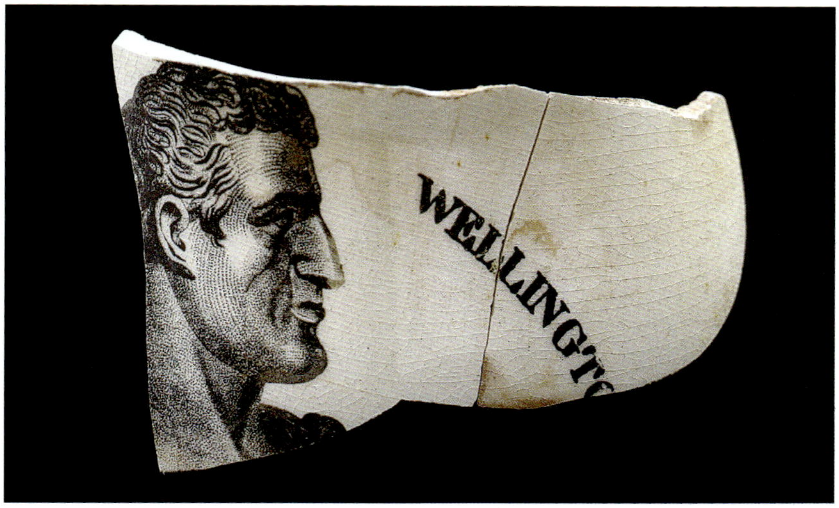

Figure 22 Pitcher fragment, England, ca. 1815. Creamware. This pitcher celebrates Wellington's defeat of Napoleon at Waterloo, which ended twenty-three years of war between Britain and France.

troops were victorious over Napoleon. After this naval defeat France was no longer a threat to Britain, and Nelson, a greatly loved British hero, was commemorated with a large assortment of souvenirs. A decade later, in June 1815, the duke of Wellington defeated Napoleon on land at the Battle of Waterloo, near Brussels. The Wellington ceramics commemorate this battle, which ended twenty-three years of war between Britain and France and exiled Napoleon to St. Helena. Wellington became a hero throughout Europe. He played a role in the War of 1812, urging Britain to accept the Treaty of Ghent, which brought to a close the war with the United States, and went on to become Britain's prime minister in 1828.

Alexandria and the War of 1812

The Embargo of 1807, placed by Jefferson on commercial trade with England, was in response to a British decree that limited American trade and required that duty be paid at British ports. Ultimately leading to the War of 1812, the embargo greatly affected Staffordshire potteries, which had enjoyed a lucrative trade with America. From 1795 to 1800 approximately ten million pieces of Staffordshire pottery were imported annually by American merchants, and although trade did not cease during the embargo, production at the potteries decreased by 60 percent and unemployment was high.[29]

After the War of 1812 began, Alexandria merchants again supplied the navy with ships' biscuits and other supplies, as they had during the Quasi-War with France. On August 23, 1814, British troops set fire to Washington. A British soldier wrote:

> [T]his was a night of dismay to the inhabitants of Washington. They were taken completely by surprise. . . . The first impulse of course tempted them to fly, and the streets were in consequence crowded with soldiers and senators, men, women, and children, horses, carriages, and carts loaded with household furniture, all hastening toward a wooden bridge which crosses the Potomac. . . . Of the senate-house, the President's palace, the barracks, the dockyard, etc., nothing could be seen, except heaps of smoking ruins; and even the bridge, a noble structure upward of a mile in length, was almost wholly demolished.[30]

Threatened with an invasion as well, and with insufficient forces to defend its city, Alexandria's Common Council surrendered to the British without resistance. British forces occupied the city for five days, from August 29 through September 2, and although the British did not burn the city, they looted its stores and warehouses. According to the terms of capitulation, "The town of Alexandria, with the exception of public works, shall not be destroyed, unless hostilities are commenced on the part of the Americans; nor shall the inhabitants be molested in any manner whatever, or their dwelling-houses entered, if the following articles are complied with. . . ."[31] Alexandria's mayor wrote: "It is impossible that men could behave better than the Britissh [*sic*] behaved while the town was in their power, not a single inhabitant was insulted or injured by them in their person or houses."[32] A prominent Alexandrian Quaker, the apothecary Edward Stabler, wrote: "Their conduct was respectful and decorous; and instead of that exultation and triumph which expands the heart of a soldier when he encounters and overcomes a force like his own, these were evidently dejected and adverse to what they were doing."[33]

Alexandria was sharply criticized and ridiculed for its surrender, however. A political cartoon published in Philadelphia in 1814 (fig. 23) shows

Figure 23 *Johnny Bull and the Alexandrians,* William Charles, 1814. Etching and aquatint. Sheet 9¾" x 13½". (Courtesy, The Lyceum: Alexandria's History Museum.) Alexandrians are shown cowering and pleading with Johnny Bull, a symbol for England. The cartoon ridicules Alexandria's 1814 surrender to the British in the War of 1812. The Alexandrians plead: "Pray Mr. Bull don't be too hard with us—You know we were always friendly, even in the time of our Embargo!" Their words suggest that British goods might have reached the port of Alexandria during the embargo years.

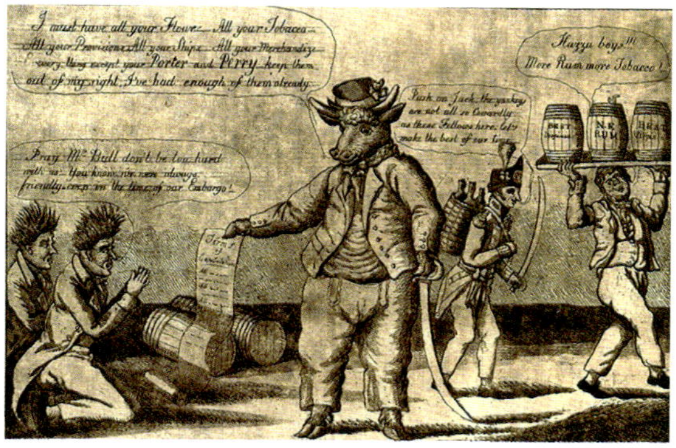

Alexandrians cowering and pleading with Johnny Bull, symbolic of England.[34] On August 31, 1814, an editorial in the *Richmond Enquirer* stated, "In what terms can we express our indignation against the conduct of the citizens of Alexandria? Thanks be to the Almighty God: that this degraded town no longer forms a part of the state of *Virginia*!"[35]

When the Treaty of Ghent was signed, on Christmas Day 1814, Alexandria had much to celebrate. With the end of the War of 1812, America's continuing independence was assured and the seas were open to trade with all nations. A number of ceramics celebrating the end of the war and honoring its heroes have been found at archaeological excavations in Alexandria, many at tavern sites.

Arell's Tavern, in particular, used many patriotic pieces, including a set of small shell-edged plates with an "even scallop" pattern and depicting the American eagle (fig. 24), a motif that appears on wares made before the

Figure 24 Plate, England, ca. 1815–1830. Pearlware. D. 7". The even scallop rim dates from ca. 1810–1835. Because of the limited palette of colors available for underglaze pearlware decoration, the shield is painted in orange and blue enamels. A set of these plates, marked with an impressed "B," was found at Arell's Tavern. An 8" blue shell-edged eagle plate was found at McKnight's Tavern, whose location was marked by the shop's sign, a spread eagle.

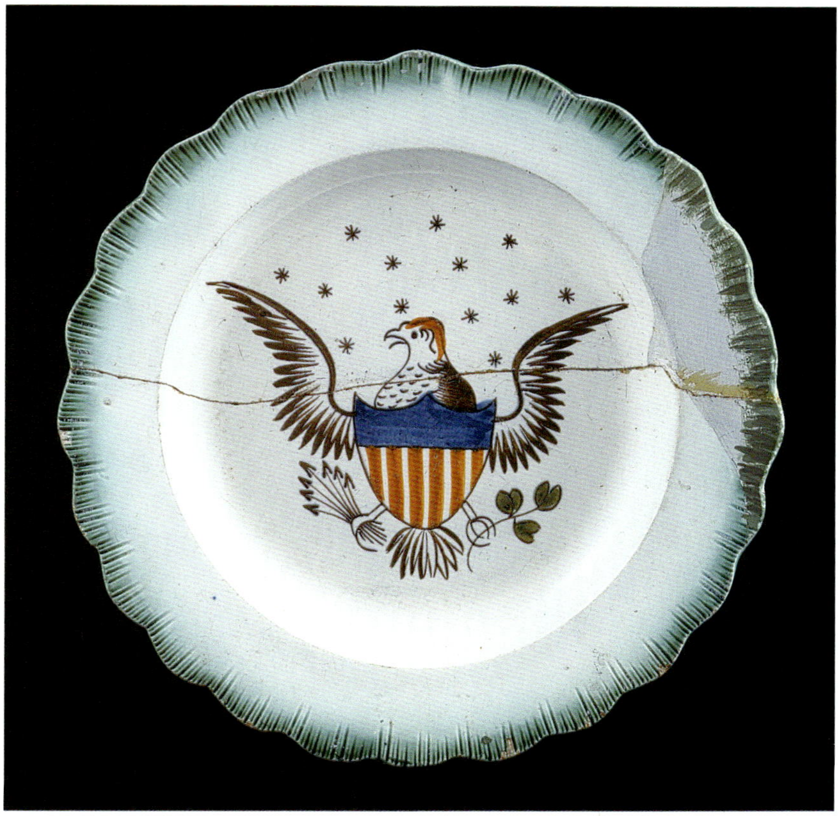

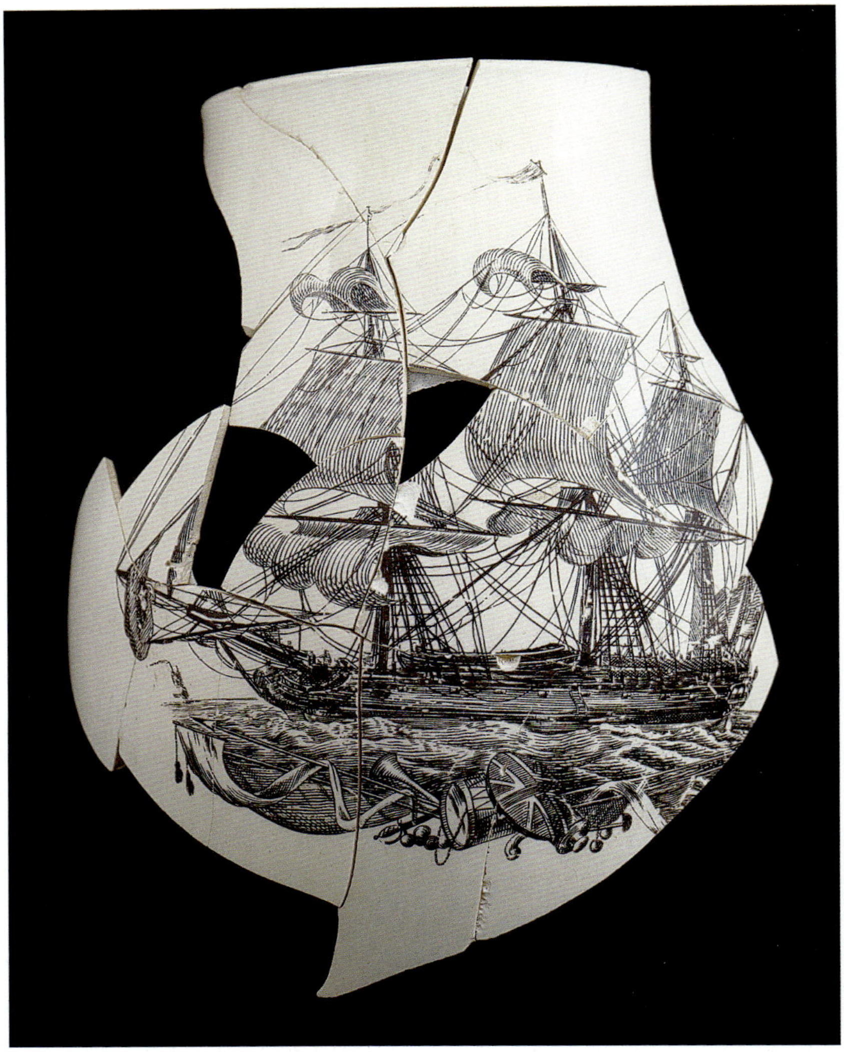

Figure 25 Pitcher fragment, England, ca. 1815. Creamware. This fragment, found at the site of McKnight's Tavern, displays a merchant ship. The ship motif was common on creamware pitchers when the War of 1812 ended and trade was restored. It was often used in conjunction with commemorative patterns.

Embargo of 1807 (see, for example, figs. 15–18) but was commonly used following the end of the War of 1812.[36] These dishes were popular among the lower and middle classes from about 1815 to 1830[37] and therefore were appropriate for Arell's, a working-class tavern. Another blue shell-edged eagle plate was found at McKnight's Tavern, which was known by its sign of the spread eagle.

Merchant ships were also a common motif on creamware pitchers, celebrating the resumption of trade following the War of 1812 and the prosperity it was expected to bring. McKnight's Tavern, for example, had a creamware pitcher showing a merchant ship under sail (fig. 25), a motif often used on the reverse of other commemorative patterns.

Many variations of the popular print *Peace, Plenty and Independence* appeared on creamware and pearlware pitchers made for the American market by the Staffordshire potters Enoch Wood and Sons, Herculaneum, and others. The print depicts the spread eagle perched atop a cannon, beneath which is a circle with the slogan "Peace, Plenty and Independence." There is a female figure at right, Peace, who torches the tools of war; a figure at left, Plenty, holds a sheaf of wheat and a cornucopia.

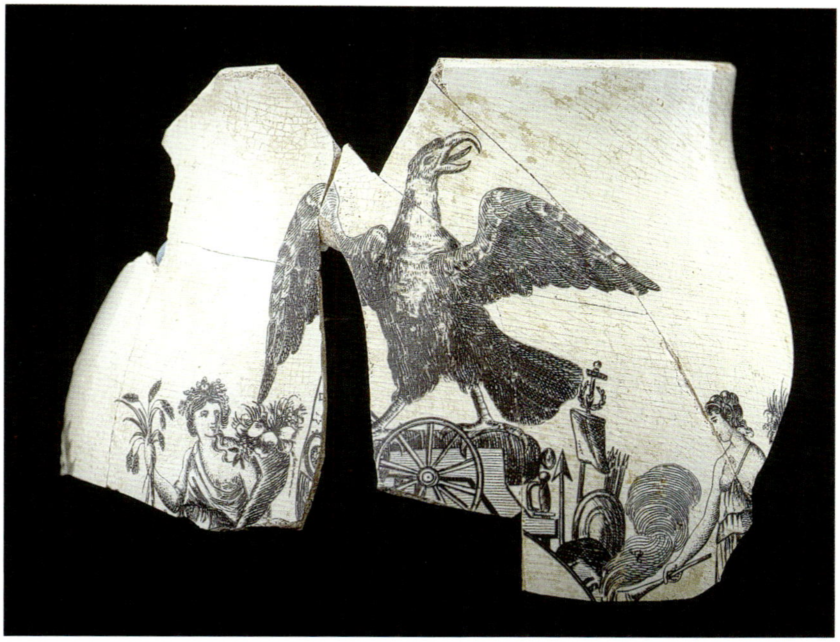

Figure 26 Pitcher fragment, England, ca. 1815. Creamware. These fragments from Arell's Tavern show the upper portion of a pattern with the spread eagle motif perched atop the cannon. The female figure at right, representing Peace, is torching the tools of war. The figure at left, representing Plenty, holds a sheaf of wheat and a cornucopia. See fig. 27 for the decoration that would appear beneath the cannon. Additional fragments of this pitcher have the names Georgia, Massachusetts, and New Jersey on cartouches beneath the spout.

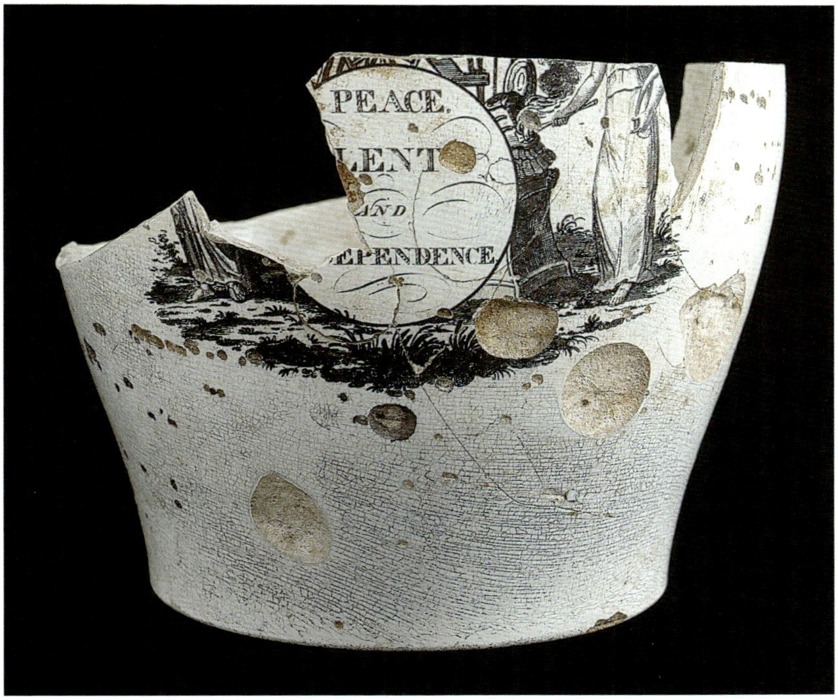

Figure 27 Pitcher fragment, England, ca. 1815. Creamware. This pitcher base was found in a privy associated with Gadsby's Tavern. The pattern is similar to the example from Arell's Tavern illustrated in fig. 26. The slogan "PEACE, PLENTY AND INDEPENDENCE" appears at the bottom. Another fragment of this pitcher, from beneath the spout, shows the American eagle carrying in its beak a ribbon with the motto "E PLURIBUS UNUM."

Creamware pitchers were found at Arell's Tavern and Gadsby's Tavern with representations of Peace, Plenty, and Independence. In cartouches under the spout of the pitcher from Arell's (fig. 26) appear the names of Georgia, Massachusetts, and New Jersey. The example from Gadsby's (fig. 27), which was produced at another pottery, has a version of the Great Seal under the spout with the eagle carrying a ribbon bearing the motto "E PLURIBUS UNUM."

Gadsby's Tavern also had a pearlware transfer-printed serving dish, attributed to James and Ralph Clews, circa 1818–1834, which was part of a

Figure 28 Serving dish, attributed to James and Ralph Clews, Staffordshire, ca. 1818–1834. Pearlware L. 12½". The America and Independence series depicts a variety of generic landscapes. This serving dish, recovered from the privy of Gadsby's Tavern, shows an unidentified castle. The states—New Jersey, Pennsylvania, Maryland, New York, Delaware, Virginia, New Hampshire, Massachusetts, Rhode Island, Connecticut, North Carolina, South Carolina, Georgia, Vermont, and Kentucky—are named in the border. Two figures flank the scenic view, standing on plinths bearing the words "AMERICA AND / INDEPENDENCE." A blindfolded figure labeled "JUSTICE" holds a portrait of Washington. A kneeling figure holds the Liberty Cap.

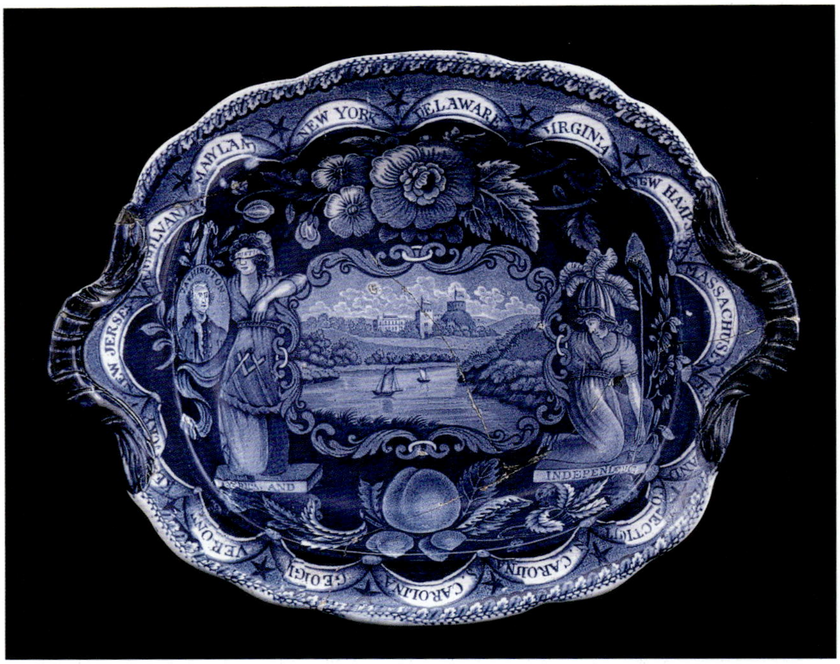

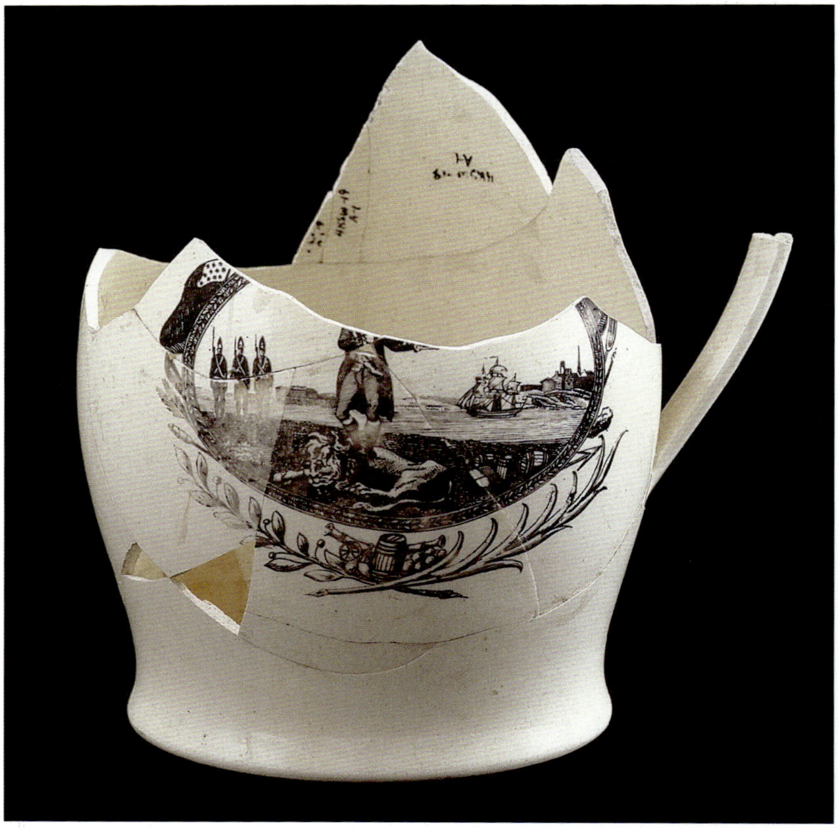

Figure 29 Pitcher fragment, England, ca. 1815. Creamware. This pitcher, found in a privy on a residential site at 414 King Street, commemorates the end of the War of 1812. The transfer print depicts a soldier standing on a lion—representing Britain—and gesturing toward a merchant ship under sail. American flags and weapons of war ornament the cartouche. A hunting scene is on the reverse.

series known as America and Independence (fig. 28). Despite the generic English landscapes, the America and Independence series has strong American patriotic content.

From nearby residential sites, archaeologists recovered pitchers with representations of Virtue and Valour and Peace and Plenty, and others honoring the war heroes General Zebulon Pike and Captain Jacob Jones. An Alexandria creamware pitcher (fig. 29) found in a privy on a residential site depicts a soldier standing on a lion (Britain) and gesturing toward a merchant ship under sail. American flags and weapons of war ornament the cartouche. Many different versions of this print appear on pitchers, bowls, and plates, and each usually includes the words "By Virtue and Valour We Have Freed Our Country, Extended Our Commerce, and Laid the Foundation of a Great Empire."[38]

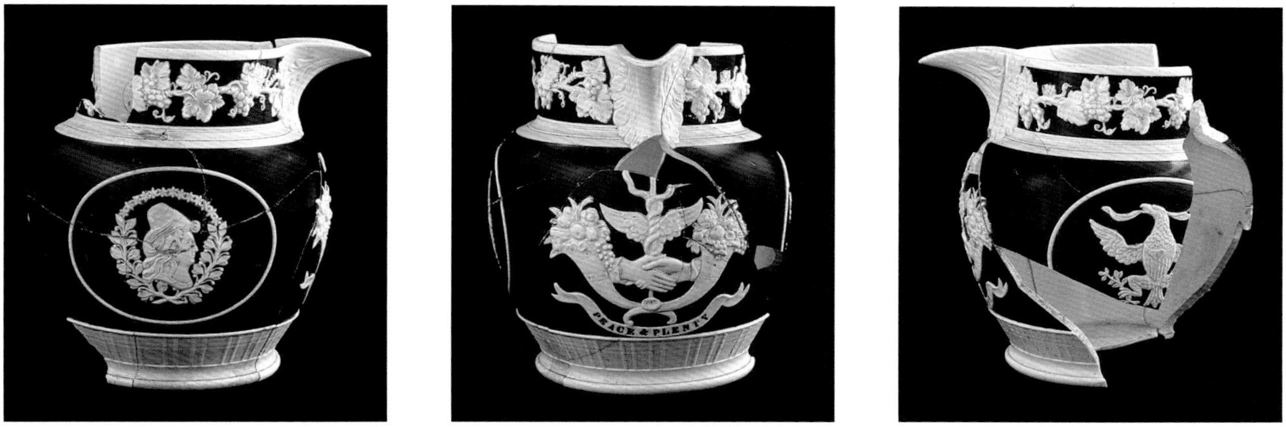

Figure 30 Pitcher, England, ca. 1815. "Cameo jasper" refined stoneware. H. 6". This pitcher has applied relief decoration on a white stoneware body against a dark brown ground. The same pattern is also found on a blue or tan ground. The Great Seal is depicted on one side, and Miss Liberty on the other. Peace is represented by the caduceus and hands clasped in friendship, Plenty by the overflowing cornucopia. This pitcher celebrates the Treaty of Ghent, which was signed on Christmas Day 1814.

On the stoneware Peace and Plenty pitcher (fig. 30), hands clasped in friendship and the caduceus represent Peace, and the overflowing cornucopia represents Plenty.[39] The American Eagle is depicted on one side, and Miss Liberty on the other. This depiction of Liberty is known as the "capped bust to right" found on gold coins from about 1795 to 1806.[40] The same pattern is also found on a blue or tan ground.[41] This pitcher celebrates the Treaty of Ghent, which was signed on December 25, 1814, but similar pitchers and mugs were made to celebrate the Battle of Waterloo, which took place six months later. On these pieces the words "Peace and Plenty" were replaced with "La Belle Aliance," a reference to the meeting between the duke of Wellington and Marshal Blücher at the inn of that name. On the sides are busts of Wellington, or Wellington and Blücher, replacing the Eagle and Liberty.[42]

Rembrandt Peale's portraits *Jones of the Macedonian* and *General Pike,* first published in 1813,[43] adorn a creamware pitcher with a buff-colored ground

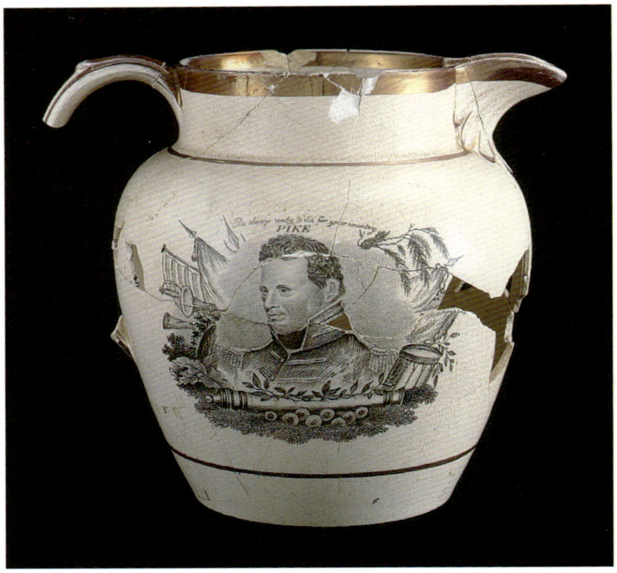

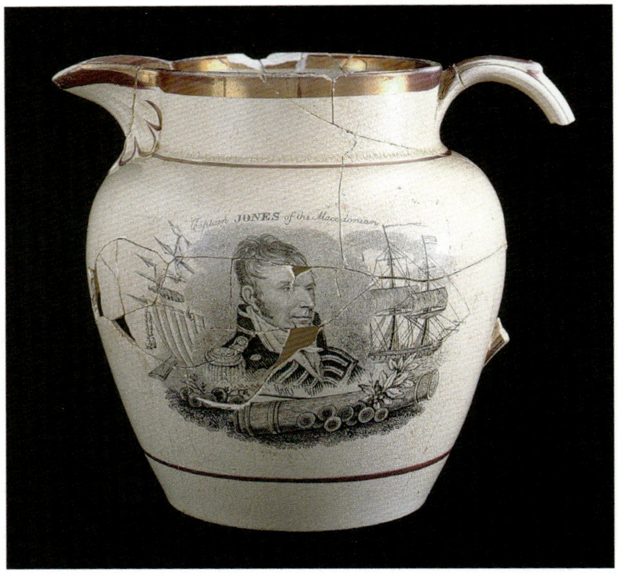

Figure 31 Pitcher, attributed to Enoch Wood, Staffordshire, ca. 1815. Creamware with buff-colored ground. H. 6½". On one side appears Rembrandt Peale's likeness of Captain Jacob Jones, and on the other that of General Zebulon Pike, portraits that were first published in the *Analectic Magazine* in 1813.

attributed to Enoch Wood (fig. 31). Jacob Jones was appointed midshipman in 1799 and served during the Quasi-War with France. In the War of 1812, as commander of the *Wasp* he captured the British brig *Dolphin* and defeated the British warship *Frolic*. After he was commissioned captain, in 1813, Jones commanded the *Macedonian,* which in 1815, after the war, he commanded against the Barbary powers in Algiers.[44] General Zebulon Pike was an explorer, best known for his discovery of Pike's Peak in Colorado. Pike led expeditions on behalf of the U.S. Army to explore the headwaters of the Mississippi River and the southwestern borders of the Louisiana Purchase. A brigadier general in the War of 1812, Pike commanded American troops at the Battle of York (now Toronto) on April 27, 1813. As his troops entered the fort, the British blew up a stone powder magazine. The rain of stone and timber killed Pike and many of his men. Still, the Americans captured the fort and began to take control of the Great Lakes.[45]

"*A La Fayette* this, *and a La Fayette* that"

Lafayette served on George Washington's staff and helped to defeat Cornwallis at Yorktown. Cornwallis surrendered to American and French allies under Washington in what is considered to be the definitive battle of the Revolution. Lafayette returned to France and the French Revolution but was forced to flee during the Napoleonic era; he returned to public life in France after Napoleon's defeat.

General Lafayette's triumphal tour of America in 1824, after a forty-year absence, was celebrated throughout the country. As the last surviving general of the Revolutionary War, Lafayette was invited by President James Monroe to be "the Nation's Guest." He attended parades, ceremonies, and receptions in all twenty-four states. Lafayette visited Alexandria five times,[46] but his most important visit began on October 19, 1824, when a grand reception and celebration were held in his honor.

> [T]he reception and welcome of Lafayette . . . in Alexandria were pronounced by John Q. Adams as equaling any, in any part of the country, in design and execution. . . . Lafayette was much interested in the many associations connected with Washington surrounding Alexandria, and spoke of it as Washington's town, which he well remembered.[47]

For almost a month Lafayette resided at 301 South St. Asaph Street, in an imposing brick house lent to him by Mrs. Thomas Lawrason (fig. 32).[48] Another contemporary, the well-known Quaker educator Benjamin Hallowell, wrote in his diary:

> Washington street in Alexandria is one hundred feet wide, each sidewalk being eighteen feet. There were three spans of arch erected in honor of La Fayette across it, just north of King street, two of eighteen feet span over the sidewalks, and the central one sixty-four feet span. They made a very handsome appearance. Appropriate mottos were on each side of each arch. I remember the two that were on the central arch. On the side of his approach was, "Welcome La Fayette! A nations's gratitude thy due!" On the other was an extract from a speech he had made in the Paris Tribune, "For a nation to be free, it is sufficient that she *wills* it."
>
> Colonel Mountford, keeper of the Alexandria Museum, had fastened one of his live eagles on the crown of the arch, with its head in the direction of the approach of the procession, and he and others said that just as the General was passing under the arch, the eagle rose, spread out his wings to their full extent, then gracefully folding them, resumed his former position.[49]

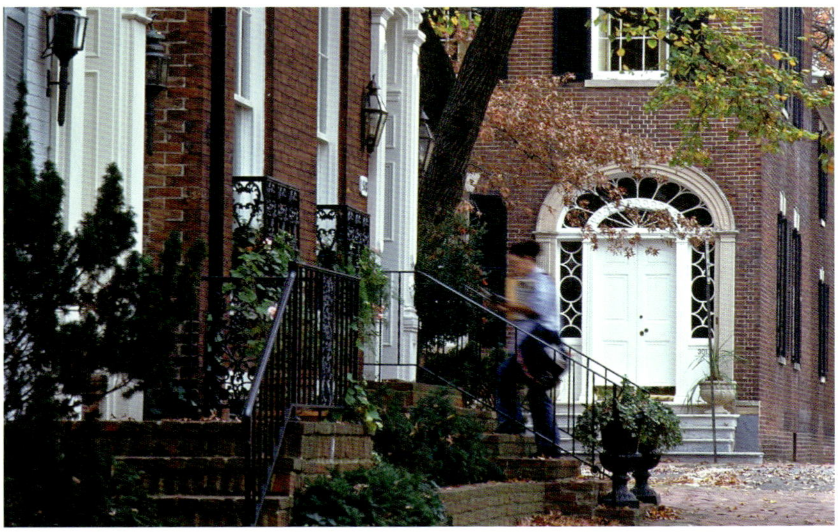

Figure 32 Lafayette House, 301 South St. Asaph Street, Alexandria, Virginia. In October 1824 Mrs. Thomas Lawrason lent her home to Lafayette during his visit to Alexandria. The house continues to be a private residence. (Courtesy, Office of Historic Alexandria; photo, Eric Kvalsvik.)

Another account of the procession describes "a fully rigged ship hung beneath the central span of the arch and the whole . . . decorated with cedar, laurel and oak, set off by a liberty cap."[50] Regarding Colonel Mountford and the eagle, the "Old Bachelor" adds:

> His happiest and busiest days were about the time of the reception of Lafayette here, when he was employed at the arch on Washington street, for which he procured the "live eagle," which afterwards gave him so much trouble, but which he would never give up or suffer to go into other hands, but kept it "flourishing" as long as he lived.[51]

Numerous Lafayette souvenirs were available throughout the country in anticipation of Lafayette's visit, and they continued to be sold for more than a year. In September 1824 the *Alexandria Gazette* advertised "La Fayette badges" for sale at J. H. Runnell's hat and fancy store on King Street.[52] On July 14, 1825, another King Street merchant, Robert H. Miller, offered "China cups and saucers, Tea plates & snuff boxes, Imitation China pitchers, Mugs and bowls, Lustre pitchers of all sizes, mugs and cans With a drawing of La Fayette & the surrender of Cornwallis. Executed expressly for him, from a drawing sent out."[53] These proved to be so popular that in March 1826 Miller offered "a further supply of Lafayette Ware" along with "tea-sets gold edge and view of Mount Vernon."[54] A prominent member of Alexandria's Quaker community, as well as the son and grandson of successful merchants, Miller was one of Alexandria's foremost businessmen and most influential citizens.[55]

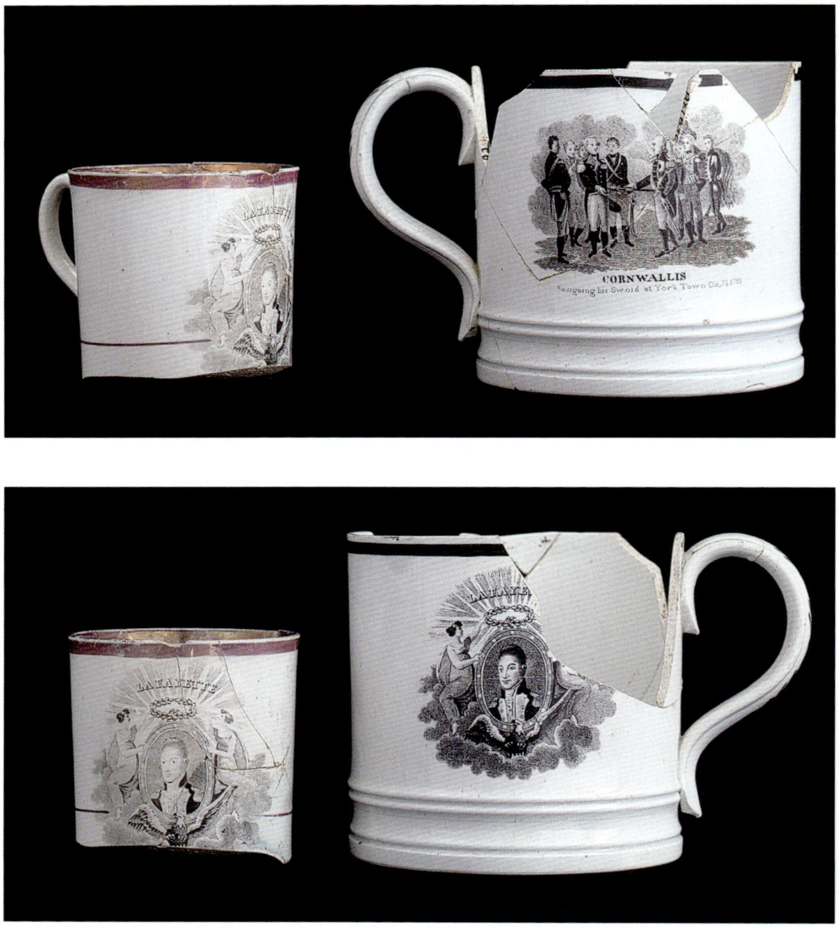

Figure 33 Coffee can and mug, England, ca. 1825. Pearlware. H. of coffee can 2⅛"; H. of mug 3⅛". The engraving *Lafayette Crowned at Yorktown* appears on both vessels. The engraving *Cornwallis Resigning His Sword at York Town, Oct. 19, 1781,* is on the reverse of the mug.

Archaeologists recovered a Lafayette and Cornwallis pearlware mug (fig. 33, right) from a well in which it had been discarded by the family of Ann Buckland, who lived at 104 South St. Asaph Street from 1810 to 1834.[56] Several examples of ceramics marked "Manufactured for Robt H. Miller / Alexandria D·C" have been found in Alexandria (see figs. 34, 36), and although this mug is unmarked, it fits the description in Miller's 1825

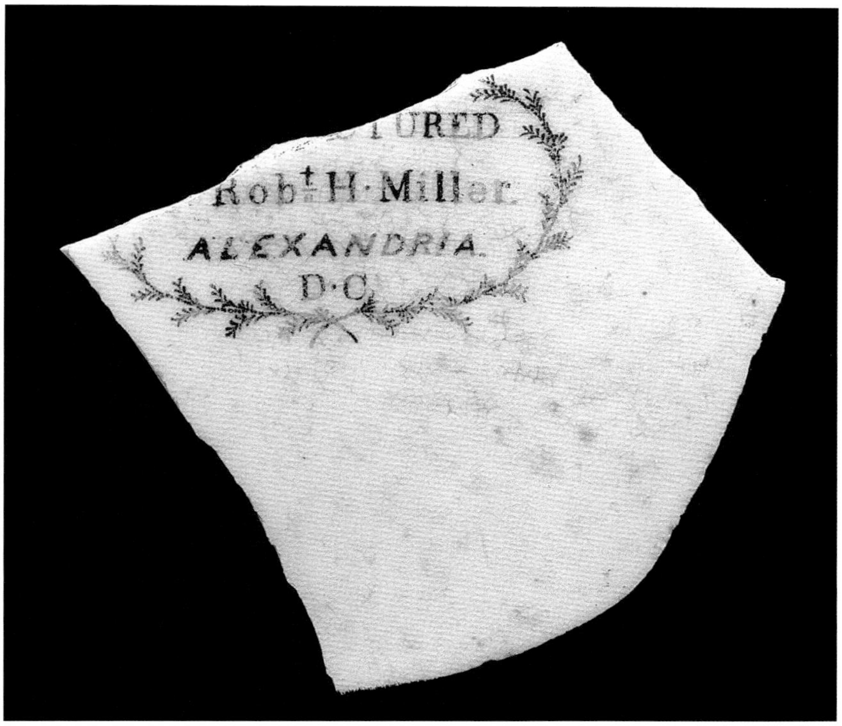

Figure 34 "MANUFACTURED / for Robt H·Miller / ALEXANDRIA. / D·C." Importer's mark, England, ca. 1830. Whiteware. Robert H. Miller was one of Alexandria's most successful businessmen. In 1822 he opened a shop at 317 King Street, on the Market Square, having traveled to England to select the first stock of china and earthenware. Miller's advertisements in the *Alexandria Gazette* state that the Lafayette/Cornwallis wares were "executed expressly for him, from a drawing sent out" (July 14, 1825), and that the Harrison campaign wares are from "designs sent out to the Potteries by himself" (September 19, 1840). In fact the designs are stock prints, but he may have selected the prints and vessel forms from a catalog. This mark is from the base of an unknown ceramic vessel that was discarded about 1830 at 418–422 S. Royal Street (44AX30), in the historic free black neighborhood known as Hayti.

advertisement. The mug is adorned with an engraving, *Lafayette Crowned at Yorktown,* depicting two angels holding the wreath of laurels with which Lafayette was crowned at Yorktown in 1824. On the reverse is *Cornwallis Resigning His Sword at York Town, Oct. 19, 1781,* which originally illustrated the epic patriotic poem "The Columbiad," published by Joel Barlow in 1807.[57] This print was also used on earlier creamware.[58]

A small coffee can with a similar image of *Lafayette Crowned at Yorktown* was found in a well associated with Gadsby's Tavern, one of the places at which Lafayette is known to have dined. The coffee can is of pearlware with a black overglaze transfer print and a purple luster rim (see fig. 33, left).

Benjamin Hallowell wrote about the ubiquitous Lafayette souvenirs in a letter to his uncle, as well as in his diary. Hallowell and his wife were returning from their wedding, which had taken place the day before Lafayette's visit to Alexandria:

> Each lover of liberty surely must get
> Something in honor of La Fayette.
> There's a La Fayette watch-chain, a La Fayette hat,
> A La Fayette *this,* and a La Fayette *that.*
> But I wanted something as lasting as life,
> And took to myself a La Fayette wife.[59]

Alexandria and Presidential Campaigns

Two American heroes of the War of 1812, Andrew Jackson and William Henry Harrison, went on to become president of the United States—in 1829 and 1841, respectively—and the Alexandria merchant Robert H. Miller imported campaign souvenirs for both of their elections.

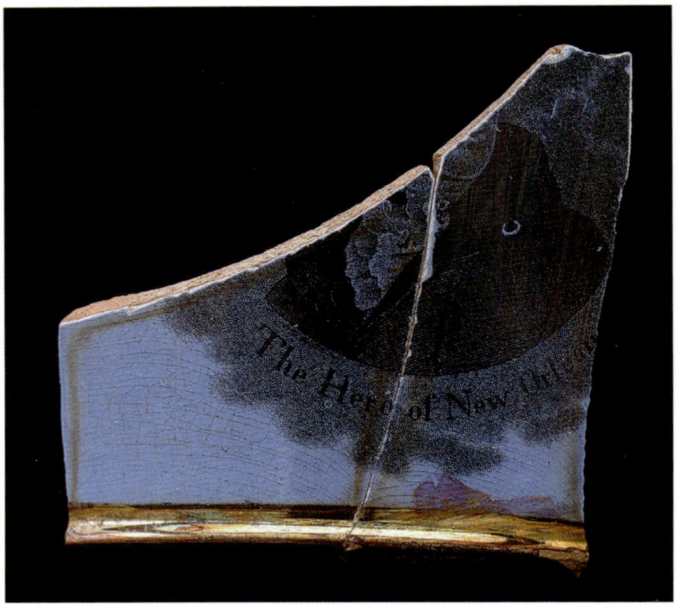

Figure 35 Pitcher fragment, Enoch Wood and Sons, England, ca. 1828. Copper luster, red stoneware body, white slip interior. The Andrew Jackson, Hero of New Orleans motif is usually printed on a canary or blue band, but has also been seen on rust-colored bands. It is also found on a round form, although the slant-sided pitcher is more common. Ellouise Baker Larsen attributes the portrait to J. Wood, engraved by C. G. Childs (Philadelphia, 1828). Copper luster was made from about 1823 to 1845, so such pitchers were intended to celebrate Jackson's presidential election, long after the Battle of New Orleans.

Andrew Jackson

On January 3, 1829, the *Alexandria Gazette* advertised: "Robert H. Miller has just received per Pioneer, direct from Liverpool, a further supply of EARTHENWARE & CHINA; among which are . . . plain and lustred pitchers, mugs, twiflers and muffins with an excellent portrait of Gen. JACKSON. . . ." One such "lustred" pitcher, which had been made for Jackson's presidential election (fig. 35), was found in a privy at 522–524 King Street.[60] This property was occupied by a series of short-term renters, one of whom was Lewis McKenzie, a grocer and Union sympathizer who became Alexandria's mayor during the Civil War.

Andrew Jackson became the seventh president of the United States on March 4, 1829, after an unsuccessful attempt four years earlier. Born on the frontier and known as "Old Hickory," he was hailed as a symbol of the common man. He was nationally known for leading American troops to victory in the Battle of New Orleans, a battle fought on January 8, 1815, two weeks after the Treaty of Ghent was signed. In the days of slow communication, word that the treaty had been signed in Europe did not reach troops in America, and the Battle of New Orleans was the decisive battle ending the war. The American public heard about the battle before it heard about the treaty, and so hailed Jackson for his role in ending the war. His popularity as a military hero helped him to gain the presidency.

In 1833, during his second term, Jackson was aboard the presidential steamer in Alexandria's harbor when Lieutenant Robert Beverly Randolph, a naval officer who had been dismissed for misconduct, went on board and assaulted Jackson, striking him in the face (or, by another account, seizing his nose). Randolph was held and punched in the ribs, but in the confusion his friends helped him to escape and provided him with a horse. One of them was said to be the son-in-law of John Gadsby, former owner of Gadsby's Tavern. Randolph was not captured for several months and the charges eventually were dismissed.[61]

William Henry Harrison

Harrison, like Jackson, was a nationally known military hero. He served in a campaign against the Indians, which opened up the Ohio frontier. He served as secretary of the Northwest Territory and, later, as governor of the Indiana Territory. Harrison worked to obtain Indian lands and met opposition from Tecumseh. Harrison is remembered for fighting Tecumseh and the Indian confederacy at the Battle of Tippecanoe in 1811. In the War of 1812 he defeated British and Indian forces near Lake Erie. His old nemesis Tecumseh was killed in this battle, which ended serious Indian resistance in the Northwest.

Harrison was cast as the common man, as was Jackson, although he was a candidate of the conservative Whig Party, the opponents of Jacksonian democracy. Harrison was the son of an aristocratic Virginia planter, but his military and political career in the Northwest enabled him to be presented as a simple frontiersman, in contrast to the aristocratic Martin Van Buren. With the "Log Cabin" campaign and the slogan "Tippecanoe and Tyler Too," Harrison was the first presidential candidate to conduct a public campaign. Harrison, said to have caught cold at his inauguration, developed pneumonia and died after only one month in office, ending the hopes of the Whig Party.[62]

Whereas the ceramic pitchers commemorating Andrew Jackson were imported from England by R. H. Miller after Jackson was elected president, the Harrison wares were souvenirs distributed as part of his election campaign. An advertisement published in the *Alexandria Gazette* on September 19, 1840, read in part: "N.B.— By an early arrival, R. H. M. expects supplies of ware with Harrison and Log Cabin engravings, from designs

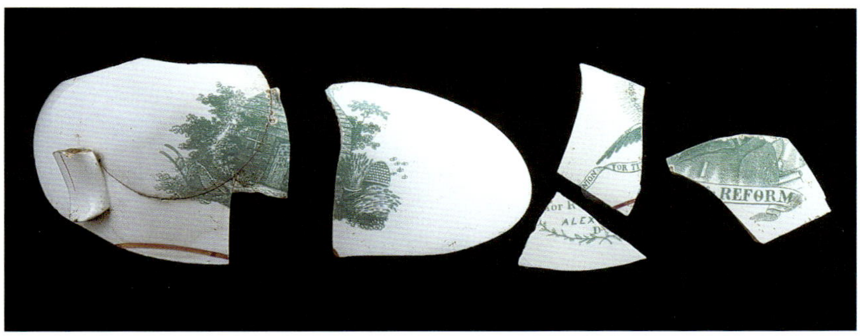

Figure 36 Pitcher sherds, England, 1840. Whiteware. These fragments of a small pitcher, made for Harrison's "Log Cabin" campaign, were found in a well associated with a doctor's office (among the other artifacts were tongue depressors, syringes, and eyeglasses). Only a few pieces were found, but they include a portion of the base with part of the mark "Manufactured for Robt H. Miller, Alexandria DC." The log cabin depicted in the overglaze green transfer print has a sign on it that reads "TO LET IN 1841," the year Harrison went to the White House. On the reverse is a portrait of Harrison and the motto "HARRISON AND REFORM." Under the spout is an eagle holding a scroll bearing the motto "UNION FOR THE SAKE OF THE UNION."

sent out to the Potteries by himself. Whig merchants will be supplied upon the 'Credit System' at reasonable prices, in time to celebrate the approaching triumph of correct principles."

Fragments of a Harrison pitcher (fig. 36) with an overglaze green transfer print were recovered from a well on Market Square containing artifacts relating to a doctor's office. The pitcher is marked "Manufactured for Robt H. Miller, Alexandria DC."[63] Robert Miller's son Warwick recalled:

> The Whig campaign of 1840 was a memorable time; it seemed as if the whole country was crazy over the log cabin candidate General Harrison and John Tyler who was nominated at Harrisonburg. Uncle John Janney was a delegate from Virginia and came very near being nominated for

> Vice President but John Tyler got it and became President after Harrison's death in a few months and then betrayed the party that nominated him. Alexandria was then in the District but there was as much excitement as though they had votes.[64] A glee club was started and was so noted that they were invited to attend meetings all over that region. Was a member of the junior club—my only appearance as a vocalist. I think I must have ruined my voice at that time shouting!!
>
> After the election there was a jollification meeting and a most remarkable one for the orators assembled: Daniel Webster, William Preston of South Carolina, Waddy Thompson, Ogden Hoffman, Henry A. Wise, George H. Proffitt and others winding up with Uncle John Janney. All these were noted orators; Webster was at his best under the influence of Uncle Phineas Janney's old Port wine of which he drank tumbler after tumbler.[65]

Harrison commemorative wares, including a plate and a cup and saucer, were also found in recent excavations at the site of the Benjamin Delaney mansion on Shuter's Hill. The mansion burned in 1842, just one year after the election. The Harrison plate fragment is similar to the pitcher, with a green overglaze transfer-printed portrait. The three small fragments from a Harrison teacup and saucer have black underglaze transfer prints. The Alexandria fragments are whiteware but correspond to a cup and saucer illustrated in David and Linda Arman's *Anglo-American Ceramics* that are listed as Staffordshire porcelain. On the saucer is a portrait of Harrison flanked by flags; below are the words "Harrison and Reform" and, above, "Our Country's Hope."[66] In a reference to Harrison's so-called log cabin upbringing promoted in his presidential campaign, the cup shows on one side a log cabin with the words "To Let in 1841."

Alexandria and the Civil War: The Marshall House Incident

On May 23, 1861, when Virginia voted to secede from the Union, James A. Jackson, a notorious secessionist and the proprietor of the Marshall House Hotel, raised a Confederate flag on the hotel's roof (fig. 37). The next day, Colonel Elmer Ellsworth, commander of the New York Fire Zoaves and well known throughout the country before the war, led his federal troops into Alexandria to occupy the city. In the ensuing incident Jackson became the first civilian casualty and Ellsworth the first martyr for the Union.

In the days and months leading up to the incident, Alexandria was in turmoil, fearing that the Union would be dissolved and that chaos would result. On election day in 1860 the editor of the *Alexandria Gazette* wrote:

> Let every friend of the Union and the Constitution vote this day for BELL AND EVERETT . . . Union men, one and all, let us urge you to go to the polls today, and should the Union be dissolved, you will have the proud satisfaction of having done your duty. And should our ticket be successful, you have the greater satisfaction of seeing our country delivered from the Disunionists of the South and the Abolitionists of the North.[67]

Alexandria voted overwhelmingly for the Constitutional-Unionist candidate John Bell, with only two votes cast for Abraham Lincoln. The *Gazette*'s view on Lincoln's election was that "the result was expected as well as dreaded. . . . The crisis is upon us and we must continue to do our duty to our country!"[68]

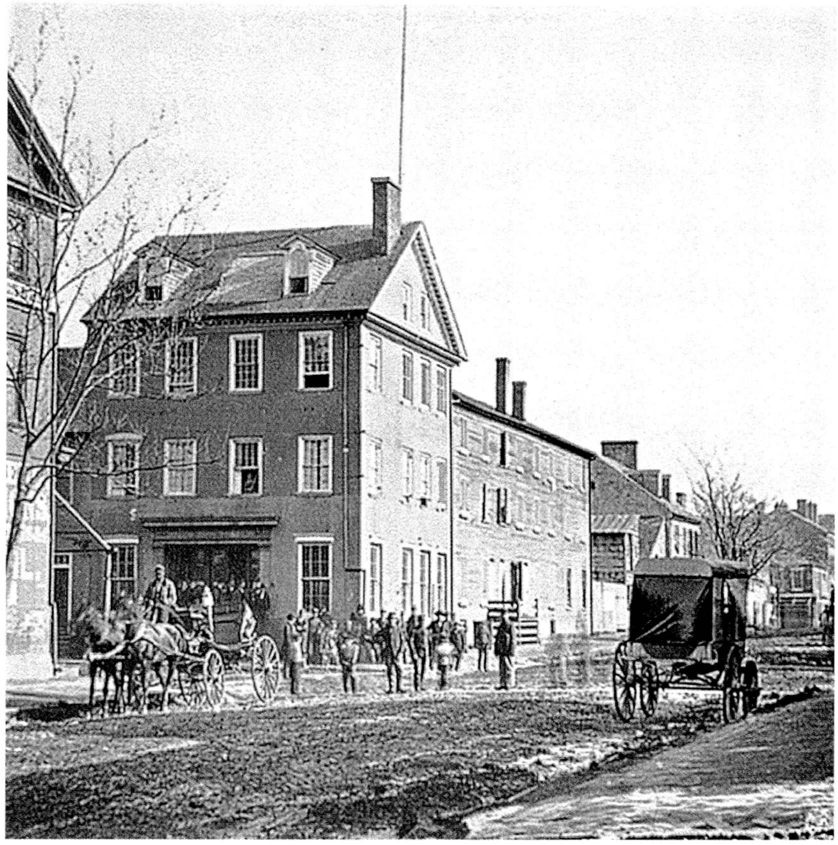

Figure 37 The Marshall House Hotel, King and Pitt Streets, during the Civil War. (Courtesy, Library of Congress, Prints and Photographs Division, LC-B8171-2294 DLC.)

The Virginia state convention passed an ordinance of secession from the Union on April 17, 1861. That same day, James A. Jackson raised the Confederate flag at the Marshall House Hotel, at the southeast corner of King and Pitt Streets. The *Alexandria Gazette* reported:

> The announcement that a Confederate States' flag would be hoisted on the flag staff at the Marshall House, collected quite a large assemblage in that neighborhood yesterday afternoon. The flag was hoisted about 4 o'clock and greeted with cheers and a salute of Artillery. . . . Several American flags, in addition to those already up, were raised yesterday at private residences in the city.[69]

On May 23 Virginians voted to ratify the ordinance of secession—with Alexandria voting 958 for and 48 against. Alexandria was occupied by federal troops the next day. As Colonel Ellsworth led his troops up King Street to seize the telegraph office, he saw the Confederate flag flying atop the Marshall House. Ellsworth entered the hotel and went up to the roof, where he seized the flag. On his way down the stairs, Jackson approached him with a shotgun and killed him. He also fired at Private Brownell, but missed. Jackson, killed by Brownell, became the first civilian casualty of the war. On June 15, 1861, *Harper's Weekly* published this graphic account "from the pen of Mr. House, the [New York] *Tribune* correspondent, who actually had his hand on Colonel Ellsworth's shoulder when Jackson shot him":

> On entering the open door, the Colonel met a man in his shirt and trowsers, of whom he demanded what sort of a flag it was that hung above the roof. The stranger, who seemed greatly alarmed, declared he

knew nothing of it, and that he was only a boarder there. Without questioning him further the Colonel sprang upstairs, and we all followed to the topmost story, whence, by means of a ladder, he clambered to the roof, cut down the flag with Winser's knife, and brought it from its staff. There were two men in bed in the garret, whom we had not observed at all when we entered, their position being somewhat concealed, but who now rose in great apparent amazement, although I observed that they were more than half dressed. We at once turned to descend, Private Brownell leading the way, and Colonel Ellsworth immediately following him with the flag. As Brownell reached the first landing-place, or entry, after a descent of some dozen steps, a man [Jackson] jumped from a dark passage, and hardly noticing the private, leveled a double-barreled gun square at the Colonel's breast. Brownell made a quick pass to turn the weapon aside, but the fellow's hand was firm, and he discharged one barrel straight to its aim, the slugs or buckshot with which it was loaded entering the Colonel's heart, and killing him at the instant. I think my hand was resting on poor Ellsworth's shoulder at the moment. At any rate, he seemed to fall almost from my own grasp. He was on the second or third step from the landing, and he dropped forward with that heavy, horrible, headlong weight which always comes of sudden death inflicted in this manner. His assailant had turned like a flash to give the contents of the other barrel to Brownell, but either he could not command his aim or the Zouave was too quick with him, for the slugs went over his head, and passed through the panels and wainscot of a door which sheltered some sleeping lodgers. Simultaneously with this second shot, and sounding like the echo of the first, Brownell's rifle was heard, and the assassin staggered backward. He was hit exactly in the middle of the face, and the wound, as I afterward saw it, was the most frightful I ever witnessed. Of course Brownell did not know how fatal his shot had been, and so, before the man dropped, he thrust his sabre bayonet through and through the body, the force of the blow sending the dead man violently down the upper section of the second flight of stairs, at the foot of which he lay with his face to the floor. . . . The first thing to be done was to look to our dead friend and leader [Ellsworth]. . . . Winser and I lifted the body with all the care we could apply, and laid it upon a bed in a room near by. The rebel flag, stained with his blood and purified by this contact from the baseness of its former meaning, we laid about his feet.

A coroner's inquest found that Jackson had died in defense of his private property. When the troops learned of Ellsworth's death, they talked of burning the town, but calm was restored. Work on Fort Ellsworth, part of a ring of Union forts known as the Defenses of Washington, began two days later. The outlines of the fort's earthworks are still visible on Shuter's Hill, on upper King Street in Alexandria.[70]

The Marshall House incident had a great effect on Alexandria and the nation. Jackson and Ellsworth were regarded as heroes and martyrs by their respective constituents. Union soldiers as well as Confederate supporters visited the empty hotel and took souvenirs of the event, including fragments of the flagpole and even the stair banisters and flooring.[71] A framed fragment of the Marshall House's Confederate flag and the large wooden *O* from the hotel's sign are preserved in the collection of Alexandria's Fort Ward Museum. Another souvenir of this incident was the Ellsworth pitcher (fig. 38), produced later that year in Trenton, New Jersey.[72] The pitcher, which illustrates the incident in molded relief, labels Jackson "the Traitor." The Ellsworth pitcher has not been found archaeologically,

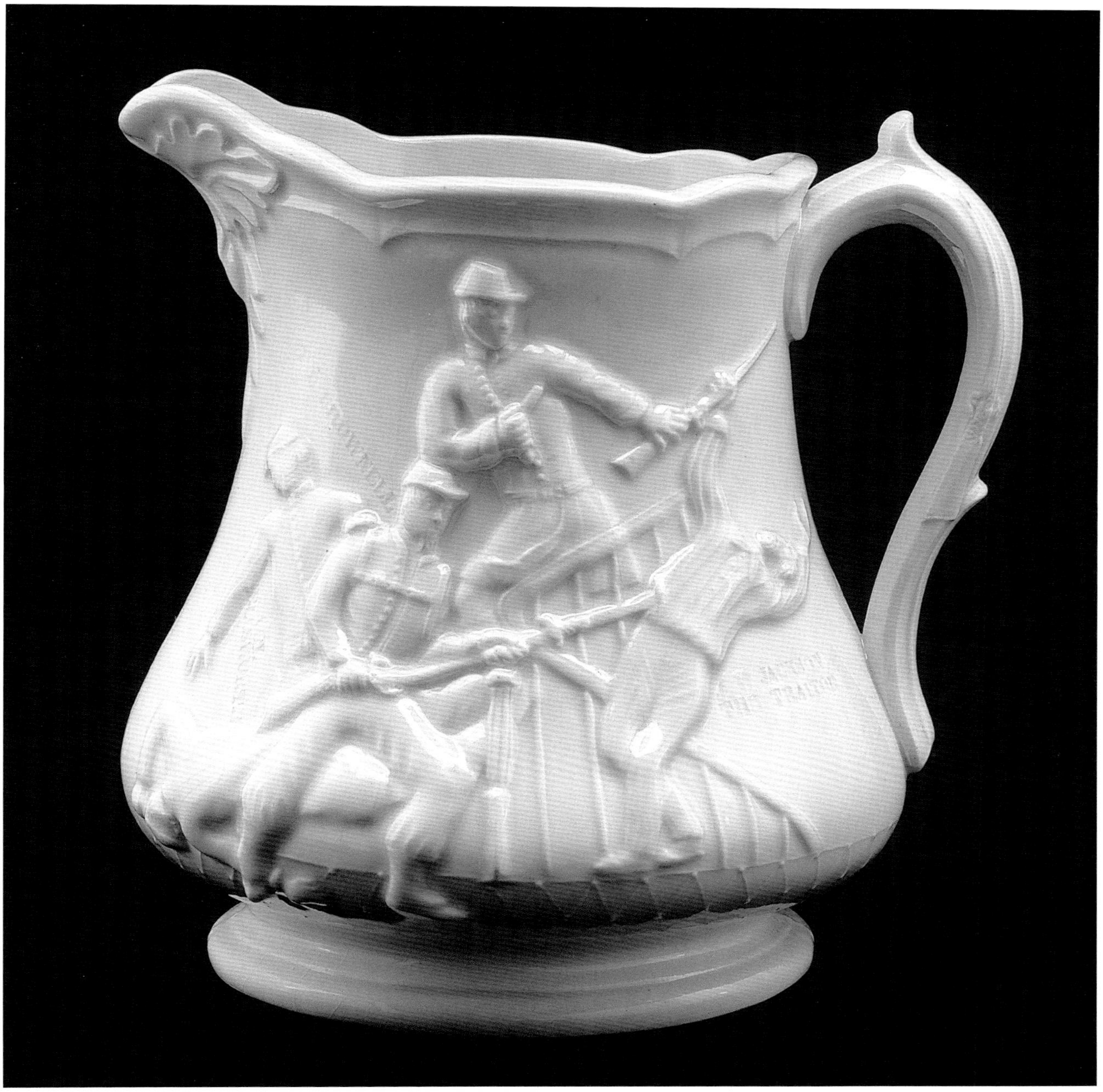

Figure 38 Ellsworth pitcher, attributed to Millington, Astbury and Poulson, Trenton, New Jersey, 1861. Whiteware. H. 9". (Courtesy, The Lyceum: Alexandria's History Museum.) This famous pitcher memorialized Colonel Elmer Ellsworth, the first Union martyr, who was killed in Alexandria on May 24, 1861. The design is based on a sketch by Corporal (formerly Private) Brownell. The molded design shows the figures of the slain Colonel Ellsworth on the left. Brownell, who avenged Ellsworth's death by killing Jackson, is shown in the center, with "J. W. JACKSON THE TRAITOR" identified on the stairs. A Union eagle and Confederate serpent are depicted on the reverse, with a South Carolina palmetto supporting the early Confederate flag. The United States flag is supported by rifles.

but an antique example is preserved in the collection of The Lyceum: Alexandria's History Museum. Alexandria remained an occupied city for the remainder of the war. More than half of Alexandria's men left town, either to fight or to escape the occupation. In 1863 the *New York Herald* reported that Alexandria "has essentially become a military city. Its streets, its docks, its warehouses, its dwellings, and its suburbs have been absorbed to the thousand uses of war . . . . Alexandria is filled with ruined people; they walk as strangers through their ancient streets, and their property is no longer theirs to possess. . . ."[73]

Alexandria Landmarks

In the eighteenth and early nineteenth century, the Staffordshire Potteries usually designed commemorative wares that were for sale to interested merchants. By the 1820s, however, china merchants such as Robert H. Miller were commissioning designs from the potteries. These continued to be, in many cases, stock designs, but they were made to order and marked with the merchant's name. Examples of commissioned designs are the Lafayette and Harrison wares made for Robert H. Miller in the 1820s and 1840s (see figs. 34, 36).

Queen Victoria's Jubilee, in 1897, may have provided the impetus for a revival in souvenir wares, which ultimately led to a renewed interest in American landmarks. Many of the wares for the American market were made in Austria and Germany (figs. 39–42), as well as in the United States, rather than in England. Another factor in the revival was the development of new printing techniques, including the full-color process known as decalomania, introduced in the 1890s. None of these later souvenir wares

Figure 39 Pitchers, Austria, ca. 1897–1908. Porcelain. *Left:* H. 6½"; *right:* H. 3½". (Courtesy, The Lyceum: Alexandria's History Museum. Gift of Ken Turino.) They differ in shape, but these porcelain pitchers share the illustration of Christ Church, Alexandria, and the marks on the base (see fig. 40). Other scenes of churches and historic sites were made for retailers throughout the country. These pitchers were made for E. J. Miller and Co., at 119 King Street, Alexandria.

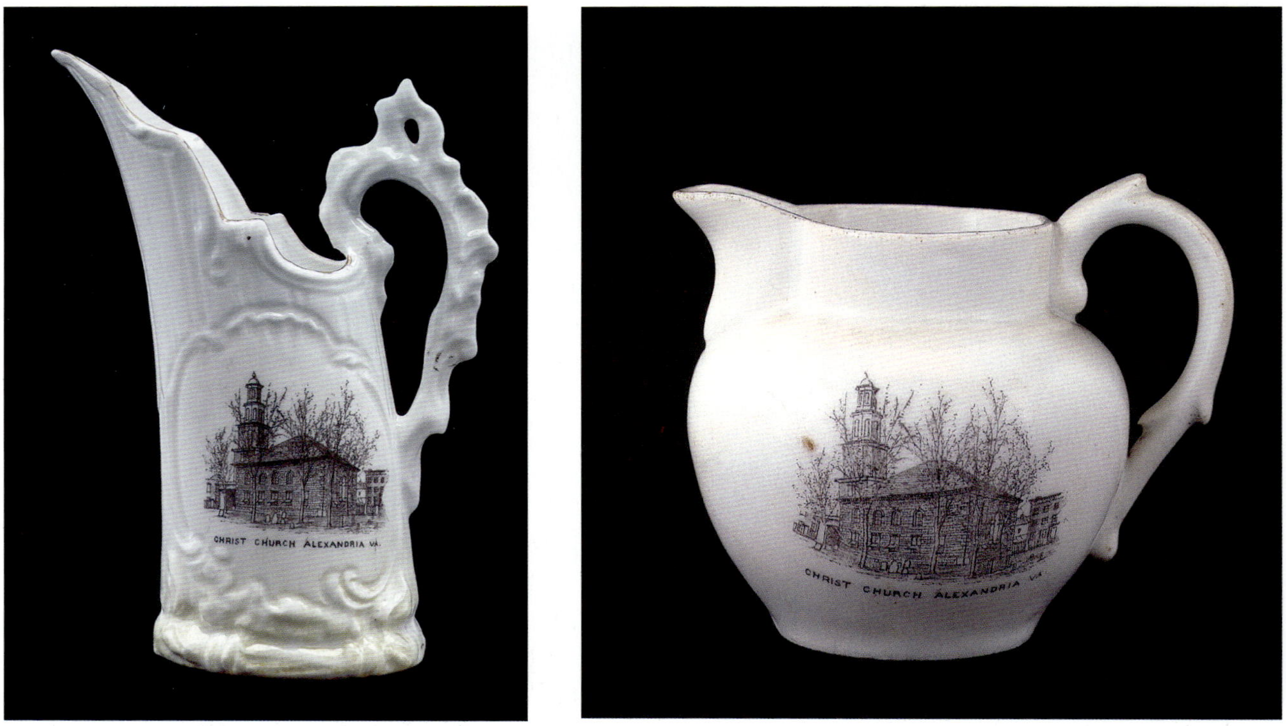

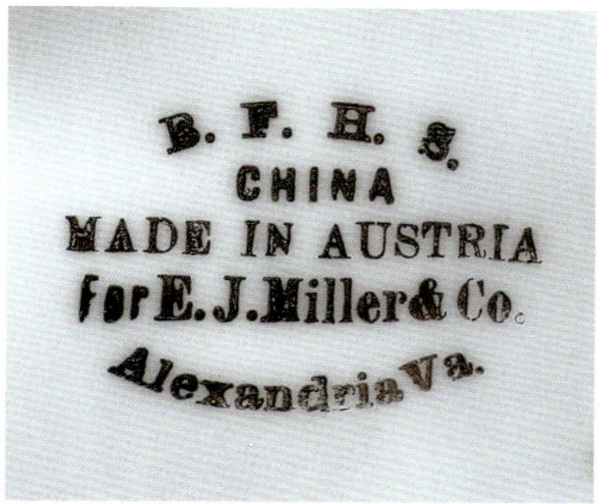

Figure 40 Detail of the mark used on the pitchers illustrated in fig. 39. Benjamin F. Hunt and Sons (B.F.H.S.) was a manufacturer and importer of porcelain from Austria at the turn of the century.

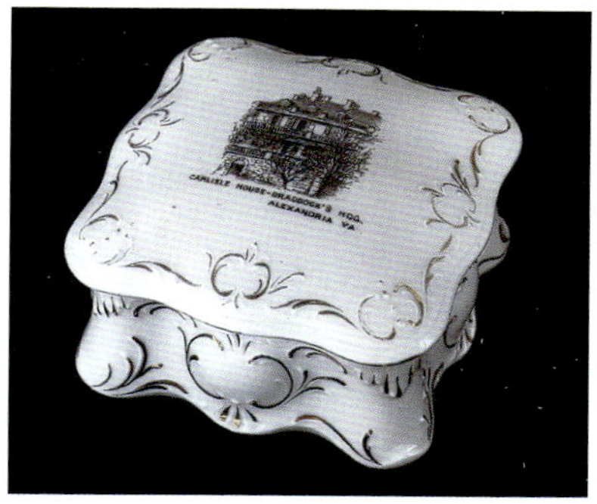

Figure 41 Lidded box, Austria, ca. 1897–1908. Porcelain. W. 4". (Courtesy, Alan Darby.) The box is inscribed "CARLISLE HOUSE—BRADDOCK'S HDQ., ALEXANDRIA VA," and the markings on the bottom are the same as those illustrated in fig. 40.

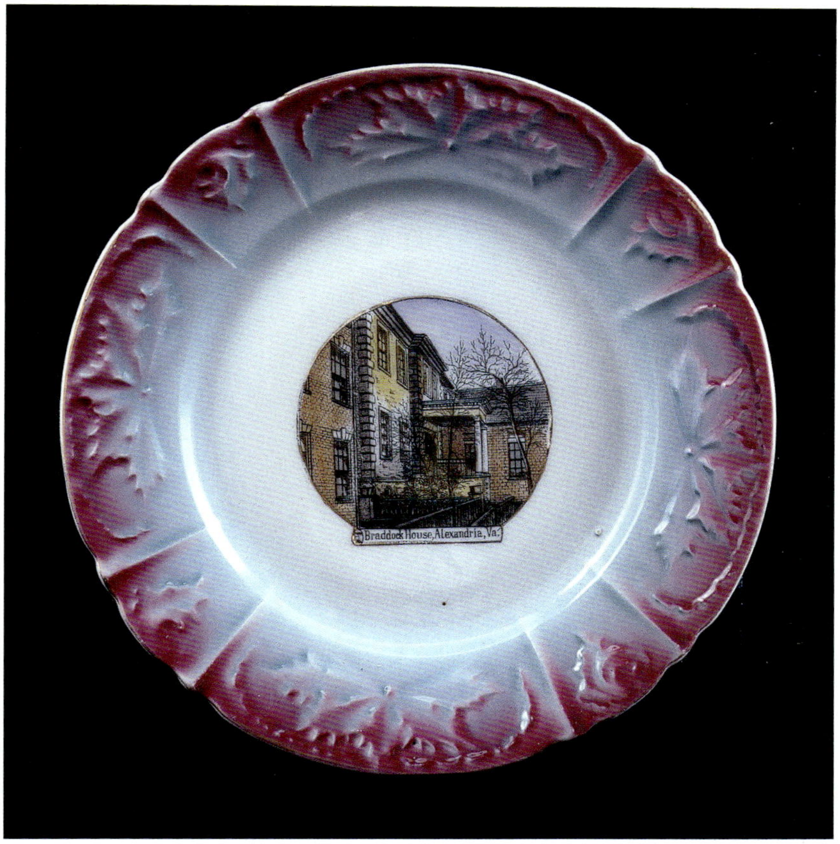

Figure 42 Plate, Germany, early twentieth century. Whiteware. D. 7⅝". (Courtesy, The Lyceum: Alexandria's History Museum. Bequest of William R. Adam.) General Braddock headquartered at John Carlyle's house for three weeks in 1755, during the French and Indian War. This souvenir plate is illustrated with a postcard view of the Carlyle House. For many years the historic mansion was surrounded by a large hotel, which was at one time known as the Braddock House. C. E. Wheelock imported both postcards and china from Germany, and the company marketed plates with local postcard views directly to stores around the country. This plate was made for R. E. Knight in Alexandria. Knight's Stationery, Toy and Souvenir store opened at 621–625 King Street in 1906. Souvenir china was imported from Germany from about 1890 to 1930.

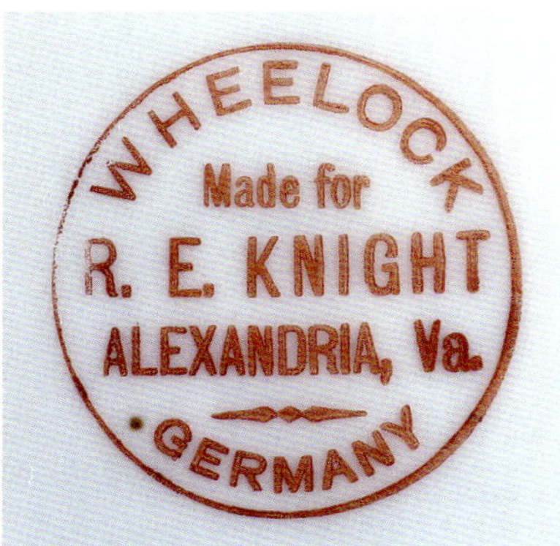

Figure 43 Detail of the mark used on the bottom of the plate illustrated in fig. 42.

has been found archaeologically in Alexandria, but several views of the city's landmarks, including historic Christ Church and the Carlyle House (once known as the Braddock House), were immortalized in ceramics around the turn of the century.

At the turn of the twentieth century, manufacturers and importers continued to create series of commemorative wares, marketing them to local merchants or directly to the historic properties. Both the Christ Church pitchers and the later Methodist Church plate are part of a series of churches from around the country. The Braddock House plate is from a series of postcards that C. E. Wheelock imported from Germany along with the ceramic wares. The Christ Church pitchers and Braddock House box were made for Robert's son E. J. Miller, and the Braddock House plate was made for R. E. Knight.[74] Eventually, the historic sites themselves began to

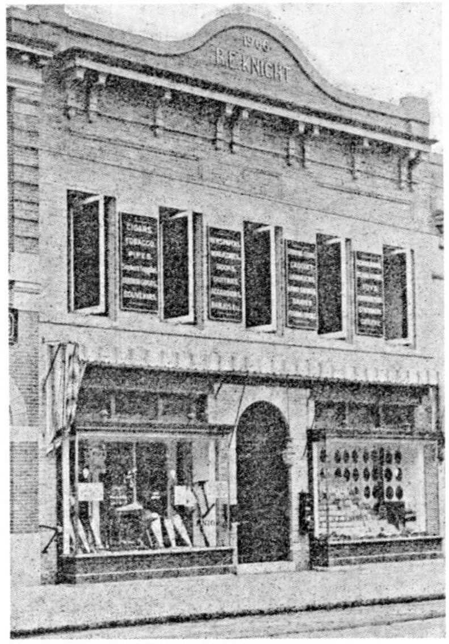

Figure 44 "R. E. Knight's new Stationery, Toy and Souvenir stores," 621–625 King Street, 1906. (Courtesy, Alexandria Archaeology Museum.) The store opened in 1906 and was illustrated in a 1907 souvenir booklet. Alexandria Archaeology's offices were located in this building in the early 1980s.

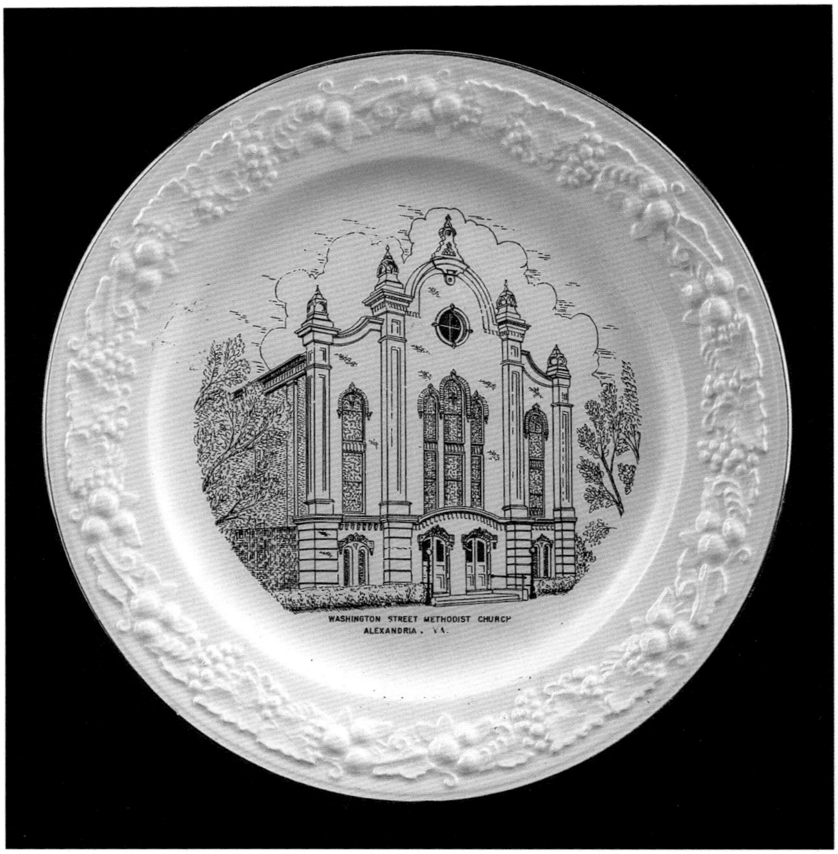

Figure 45 Plate, Homer Laughlin China Company, Newell, West Virginia, 1952. Whiteware. D. 9¾". (Courtesy, The Lyceum: Alexandria's History Museum.) These plates likely were ordered for the centennial of the Washington Street Methodist Church, Alexandria, which was built in 1850–1851, as the numbers and letters in the green-printed maker's mark date this plate to February 1952.

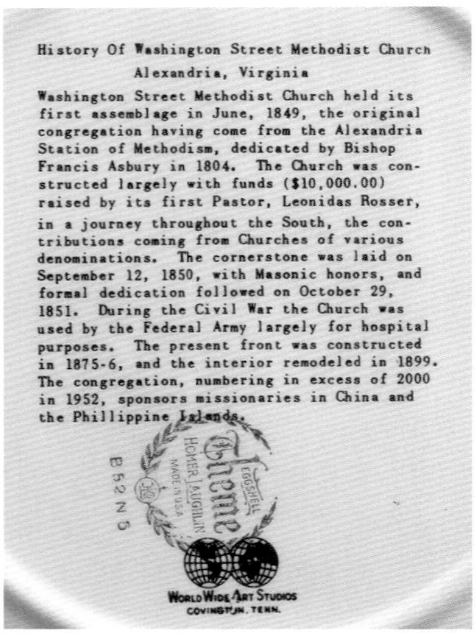

Figure 46 Detail of the mark used on the bottom of the plate illustrated in fig. 45.

commission the wares, as with the Methodist Church plate from the 1950s and the more recent Masonic Memorial plate (figs. 44–48).

Specially made commemorative ceramics continue to be popular. The mug shown in figure 49, for example, was commissioned by the City of Alexandria's 250th Birthday Commission to celebrate the 1999 anniversary of the city's founding. The logo shows a three-masted schooner under sail, a reference to Alexandria's maritime history. The logo was designed "to

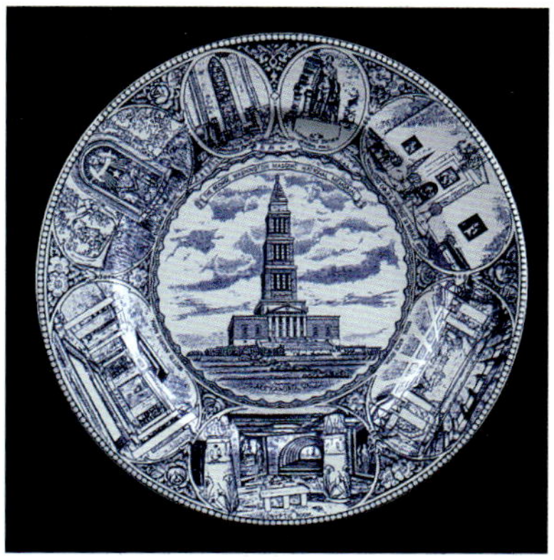

Figure 47 Plate, J. H. Weatherby and Sons, Staffordshire, ca. 1990s. Whiteware. D. 9⅞". (Courtesy, The Lyceum: Alexandria's History Museum. Gift of Monta Lee Dakin.) This plate depicts the George Washington Masonic National Memorial in Alexandria, Virginia, which was built between 1922 and 1932. The memorial served as the home for Alexandria Lodge 22, of which George Washington was a member, serving as master from 1788 to 1789.

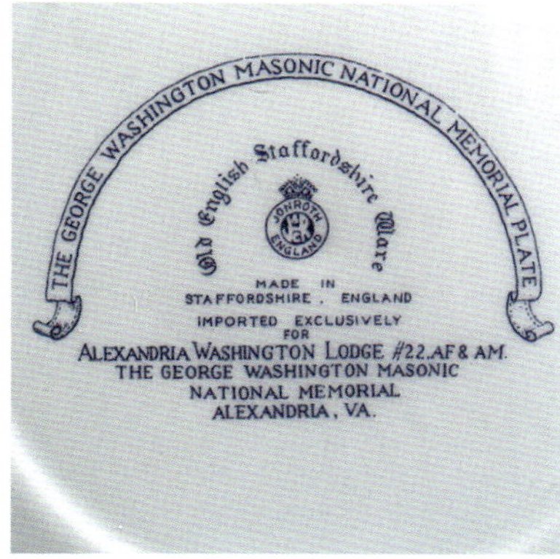

Figure 48 Detail of the mark used on the bottom of the plate illustrated in fig. 47.

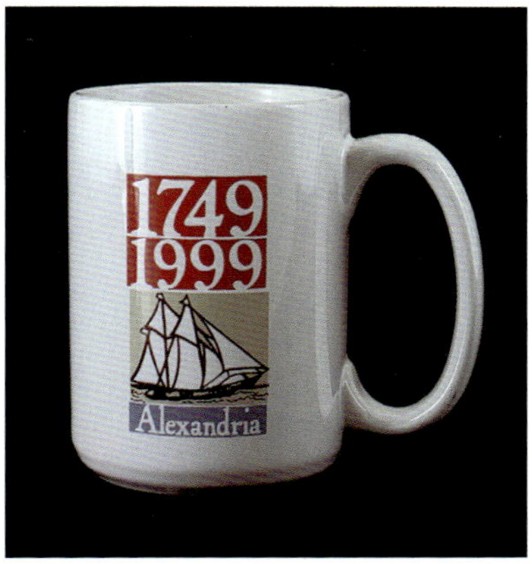

Figure 49 Mug, 1999. Whiteware. H. 3¾". This mug was commissioned by the City of Alexandria's 250th Birthday Commission to celebrate the 1999 anniversary of the city's founding. The logo shows a three-masted schooner under sail, honoring the city's maritime history.

convey energy, optimism and strength for the future of the City." Undoubtedly, discarded fragments of these mugs will one day end up in the local soils to join the other commemorative ceramics that constitute Alexandria's archaeological record.

ACKNOWLEDGMENTS

For sharing their vast knowledge of Alexandria's history, I would like to thank T. Michael Miller, Research Historian for the City of Alexandria; Jim Mackay, Director of The Lyceum; and Timothy J. Dennée, a former employee of Alexandria Archaeology. Thanks also go to curator Kristin Lloyd for providing access to Alexandria commemorative wares in the collection of The Lyceum. I am grateful to Jean Federico, Director of the Office of Historic Alexandria, for her support of this project and for allowing me to use Eric Kvalsvik's photographs of Alexandria's historic sites, which were originally published in William Seale, *A Guide to Historic Alexandria* (Alexandria, Va.: City of Alexandria 250th Anniversary Commission, 2000). And, as always, the support, assistance, and editorial comments of Pamela J. Cressey, City Archaeologist, and all the staff of the Alexandria Archaeology Museum were invaluable.

1. The reminiscences of this anonymous author, who visited Alexandria after a thirty-year absence, were serialized in the Alexandria newspaper *The Local News* in 1861–1862 under the title "Reminiscences of an Old Bachelor" and are reproduced in T. Michael Miller, *Manuscripts of an Old Bachelor: Reminiscences of Alexandria* (Alexandria, Va., 1987), p. 30.

2. *The Great Seal of the United States* (Washington, D.C.: Bureau of Public Affairs, U.S. Department of State, U.S. Government Printing Office, July 1980), p. 6.

3. James D. Munson, *Col. John Carlyle, Gent., 1720–1780: A True and Just Account of the Man and His House* ([Alexandria, Va.]: Northern Virginia Regional Park Authority, 1986), pp. 62–71.

4. Ibid., p. 67.

5. Ibid., pp. 71–73.

6. In his *Guide to Artifacts of Colonial America* (New York: Alfred A. Knopf, 1969), pp. 115–17, Ivor Noël Hume points out that a similar plate was advertised in the *Boston Gazette* on November 13, 1758.

7. Rear Admiral Charles Holmes to unknown addressee, September 18, 1759, Special Collections Department, University of Waterloo Library, Ontario, Canada. For a reproduction of this letter, see www.lib.uwaterloo.ca/discipline/SpecColl/archives/holmes/holmes.html, accessed February 18, 2005.

8. Ellouise Baker Larsen, *American Historical Views on Staffordshire China,* 3rd ed. (New York: Dover Publications, 1975), p. 201, fig. 505.

9. *The Death of Wolfe* has been called "probably the most famous of all historical paintings"; C. P. Stacey, "Benjamin West and 'The Death of Wolfe,'" *National Gallery of Canada Bulletin* 7, vol. IV, no. 1 (1966), available at http://collections.ic.gc.ca/bulletin/num7/stacey1.html, accessed April 4, 2005.

10. Ivor Noël Hume, *All the Best Rubbish* (New York: Harper and Row, 1974), p. 153; idem, *If These Pots Could Talk: Collecting 2,000 Years of British Household Pottery* (Hanover, N.H.: University Press of New England for the Chipstone Foundation, 2001), p. 264. Noël Hume first identified this Alexandria sherd as depicting Lord Rodney and notes that sherds commemorating the British hero were also found, incongruously, at the French Fortress Louisburg in Nova Scotia. According to John May and Jennifer May, *Commemorative Pottery, 1780–1900: A Guide for Collectors* (New York: Charles Scribner's Sons, 1972), p. 81, Rodney souvenirs may have been produced as early as 1768, when he contested the seat from Northampton. While Rodney's involvement in the French and Indian War would provide a better pretext for the presence of these wares in the American colonies, this is too early a date for the Alexandria sherd.

11. *Alexandria Gazette,* September 16, 1784.

12. *Alexandria Gazette,* February 7, 1857, cited in T. Michael Miller, *Pen Portraits of Alexandria, Virginia, 1739–1900* (Bowie, Md.: Heritage Books, 1987), p. 41.

13. Miller, *Manuscripts of an Old Bachelor,* pp. 30–32. In the nineteenth century the museum was located in the town hall and operated by Alexandria Lodge 22. The building was destroyed by fire in 1871, but this letter—and other items that survived from the museum's collection—are now in the collection of the George Washington National Masonic Memorial, Alexandria.

14. *Virginia Herald* (Fredericksburg), December 31, 1799, Papers of George Washington, Alderman Library, University of Virginia.

15. Harold Vedeler, "A History of the Old Presbyterian Meeting House," April 1996, www.opmh.org/history.htm, accessed March 8, 2005.

16. *Alexandria Times and District of Columbia Advertiser,* December 20, 1799, from Gay Montague Moore, *Seaport in Virginia—George Washington's Alexandria* (Richmond, Va.: Garrett and Massie, 1949).

17. A similar pitcher, from the collection of the Mattatuck Museum, Waterbury, Connecticut, is illustrated in Alan Smith, *Illustrated Guide to Liverpool Herculaneum Pottery, 1796–1840* (New York: Praeger Publishers, 1970), fig. 50. These transfer prints were combined with other illustrations on a larger pitcher from the same factory, now in the Philadelphia Museum of Art; see Smith, *Illustrated Guide to Liverpool Herculaneum Pottery,* fig. 49.

18. Wedgwood had made earlier busts of Washington, all based on a sculpture by Houdon taken from a life cast. British ceramics expert Terrence Lockett dated this piece for the Historic Alexandria museums. The bust is marked inside the base with the letters XCD. According to Godden, the third letter refers to the year of manufacture, with D denoting 1875. Geoffrey A. Godden, *Encyclopaedia of British Pottery and Porcelain Marks* (New York: Bonanza Books, 1964), pp. 658–59.

19. For more on the use of designs inspired by the Treaty of Amiens, see Noël Hume, *If These Pots Could Talk,* pp. 237–39.

20. Terence A. Lockett, "Pearlware: Origins and Types (Part 2)," www.thepotteries.org/features/pearlware2.htm, accessed February 18, 2005.

21. Diana Edwards Roussel, *The Castleford Pottery, 1790–1821* (Wakefield, West Yorkshire, Eng.: Wakefield Historical Publications, 1982), pp. 33, 42–43, and pls. 90–93. The "S & Co." impressed marks seen on similar vessels are attributed to Sowter and Company, Mexborough, Yorkshire. Sowter and Co., which made feldspathic stoneware and pearlware, operated the pottery from 1800 to 1811.

22. Ibid., p. 85.

23. Illustrated in David Arman and Linda Arman, *Anglo-American Ceramics,* vol. 1, *Transfer Printed Creamware and Pearlware for the American Market 1760–1860* (Portsmouth, R.I.: Oakland Press, 1998), p. 168.

24. Smith, *Herculaneum Pottery,* fig. 56, shows the *Emblem of America* engraving paired with *The Coopers Arms,* now in the collection of the National Museum of American History, Smithsonian Institution.

25. Miller, *Pen Portraits,* p. 32, citing Marquis de Chastellux, *Travels in North America in the Years 1780, 1781 and 1782,* translated by Howard C. Rice Jr., 2 vols. (Chapel Hill: University of North Carolina Press, 1963), 2: 583 (July 19, 1782).

26. "The QuasiWar with France: First Foreign Fight," *VFW Magazine* (June/July 1998): 40; Michael J. Crawford and Christine F. Hughes, "Historical Overview of the Federalist Navy, 1787–1801," in *Reestablishment of the Navy, 1787–1801: Historical Overview and Select Bibliography,* Naval History Bibliographies, No. 4 (Washington, D.C.: Department of the Navy, Navy Historical Center, 1997), available at www.history.navy.mil.biblio/biblio4/biblio4.htm, accessed April 4, 2005.

27. Donald G. Shomette, *Maritime Alexandria: An Evaluation of Submerged Cultural Resource Potentials at Alexandria, Virginia,* report prepared for Alexandria Archeology, 1985, pp. 104–6, citing *Naval Documents Related to the Quasi-War between the United States and France,* 7 vols. (Washington, D.C.: U.S. Government Printing Office, 1935–1938), 1: 77–78, 2: 364.

28. T. Michael Miller, Pamela J. Cressey, and Timothy J. Dennee, "Discovering the Decades: Alexandria Archaeology Looks Back at 250 Years of Alexandria History; The 1790s," *Alexandria Archaeology Volunteer News* 16, no. 10 (October 1998): 7. Invoices for cargo seized from the Alexandria-registered ships are in the collections of the National Archives.

29. Arman and Arman, *Anglo-American Ceramics,* pp. 5–7. When trade resumed, in 1815, the potteries exported more than 22 million pieces to America. Neil Ewins, "'Supplying the Present Wants of Our Yankee Cousins . . .': Staffordshire Ceramics and the American Market, 1775–1880," *Journal of Ceramic History* 15 (1997): 6.

30. George Robert Gleig, "A British Account of the Burning of Washington" (Washington, D.C.: National Center for Public Policy Research, 2004), available at www.nationalcenter.org/BritishBurnWashington1814.html, accessed March 2, 2005. Originally published in G. R. Gleig, *A Narrative of the Campaigns of the British Army at Washington and New Orleans, under Generals Ross, Pakenham, and Lambert, in the Years 1814 and 1815: With Some Account of the Countries Visited* (London: J. Murray, 1821).

31. Joseph F. Skivora, "The Surrender of Alexandria in the War of 1812 and the Power of the Press," *Northern Virginia Heritage* (June 1988): 9–14.

32. Ibid., p. 10, citing Mayor Charles Simms in a letter to his wife, Nancy, dated September 3, 1814.

33. William Stabler, *A Memoir of the Life of Edward Stabler, Late of Alexandria in the District of Columbia, with a Collection of His Letters* (Philadelphia: John Richards, 1846), p. 56, citing a letter to "My Dear Friends" (other Alexandria Quakers) dated September 21, 1814.

34. The cartoon was drawn and etched by William Charles, a British caricaturist who lived and worked in America from 1805 until his death in 1820. The satire, which Charles also published and sold, was either a companion piece or a sequel to *John Bull and the Baltimoreans*; both were registered for copyright on October 21, 1814. (I would like to thank James Mackay, director of The Lyceum: Alexandria's History Museum, for this information).

35. John S. Williams, *History of the Invasion and Capture of Washington, and of the Events which Preceded and Followed* (New York: Harper and Brothers, 1857), pp. 282–97. Alexandria had become part of the District of Columbia in 1801; it remained so until 1847, when it was retroceded to Virginia.

36. Arman and Arman, *Anglo-American Ceramics,* p. 6.

37. Robert Hunter and George L. Miller, "English Shell-Edged Earthenwares," *Antiques* 165, no. 3 (March 1994): 432–33.

38. Arman and Arman, *Anglo-American Ceramics,* pp. 72–73.

39. On Roman coins Felicitas, the Roman goddess of good fortune, is depicted holding a cornucopia to represent prosperity and a caduceus to represent peace. "Ancient Coins at the Museum of Antiquities: Trajan (AD 98 to 117)" (Saskatoon, Saskatchewan, Canada: Museum of Antiquities, University of Saskatchewan), www.usask.ca/antiquities/coins/trajan.html, accessed February 19, 2005.

40. The capped bust is found on quarter eagle ($2.50), half eagle ($5.00), and eagle ($10.00) gold pieces. On the reverse of the coins are various depictions of the American eagle. R. S. Yeoman, *A Guide Book of United States Coins,* 51st ed. (Racine, Wis.: Golden Books Publishing Co., 1998), pp. 191–211.

41. Arman and Arman, *Anglo-American Ceramics,* p. 239.

42. May, *Commemorative Pottery,* pp. 110–11.

43. Larsen, *American Historical Views,* pp. 230, 269–70, and 275.

44. James L. Mooney, ed., *Dictionary of American Naval Fighting Ships,* 9 vols. (Washington, D.C.: Navy Department, Office of the Chief of Naval Operations, Naval History Division, 1859–1991), 3: 484–85.

45. "Pike, Zebulon Montgomery," *Britannica Concise Encyclopedia,* accessed October 15, 2005, from Encyclopaedia Britannica Premium Service, www.britannica.com/ebc/article-9375263.

46. T. Michael Miller, "General Lafayette and His Many Visits to Alexandria, Virginia," memorandum, July 15, 1987, Alexandria Library Lloyd House. Lafayette lodged at Gadsby's Tavern in 1777 (Dorothy Kabler, *A Story of Gadsby's Tavern* [Alexandria, Va.: D. Kabler, 1952], p. 13); wrote a letter to Jefferson from Alexandria in 1781 (*Writings of Thomas Jefferson,* edited by Albert Ellery Bergh, 20 vols. [Washington, D.C.: The Thomas Jefferson Memorial Association, 1905], 5: 523); and attended a reception in his honor at Lomax's Tavern in 1784 (*Alexandria Gazette,* August 2, 1784, p. 2). When he visited America in 1824, Lafayette spent almost a month in Alexandria, beginning in October (*Lafayette, Guest of the Nation: A Contemporary Account of the Triumphal Tour of General Lafayette through the United States in 1824–1825 as Reported by the Local Newspapers,* edited by Edgar Ewing Brandon, 3 vols. [Oxford Historical Press, 1951–1957], 3: 33–36). In December he dined there with Mrs. Henry Lee (*Alexandria Gazette,* December 24, 1824, p. 3).

47. Miller, *Manuscripts of an Old Bachelor,* p. 32.

48. Moore, *Seaport in Virginia,* pp. 240–41. In gratitude, the Alexandria City Council presented Mrs. Thomas Lawrason with a silver cup.

49. Benjamin Hallowell, *Autobiography of Benjamin Hallowell, Written . . . in the Seventy-sixth Year of His Age,* 2nd ed. (Philadelphia: Friends' Book Association, 1884), pp. 99–101.

50. Moore, *Seaport in Virginia,* pp. 240–41.

51. Miller, *Manuscripts of an Old Bachelor,* pp. 52–53.

52. T. Michael Miller, *Portrait of a Town: Alexandria, District of Columbia (Virginia), 1820–1830* (Bowie, Md.: Heritage Books, 1995), p. 211.

53. *Alexandria Gazette,* September 14, 1825.

54. *Alexandria Gazette,* March 4, 1826, p. 3:

> EARLY SPRING SUPPLY
> ROBERT H. MILLER has just received from Liverpool, per ship Shenandoah,
> Sixty-two Crates & Hogsheads
> of Earthenware and China
> being part of his spring supply—among which are
> Ironstone China table services
> Blue printed (Liverpool) do
> China pitchers and tea-sets, richly gilt
> Do tea-sets gold edge and view of Mount Vernon
> Do Do plain, with do and common patterns
> Which, with the usual articles in his line, such as edged
> plates and dishes, enamelled and printed bowls, &c &c.
> a further supply of *Lafayette Ware,* makes his assortment very handsome.
> He has also lately received a superior assortment of
> Rich Cut Glass, viz: . . .

55. Warwick Price Miller, *Reminiscences of Warwick P. Miller of Alexandria, Virginia, 1896* ([Alexandria, Va.]: Lloyd House, Alexandria Library, 1981), p. 3. Robert Miller had opened a shop at 317 King Street, on the Market Square, in 1822, after traveling to England to select the first stock of china and earthenware. His brother was sent to St. Louis in 1835 to establish a

branch store, N. E. Janney and Brothers. An ironstone wash basin found at a mid-nineteenth-century residential site at 809 Duke Street bears the impressed mark "PEARL WARE J. HEATH" and the printed mark "MANUFACTURED FOR R-H MILLER & CO ST. LOUIS M / IRONSTONE CHINA / J. HEATH." The Heath mark dates to 1845–1853, according to Godden, *Encyclopaedia,* p. 318. Robert retired in 1865, but his son Elijah Janney Miller entered the business in 1856, and the company continued into the early twentieth century.

56. Philip Terrie, "A Social History of the 500 Block, King Street in Alexandria, VA" (1979), unpublished manuscript in the collection of the Alexandria Archaeology Museum.

57. Larsen, *American Historical Views,* p. 263. An oval pearlware teapot with the Lafayette and Cornwallis prints is reproduced in Arman and Arman, *Anglo-American Ceramics,* p. 12, fig. 2.4.

58. May, *Commemorative Pottery,* p. 82.

59. Hallowell, *Autobiography,* pp. 99–101.

60. Arman and Arman, *Anglo-American Ceramics,* pp. 51, 112; Larsen, *American Historical Views,* p. 229.

61. *Alexandria Gazette,* May 7, 1833, and December 29, 1854, cited in T. Michael Miller, *Murder and Mayhem: Criminal Conduct in Old Alexandria, Virginia, 1749–1900* (Bowie, Md.: Heritage Books, 1988), pp. 184–87. The incident is also described in Dorothy H. Kabler, *The Story of Gadsby's Tavern* (Alexandria, Va.: Printed by Newell-Cole Co., 1952), p. 49.

62. "The Whitehouse: Presidents' Hall, William Henry Harrison," reprinted from Frank Burt Freidel, *The Presidents of the United States of America,* 14th ed. (Washington, D.C.: White House Historical Association with the cooperation of the National Geographic Society, 1995), available at www.whitehouse.gov/history/presidents/wh9.html, accessed March 6, 2005.

63. Robert H. McCauley, "American Importers of Staffordshire," *Antiques* (June 1944): 295. McCauley illustrates a similar pitcher with the Robert H. Miller mark, printed in mulberry. Larsen (*American Historical Views,* p. 266) also describes a Harrison pitcher marked "Manufactured for Robert H. Miller, Alexandria, D.C."

64. Alexandria was part of the District of Columbia from 1801 until 1847. Residents of the District of Columbia did not have votes in the electoral college until 1961, when the twenty-third constitutional amendment was ratified.

65. Miller, *Reminiscences of Warwick P. Miller,* pp. 17–18.

66. The Armans (*Anglo-American Ceramics,* p. 103) illustrate a green-printed pitcher with the *Harrison and Reform* print but without the eagle beneath the spout.

67. "Rally: Conservative Men, Whigs, Union Men!" *Alexandria Gazette,* November 6, 1860, p. 2.

68. William B. Hurd, *Alexandria, Virginia, 1861–1865* (Alexandria, Va.: Fort Ward Museum, 1970), pp. 3–4.

69. "Raising of Confederate States Flag," *Alexandria Gazette,* April 18, 1861, p. 3.

70. Hurd, *Alexandria, Virginia, 1861–1865,* pp. 3–7.

71. James Barber, *Alexandria in the Civil War* (Lynchburg, Va.: H. E. Howard, 1988), p. 17.

72. David J. Goldbert, "Charles Coxon: Nineteenth-Century Potter, Modeler-Designer, and Manufacturer," *American Ceramic Circle Journal* 9 (1984): 44–49.

73. Hurd, *Alexandria, Virginia, 1861–1865,* p. 6.

74. Laurence W. Williams, *Collector's Guide to Souvenir China: Keepsakes of a Golden Era* (Paducah, Ky.: Collector Books, 1998). Knight's store is illustrated in Alexander J. Wedderburn, *Souvenir Virginia Tercentennial, 1707–1907, of Historic Alexandria, Virginia* (Alexandria, Va.: A. J. Wedderburn, 1907), p. 56.

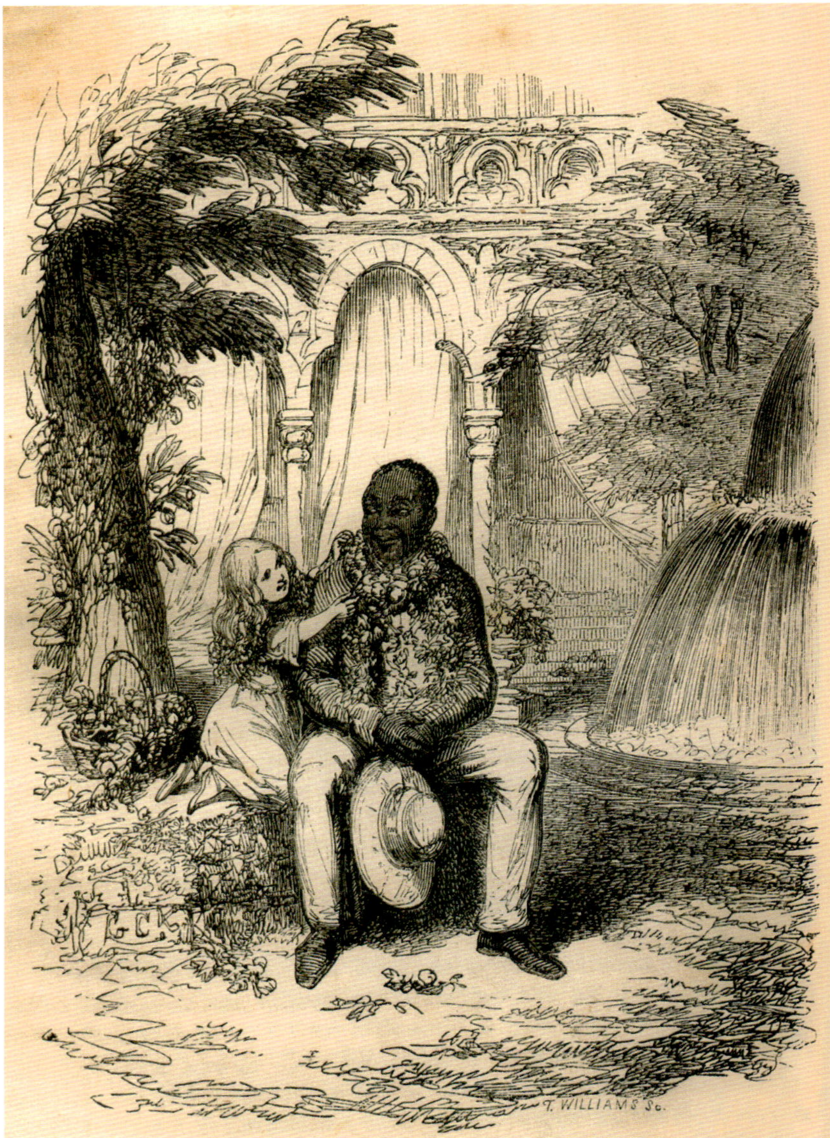

Figure 1 George Cruikshank, "Eva Dressing Uncle Tom," wood engraving, from Harriet Beecher Stowe, *Uncle Tom's Cabin,* London, 1852. (Courtesy, Harriet Beecher Stowe Center.) Staffordshire manufacturers often replicated the relative scale and pose of Eva and Tom in their figural reproductions based on characters in Stowe's novel.

Jill Weitzman Fenichell

Fragile Lessons: Ceramic and Porcelain Representations of *Uncle Tom's Cabin*

▼ HARRIET BEECHER STOWE'S *Uncle Tom's Cabin* (1851) was not just a novel; it was a cultural phenomenon. Its influence spread across America and Europe through political discourse, theatrical interpretations, and a host of visual and material representations. Now, more than 150 years after its initial publication, it is inseparable from the fabric of American culture and mythology. Of the multitude of objects produced in the wake of the novel's publication, some of the most interesting are the ceramic and porcelain figures, vases, and decorative plates depicting scenes from the novel. These objects, produced primarily for use as decoration in Victorian parlors, speak to the varied ways *Uncle Tom's Cabin* functioned in the domestic lives of their owners and teach us much about the objectification of the reader's experience of the novel in mid-nineteenth-century America.

Although manufacturers made thousands of Uncle Tom items, more than 75 percent of the objects portray two specific scenes in the book. Because these scenes crystallize the moral and familial dramas at the novel's center, it is no surprise that they were reproduced so extensively. The first, representing the highest "moral moment," is the scene in which Evangeline, or Eva, "dresses" Uncle Tom in garlands of flowers (fig. 1): "every one of his buttonholes stuck full of cape jessamines . . . Eva, gayly laughing . . . hanging a wreath of roses around his neck."[1] In this brief interlude of spontaneous love and affection, Tom's childish simplicity makes him Eva's emotional equal. The old slave, holding a Bible in his left hand, is now a literate and God-fearing man, embraced by the loving daughter of his owner. The garland of flowers around Tom's neck strikes another religious note, reminding the viewer of the Lamb of God, similarly garlanded, who also sacrificed his life for others.

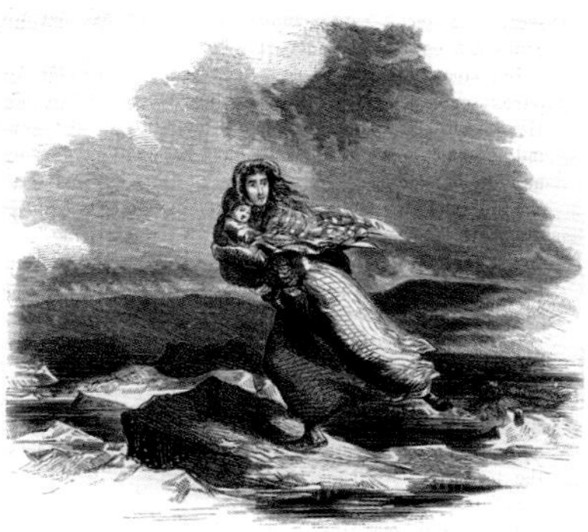

Figure 2 Hammatt Billings, "Eliza on the Ice Floes," wood engraving, from Harriet Beecher Stowe, *Uncle Tom's Cabin, the Illustrated Edition,* Boston, 1853. (Courtesy, Harriet Beecher Stowe Center.) Billings's rendering of Eliza as a Caucasian helped to elicit sympathy for the character's plight. In Staffordshire, Eliza is also shown as white, whereas French ceramic manufactories tend to show her skin color more accurately.

Conversely, the lowest moral and greatest melodramatic moment occurs as Eliza crosses the frozen Ohio River with Baby Harry in her arms (fig. 2). "Nerved with strength such as God gives only to the desperate,"[2] she runs away from slave trappers, away from a life of misery in Kentucky toward life as a free woman in Canada. This second scene reminded contemporary viewers that slave owners separated family members from one another for commercial reasons and sent bounty hunters across state lines to find escaped slaves and return them, as chattel, to their owners.

These two moments are beautifully shown in a pair of elegant figural vases, full of movement and emotion, produced at Limoges for export to America, about 1852–1853 (fig. 3).[3] The dramatic poses and coloration of the

Figure 3 "Tom et Evangeline" and "La Fuite d'Elise," Limoges, ca. 1852–1853. Biscuit- and glazed hard-paste porcelain with polychrome enamels and gilding. H. 11¾". (Courtesy, Harriet Beecher Stowe Center; photo, author.)

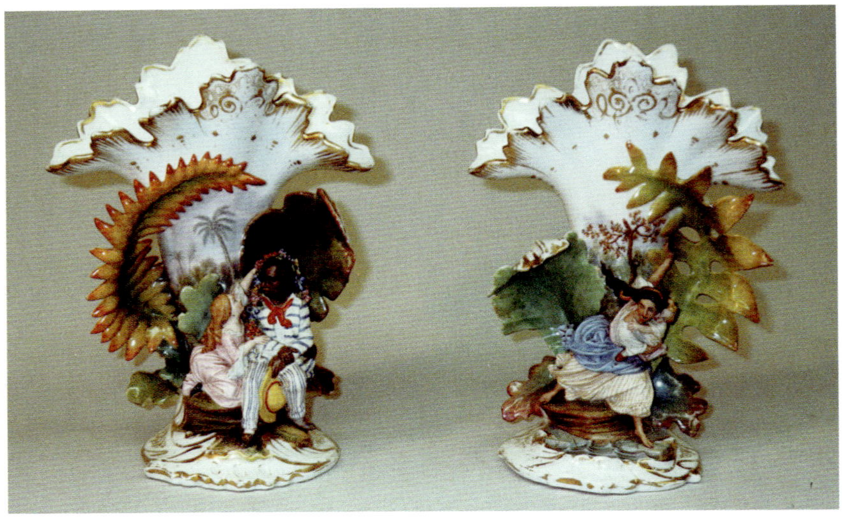

Figure 4 Charles Bour, "Tom et Evangeline," lithograph, from Harriet Beecher Stowe, *La Case de l'Oncle Tom,* Paris, 1853. (Courtesy, Harriet Beecher Stowe Center, 68.288.Pr.F3.)

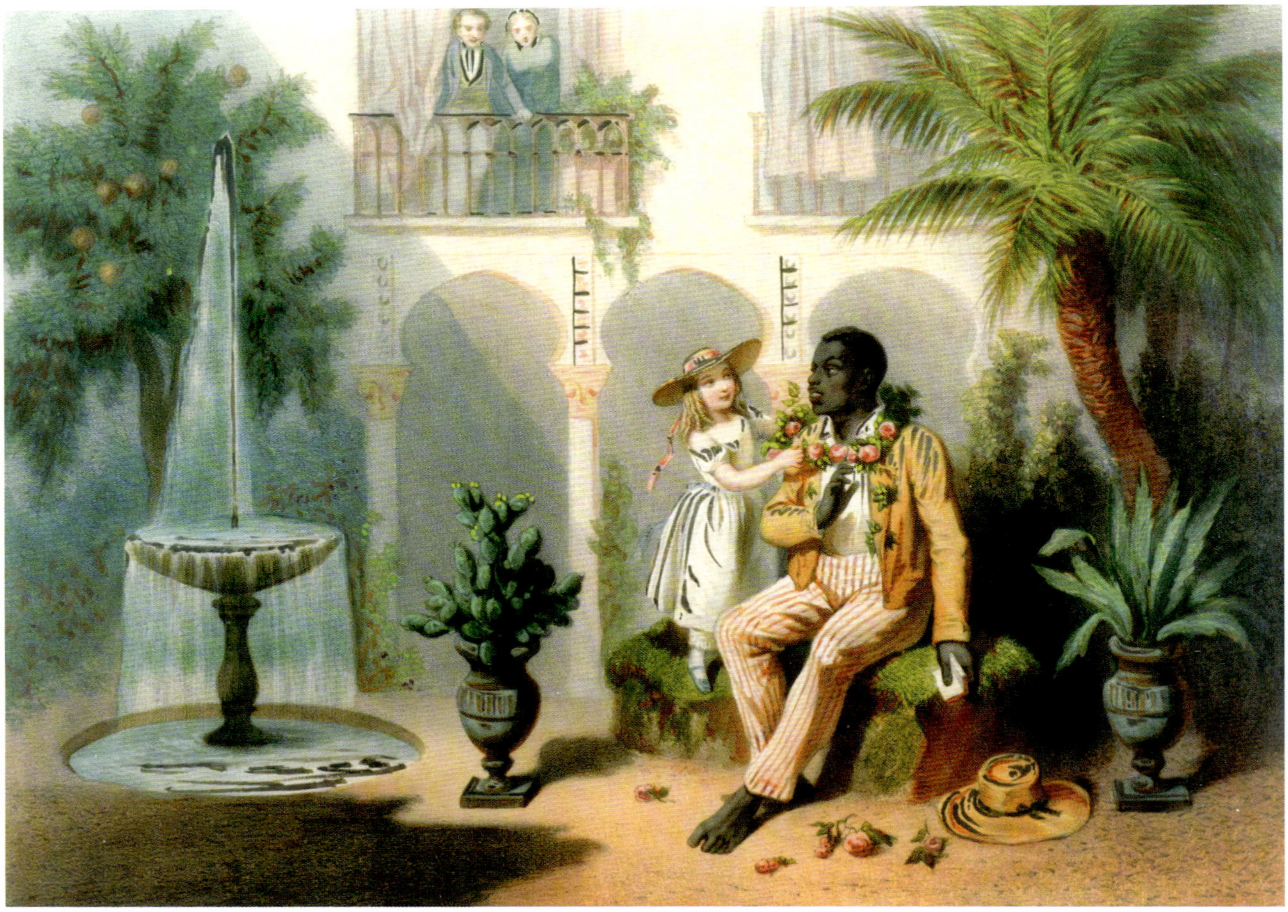

vases probably derive from a pair of prints published in the first French edition of *Uncle Tom's Cabin, La Case de l'Oncle Tom*.[4] The scene of Eva and Tom (fig. 4) is by Charles Bour (act. 1844–1880), who exhibited paintings in the Paris Salon and made lithographs of competition entries. The second image, of Eliza's flight (fig. 5), is signed by Bettannier, about whom little is known. Both prints share an interest in bright polychromy and imaginative foliage that are present as well on the vases.

The fanciful and heavy foliage creeping around the vases and the awkward position of Uncle Tom's legs may derive from an additional source, the sixth illustration of the 1853 children's edition of Stowe's novel, called *Pictures and Stories from Uncle Tom's Cabin* (fig. 6).[5] This image emphasizes Tom's exotic, nearly Asian features, and his awkwardness before social superiors (that is, Caucasian plantation owners). His deference is underscored by his holding his hat rather than a Bible.

The figural work on these vases is of biscuit porcelain, which contrasts with the glassy glaze and bright, mercury-flashed gilding on the very high Victorian flat-backed frill vases behind the figures. The contrast of textures echoes the contrast of movement, particularly in the Eliza and Harry vase, which looks as if the figures are trying to escape from the front of the vase, running away as we do in nightmares, our feet stuck in concrete. Visual tension underlines dramatic motion. In addition, the application of over-

Figure 5 Bettannier, "La Fuite d'Elise," lithograph, from Harriet Beecher Stowe, *La Case de l'Oncle Tom,* Paris, 1853. (Courtesy, Harriet Beecher Stowe Center, 68.288.Pr.F3.)

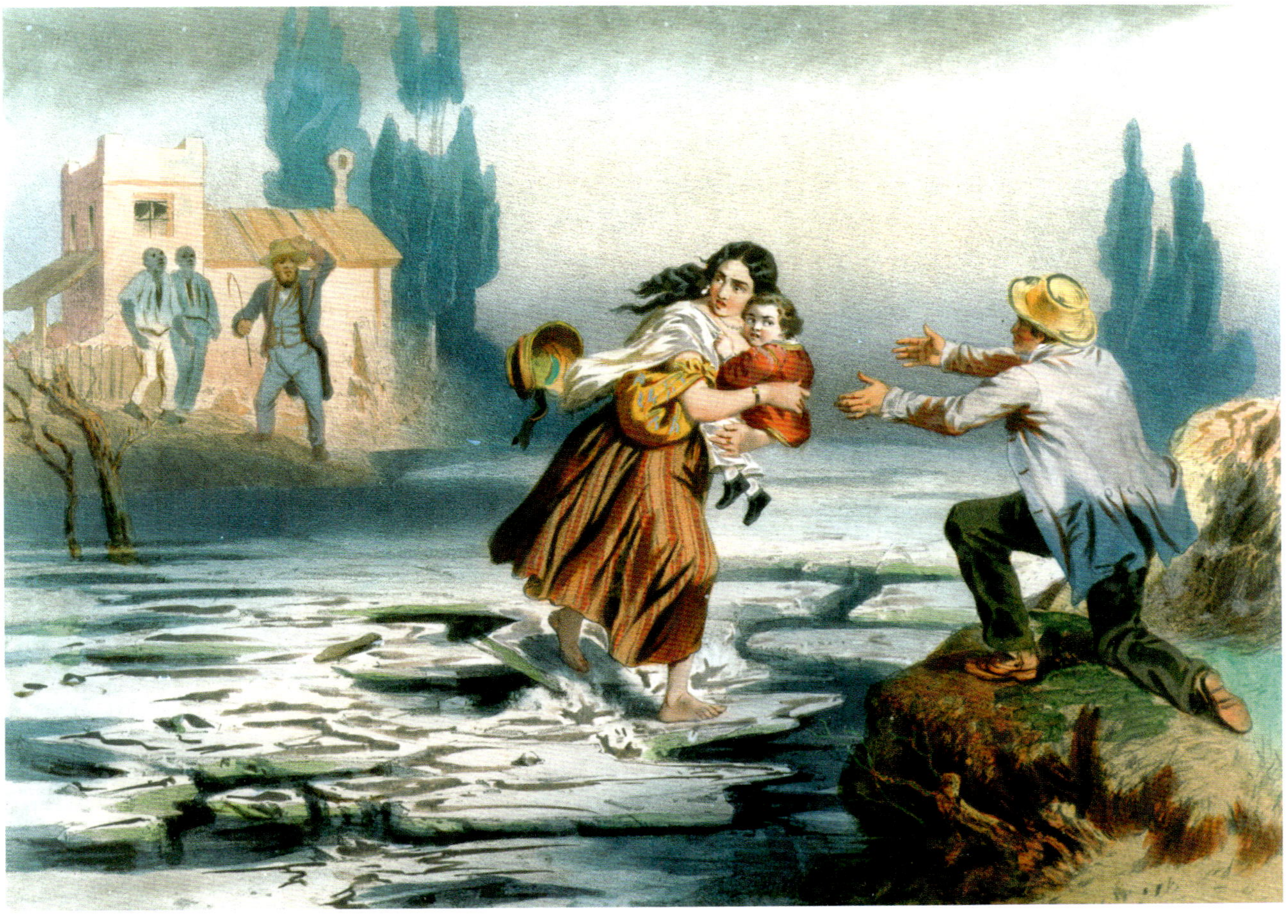

glaze polychrome enamels to the surface changes the tenor of the sculptures. In unglazed biscuit and gold, they are like refined marble statues, projecting virtue and moral aspirations. The polychrome enamels make the figures look more human, and they project the hopes and fears of the characters they depict. In underlining the melodrama of the various scenes, the coloration enabled merchants who wanted to take advantage of the fad for figures based on Stowe's novel to sell commercial goods in quantity.

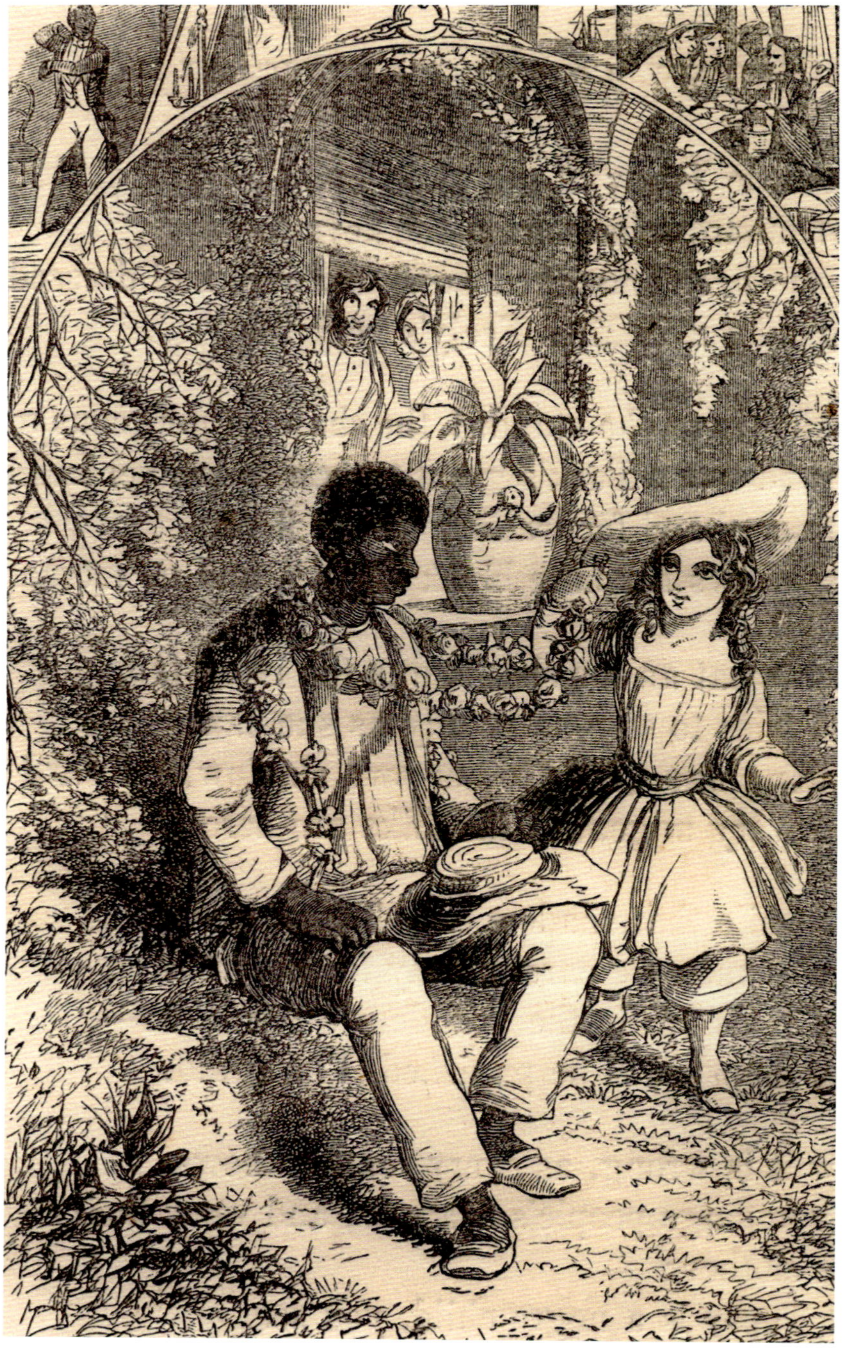

Figure 6 Anonymous, "Eva and Uncle Tom," from Harriet Beecher Stowe, *Pictures and Stories from Uncle Tom's Cabin,* Boston, 1853. (Courtesy, Harriet Beecher Stowe Center.) This was a juvenile edition of Stowe's novel. Note the shape of Tom's eyes and his unusually expressive face. The style of illustration is orientalist and idealized, rather than relying on caricature and overexaggerated features. This approach to depicting negroid features typifies the French decorative style, whether on vases or tablewares.

Research to date has identified two distinct models: a smaller model, at 11¾ inches (fig. 8), and a larger example, at 19 inches high (fig. 9). Both are found decorated in different ways, varying from undecorated biscuit and elaborate gold patterning on the vase forms to overglaze polychrome enameling with extensive gilding. Four pairs of large vases and a single example missing its mate have been identified: one pair in biscuit with gilt highlights in the collection of the Newark Museum, Newark, New Jersey (see fig. 9); a single vase of Uncle Tom, also with gilt decoration decorated with gold only, now in a private collection; a pair decorated in biscuit and matte blue with gold, auctioned at Christie's East, New York, October 27,

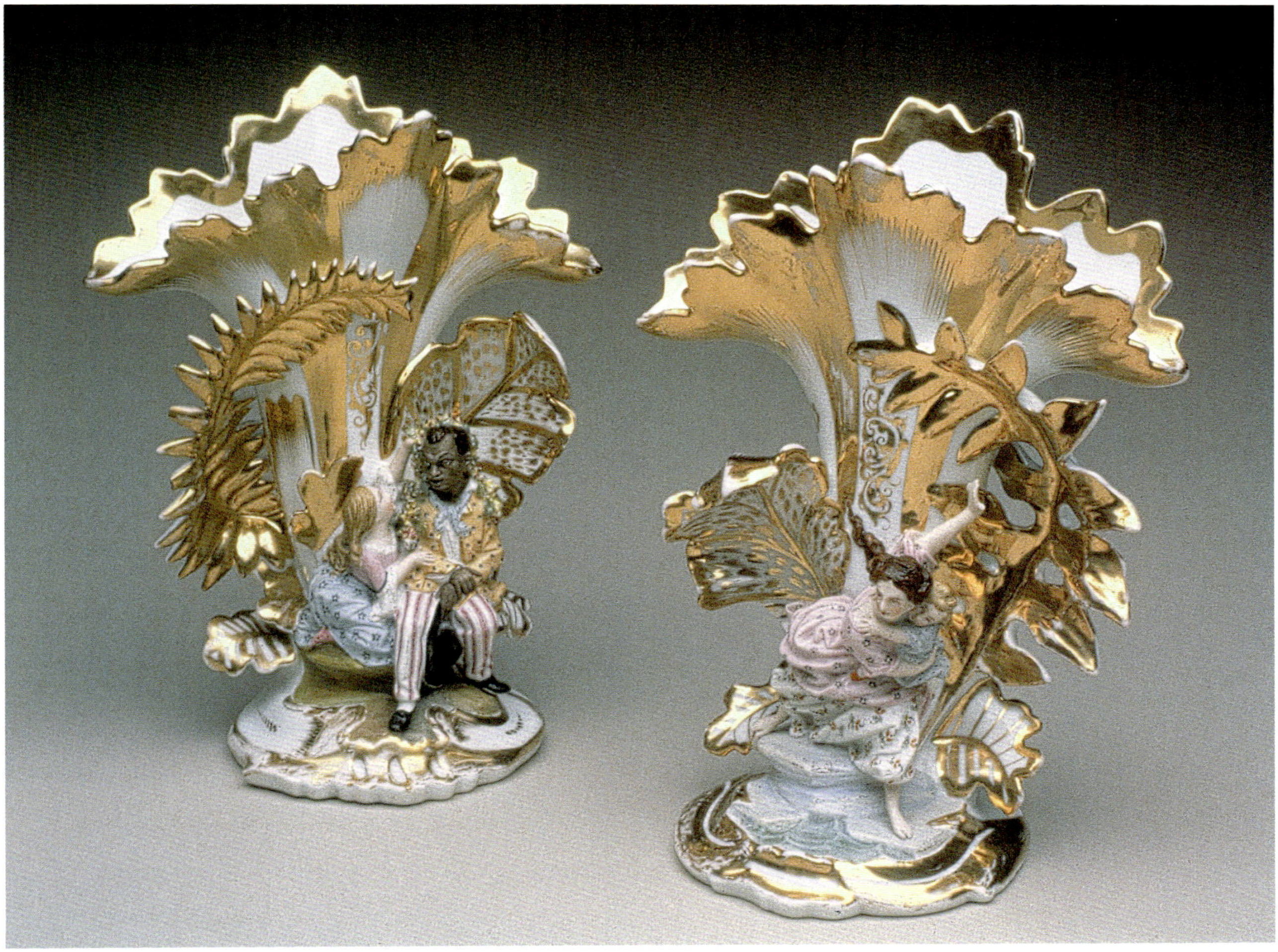

Figure 7 "Tom et Evangeline" and "La Fuite d'Elise," Limoges, ca. 1850–1860. Biscuit- and glazed hard-paste porcelain with polychrome enamels and gilding. H. 11¾". (Courtesy, Museum of Fine Arts, Houston; Bayou Bend Collection, museum purchase in memory of Genevieve Lydian Dutton Gundry by her son Jas A. Gundry, acc. nos. B2000.6.1, 6.2.) Although these figures share similar pastel palettes, the gilding on each of the pairs differs dramatically.

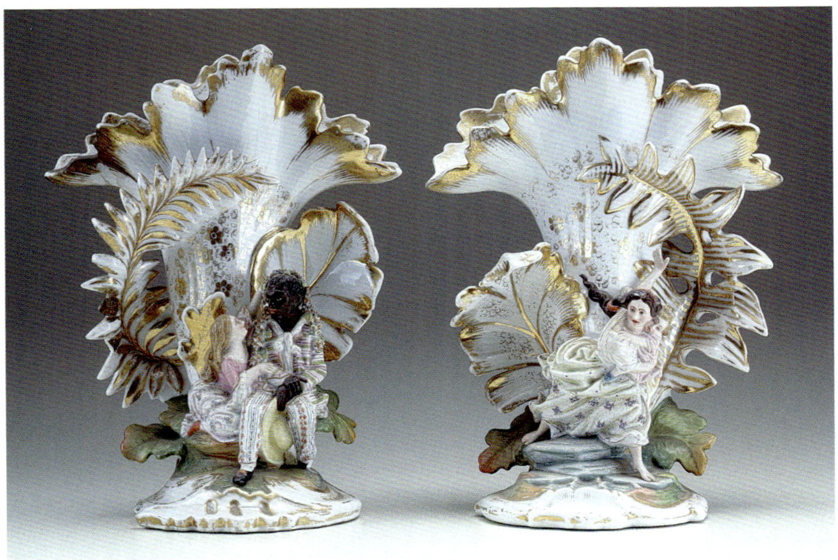

Figure 8 "Tom et Evangeline" and "La Fuite d'Elise," Limoges, ca. 1852–1853. Biscuit- and glazed hard-paste porcelain with polychrome enamels and gilding. H. 11¾". (Courtesy, Rex Stark; photo, Gavin Ashworth.) These and several other pairs of vases are decorated with fanciful enameling, but no two pairs have surfaced with identically painted features.

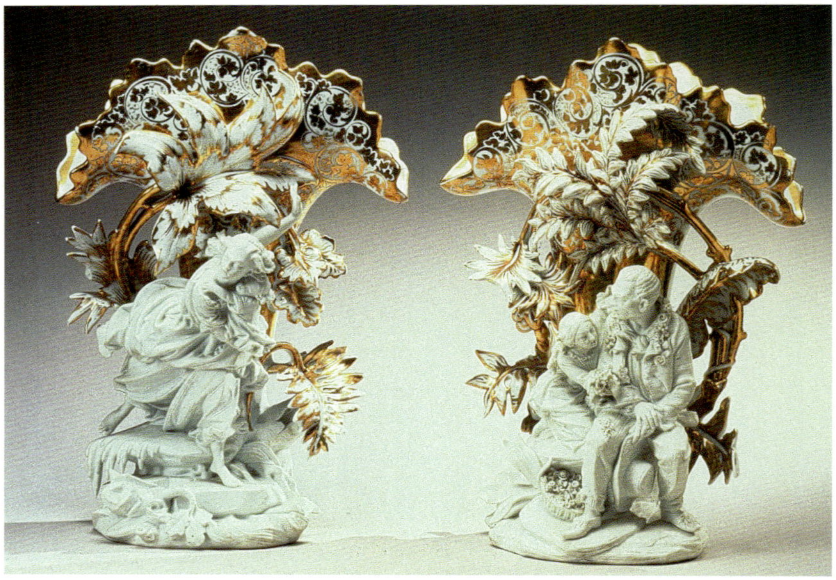

Figure 9 "Tom et Evangeline" and "La Fuite d'Elise," Limoges, ca. 1852–1853. Biscuit- and glazed hard-paste porcelain and gilding. H. 19". (The Newark Museum, Newark, New Jersey, Purchase 1968, Mrs. Parker O. Griffith Fund, 68.106 a,b.)

1998; and two pairs with extensive polychrome decoration and gilding in the collections of the Harriet Beecher Stowe Center, Hartford, Connecticut, and the Museum of the City of New York, respectively.

Four pairs of the small vases have also been identified: a pair in biscuit with gold highlights from the inventory of Gary and Diana Stradling, New York dealers who specialize in American ceramics; and three pairs decorated with polychrome enamels and gilding, one at Bayou Bend, The Museum of Fine Arts, Houston, Texas (fig. 7), one in a private New England collection, and one at the Harriet Beecher Stowe Center (see fig. 3).

Although the two sizes of vases are modeled differently from each other, all of the large vases are identically modeled, as are all of the small vases. This is typical of Limoges wares and speaks to the desire of manufacturers to reach diverse segments of the purchasing public. No two pairs of vases have been found with identical decoration, a fact explained by the desire of American merchandisers to present consumers with a wide array of choices. As has been noted by historians of the period 1850–1860, retailers of luxury goods had workrooms above their showrooms where special orders were executed, including those for porcelains.[6] As proof, the *Cosmopolitan Art Journal* illustrated the premises of the New York retailers Haughwout and Dailey of 540 Broadway at Broome Street (fig. 10). In their upstairs workrooms, more than one hundred "mostly English" china painters and several female gilders, imported from Staffordshire and elsewhere, labored to decorate items chosen by the customers downstairs.[7]

Overglaze enameling work was done after the modeling factory shipped the vases in the white, as undecorated ceramics were taxed and dutied at a much lower rate than those with elaborate details.[8] Museum curators and ceramics dealers have often misattributed these figures to Bohemia. It is now clear, however, that the vases were produced in Limoges and sent undecorated to New York, where they were decorated by Bohemian, French, and English painters and then sold. As the painters had been taught enameling and gilding in distinctive European markets, the figures reflect the background of their decorators. For instance, the vases with matte blue and gilding were decorated by an enameler trained in Limoges, where that style of decor was considered chic. The vases with enamels of intense hue and with fanciful water and plant colorations would have been painted by a decorator from Staffordshire accustomed to enameling Staffordshire figurines. The vases with very precise costume details and distinct orientalist landscapes, with attention paid to minutiae and having large areas of picked out and flashy gilding, were enameled by Bohemian workers.

The decorating on all the Uncle Tom vases may have been done at the same establishment because certain decorative elements appear more than once on the various pairs. For instance, two pairs were painted with trompe l'oeil orientalist landscapes on the tapered bases of the vases; these scenes lend an overall effect of exotic three-dimensionality to the figural work. Several pairs share an identical gilding motif, a wide band of scrolling incurvate oval forms, which appears on the back of some vases and as a decorative feature on the front of the vase section of others. The distinctions allow us to act as detectives, trying to attribute certain motifs and gilding to specific hands or schools of decoration. For instance, a pair of figural vases at the Peabody Essex Museum in Salem, Massachusetts, flamboyantly painted, yet with very detailed costumes and gilt scrolls, were almost certainly painted in the same workshop as the pair of vases owned

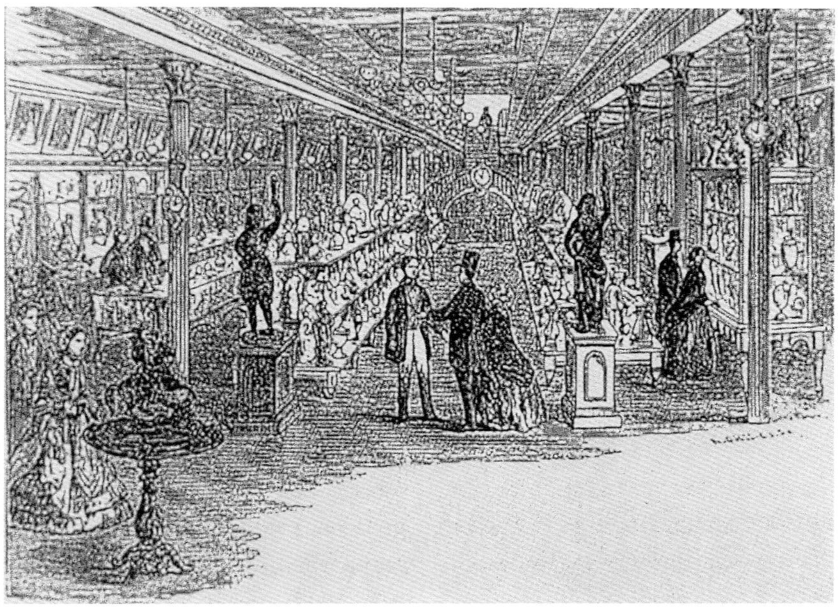

Figure 10 Nathaniel Orr and Company, "Saleroom, Main Floor, Haughwout Building, New York," wood engraving, from *Cosmopolitan Art Journal* 3 (June 1859). (Courtesy, the author.) This view of the ground-floor premises of the retailing emporium of Haughwout and Dailey at 540 Broadway shows a tantalizing survey of stylish porcelain and glass available for purchase in the 1850s. Haughwout's hired women—mostly from Staffordshire but also from porcelain towns in eastern France and western Germany—to work as decorators in the custom workshop upstairs from the main selling floor. Many of the *Uncle Tom's Cabin*-motif French vases might have acquired their unusual coloration and gilding styles from the women decorators working at this establishment.

by the Museum of the City of New York and the small pair of vases now owned by the Harriet Beecher Stowe Center (see fig. 3).

It may seem surprising that these ceramics illustrating scenes from an American novel were made in Europe, but it is necessary to remember that *Uncle Tom's Cabin* was very popular overseas, selling hundreds of thousands of copies in England and France alone. Indeed, without this transatlantic connection, these objects could not have been made, because no American ceramic manufactory could create the quantities of inexpensive objects demanded by the popularity of the novel.[9] In Europe, by contrast, literacy rates were much higher than in America, and factories that had experience making souvenir-type wares were well established. The combination made Stowe's novel and items dramatizing its characters and moments readily available.

The five potting towns that make up Stoke-on-Trent in Staffordshire were a natural place to create a plethora of *Uncle Tom's Cabin* trinket wares, first, because there was already an English audience long accustomed to purchasing items with literary connections, and second, because the existing industrial base was large enough to support multiple factories producing similar items quickly and cheaply.[10] Furthermore, Stoke was ideally located near Liverpool and its huge port to facilitate nationwide distribution.

It therefore is not surprising to learn that Staffordshire manufacturers produced thousands of items based on *Uncle Tom's Cabin* (fig. 11). These include figures that commemorate Uncle Tom meeting Eva on the boat; Eva with Topsy; Eva crowning Uncle Tom with a garland of flowers; Eva and Uncle Tom on her or his own scroll base. According to P. D. Gordon Pugh, at least eighteen different models are known, including ones of Aunt Chloe and George and Eliza Harris. Some of the figures are as small as 1½ inches high; others are closer to 10 inches.[11] Such figurines turn up at public sales in England far more often than in the United States.

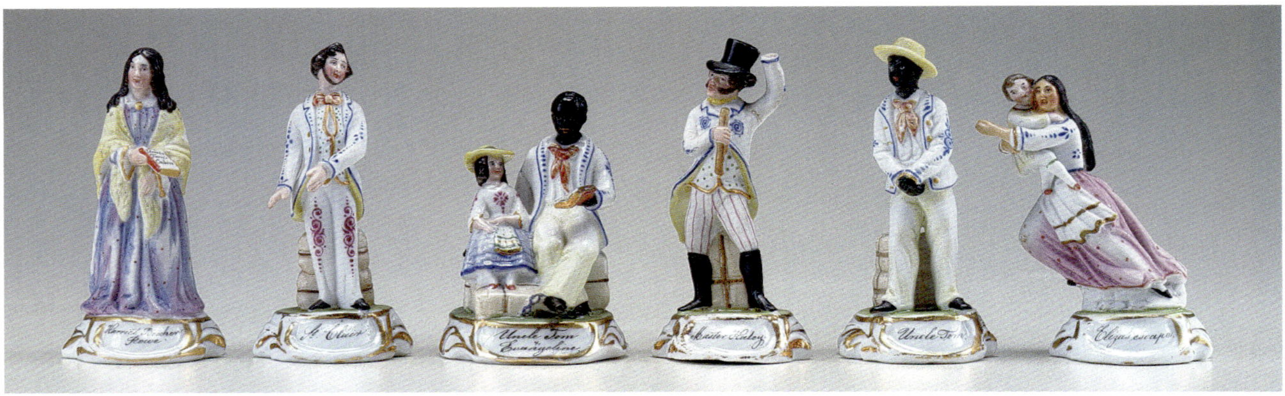

Figure 11 Figurines based on characters from *Uncle Tom's Cabin,* Staffordshire potteries, Stoke-on-Trent, ca. 1852–1860. Bone china. H. of tallest approx. 8". (Courtesy, Rex Stark; photo, Rex Stark.)

The Dudson factory, in the nexus of ceramic manufacturing at Stoke-on-Trent, is the only one that has been clearly identified as making *Uncle Tom*–related figures, based on factory records dating to the early 1850s. These list "Eva, Uncle Tom, Boat," which had the typically stylized three-dimensional appearance of other flatback figures but with the addition of the "ebonized" figure of Uncle Tom placed next to the petite figure of Eva (figs. 12, 13). A Dudson family member who was present at excavations on the site of the old Dudson factory has identified fragments of a figurine of "Eva and Uncle Tom" in a banana-shaped boat, of which only two examples are known.[12]

By far the most popular figural group shows a very small Eva seated beside a huge figure of Uncle Tom. She is variously helping Uncle Tom read his Bible in a bower or "dressing" him with a garland of flowers, and

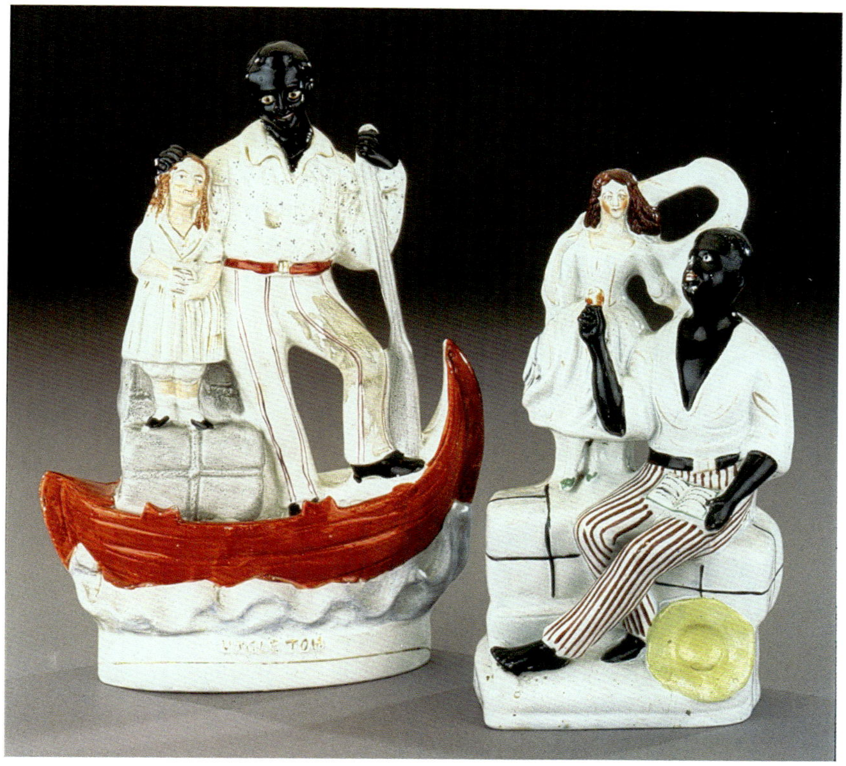

Figure 12 *Left*: "Eva, Uncle Tom, Boat," Dudson factory, Stoke-on-Trent, ca. 1852–1855. Whiteware. *Right*: "Eva and Uncle Tom on the Boat," unknown factory, Stoke-on-Trent, ca. 1852–1855. Whiteware. H. of tallest 7". (Courtesy, Christie's, London.) "Eva, Uncle Tom, Boat" is the only figural group whose attribution to a specific Staffordshire factory producing earthenware figures based on *Uncle Tom's Cabin* is confirmed.

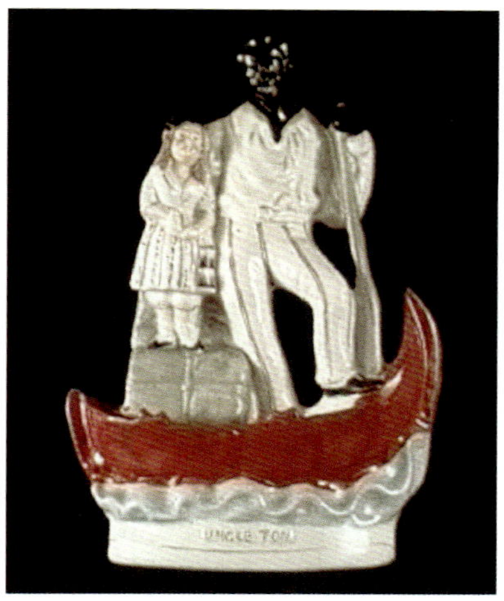

Figure 13 Figural group, "Eva, Uncle Tom, Boat," Dudson factory, Stoke-on-Trent, ca. 1852–1855. Whiteware. H. 7". (Courtesy, Harriet Beecher Stowe Center, 81.18.)

Tom holds either his hat or a Bible in his left hand. The idea for the group comes from the 1852 London edition of *Uncle Tom's Cabin,* illustrated with twenty-seven woodcuts by the famed caricaturist George Cruikshank.[13] The headpiece shows the two figures in positions very similar to the figural group, seated in a bower. The image had strong resonance for the English, because bowers formed part of their own gardens, and bower groups made in salt-glazed stoneware and glazed earthenwares were made as early as the 1740s. The *Uncle Tom's Cabin* groups hark back to that eighteenth-century tradition.[14]

Whether depicting Eva and Uncle Tom, Eva and Topsy, or George and Eliza, all these Staffordshire figural groups are loosely and quickly modeled, and even more quickly decorated with overglaze enamels. The palette of the figures is typically bright, and the figure of Uncle Tom is usually lustrously black, with exaggerated lips and mouth. All of the figures share flat backs, abstractly painted surfaces, idealized faces, and the exploitation of "exotic" or "foreign" negroid features. By mass-producing quantities of such wares and decorating them minimally, manufactories such as Dudson's could capitalize on momentary fads among the newer consuming classes with their reasonably priced wares.

Staffordshire manufacturers also had a long history of producing so-called ABC plates, that is, plates with borders and transfer-printed with images above brief mottos. Intended not for use but for playful instruction, they were meant to stimulate youngsters to think good religious and moral thoughts. Many such plates have been found reproducing scenes from *Uncle Tom's Cabin,* most of which are transfer-printed with direct copies of Cruikshank's illustrations for the 1852 edition, such as "Eva Dressing Uncle Tom," shown on a mug, and a plate with a visual interpretation of Uncle Tom's cabin, true to Stowe's novel right down to the portrait of George Washington hanging over the mantel (figs. 14, 15). Other plates show signs of being

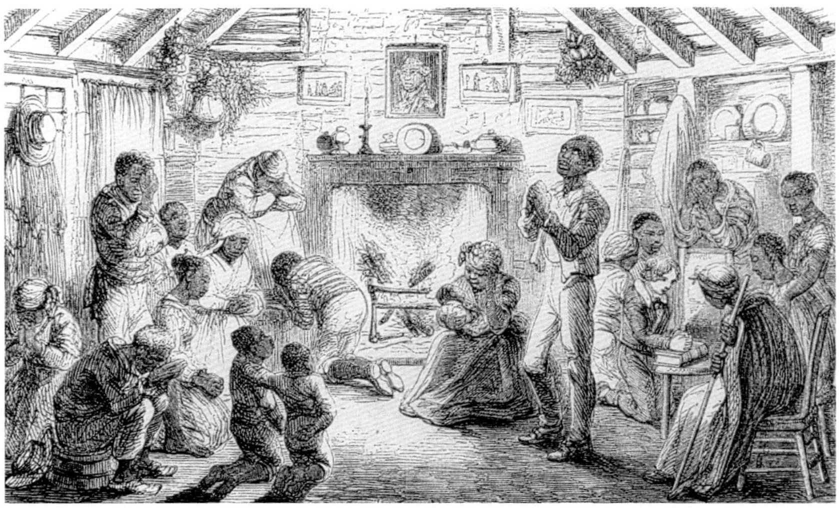

Figure 14 George Cruikshank, "Prayer Meeting in Uncle Tom's Cabin," from Harriet Beecher Stowe, *Uncle Tom's Cabin,* London, 1852.

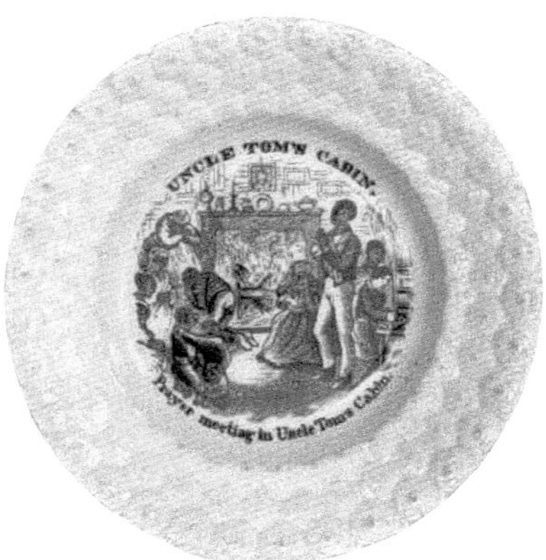

Figure 15 Plate, "PRAYER MEETING," unknown factory, Stoke-on-Trent, ca. 1852–1855. Molded earthenware transfer-printed in warm black enamel. D. 7⅝". (Courtesy, Harriet Beecher Stowe Center, 81.34.) The image of the "Prayer Meeting" is taken directly from the Cruikshank illustration shown in fig. 14. The mantel decor is typical of aspirational middle-class taste, not a true slave home interior.

copies of copies of copies, evident because of the thick outlines and wavy printed backgrounds on the faces of the characters, seen, for example, in a plate illustrating "Eva and Topsy." Manufacturers enjoyed making these plates with variegated shoulder molding because it enabled them to charge a little extra for the plates without doing anything complicated.[15]

A favorite topic for the ABC treatment repeats the imagery found on the French vases discussed earlier: Eliza's crossing of the Ohio, a small moment in the book but one whose drama is beautifully encapsulated in Cruikshank's illustration (fig. 16). In a relatively early version, circa 1853, we have "Eliza Crossing the Ohio on the Freezing Ice"; a later version with a more degraded print bears a new caption: "Lizzy's Bridge: The Article Exhibits the Dominion of Maternal Love."[16]

"The Death of Uncle Tom" is not often seen, but its imagery recalls an earlier ceramic item made by the antislavery advocate Josiah Wedgwood. As early as 1780, Wedgwood introduced little plaques that looked as

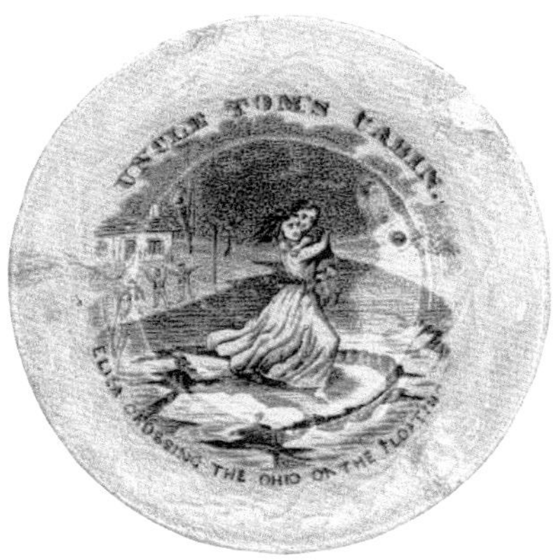

Figure 16 Plate, "UNCLE TOM'S CABIN / ELIZA CROSSING THE OHIO ON THE FLOATING ICE," unknown factory, Stoke-on-Trent, ca. 1855. Earthenware with molded rim, transfer-printed in warm black enamel. D. approx. 5". (Courtesy, Harriet Beecher Stowe Center, 81.28.) This plate shows a revised illustration for *Uncle Tom's Cabin.* Note that Eliza and Baby Harry are running toward the right side of the plate, whereas the standard orientation shows them running toward the left. The plate imagery is highly degraded.

though they were made in ancient Greece. They clearly depict a slave on bended knees, his arms shown in prayer to an unseen God, below the motto of the British Anti-Slavery Society, "AM I NOT A MAN AND A BROTHER?" Such items were mounted in gold and worn as jewelry, such as a stick pin. The slave's image, representing a repressed cry for dignity and recognition, was often copied, albeit with some changes, by other Staffordshire manufacturers, using the new technology of lithography to create transfer-prints. Early-nineteenth-century drabware cups and saucers and unusual bone china wares such as a souvenir shoe and chamberstick, both made about 1835, serve as similar small antislavery tokens. Here the motto from the Wedgwood era has been changed and now reads, "REMEMBER THEM THAT ARE IN BONDS."[17]

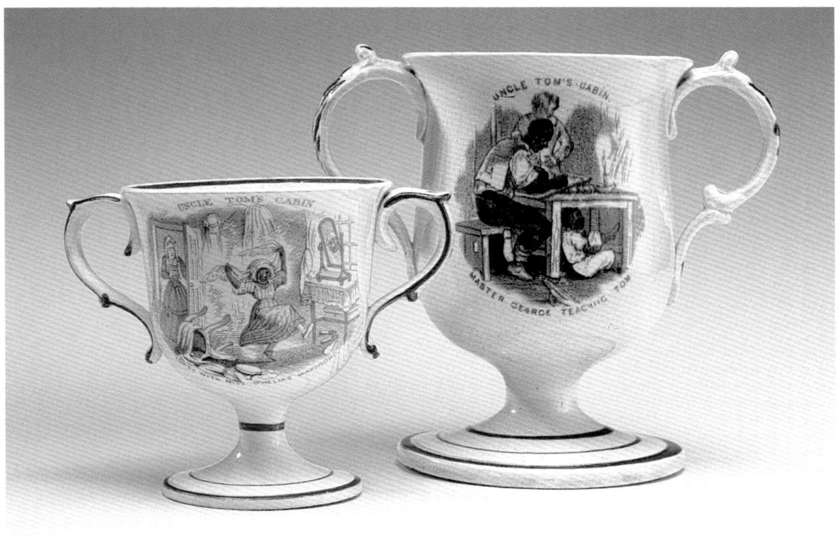

Figure 17 Two-handled cups, probably Staffordshire, ca. 1854–1858. Whiteware. H. of tallest 6". (Courtesy, Rex Stark; photo, Gavin Ashworth.) Compared to plates, these unusual hollow-bodied wares were relatively expensive to produce. On the smaller cup is depicted Topsy, who, while supposedly working at Olivia's house, is being mischievous. "MASTER GEORGE TEACHING TOM" appears on the trophy cup. Both images are from Cruikshank's illustrations for the first English edition of Stowe's novel. The message on the trophy cup would have stirred the emotions of abolitionist sympathizers, who were outraged that teaching slaves to read and write was illegal in the South.

Figure 18 Mug, probably Staffordshire, ca. 1852–1855. Whiteware. H. 2½". (Courtesy, Rex Stark; photo, Gavin Ashworth.) Some of the Uncle Tom prints used by ceramic factories provide wonderfully detailed views of mid-nineteenth-century Victorian interiors. George Cruikshank's "DEATH OF ST. CLARE" has been transposed from the English edition of Stowe's novel to this child's cup. The louche plantation owner is surrounded by fine furniture and other talismans of wealth.

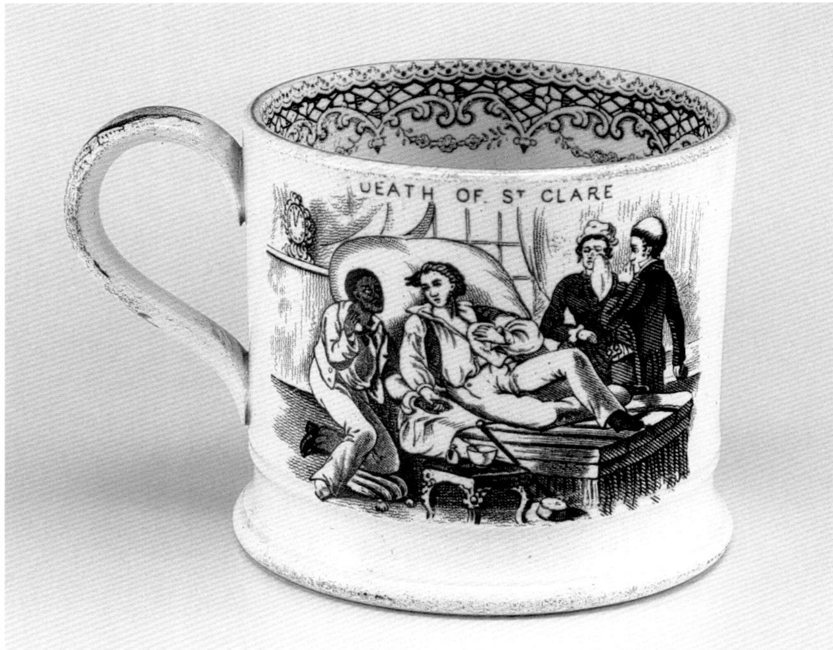

Figure 19 Plate, attributed to J. Vieillard et Cie., Bordeaux, ca. 1855. Whiteware. D. 8". (Courtesy, Rex Stark; photo, Gavin Ashworth.) This plate illustrates the scene "Uncle Tom Refuses to Hit Legree" (*L'Oncle Tom Refuse de Battre Legree*). The international success of Stowe's novel was fueled by worker literacy movements on the Continent. Because Stowe capitalized on the popularity of the sentimental novel, women bought her book; men bought it for the action within; and religious groups and parents approved of its high moral tone. The novel was printed in more languages and in more countries at mid-century than any book other than the Bible.

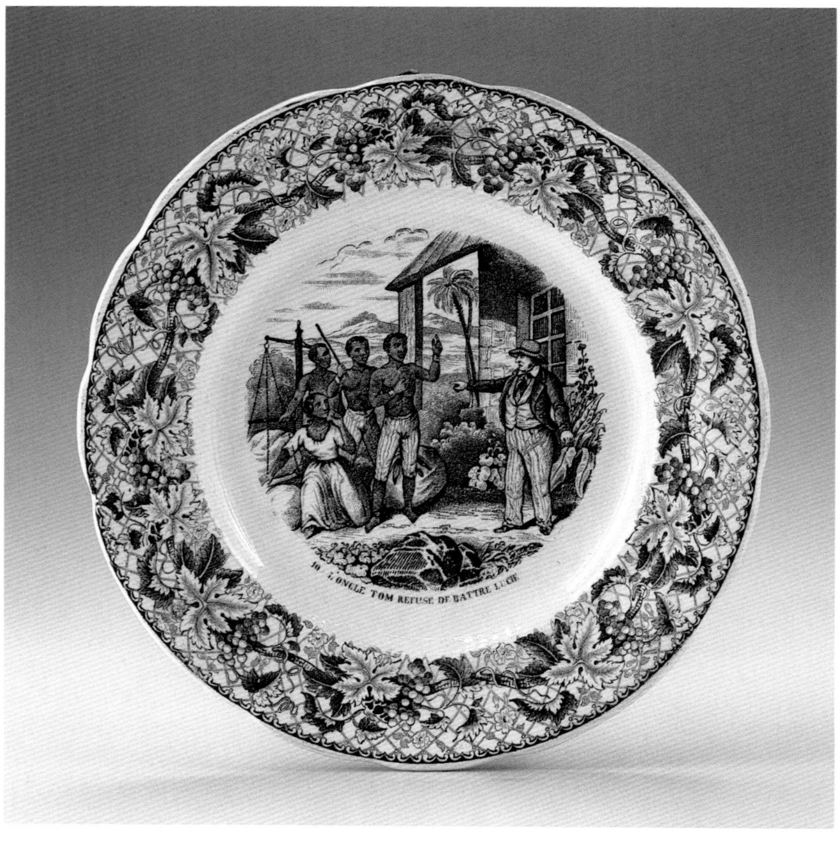

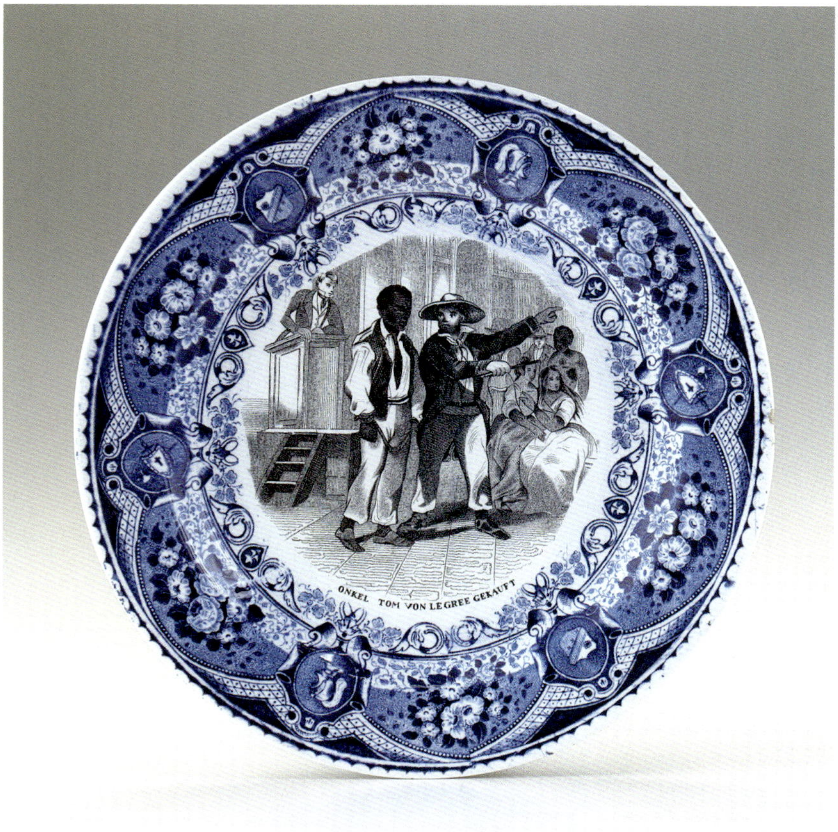

Figure 20 Plate, Schramburg, Germany, ca. 1855. Whiteware. D. 8". (Courtesy, Rex Stark; photo, Gavin Ashworth.) This two-color transfer-printed plate illustrates "Uncle Tom Leaving with Legree" (*Onkel Tom mit Le Gree Gereitt*).

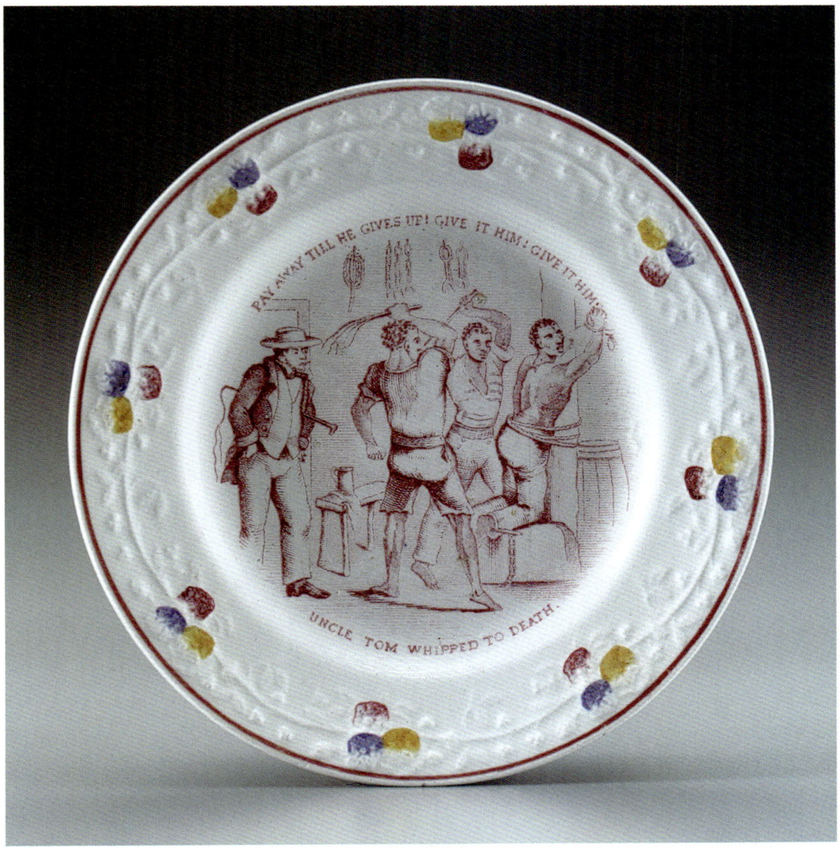

Figure 21 Plate, probably Staffordshire, ca. 1855. Earthenware. D. 6¼". (Courtesy, Rex Stark; photo, Gavin Ashworth.) This molded polychrome-painted and transfer-printed plate depicts "THE DEATH OF UNCLE TOM," based on Cruikshank's illustration.

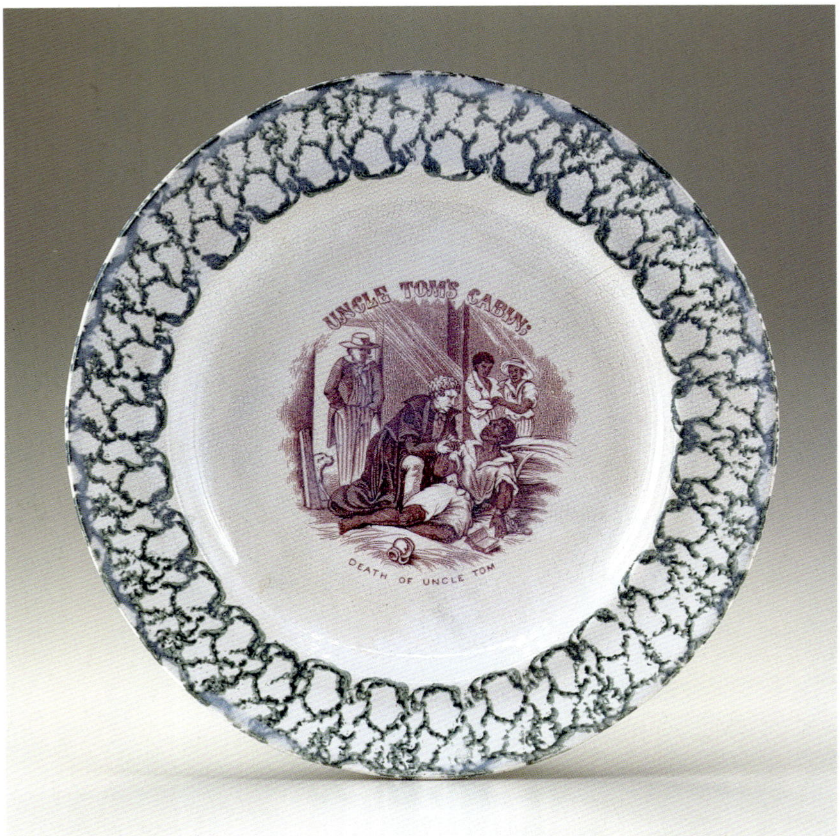

Figure 22 Plate, probably northern England or Scotland, ca. 1855. Whiteware. D. 7". (Courtesy, Rex Stark; photo, Gavin Ashworth.) This earthenware polychrome transfer-printed and painted plate illustrates the "DEATH OF UNCLE TOM." The plate was meant to appeal to a person of higher social standing than the example shown in fig. 21. Here, the teal blue sponged border frames the puce image of Tom lying on his deathbed.

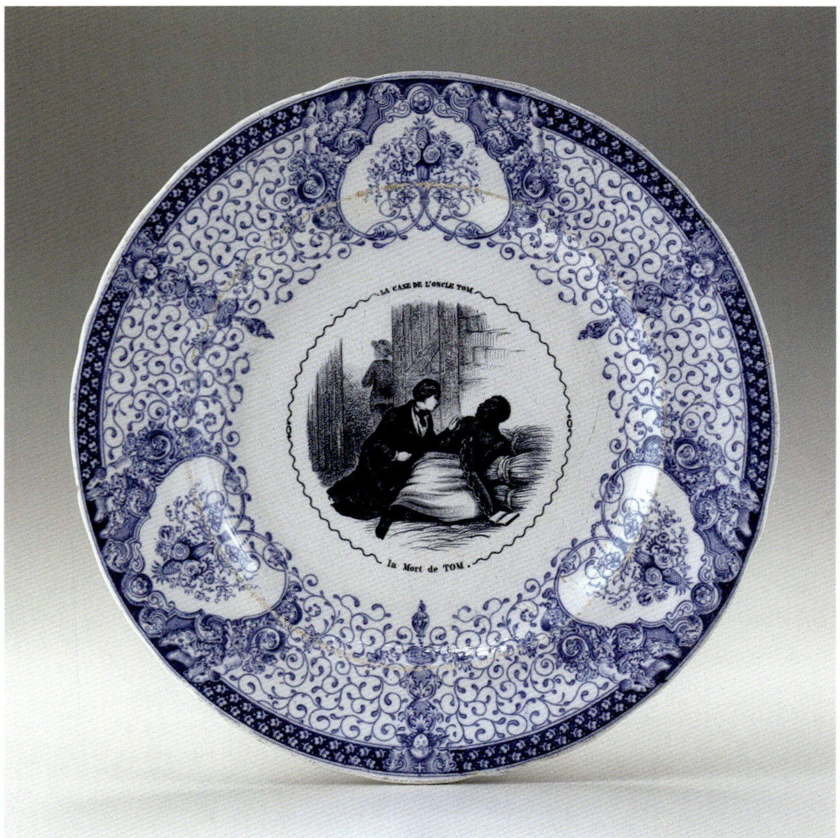

Figure 23 Plate, J. Vieillard et Cie., Bordeaux, ca. 1855. Whiteware. D. 8". (Courtesy, Rex Stark; photo, Gavin Ashworth.) This plate has an even more delicate rendering of the tragic death of Uncle Tom. The elaborate printed blue border is quite wide, diminishing the visual impact of the central image. Uncle Tom's death, illustrated in a warm, charcoal black with delicate detail, is based on an entirely different print source from the previous images, Girard's steel-cut engravings for the first French edition of *Uncle Tom's Cabin* (1853).

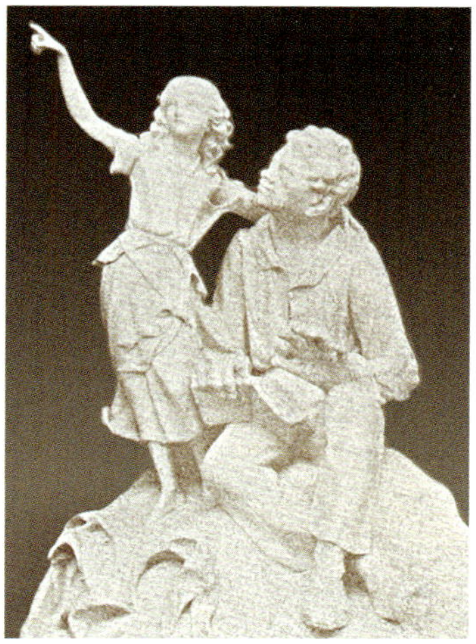

Figure 24 "Eva and Uncle Tom," parian sculpture from the Royal Worcester production catalogue, ca. 1853–1855. Salt print photograph. 3⅛" x 2". (Courtesy, The Worcester Porcelain Museum.) Wendy Cook, curator at the museum, has seen only one example of this figural group, and it was painted with polychrome enamels.

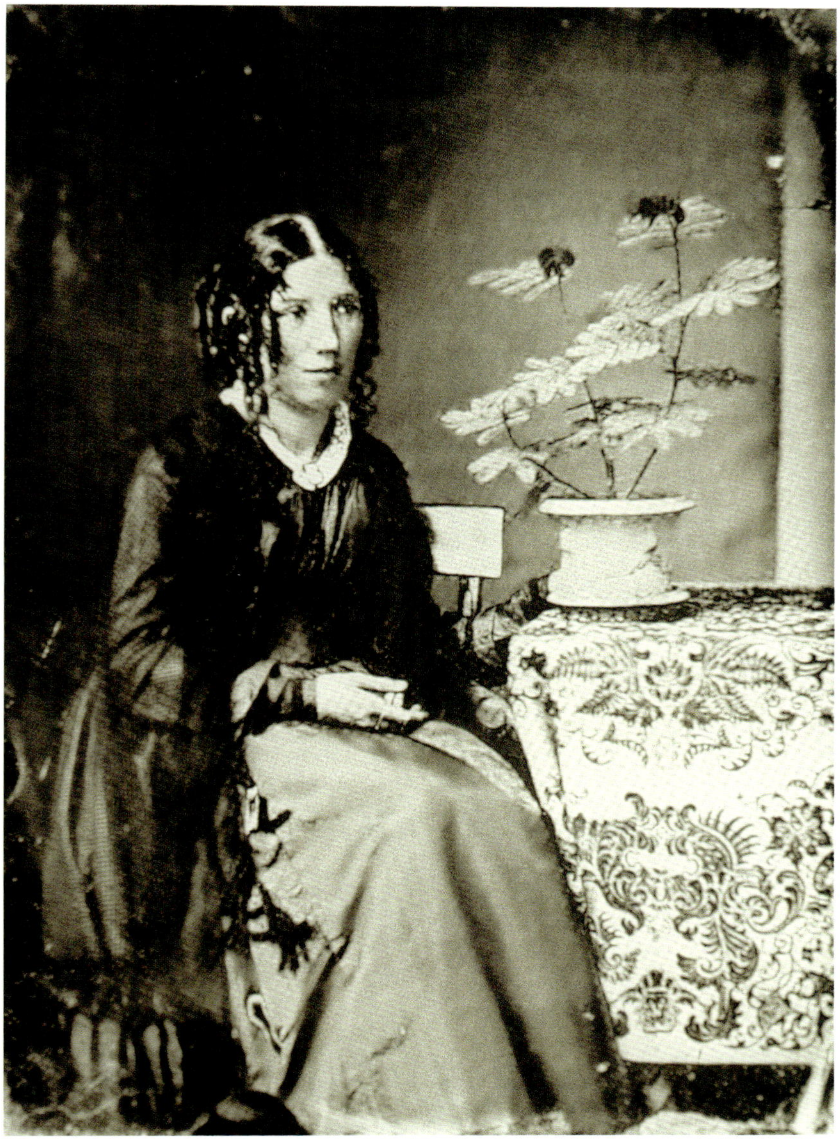

Figure 25 Southworth and Hawes, *Harriet Beecher Stowe.* Quarter-plate daguerreotype, ca. 1850. 4¼" x 3¼". (Metropolitan Museum of Art, New York, Gift of I. N. Phelps Stokes, Edward S. Hawes, Alice Mary Hawes, and Marion Augusta Hawes, 1937, acc. no. 37.14.40.)

The 1852 publishing contract between John Jewett of Boston and Harriet Beecher Stowe gave Stowe only 10 percent of the novel's overall sales, but Jewett agreed to "invest" another 10 percent in advertising the book. He agreed to "employ agents every where, to spare no pains nor expense nor effort to push the book into an unparalleled circulation." The contemporary historian Francis Lieber wrote, "It [Stowe's novel] will enter largely into exhibitions of paintings and possibly statuary. It will have its music."[18]

The tantalizingly vague mention of "statuary" based on the novel allows us to imagine marble figural groups based on illustrations from *Uncle Tom's Cabin,* which would later be translated into parianware ceramics (fig. 24).[19] It is uncertain what parianware models of the figural group survive; a preparatory drawing of *Eva and Uncle Tom* dating to about 1856–1857 is recorded in the factory archives of the manufactory at Royal Worcester.[20] Stowe herself (fig. 25) was the subject of a commemorative mug, which bears the legend "Mrs. Harriet Beecher Stowe / AUTHOR OF UNCLE

Figure 26 Mug, Staffordshire, ca. 1855–1860. Earthenware. H. approx. 8". (Courtesy, Rex Stark; photo, Gavin Ashworth.) Transfer-printed in warm medium blue, the legend reads "Mrs. Harriet Beecher Stowe / AUTHOR OF UNCLE TOM'S CABIN."

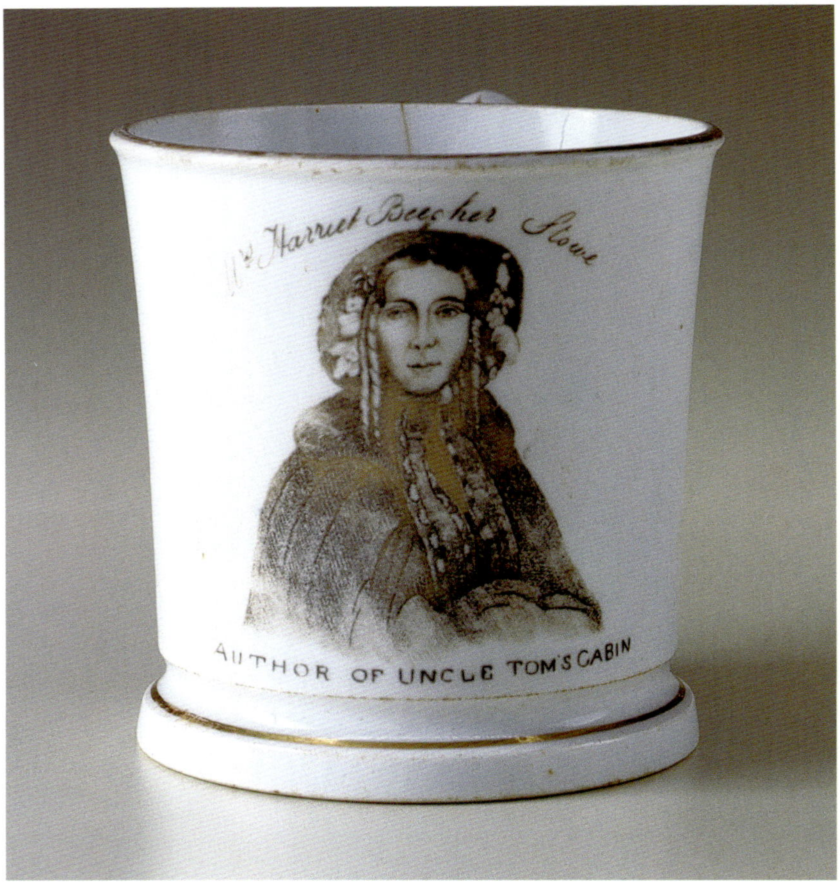

TOM'S CABIN" (fig. 26). Stowe's image dominates the central reserve, presenting her as a national—or international—hero surrounded by figures from *Uncle Tom's Cabin* as if they were supporting figures in her personal crusade against slavery. It is an amazing artifact, and one that advertises Stowe's household name recognition across international boundaries, the first time that a literary figure and a woman, no less, was so commemorated.

ACKNOWLEDGMENTS

The author would like to thank Alice Cooney Frelinghuysen, Anthony W. and Lulu C. Wong Curator of American Decorative Arts, The Metropolitan Museum of Art, for the inspiration to examine more closely these fascinating figural groups, and Leah C. Dilworth, chair of the English Department at Long Island University, with whom this essay was first imagined. A version of this essay was presented with Dilworth at the Southern American Studies Association Conference, February 2001.

1. Harriet Beecher Stowe, *Uncle Tom's Cabin* (New York: Bantam, 1981), p. 176.

2. Ibid., p. 58.

3. Although unmarked, the vases can be from no place other than Limoges. Many Limoges factories made decorative wares that featured figures either from popular contemporary literature and drama or of newsworthy appeal such as politicians, preachers, sportsmen, monarchs, and so forth. These were meant to be of momentary fashionable interest, but they were better looking than very cheap fairings. Limoges, more than factories in Paris or Bohemia, produced

vases in two sizes. Alice Cooney Frelinghuysen discusses a pair of similar vases from the collection of the Newark Museum, Newark, New Jersey, in her essay "Empire City Entrepreneurs: Ceramics and Glass in New York City," in *Art and the Empire City: New York, 1825–1861,* edited by Catherine Hoover Voorsanger and John K. Howat, exh. cat. (New York: Metropolitan Museum of Art, 2000), p. 330, no. 252, color ill. p. 537. Frelinghuysen cites Limoges rather than Paris as the likely place where these vases were made.

4. Harriet Beecher Stowe, *La Case de l'Oncle Tom,* translated by Louise Swanton-Belloc (Paris: Charpentier, 1853). As Adolphus M. Hart notes in *Uncle Tom in Paris,* the French rejected the high moral tone of Stowe's novel, preferring to show the bloodthirsty melodramatic elements of the book; Hart, *Uncle Tom in Paris; or, Views of Slavery Outside the Cabin* (Baltimore: William Taylor and Company, 1854). Even popular American theatrical interpretations relied on these same elements to draw crowds into their halls.

5. Harriet Beecher Stowe, *Pictures and Stories from Uncle Tom's Cabin* (Boston and London: John P. Jewett and Company, 1853).

6. See Dell Upton, "Inventing the Metropolis: Civilization and Urbanity in Antebellum New York," in *Art and the Empire City: New York, 1825–1861,* edited by Catherine Hoover Voorsanger and John K. Howat, exh. cat. (New York: Metropolitan Museum of Art, 2000),, pp. 3–45.

7. See ibid., fig. 16, for the illustration of the Haughwout salesroom; the details about the company's decorating staff can be found in Margaret Brown Klapthor et al., *Official White House China, 1789 to the Present,* 2nd ed. (New York: Harry N. Abrams, 1998), p. 77.

8. Klapthor, *Official White House China,* p. 82, says that two services decorated for Abraham Lincoln were completed with imported blanks from various French factories.

9. Alice Cooney Frelinghuysen, *American Porcelain, 1770–1920,* exh. cat. (New York: Metropolitan Museum of Art, 1989), suggests that the factories founded in America before the last third of the nineteenth century were prevented from enjoying widespread success through a combination of technical, marketing, and management flaws in their businesses.

10. The Bow factory had been making porcelain figures based on contemporary prints of famous actors and actresses since the 1750s, and the Derby factory made twelve bone china representations of William Coombs's hilarious buffoon Dr. Syntax based on differing episodes in his travels.

11. P. D. Gordon Pugh, *Staffordshire Portrait Figures and Allied Subjects of the Victorian Era* (London: Barrie and Jenkins, 1970).

12. The plethora of figures and the Dudson find were discussed via email at length with Miranda Goodby, beginning on January 5, 2001.

13. Harriet Beecher Stowe, *Uncle Tom's Cabin* (London: J. Cassell, 1852).

14. Obviously, bowers were not and are not exclusively English, but the English held to their garden traditions and associated an orderly garden to an orderly, morally correct life.

15. See www.iath.virginia.edu/utc for a fairly complete list of plates found with *Uncle Tom* scenes.

16. See ibid. This latter image particularly helped stir moral outrage against slave-holding and assisted Stowe's abolitionist cause. Stowe herself said she wrote *Uncle Tom's Cabin* because the true crime of slaveholders lay in their dividing families and destroying homes. The cabin in the novel is a symbol of good, holy, moral love despite the ignorance of its occupants.

17. For a full discussion of these items, see Sam Margolin, "'And Freedom to the Slave': Antislavery Ceramics, 1787–1865," *Ceramics in America,* edited by Robert Hunter (Hanover, N.H.: University Press of New England for the Chipstone Foundation, 2002), pp. 80–109.

18. Quoted in Joan D. Hedrick, *Harriet Beecher Stowe: A Life* (New York: Oxford University Press, 1994), p. 224.

19. "Inexpensive Parian-ware sculptures had become favorite domestic adornments by this period and had proliferated on the market; one advertisement run repeatedly by a Maiden Lane proprietor boasted of more than 400 selections"; Thayer Tolles, "Modeling a Reputation: The American Sculptor and New York City," in *Art and the Empire City: New York, 1825–1861,* edited by Catherine Hoover Voorsanger and John K. Howat, exh. cat. (New York: Metropolitan Museum of Art, 2000), p. 161.

20. See Paul Atterbury and Maureen Batkin, eds., *The Parian Phenomenon: A Survey of Victorian Parian Porcelain Statuary and Busts* (Somerset, Eng.: Richard Dennis, 1989), which includes this wonderful sketch from the factory archives. Although the image is not very distinct, it clearly shows Eva in an almost angelic state on Uncle Tom's knee. For the pitcher, see www.iath.virginia.edu/utc.

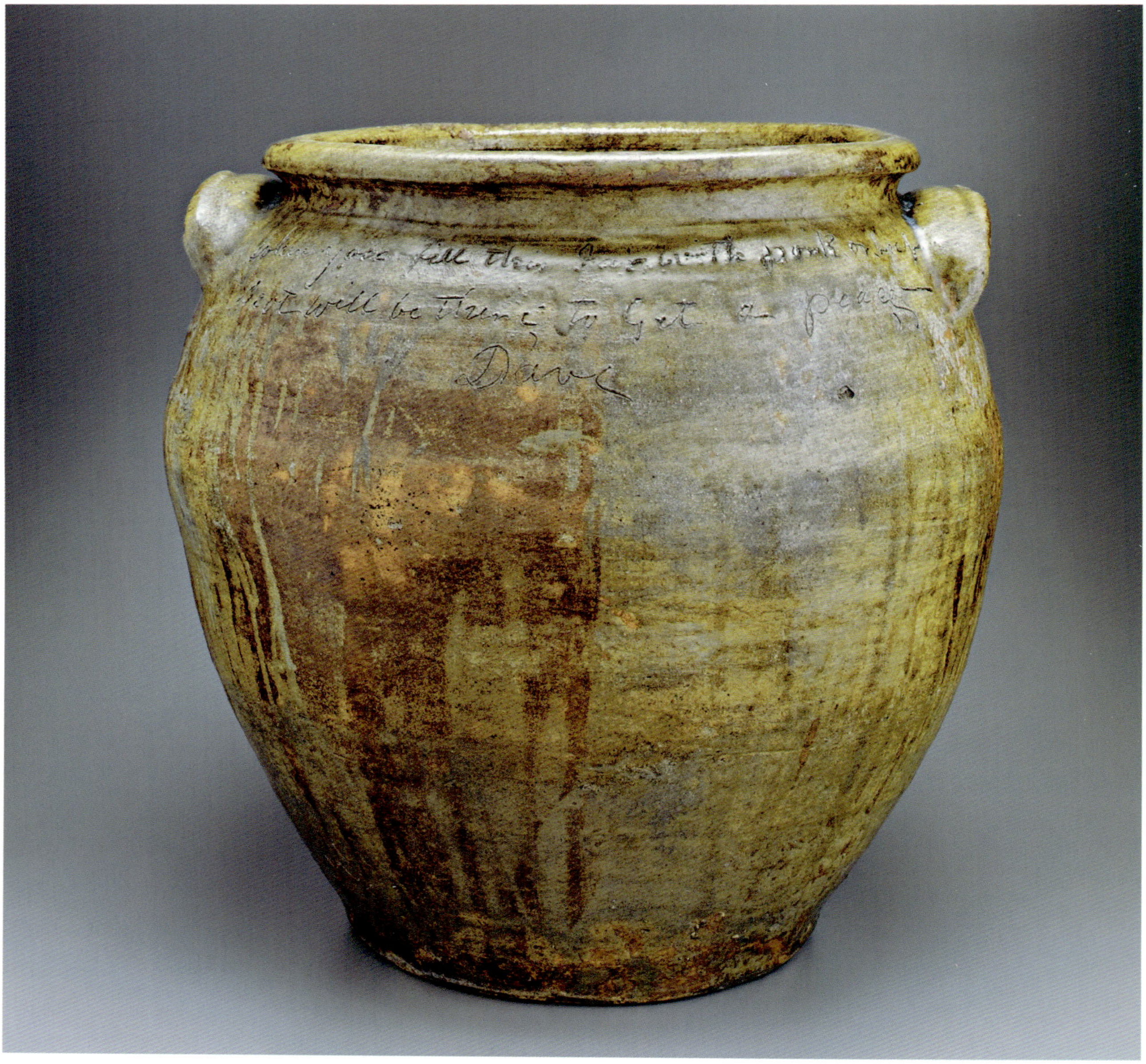

Figure 1 Storage jar, Lewis Miles pottery, 1858. Alkaline-glazed stoneware. H. 25⅝". Incised in script at shoulder: "This jar is to Mr Segler who keeps the bar in orange burg / for Mr Edwards a Gentleman—who formly kept Mr thos bacons horses / April 21 1858"; on opposite shoulder: "When you fill this jar with pork or beef / Scot will be there to get a peace,— / Dave" and twenty-five short slashes (two rows of ten and one of five). (Courtesy, Arthur Goldberg; unless otherwise noted, all photos by Gavin Ashworth.)

Arthur F. Goldberg
and
James P. Witkowski

Beneath His Magic Touch: The Dated Vessels of the African-American Slave Potter Dave

▼ DAVE, AN ENSLAVED African-American potter, made alkaline-glazed stoneware vessels in the nineteenth century in the Edgefield District of South Carolina. His ceramics, especially his large pots, are very expressive and powerful, but his contribution goes beyond the vessels themselves. He is also known for his literacy, at a time when slave literacy was not permitted. It is expressed in the dates, signatures, and poetry incised in script on many of his pots. Dave's pottery, valued for its functionality when first produced, is now appreciated for its greater value as the historical record of a slave who became an American icon.

This article reviews all of the known dated stoneware vessels that were signed by Dave or are attributed to him. His practice of dating pieces in script to such extent—with month, day, and year—as well as incising poems and inscriptions is unique in the American stoneware tradition and affords an opportunity to analyze his output in a systematic fashion. The most extensive published to date, this listing was compiled by examining pieces in private and public collections, interviewing collectors and curators, and reviewing publications and auction records. A sufficient number of examples exist to allow certain conclusions to be made as to where the objects were produced and the evolution of their forms.

These dated vessels are examined within a coherent chronology of Dave's life, which includes a review of the potter's various masters as well as a discussion of the social and political issues in Edgefield at the time. Also included is a general discussion of alkaline-glazed stoneware, which became a uniquely Southern ceramic tradition, and an updated compilation of his poems and inscriptions.

Chronology

The biographical record that follows has been constructed from a review of legal documents, as well as from the written record incised by Dave onto his pots.[1] Dave's ownership passed between brothers, from father to son, and among in-laws, and indeed, most of his masters are believed to be related. Central figures in his life were two brothers, Dr. Abner Landrum and Reverend John Landrum (the codevelopers of alkaline-glazed stoneware in the South), Harvey Drake, and his last master, Lewis Miles, who permitted a full expression of Dave's poetic abilities.

Dave was born into slavery in South Carolina about 1800 and was owned by one man, Harvey Drake, until Drake's death in December 1832. A mortgage agreement dated June 13, 1818, between Harvey Drake, Abner

Figure 2 Reverse side of the storage jar illustrated in fig. 1.

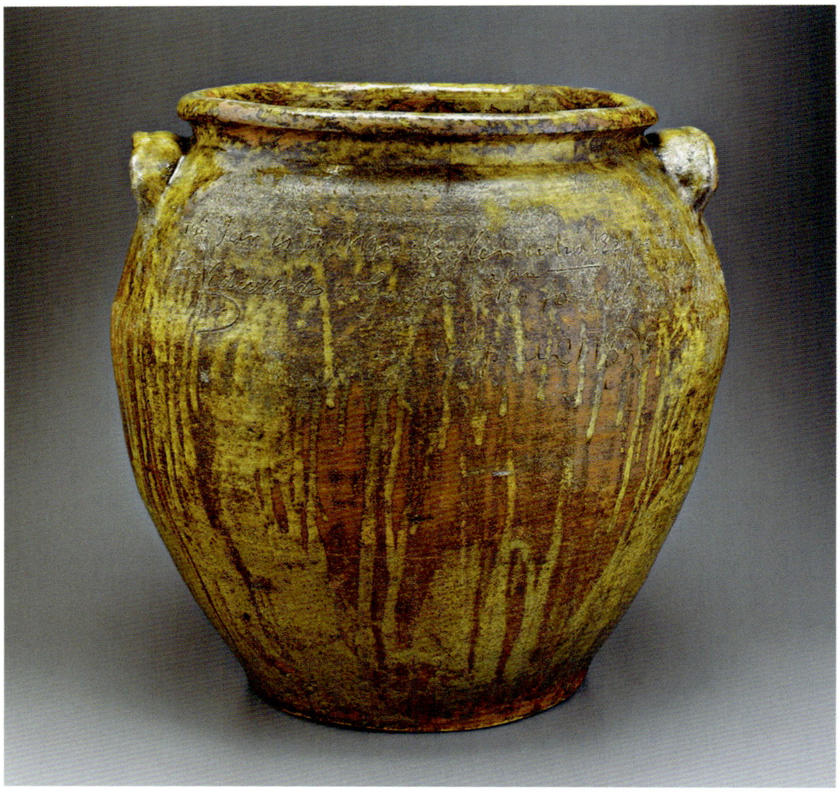

Landrum, and Eldrid Simkins, mentions "two negroes Slaves Dave a boy about 17 years old Country born and Eliza a girl about 18 or 20 years old both my property and my indisputed possession."[2] An agreement between Amos Landrum (brother of Abner and John) and Eldrid Simkins, dated October 26, 1818, settled the mortgage debt so that Dave would remain in Drake's possession. It also identifies Harvey Drake, Abner Landrum, and Amos Landrum as "copartners in the business of a pottery establishment in Pottersville," confirming Harvey Drake's importance as a figure in the early development of alkaline-glazed stoneware in the Edgefield District and his association with the Pottersville pottery.[3]

The Pottersville pottery, located about a mile from the Edgefield courthouse, was initially established by Abner Landrum about 1815.[4] Dr. Landrum was a physician, businessman, scientific farmer, newspaper editor, and publisher. He displayed his worldly knowledge and enthusiasm for ceramics by naming three of his sons—Wedgwood, Pallisey, and Manisis—after well-known European potters; another son, Linnaeus, conveyed his interest in botany. The Pottersville pottery was very successful, and in his *Statistics of South Carolina* (1826) Robert Mills credits Abner Landrum with the manufacture of stoneware "which by his discoveries is made stronger, better and cheaper than any European or American ware of the same kind."[5] In November 1828 Abner Landrum sold his interest in the pottery to Harvey Drake and his brother, Reuben. A partnership agreement between Harvey and Reuben Drake dated July 1, 1830, states: "The following property belongs to Harvey Drake . . . the servants—Dave, Lidy and her children. . . ."[6]

Harvey Drake died in 1832, and the January 31, 1833, estate settlement states that "1 Negro man named Dave [was sold] to Drake and Gibbs for $400."[7] Reuben Drake and Jasper Gibbs both owned the Pottersville pottery at various times. Drake maintained an interest in the enterprise until September 1836; he moved to Louisiana in 1837. Jasper Gibbs sold his interest in the pottery in October 1839.[8] No dated vessels from the pottery are known to us from 1838.

Figure 3 Detail of a portion of the poem and Dave's signature on the shoulder of the storage jar illustrated in fig. 1.

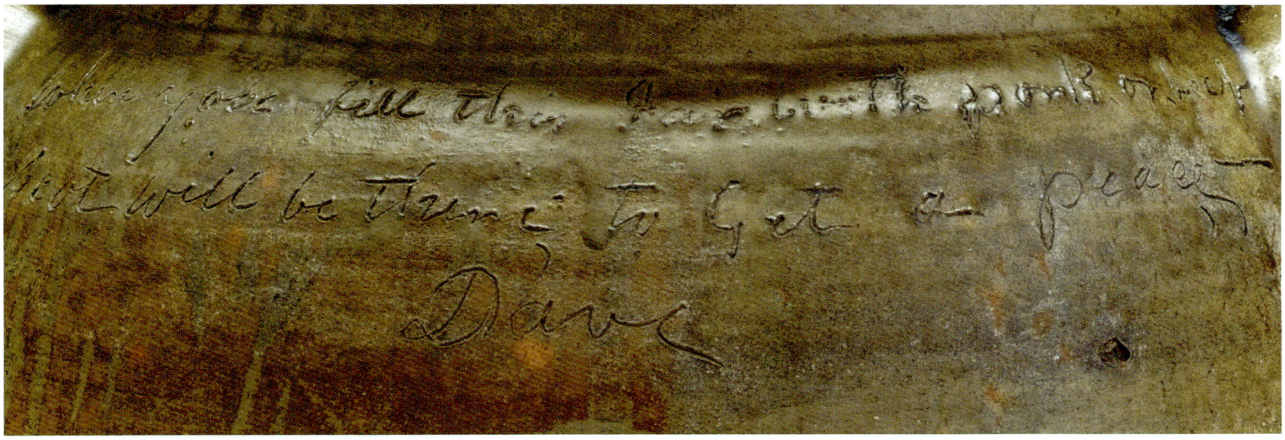

Figure 4 Detail of the rim and one of the large lug handles of the storage jar illustrated in fig. 1. Note blue rutile between the rim and handle.

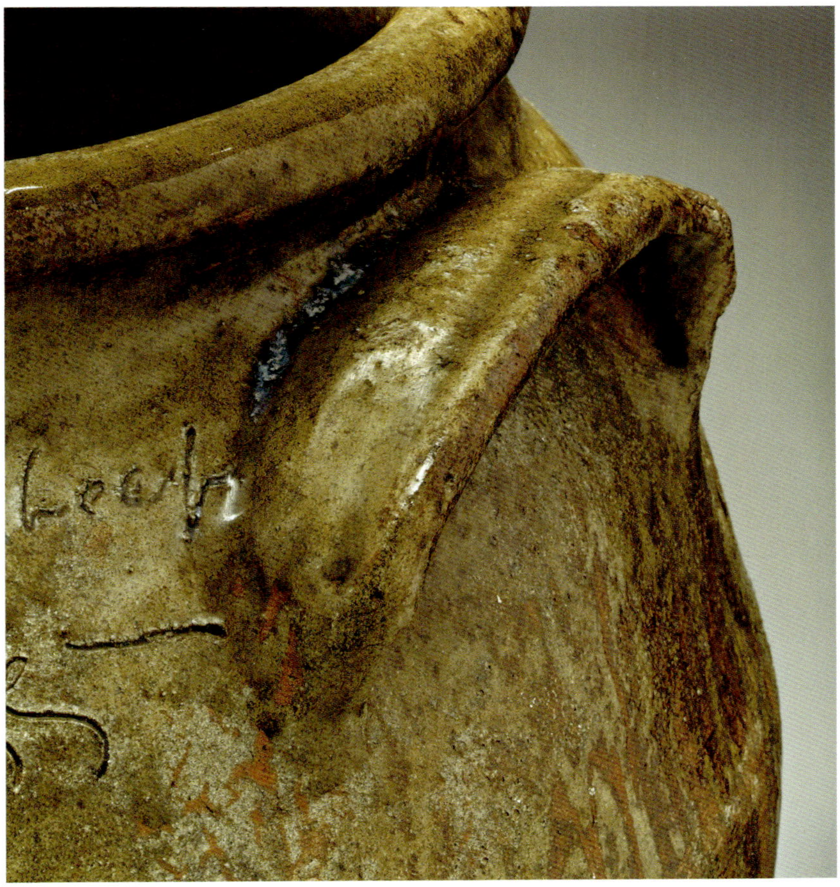

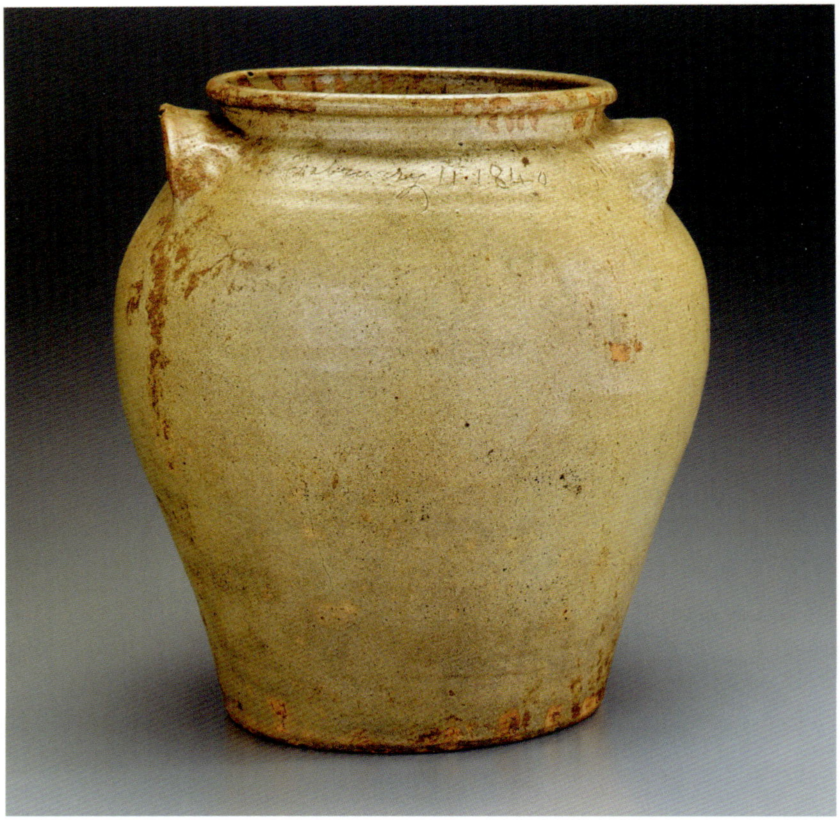

Figure 5 Jar, Lewis Miles pottery, 1840. Alkaline-glazed stoneware. H. 14½". Incised in script at shoulder: "Mr. L. Miles Dave"; on opposite shoulder: "February 11th · 1840" plus two slash marks and five punctates. (Courtesy, Arthur Goldberg).

An article of reminiscences about "Old Pottersville and Dr. Landrum," published in the *Edgefield Advertiser* on May 11, 1859, not only confirms Dave's importance at Pottersville but also sheds light on how he was viewed by adults and children in the Edgefield community:

> We cheerfully allow the old *ville* all the ancient glory claimed for it. Although not quite aged enough to go back distinctly to those stirring days when she was the *Cincinnati* of Edgefield, we can vividly recall the times when she was the Juggery of South Carolina. The first sight of that furnace nearly thirty years ago,—can we ever forget it? Have we ever been able to think of any other kind of furnace when reading of *Shadrack, Meschuck* and *Abednego*? Do we still mind how the boys and girls used to think it a fine Saturday frolic to walk to old Pottersville and survey its manufacturing peculiarities? to watch old Dave as the clay assumed beneath his magic touch the desired shape of jug, or jar, or crock, or pitcher, as the case may be?—No, we are not unmindful of the old times of old *Age*field; And we would not have old Pottersville omitted when our history comes to be written out, nor Dr. Landrum, nor the *Hive,* nor anything good or clever that ever adorned that once prominent locality.[9]

Presumably Dave continued to work in Pottersville until he came under the ownership of Lewis Miles, who, it is thought, owned him from about 1840 to 1843. It is not known how Miles came to own Dave, but Miles's association with the Landrum family is well documented. Lewis Miles married Dr. Abner Landrum's daughter; after her death, he married Reverend John Landrum's daughter, Sarah.[10] A jug dated January 27, 1840, and signed "Dave" and "L. Miles" is the earliest known evidence of their association, and a jar dated July 31, 1840, bears a poem that reads, in part:

". . . Dave belongs to Mr. Miles. . . ." Additional related information comes from the 1840 census, which indicates that Lewis Miles owned one male slave between thirty-six and fifty-five years old (Dave would have been forty years old at that time).[11] A jar signed "Dave" and "Mr. L. Miles," dated May 16, 1843, is the last known signed and dated vessel until January 27, 1849.

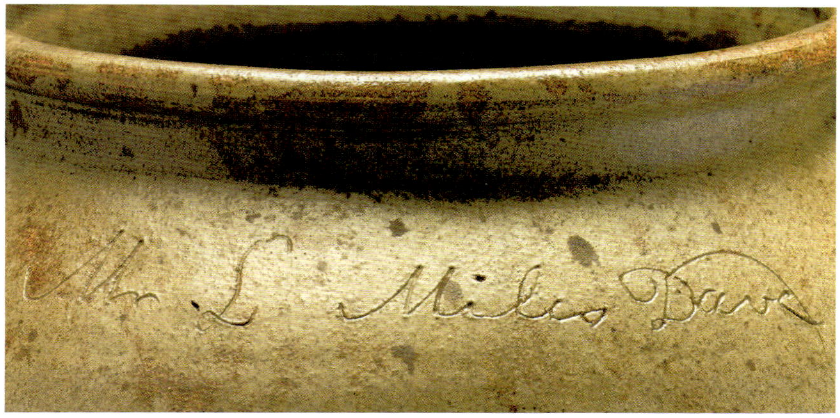

Figure 6 Detail of the inscription and signature on the shoulder of the storage jar illustrated in fig. 5.

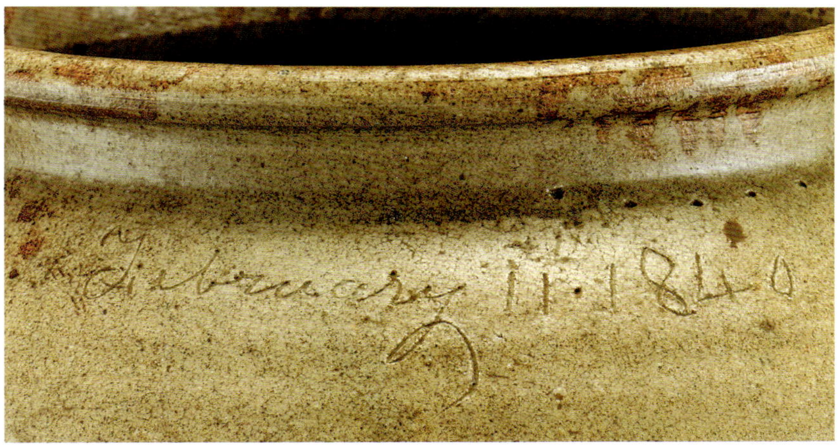

Figure 7 Detail of the incised date on the shoulder and the punctates under the rim of the storage jar illustrated in fig. 5.

Evidence suggests that from about 1843 to 1847 Dave was owned by Reverend John Landrum, Lewis Miles's father-in-law, who as early as 1817 had operated a pottery in the Big Horse Creek section of the Edgefield District, about twelve miles south of Pottersville.[12] How Dave came to be owned by Reverend Landrum is not known. The inventory of the reverend's estate, dated January 22, 1847, lists "1 Boy Dave."[13] The estate sale, on February 22, 1847, records the sale of a "Man Dave" to Reverend Landrum's son, Benjamin Franklin (B. F.) Landrum, who operated a pottery in the Horse Creek area, adjacent to Lewis Miles's plantation at that time. B. F. Landrum owned Dave from about 1847 to 1849.[14]

There are no known vessels signed Dave that date from October 1843 through January 1849, when he undoubtedly worked at both of the Landrum potteries in the Horse Creek area. Archaeological investigations at these pottery sites by Joe and Fred Holcomb in the 1970s yielded sherds with "incised letters, words, decorations, or marks similar to those on

Dave's extant vessels." There were also sherds similar in form to those on Dave's double-collared ring necks.[15] In 1994 Carl Steen found sherds at the B. F. Landrum site with characteristics of Dave-made vessels.[16]

From 1849 until he was freed in 1865 Dave was a potter at the Lewis Miles Stoneware Factory at Stoney Bluff Plantation on Big Horse Creek. Miles's second ownership of Dave, beginning in 1849, is believed to have resulted when Sarah Miles (Reverend John Landrum's daughter and Lewis Miles's wife) petitioned for a new executor of her father's estate, which somehow resulted in Miles's ownership of Dave.[17] Ending a period of six years with no known marked vessels is a jar dated January 27, 1849, and signed "Lm." A poem from May 13, 1859, "Made at Stoney Bluff. . . ," commemorates the pottery. The 1850 census indicates Miles owned a fifty-year-old male slave; the 1860 census shows his ownership of a fifty-six-year-old male slave, presumed to be Dave.[18]

More than one hundred known vessels are dated between 1849 and 1864, and signed "Dave" and "Lm," and all are believed to have been made at Miles's Stoney Bluff Plantation pottery. The Census of Manufactures for 1860 provides information on the scale of this operation, with annual production recorded at fifty thousand gallons.[19] Lewis Miles's personal involvement in the stoneware production, as well as Dave's dry wit and his ability to tease his owner, is revealed in the inscription on the jug signed by Dave on June 28, 1854: "Lm says this handle will crack."

Dave's last known poem, dated May 3, 1862, is charged with political, religious, and social meaning, given the context of slavery and civil war: "I made this Jar all of cross / If you don't repent, you will be lost." The

Figure 8 Storage jar, Lewis Miles pottery, 1861. Alkaline-glazed stoneware. H. 17¼". Incised in script on shoulder: "Lm May 7. 1861 Dave"; on opposite side: two slash marks. (Courtesy, Arthur Goldberg.)

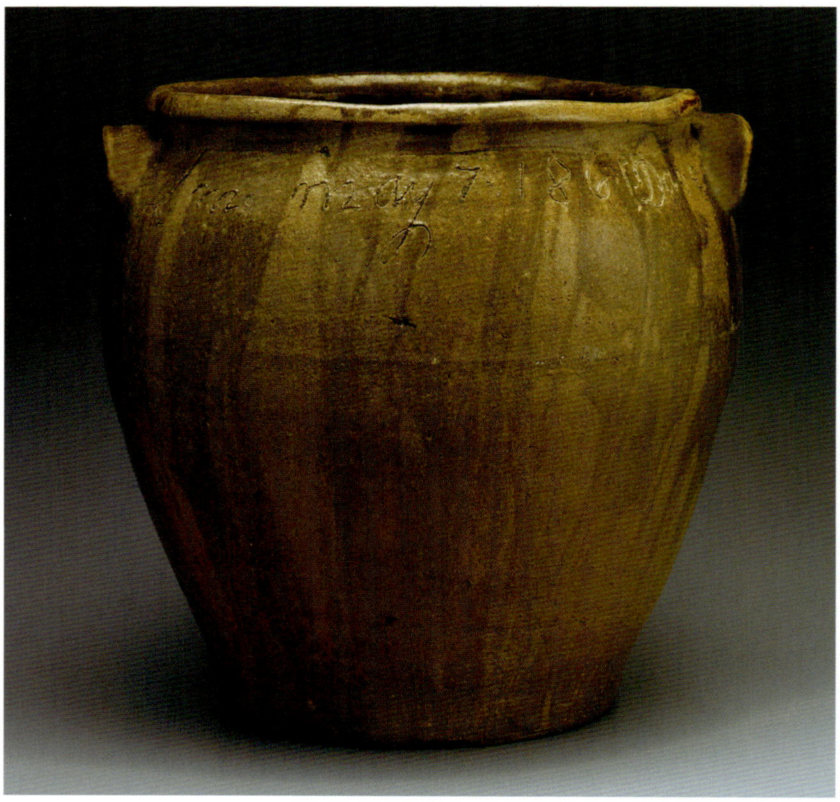

last known dated stoneware vessel inscribed "Dave" and "Lm" is dated March 31, 1864.

There is evidence that Dave lost one of his legs in an accident but continued to work as a potter. In interviews conducted by researchers from the Charleston Museum in 1930, an eighty-five-year-old former slave pottery worker, Carey Dickson, recalled:

> Of course I knew Dave. I know all about him. He used to belong to old man Drake and it was at that time that he had his leg cut off. They say he got drunk and layed on the railroad track. Later Dave went to Mile's Mill. After Dave was crippled he had Henry Simkins, who was crippled in the arms, to drive the wheel for him.[20]

Another potter, George Fletcher, also remembered being told "about an old potter there [Lewis Miles's pottery], a one-legged negro named Dave. He was said to be mighty good."[21]

Upon obtaining his freedom after the Civil War, Dave changed his name to David Drake, combining a more dignified first name with the surname of his original known owner. The 1870 Federal Census for South Carolina shows a David Drake, age seventy, occupation "turner," living in the household of Mark Jones, age thirty-five, whose occupation is also listed as "turner." All of the occupants of this household, including Jones's five children, are listed as having been born in South Carolina (fig. 11).[22] Notably, a jar dated March 10, 1859 (see figs. 33, 34), is signed "Mark and Dave," tantalizing evidence that Dave might have passed on his skills as a potter to a future generation. Since the 1880 Federal Census for South Carolina does not list a David Drake in the Edgefield District,[23] it is presumed that he

Figure 9 Reverse side of the storage jar illustrated in fig. 8.

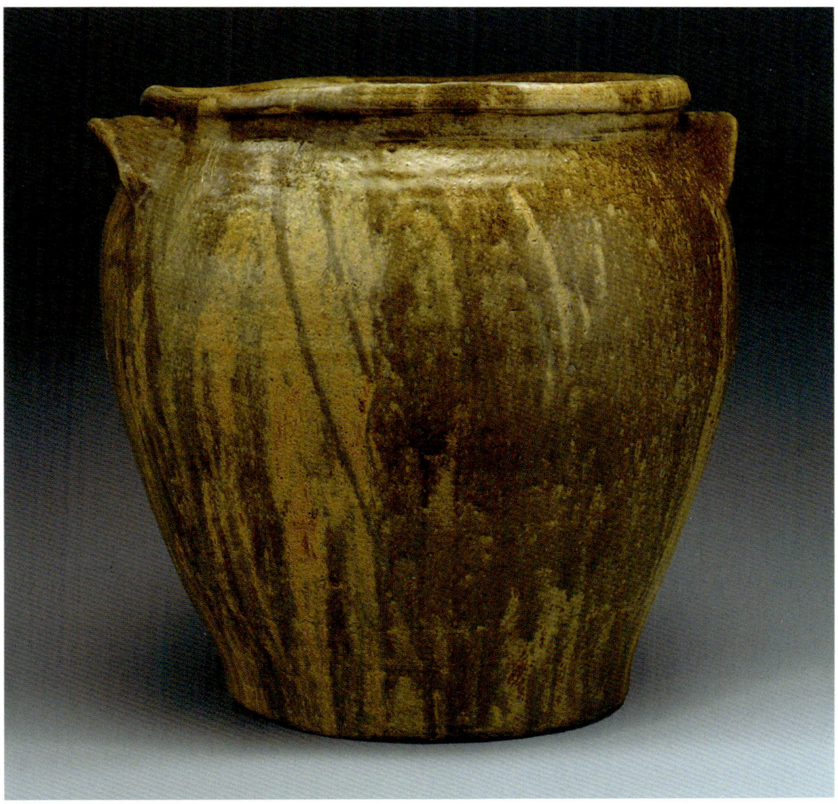

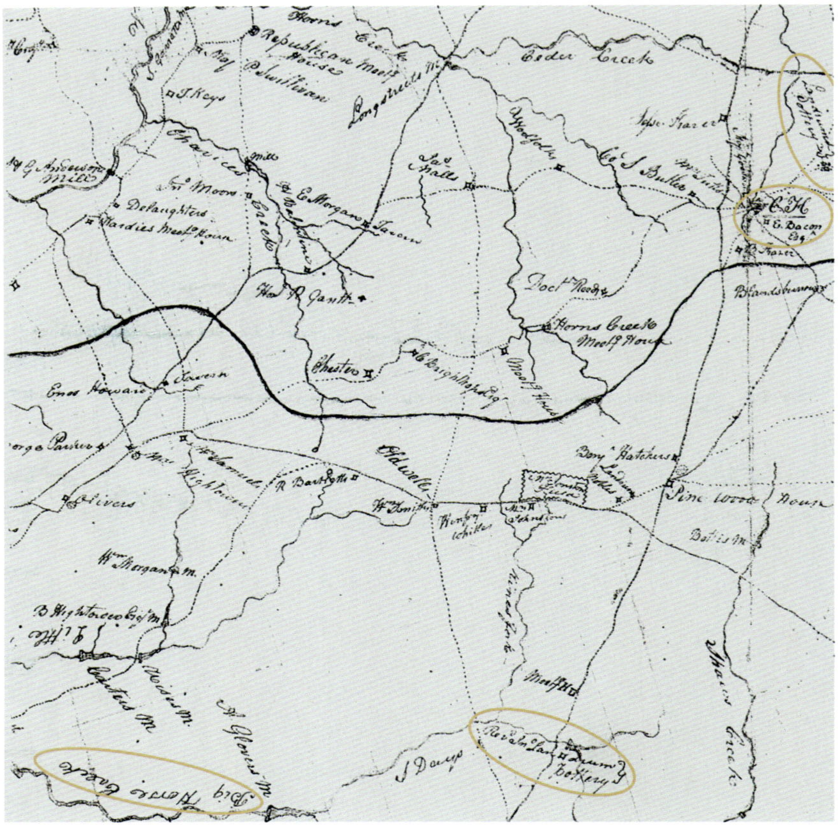

Figure 10 Detail, "A Map of the District of Edgefield," December 1816, by Thomas Anderson, D.S. (Courtesy, South Carolina Department of Archives and History, Columbia.) This map was later revised and published in the *Robert Mills Atlas of the State of South Carolina* (1825). The circled areas show the [Abner] Landrum pottery, CH [Edgefield Courthouse], the Rev. Jn. Landrum pottery, and Big Horse Creek.

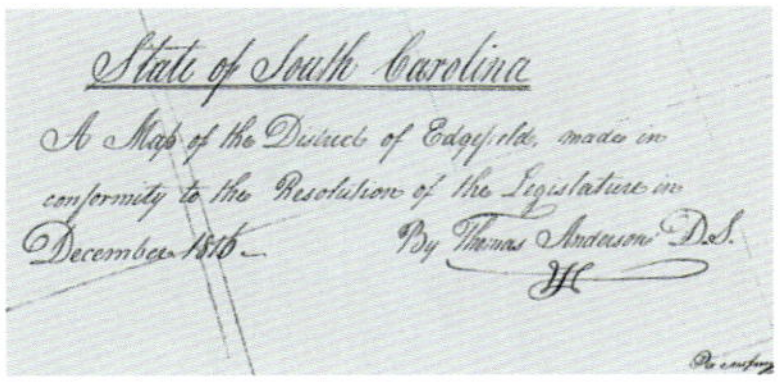

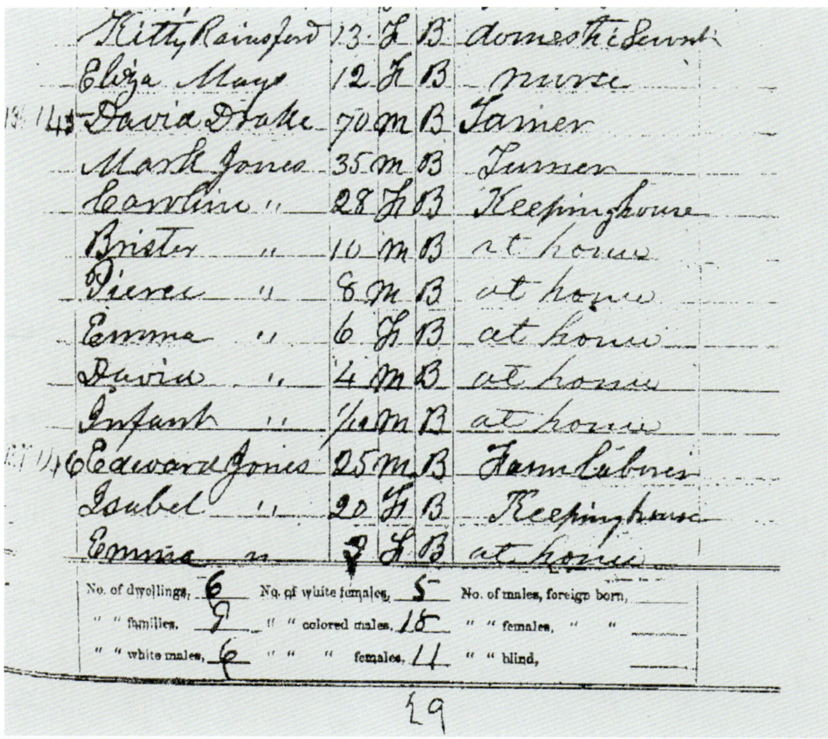

	Name	Age	Sex	Color	Occupation
	Kitty Rainsford	13	F	B	domestic Servant
	Eliza Mays	12	F	B	nurse
145	David Drake	70	M	B	Turner
	Mark Jones	35	M	B	Turner
	Caroline "	28	F	B	Keeping house
	Brister "	10	M	B	at home
	Pierce "	8	M	B	at home
	Emma "	6	F	B	at home
	David "	4	M	B	at home
	Infant "	1/12	M	B	at home
146	Edward Jones	25	M	B	Farm Laborer
	Isabel "	20	F	B	Keeping house
	Emma "	3	F	B	at home

No. of dwellings, 6 No. of white females, 5 No. of males, foreign born,
" " families, 9 " " colored males, 18 " " females, " "
" " white males, 6 " " females, 11 " " blind,

29

Figure 11 Detail of the 1870 Federal Census, Population Schedule, South Carolina, showing Dave and Mark Jones and his family in the same household.

died sometime between 1870 and 1880. After David Drake's death, Mark Jones and one of his sons worked at another Edgefield District pottery, continuing the stoneware legacy.[24]

Dated Vessels

Table 1 lists the known dated stoneware vessels signed by or attributed to Dave—169 as of this writing. The table, which is organized chronologically,

TABLE 1. Known Dated Vessels Signed by or Attributed to the Slave Potter Dave

Size given is height in inches. Jug(2) indicates a two-handled jug; jar(4) indicates a four-handled storage jar.

	Date	Vessel	Size	Dave	Miles	Marks	Day	Notes	Museum*
1	5-23-1829	Jar	13			numerous punctates	Sat.		
2	10-22-1829	Jar	12			], numerous punctates	Thur.		
3	3-17-1831	Jar	12			numerous punctates	Thur.		3
4	6-12-1834	Jar	19½			∩, eight punctates, squiggle	Thur.	Inscription	
5	7-12-1834	Jar	19¼			//, fourteen punctates	Sat.	Poem	13
6	11-16-1834	Jar	18½			∩ on handle	Sun.		3
7	3-29-1836	Jar	16			∪	Tues.	Poem	
8	4-12-1836	Jar	14½				Tues.	Inscription	
9	4-15-1836	Jar				//, five punctates	Fri.		
10	4-20-1836	Jar	15¼				Wed.		
11	4-26-1836	Jar	11			^	Tues.		
12	8-25-1836	Jug	14½			///	Thur.		
13	10-17-1836	Jug	14			///	Mon.		
14	11-9-1836	Jug(2)	17½			/, five punctates	Wed.	Poem	4
15	12-22-1837	Jar	14				Fri.		
16	12-2-1839	Jar					Mon.		
17	1-23-1840	Jar	12		L. Miles		Thur.		2
18	1-25-1840	Jar	14		L. Miles	//	Sat.		
19	1-27-1840	Jar	15½	Dave	Mr. Miles	//, six punctates	Mon.		
20	1-29-1840	Jug(2)	17¼	Dave	L. Miles	∩, six punctates	Wed.	Poem	
21	2-7-1840	Jar	14½	Dave	Mr. Miles/L.Miles	//, five punctates, squiggle	Fri.		
22	2-10-1840	Jar	15	Dave	Mr. L. Miles		Mon.	Poem	
23	2-11-1840	Jar	14½	Dave	Mr. L Miles	//, five punctates	Tues.		
24	2-27-1840	Jug(2)	16	Dave	Mr. L Miles	//	Thur.		14
25	4-3-1840	Jug	14		LM	//, two punctates	Fri.		
26	4-27-1840	Jar	14	Dave	Mr. L Miles		Mon.		
27	6-27-1840	Jar	14½	Dave	Mr. L Miles		Sat.	Poem	
28	7-3-1840	Jar		Dave	Mr. L. Miles	∩, punctates	Thur.		
29	7-22-1840	Jar	14	Dave	L. Miles		Wed.		3
30	7-31-1840	Jar	15¼	Dave	Mr. Miles	six punctates	Fri.	Poem	3
31	8-25-1840	Jar	15¾	Dave	Mr. L Miles	∪, six punctates	Tues.	Inscription	
32	8-26-1840	Jar	15¼	Dave	L. Miles	six punctates	Wed.	Poem	
33	11-20-1840	Jar	13	Dave	Mr. L Miles	//, four punctates	Fri.		
34	2-14-1841	Jug	14½	Dave	L. Miles	//	Sun.		
35	4-26-1842	Jug(2)	20	Dave	L. Miles	//, five punctates	Tues.		
36	5-16-1843	Jar	27¾	Dave	Mr. L Miles		Tues.	Inscription	
37	10-13-1843	Jar	16	Dave	L. Miles	//	Fri.		
38	1-27-1849	Jar	14		Lm		Sat.		
39	5-3-1849	Jar	15	Dave	Lm		Thur.		
40	8-29-1849	Jug	16		Lm	∪, three punctates	Wed.		

	Date	Vessel	Size	Dave	Miles	Marks	Day	Notes	Museum
41	10-15-1849	Jar	17	Dave	Mr. Miles	four punctates	Mon.		
42	10-27-1849	Jar	15½		Lm	//	Sat.		
43	10-29-1849	Jar			Lm	//, four punctates	Mon.		
44	10-31-1849	Jar	16	Dave	Lm	//, five punctates	Wed.		
45	11-1-1849	Jug(2)	19	Dave	L. Miles	five punctates	Thur.		
46	1-?-1850	Jar	16		Lm				
47	10-17-1850	Jar	24			∪	Thur.	Inscription	
48	3-21-1851	Jug(2)	17¾		Lm	)), punctates	Fri.		
49	5-2-1851	Jar	16½		Lm	//, 7	Fri.		
50	5-7-1851	Jar	12		Lm	//, three punctates	Wed.		
51	5-30-1851	Jar	14½		Lm	//	Fri.		
52	8-5-1851	Jar		Dave	Lm		Tues.		
53	8-13-1851	Jar	17		Lm	seven punctates	Wed.		
54	1-31-1852	Jar		Dave	Lm		Sat.		
55	2-11-1852	Jar	14½	Dave		//, seven punctates	Wed.		1
56	3-14-1852	Jug	14¾	Dave	Lm		Sun.		
57	5-18-1852	Jar	17	Dave	Lm	], seven punctates	Tues.		
58	5-19-1852	Jar	16		Lm	∩, six punctates	Wed.	Inscription	
59	7-10-1852	Jug(2)	16			∪	Sat.		
60	8-16-1852	Jar	16		Lm	/∪, seven punctates	Mon.		
61	8-31-1852	Jug(2)	16	Dave	Lm	∩/, four punctates	Tues.		
62	9-4-1852	Jar	16½		Lm	∩/, six punctates	Sat.		
63	9-6-1852	Jar			Lm	)/, five punctates	Mon.		
64	10-28-1852	Jar	15	Dave	Lm	∪, seven punctates	Thur.		
65	4-20-1853	Jar	13			four punctates	Wed.		
66	6-10-1853	Jug	14¼		Lm	∪, three punctates	Fri.		
67	10-26-1853	Jug	15	Dave	Lm	∩, three punctates	Wed.		
68	6-9-1854	Jar	13½	Dave	Lm	//, five punctates	Fri.		
69	6-28-1854	Jug		Dave	Lm		Wed.	Inscription	
70	8-12-1854	Jar	15	Dave	Lm	∩, six punctates	Sat.		
71	8-17-1854	Jar	13½	Dave	Lm	//	Thur.		
72	8-24-1854	Jar		Dave	Lm	//, three punctates, D	Thur.		
73	9-17-1854	Jug	11	Dave	Lm		Sun.		
74	10-16-1854	Jar	14½	Dave			Mon.		
75	4-5-1855	Jar	12½	Dave	Lm	//, three punctates	Thur.		
76	6-?-1855	Jar			Lm	∩, six punctates			
77	8-24-1855	Jug	15½	Dave	Lm	//, three punctates	Fri.		
78	10-?-1855	Jar		Dave	Lm	X, three punctates			
79	1-19-1856	Jar		Dave	Lm		Sat.		
80	1-23-1856	Jar	14	Dave	Lm		Wed.		
81	6-1-1856	Jug	13¼	Dave	Lm	//, two punctates	Sun.	Incised horse and rider	
82	8-29-1856	Jar	12¾		Lm	8, L, //, seven punctates	Fri.		
83	9-25-1856	Jar		Dave	Lm		Thur.		
84	10-24-1856	Jar	14½	Dave	Lm	//, six punctates	Fri.		
85	12-12-1856	Jar	14	Dave	Lm/Lm	//	Fri.		
86	2-25-1857	Jar	15	Dave	Lm	//, three punctates	Wed.		
87	3-12-1857	Jar	14		Lm	//, four punctates	Thur.		
88	3-20-1857	Jar	15	Dave	Lm		Fri.		6
89	3-24-1857	Jar	17	Dave	Lm	//, five punctates	Tues.		
90	5-30-1857	Jar	15		Lm		Sat.		
91	6-6-1857	Jar	17	Dave	Lm	//, eight punctates	Sat.		
92	6-8-1857	Jar		Dave	Lm	//, eight punctates	Mon.		

	Date	Vessel	Size	Dave	Miles	Marks	Day	Notes	Museum
93	6-11-1857	Jug	14	Dave	Lm	//	Thur.	Incised bird	
94	6-16-1857	Jar		Dave	Lm		Tues.		
95	7-1-1857	Jar	17	Dave	Lm		Wed.		
96	7-6-1857	Jar	21¾		Lm	W, ten punctates	Mon.	Inscription	
97	7-25-1857	Jar	15½	Dave	Lm	//, three punctates	Sat.		
98	8-1-1857	Jar	17		Lm	∩, eight punctates	Sat.		
99	8-8-1857	Jar	22½		Lm	//, ten punctates	Sat.		13
100	8-16-1857	Jar	19	Dave	Lm		Sun.	Poem	
101	8-18-1857	Jar	20	Dave	Lm	//, ten punctates	Tues.		
102	8-20-1857	Jug(2)	17	Dave	Lm	//	Thur.		
103	8-22-1857	Jar	19	Dave	Lm		Sat.	Poem	9
104	8-24-1857	Jar	19	Dave	Lm		Mon.	Poem	6
105	8-30-1857	Jar	19¼	Dave	Lm	//, ten punctates	Sun.		
106	9-5-1857	Jar	21	Dave	Lm	∩, twelve punctates	Sat.		2
107	9-7-1857	Jar	17¾	Dave	Lm		Mon.		
108	10-15-1857	Jar	21	Dave	Lm	∩, ten punctates	Thur.		13
109	12-17-1857	Jar	16¼	Dave	Lm	//, eight punctates	Thur.		
110	1-30-1858	Jar	20½	Dave	Lm	fourteen punctates	Sat.	Poem	
111	2-15-1858	Jar	21¼		Lm	fourteen punctates	Mon.		
112	2-20-1858	Jar(4)	26	Dave	Lm	∩, thirty punctates	Sat.	Poem	
113	2-24-1858	Jug	14½	Dave	Lm	//, three punctates	Wed.		
114	3-9-1858	Jug		Dave	Lm	//, punctates	Tues.		
115	3-12-1858	Jar	12½	Dave	Lm	X, three punctates	Fri.		
116	3-15-1858	Jar	23¼	Dave	Lm	//, eight punctates	Mon.		
117	3-31-1858	Jar	22	Dave	Lm	fifteen punctates	Wed.	Poem	4
118	4-8-1858	Jar	23	Dave	Lm		Thur.	Poem	
119	4-12-1858	Jar(4)	24½	Dave	Lm		Mon.	Poem	1
120	4-21-1858	Jar	23½	Dave		twenty-five short slashes	Wed.	Poem and Dedication	
121	6-9-1858	Jar		Dave	Lm	X	Wed.		
122	6-26-1858	Jug	15¾		Lm	//, three punctates	Sat.		
123	7-14-1858	Jug	13	Dave	Lm		Wed.		
124	7-29-1858	Jar	20¼	Dave	Lm		Thur.	Poem	
125	9-18-1858	Jar	15	Dave	Lm	//	Sat.		
126	10-26-1858	Jar	14½		Lm		Tues.		
127	11-3-1858	Jar	24¼	Dave	Lm		Wed.	Poem	8
128	12-6-1858	Jar	21				Mon.	Poem	
129	12-6-1858	Jug	13	Dave	Lm		Mon.		
130	1-13-1859	Jar	11½		Lm		Thur.		
131	1-27-1859	Jar		Dave	Lm		Thur.		
132	1-29-1859	Jar	14	Dave	Lm		Sat.		
133	3-10-1859	Jar	15	Dave	Lm	//	Thur.	Inscription	12
134	3-25-1859	Jar	21	Dave	Lm		Fri.	Poem	5
135	4-14-1859	Jar					Thur.	Poem	
136	4-18-1859	Jar					Mon.	Poem	
137	5-2-1859	Jug	14¼		Lm	///	Mon.		10
138	5-3-1859	Jar	26½	Dave	Lm		Tues.	Poem	11
139	5-13-1859	Jar(4)	28¾	Dave			Fri.	Poem and Inscription	3
140	5-13-1859	Jar(4)	25¾	Dave	Lm		Fri.	Poem and Inscription	3
141	6-15-1859	Jar	13	Dave	Lm	X	Wed.		
142	6-20-1859	Jug	13½	Dave	L+M	///	Mon.		6

	Date	Vessel	Size	Dave	Miles	Marks	Day	Notes	Museum
143	7-2-1859	Jug	14	Dave	Lm	///	Sat.		
144	7-4-1859	Jar	23½	Dave	Lm		Sun.	Poem	1
145	7-8-1859	Jar	12	Dave	Lm		Fri.		
146	7-15-1859	Jug	13½	Dave			Fri.		1
147	7-19-1859	Jug(2)	16¼				Tues.	Inscription	
148	9-6-1859	Jug	13½	Dave	Lm	///	Tues.		
149	12-21-1859	Jar	12	Dave	Lm	X	Wed.		
150	4-1-1860	Jar	13¾	Dave	Lm		Sun.		
151	4-3-1860	Jug	14½	Dave	Lm	// X	Tues.		
152	5-31-1860	Jar	15	Dave	Lm		Thur.		
153	6-2-1860	Jar	15½	Dave	Lm	X	Sat.		
154	8-2-1860	Jar	18	Dave	Lm	//	Thur.		
155	8-7-1860	Jar	16¾	Dave	Lm		Tues.	Poem	
156	11-9-1860	Jar	21½	Dave	Lm		Fri.	Poem	
157	4-25-1861	Jar	13¼	Dave	Lm	X	Thur.		
158	5-7-1861	Jar	17¼	Dave	Lm	//	Tues.		
159	6-18-1861	Jar	14	Dave	Lm		Tues.		
160	8-15-1861	Jar	15	Dave	Lm		Thur.		
161	10-10-1861	Jar	13½	Dave	Lm	X	Thur.		13
162	5-3-1862	Jar	20	Dave	Lm		Sat.	Poem	12
163	2-13-1863	Jar	15	Dave	Lm	X	Fri.		
164	3-18-1863	Jar	14¼	Dave	Lm	//	Wed.		
165	4-21-1863	Jar		Dave	Lm		Tues.		
166	7-27-1863	Jar	13¾	Dave	Lm		Mon.		
167	1-12-1864	Jar	12½	Dave	Lm		Tues.		
168	3-10-1864	Jar	13½	Dave	Lm	X, three punctates	Thur.		7
169	3-31-1864	Jar	15	Dave	Lm	///	Thur.		

*Legend for Museums

1. Atlanta History Center, Atlanta, Ga.
2. Augusta Museum of History, Augusta, Ga.
3. The Charleston Museum, Charleston, S.C.
4. High Museum of Art, Atlanta, Ga.
5. Madison-Morgan Cultural Center, Madison, Ga.
6. McKissick Museum, Columbia, S.C.
7. The Mint Museums, Charlotte, N.C.
8. Museum of Early Southern Decorative Arts, Winston-Salem, N.C.
9. Museum of Fine Arts, Boston, Ma.
10. The Museum of Fine Arts, Houston (Bayou Bend), Tex.
11. Philadelphia Museum of Art, Pa.
12. Smithsonian Institution, Washington, D.C.
13. South Carolina State Museum, Columbia, S.C.
14. Washington Historical Museum, Washington, Ga.

gives basic information about each pot: date, vessel type, overall height, signatures, inscriptions, and other marks. Also included is whether the pot has an incised poem or other type of inscription, and whether it is in a public museum collection.

The vessel types include bulbous storage jars, jars that are more cylindrical (perhaps butter churns), and single- and double-handled jugs. Other vessel forms are attributed to Dave, such as handled pitchers, but there are no known dated examples.

The writing on these pots typically is found on the vessel shoulders, but some of the more elaborate inscriptions extend to the sides. No known examples are marked on the base.[25] At an appropriate time, depending on the degree of drying, the incising was done with a very sharp pointed stylus. As occurs in drypoint etching, when the clay was softer, a burr outlining the writing was produced. This had to be done shortly after the vessel was turned. The size of the letters/numbers varies from about one quarter of an inch to over an inch and a half, and generally increased from the earliest pieces to the later.

On almost all of the vessels listed the dates include month, day, and year. A few vessels have only month and year, a few have the day inscribed before the month, and, in some cases, the month is abbreviated.[26]

The earliest known dated vessel signed "Dave" is a storage jar dated "January 27th, 1840"; the latest is a storage jar dated "March 31, 1864." Table 1 includes not only vessels signed "Dave" but also pieces that are attributed to him. The attributions are based on the distinctive cursive writing of the dates or other inscribed words, as well as the vessel characteristics (form, handle attachment, and so forth). The attributed pieces, including vessels dating from 1829 to 1836, are believed to have been made by Dave at the Pottersville pottery, where apparently he was not allowed to sign his wares. These attributed pieces include three examples with poems, almost certainly made by Dave (see fig. 36).

Many of the pots made when Dave was producing pottery for Lewis Miles bear inscriptions that acknowledge the relationship between the two men. Typically, the dated vessels have both Dave's name and "Lm" or some variant of "Miles" inscribed on them. A number have a Lewis Miles inscription but not Dave's signature, even though they were almost certainly made by Dave and therefore are included in Table 1.

TABLE 2 Bar chart showing the production of known dated vessels by year. Note the gap between 1844 and 1848, suggesting that Dave's owners in those years did not allow him to sign or date his pots.

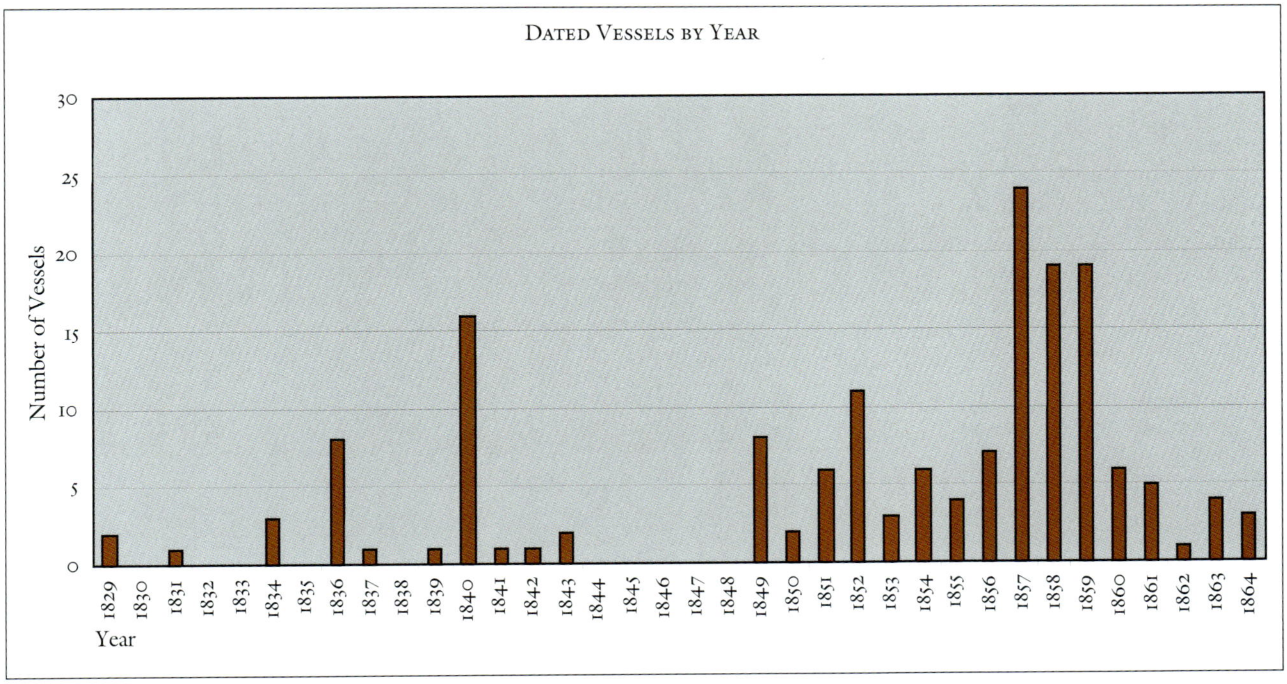

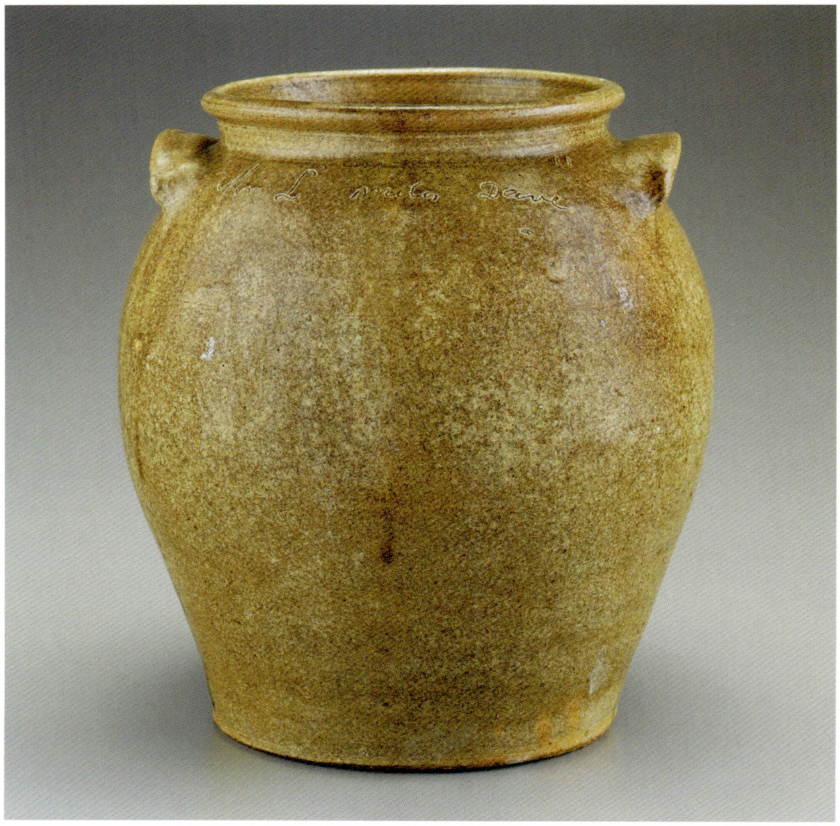

Figure 12 Storage jar, Lewis Miles pottery, 1840. Alkaline-glazed stoneware. H. 13". Incised in script at shoulder: "Mr. L miles Dave" and two slash marks; on opposite shoulder: "20th Nov 1840" and four punctates. (Courtesy, James Witkowski.)

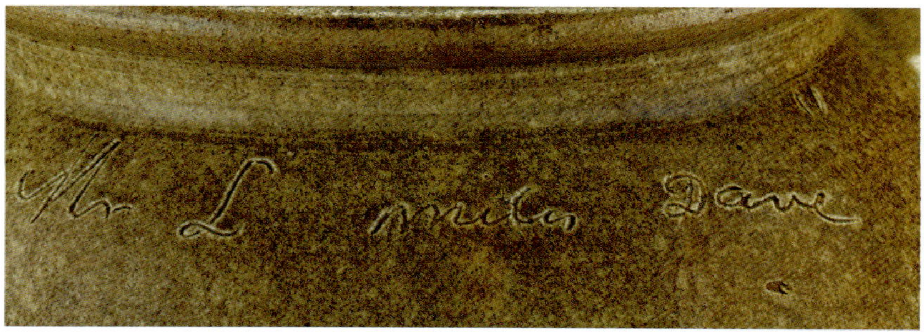

Figure 13 Detail of the inscription and signature, followed by two slash marks, on the shoulder of the storage jar illustrated in fig. 12.

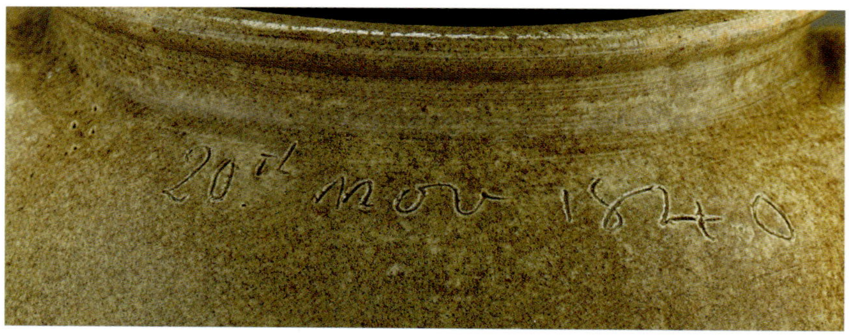

Figure 14 Detail of the four punctates and incised date on the shoulder of the storage jar illustrated in fig. 12.

The vessels of the 1829–1839 period are likely examples of Dave's work while he was at the Pottersville pottery, although they do not bear his signature. Archaeological examples of some of the marks attributed to his work and characteristically impressed at the base of the pot have been found at this site.[27] The pots themselves tend to be slightly more ovoid in shape than the later pieces and vary in height from eleven to nearly twenty inches. The glazes are generally a light yellowish green to olive green color.

The vessels dating from 1840 to 1843, when Dave was owned by Lewis Miles, all bear an inscription including some form of Miles's name: "L. Miles," "Mr. L. Miles," "Mr. Miles," and "LM." All of them are signed by Dave with few exceptions—the first known example of the group, dated "January 23, 1840," is signed "L. Miles" only. One particularly important example, from July 1840, states simply, "Dave belongs to Mr. Miles." There are five poems known from this period.

Cryptic but somewhat consistent markings appear on the shoulder of many of the pieces in this group: two parallel slash marks and a U (or reversed C), sometimes referred to as a horseshoe mark. Often they appear next to punctates, which seem to indicate vessel size in gallons (see figs. 27, 29, 30).

The size of the vessels in this group is generally comparable to that of the earlier examples, although one jar made in 1843 is over twenty-seven inches tall. The glazes, usually a light olive color, vary somewhat and are discussed in more detail in a later section of this article.

The lack of dated vessels from 1844 through 1848 (see Table 2) seems to indicate that Dave's owners—believed to have been Reverend John Landrum and, subsequently, his son, B. F. Landrum—did not allow Dave to sign and date his pots. Although no dated or signed vessels from this period have come to light, Dave undoubtedly was turning pots at the Landrum potteries on Horse Creek, and sherds found at those two sites support this conclusion.[28]

The majority of Dave's signed and dated vessels, more than one hundred, are from the years 1849 through 1864. During this period Dave started to abbreviate Lewis Miles's name as "Lm" and he used these initials exclusively from 1850 on (see fig. 23). In addition to the capacity punctates and the double slash and horseshoe markings noted earlier, an X mark (see fig. 28) and three slashes are used.

As shown in Table 2, the number of surviving pieces indicates that Dave's output peaked about 1857–1859, when he was approaching the age of sixty, and confirms that the Lewis Miles pottery was a large-scale, year-round operation.[29] Dave's poetic expression was also most prolific in this period, with at least eighteen poems. A storage jar signed "Dave" and "Lm" and dated May 13, 1859, has the incised couplet "Made at Stoney Bluff / for making lard enuff," confirming the location of the pottery at Stoney Bluff, Lewis Miles's plantation on Horse Creek (fig. 35).

The vessels produced by Dave in this period include some very large storage jars, a number over twenty-four inches in height and with capacities well over twenty gallons. The large jar shown in figures 1 and 2 was made with about five large joined coils, with its rim probably separately turned. Dave's skill at turning and coiling such large pots may have been unique in the Edgefield District, making his products much in demand as plantation utilitarian wares.

Figure 15 Jug, Lewis Miles pottery, 1858. Alkaline-glazed stoneware. H. 14½". Incised in script: "Lm Feb 24 1858 Dave"; two slash marks and three punctates. (Courtesy, James Witkowski.)

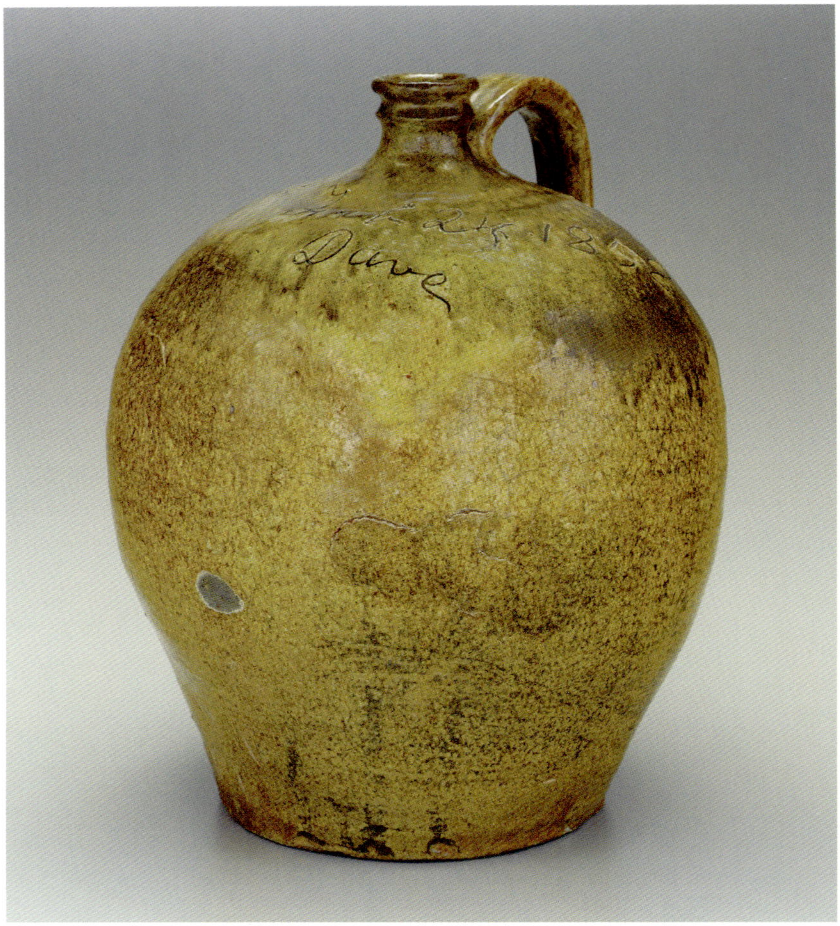

Figure 16 Detail of the handle of the jug illustrated in fig. 15. Note the top of the handle attached at the spout base, the characteristic collar, and the pronounced "thumbprint" at the base of the handle.

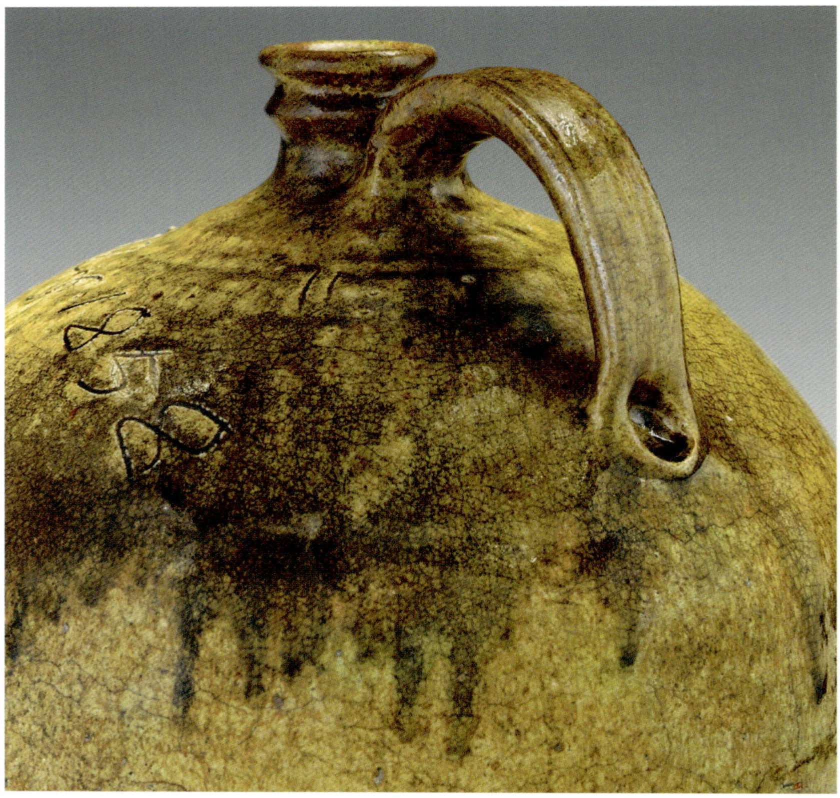

TABLE 3 Bar chart in which the dated vessels are grouped according to the day of the week in which they were produced.

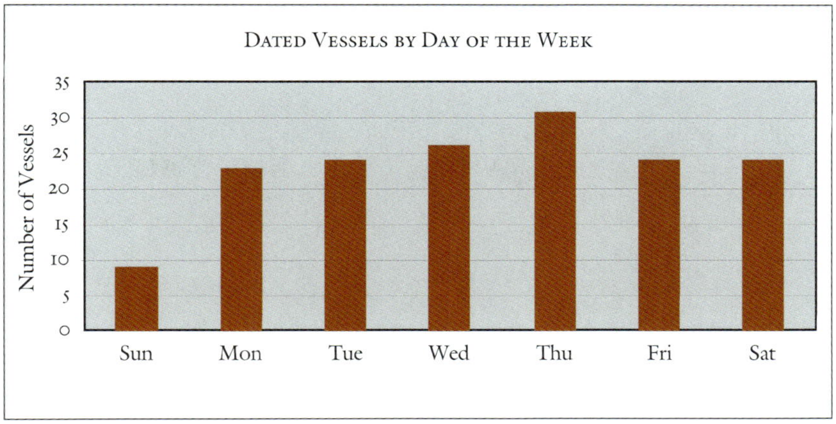

The glazes used during this period vary greatly, and their lack of uniformity might indicate experimentation with glaze materials and techniques. Many are a darker green than the earlier pieces, and some pots have a distinctive dark brown glaze. A few examples show evidence of a secondary glaze applied to the shoulder of the vessel, probably for decorative effect (see figs. 18, 22).

The dates inscribed on many of Dave's pots make it possible to determine the day of the week the vessel was produced. As shown in Table 3, Dave turned pots seven days a week. However, significantly fewer dated vessels were produced on Sundays compared to the other days of the week, a trend that holds regardless of the time period examined. This is intriguing evidence that Dave's masters might have allowed other activities or a period of rest on the Christian Sabbath.

For all of the unsigned vessels clearly attributable to Dave that bear an inscribed "Lm" and/or other marks, in general only one per day appears to have been dated.[30] Was this a practice for marking the start or end of a day's work? By the same token, we have not encountered any vessels signed by Dave that are not dated.

Characteristics of Dave's Wares

Dave made a number of changes in the pots he created over the course of nearly fifty years—most notably being the large vessels he produced in the 1850s and the different glazes—but some characteristics remained the same.

His storage jars are very substantial, typically having quite broad bases and thick walls, consistent with their function. Regardless of their size, they are described by potters today as "powerful," "strong," and with "great presence." Most have a rolled rim that could be used to tie down material covering the opening. The handles are usually "lug type" and are located quite close to the rim (see fig. 4). Some of the larger storage jars have four lug handles, most have two. There is only one known example of a storage jar with "strap" handles.[31]

Dave made both single- and double-handled jugs—the latter used for larger forms—with spouts that have "double-ringed" collars. The strap handles are usually attached quite close to the base of the spout, and many have a pronounced thumb indentation at the lower point of attachment, all characteristics that can be seen on the jug illustrated in figure 16.

Alkaline Glazes

Two of the earliest potteries in the Edgefield District are believed to have been Pottersville, owned and operated by Dr. Abner Landrum, and that owned by Reverend John Landrum on Horse Creek, both identified on the Anderson Edgefield District survey map of 1816 (see fig. 10). These men are credited with being codevelopers of the use of alkaline-glazed stoneware in Edgefield, a tradition that spread to North Carolina and the southernmost states and, eventually, westward.[32] John Burrison explains:

> Alkaline glazes are so called by researchers because they contain an alkaline (high pH or caustic base) substance—in the case of stoneware from the Deep South, wood ashes or slaked lime—as a flux or melter to reduce the maturation temperature of the glaze so that it matches that of the clay body. The wood ashes (pine from the kiln firebox or hardwood from the home fireplace) produce lye (sodium or potassium hydroxide) when mixed with water in the glaze solution. Some potters obtained lime from nearby lime-kiln operations or by burning limestone in the pottery kilns, then added water to the resultant white powder to create slaked lime (calcium hydroxide); others purchased commercial builders' lime. Variation in alkaline glaze color is largely attributable to three factors: the degree of iron present in the clay body, the amount of iron in the glaze, and the kiln atmosphere. Variation in alkaline glaze color is largely attributable to three factors: the degree of iron present in the clay body, the amount of iron in the glaze, and the kiln atmosphere.[33]

Some of the complexities of the glazing results have been discussed by ceramic scholars Charles Zug and Georgeanna H. Greer.[34] The variegated colors and their distribution on the vessels were endless. When significant reduction (reduced oxygen) occurred during the firing process, blue crystal titanium, called rutile, could form in the glaze (see fig. 4).[35]

Using local iron-bearing clays with significant kaolin content, the potters were able to produce high-fired vitrified stoneware (2200° F, 1300° C) that was impervious to liquids. In addition to the initial glaze, ashes (fly-ash) developing during the wood firing would be deposited in varying amounts on the upper portions of the vessels, melting in an unpredictable fashion onto the vessels to fuse with the glaze, particularly on the shoulders of the pieces (see figs. 16, 36). "At these temperatures the silica in the clay [or sand] will meet and fuse naturally when fluxes such as lime, soda or potash, which are naturally present in wood ashes, are added. The materials meet and harden. . . ."[36] This results in a glass surface. It is during this period that the stoneware itself matures and is vitrified.

The use of alkaline glaze and its firing techniques was one of the great contributions to American ceramics, and without apparent European antecedents. In fact, this glazing technique had its origin hundreds or perhaps thousands of years ago, in China, Japan, and Korea. So how did this rediscovery take place in such a small and relatively isolated rural community?

The answer is unclear. Perhaps the rediscovery of the use of wood ash and lime constituents of the glazes was developed from considerable exper-

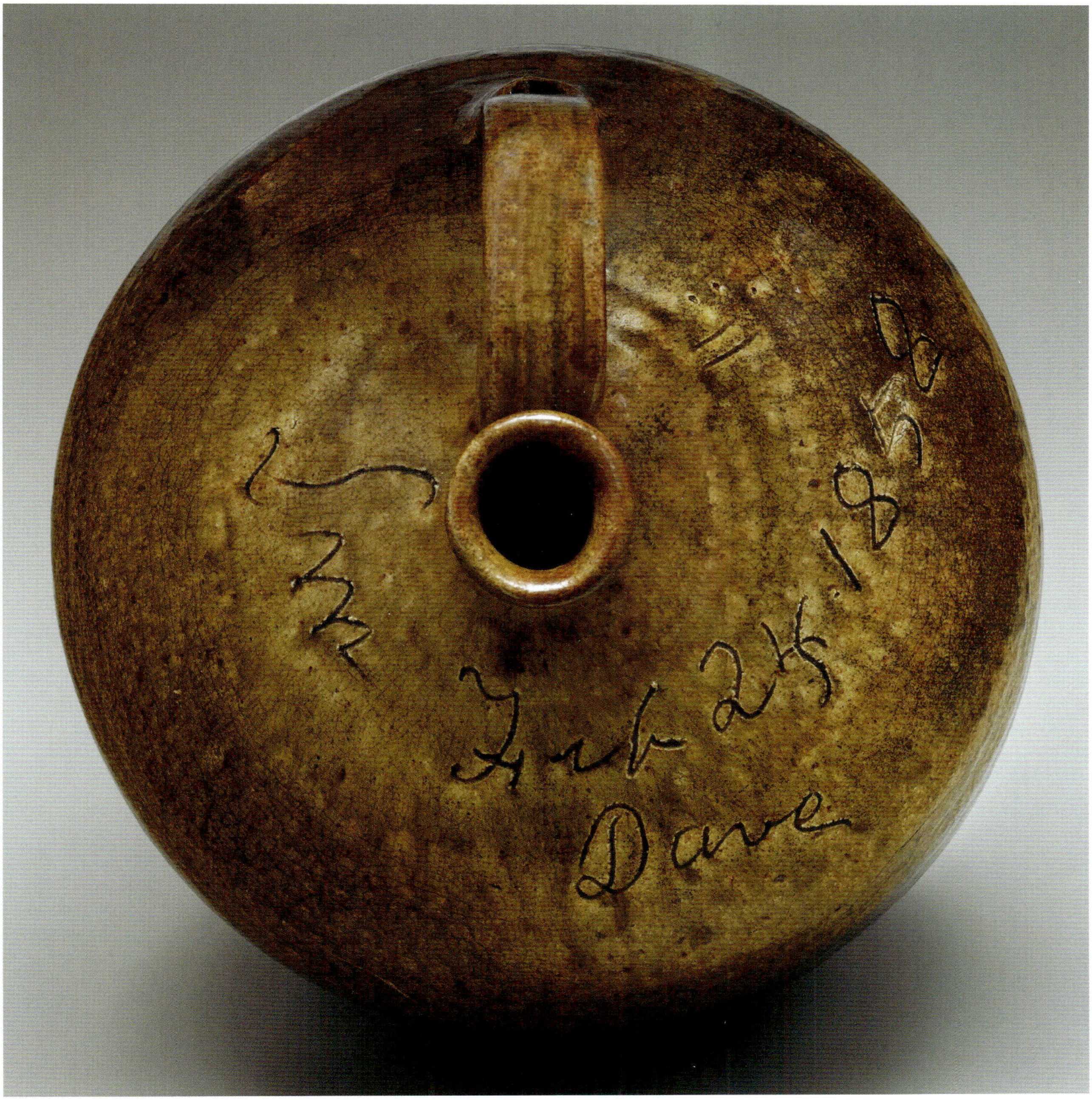

Figure 17 Top view of the jug illustrated in fig. 15.

imentation or was simply fortuitous, or perhaps the Landrums had read firsthand accounts from published eighteenth-century letters from missionaries in China.[37] Unfortunately, whereas American art museums proudly exhibit Chinese, Korean, and Japanese ash-glazed stoneware, it is rare for a major art museum in the United States to display similarly glazed Southern stoneware. These vessels are certainly of equivalent beauty, however, and deserve respect and recognition.

Glazes on Dave Vessels

The glazing and firing techniques varied at the different potteries where Dave worked during his lifetime, producing richly varied surface colors.

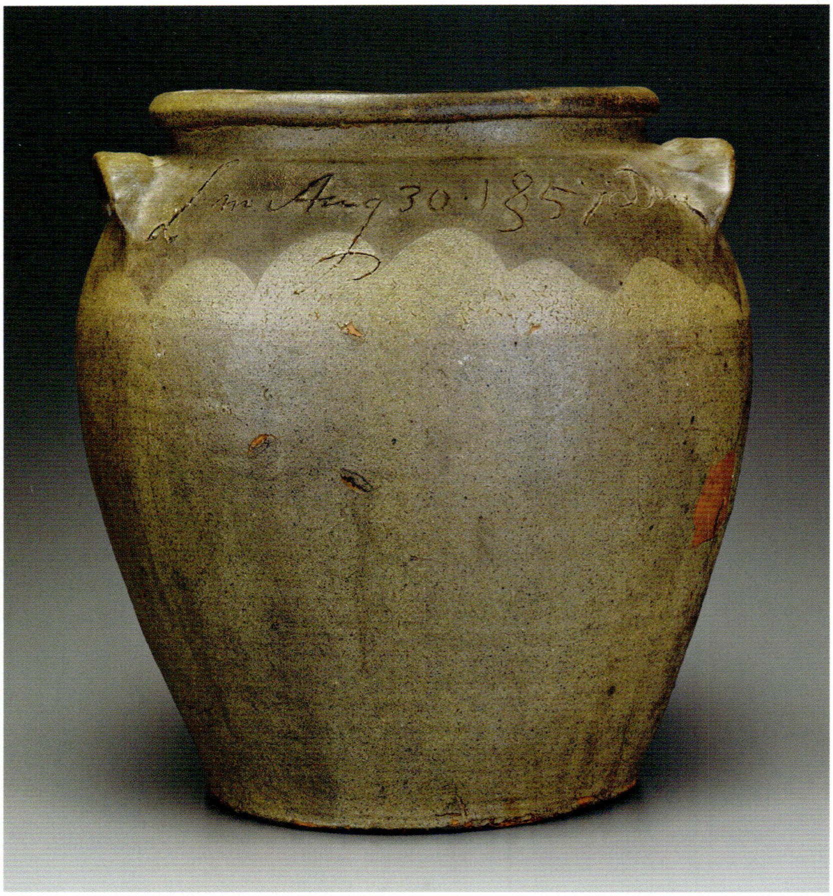

Figure 18 Storage jar, Lewis Miles pottery, 1857. Alkaline-glazed stoneware. H. 19¼". Incised in script on shoulder: "Lm Aug 30 1857 Dave"; on opposite side: two slashes with ten punctates. (Courtesy, James and Mary Smith.) Note the unusual effect produced by pouring additional glaze around the shoulder of the vessel.

When examining the glazes on his vessels, examples of all the various glazes described can be seen. The exact glazing formulas and kiln firing methods used at the potteries where he worked are not known.

It is also difficult to pinpoint when a specific vessel was made by the glaze itself, because there is such a wide range of colors and textures of alkaline-glazed stoneware. In addition to the specific glazes used, one must take into consideration the different clays and their interaction with the glazes, firing conditions (oxidation/reduction ratios, temperature, and so forth), and where the vessel was placed in the kiln. All affected glaze colors and textures as well as the fly-ash fallout. Only broad generalizations can be made about the widely varied glazes and textures on Dave's vessels.

The earliest vessel glazes of the 1830s and 1840s are typically in pleasingly warm earth tones—yellowish, green, olive, gray, tan, or brown with transitional hues. By the 1850s and through the 1860s the brown to dark brown to almost black glazes seem to have been preferred. Nevertheless, some vessels with light tan glazes and others with variations of dark olive-green to brown hues were also made (see figs. 21, 31). Rarely one finds a white drippy glaze running over the shoulder and body, as in figure 22, or patches of reddish brown.

The tactile and visual surface characteristics vary from shiny, smooth, and glossy to rough. These textural changes occurred with a thin or thick application of the glaze distributed evenly or irregularly on the vessel. We have not seen a translucent celadon-type green to bluish green glossy glaze or slip decoration on any of Dave's signed or dated vessels as seen on the wares of other Edgefield potters.[38]

In addition to the glaze colors, there are other surface additions. One of the most distinctive are fly-ash deposits, resulting from the wood-fired kiln, producing variegated colors over the shoulder. The fly-ash visually reflects the firing process, since it outlines the form above the waist of the vessel. At times there were glaze splashes, runny streaks, and drips. Occasionally patches of iridescent and luminous rutile blue (titanium oxide) can be seen, as shown in figure 4. All these unpredictable decorative effects produced a glorious array of colors, frequently melting into each other.

Literacy and Freedom of Expression

A review of the general social and political circumstances of the period and region in which Dave lived and worked serves to underscore his talent, creativity, and popularity.

The hardship, danger, and lack of dignity suffered by slaves in America, especially those in the South, are well documented. "Literate or illiterate, skilled or unskilled, African Americans lived with oppression from whites. Edgefield slaves endured savage cruelty, family separation, and sexual abuse from their masters. . . . Violence and the threat of physical punishment underlay all other devices and strategies for the control of slaves.

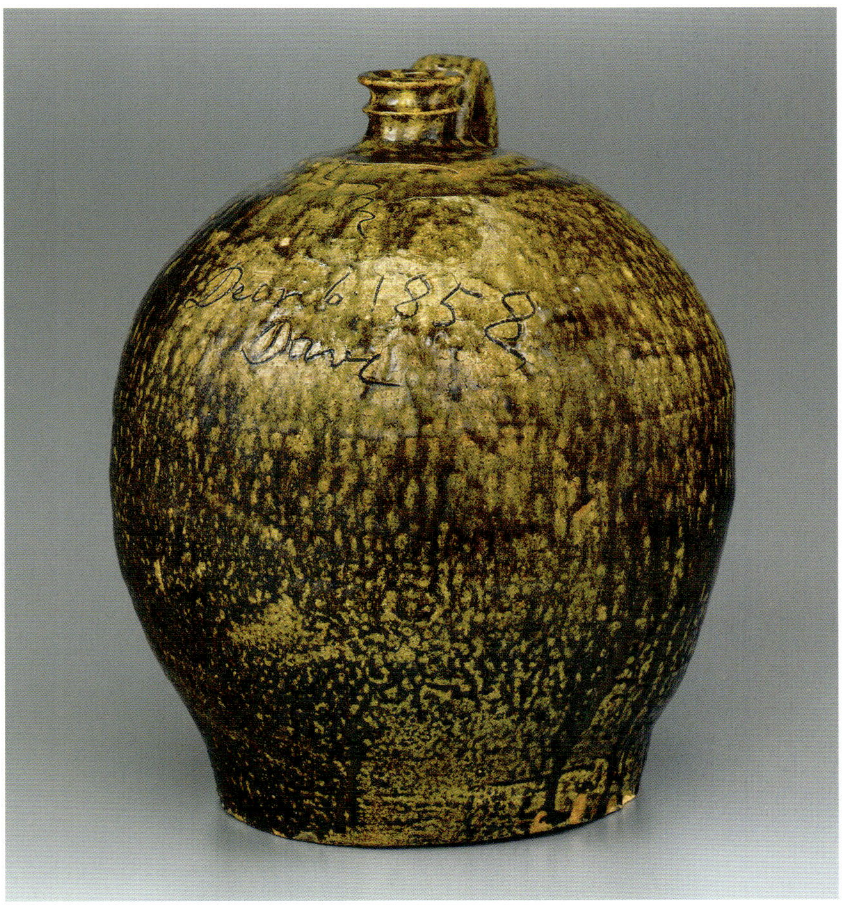

Figure 19 Jug, Lewis Miles pottery, 1858. Alkaline-glazed stoneware. H. 13". Incised in script on shoulder: "Lm Decr 6 1858 Dave." (Courtesy, Katie Green.)

Whites had absolute authority to whip slaves, whether adult or child. Slave parents had to stand aside if whites wanted to whip children. Slave children had to watch while whites humiliated and whipped fathers and mothers."[39]

Another form of control was exerted through a policy of maintaining illiteracy among slaves. Most Southern states in the 1830s had legislation prohibiting their education, and "of all of the Southern states which restricted black literacy, South Carolina [which had no public school system prior to the Civil War] passed the harshest law against teaching slaves and even free blacks to read and write."[40]

The complex relationships between slaves, their masters, religious leaders, and the general population were complicated by the diversity of opinion regarding the continued oppression of slaves, Christian moral teaching, and emancipation.[41] Four states—Georgia, North Carolina, South Carolina, and Virginia—banned the teaching of slaves and free blacks until 1865, yet those laws were seldom enforced.[42] In fact, quite a few slaves learned to read, for a variety of reasons: those with religious masters wanted them to be able to read the Bible; slave children absorbed the teachings of white children; literate slaves shared their knowledge with others; and, in some situations, literacy was necessary in order for slaves to perform their jobs.[43]

When or how Dave learned to read and write is not known, although it is likely he learned at a fairly young age. He certainly had the intelligence and ability, and by the time South Carolina's stringent anti-literacy law of

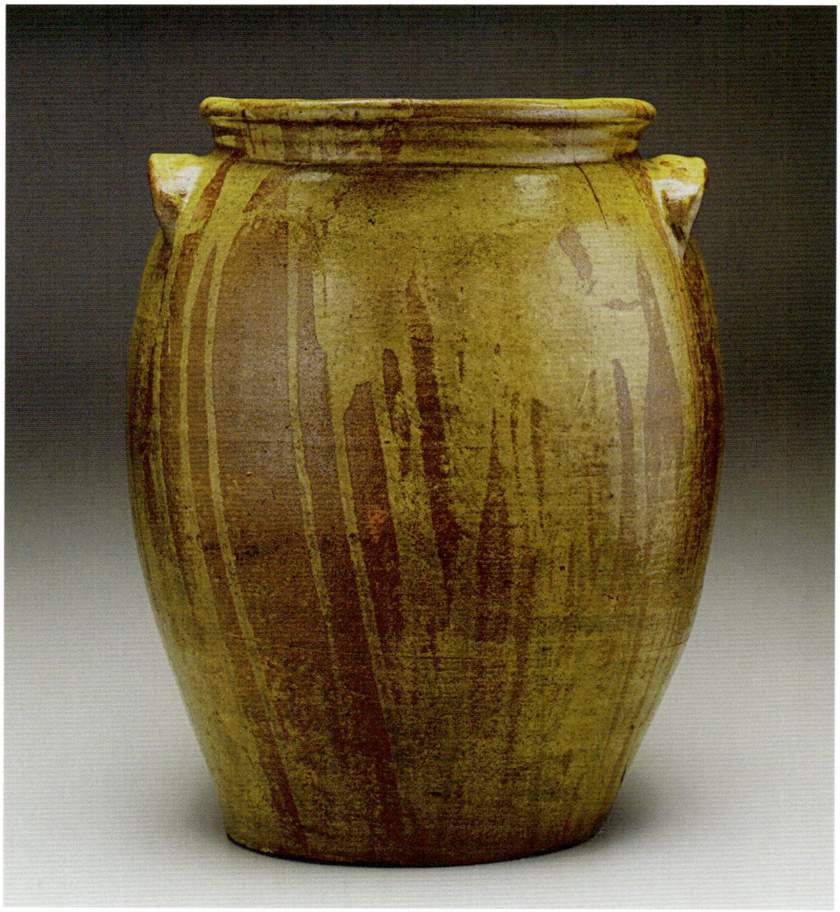

Figure 20 Storage jar, Lewis Miles pottery, 1858. Alkaline-glazed stoneware. H. 21¼". Incised in script on shoulder: "Lm Feb 15 1858" and fourteen punctates. (Courtesy, Lindsay Green.)

1834 was passed, he was an adult and already literate, as we know from the poems written on his (unsigned) jars at this time—one from 1834 and two from 1836.[44]

An editorial published in the *Edgefield Advertiser* on April 1, 1863, reads, in part:

> One day in years gone by we happened to meet DAVE POTTERY (whom many readers will remember as the grandiloquent old darkey once connected with a paper known as the Edgefield Hive) in the outskirts of his beloved hamlet. Observing an intelligent twinkle in his eye, we accosted him in one of his own set speeches: "Well, uncle DAVE, how does your corporosity seem to sagatiate"—"First rate, young master, from top to toe—I just had a magnanimous bowl-full of dat delicious old beverage, buttermilk." Who has not often felt his buttermilk as DAVE did.[45]

How Dave was "connected" to the *Hive* is not known. Many authors presume that he was a typesetter there, but offer little if any supporting evidence. Whatever the connection, the paper would have afforded Dave a significant learning opportunity. He was, as the quotation implies, treated with respect, and his grandiloquence and humor were clearly appreciated.

Dr. Abner Landrum could have played a role in teaching Dave to read and write. As a unionist—and at odds with his community in that regard—Landrum opposed anti-literacy legislation and "protected and fostered the freedom to read, write and worship for slaves and free blacks. . . ."[46] He was one of the owners of the *Edgefield Hive* from 1830 to 1831, when he departed to become editor of the state capital newspaper, renamed the *Columbia Free Press and Hive,* where his views would be more easily tolerated. This

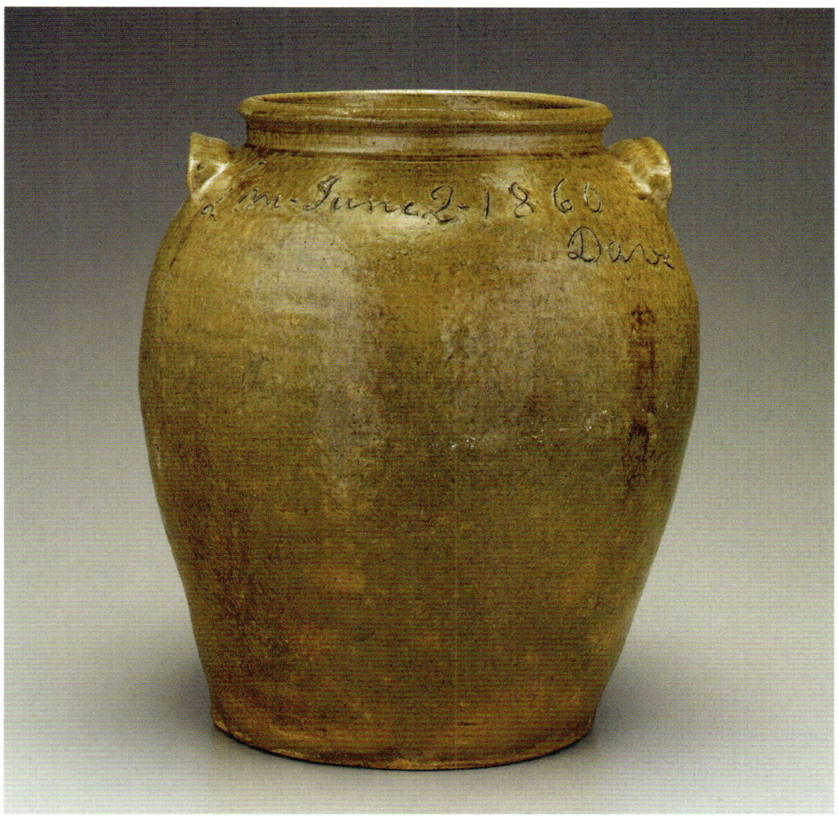

Figure 21 Storage jar, Lewis Miles pottery. Alkaline-glazed stoneware. H. 15½". Incised in script on shoulder: "Lm – June 2 – 1860 Dave"; on opposite side: "X." (Courtesy, Larry and Joan Carlson.)

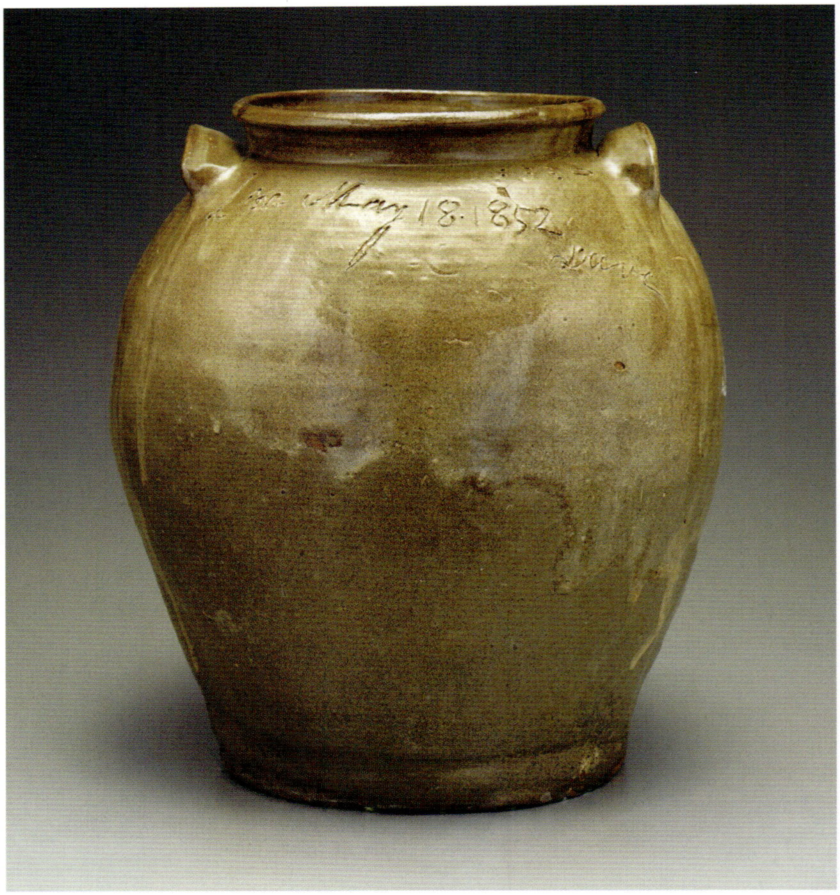

Figure 22 Storage jar, Lewis Miles pottery, 1852. Alkaline-glazed stoneware. H. 17". Incised in script: "Lm May 18 1852 Dave"; on opposite side: backward ᑕ and seven punctates. (Courtesy, Larry and Joan Carlson.)

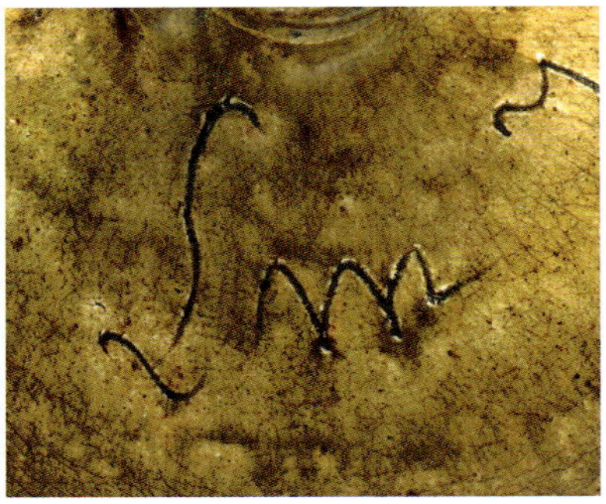

Figure 23 Detail of the inscription on the shoulder of the jug illustrated in fig. 15.

paper, the principal publication of the Republican party, was described as "the organ of the Union cause."[47] Landrum died on April 3, 1859.[48]

Dave's early potting experience at Pottersville was in close association with Dr. Abner Landrum, the Drake brothers, and Jasper Gibbs. Any of Dave's masters could have tutored him. Significantly, during this period they permitted him to write poems and words on his vessels but his name does not appear on the pots. After 1837, however, the special personal and working relationships Dave had with his masters—plus the respect of the members of his community—were such that they chose to disregard the terms of the 1834 South Carolina Slavery and Freed Persons of Color Law, knowing they could be fined and Dave could be whipped. There are even anecdotal reports of amputations of the hand or parts of digits occurring in some Southern states if a slave was known to have learned to read or write.[49]

Given Dave's relative freedom to express himself on his pots, what could account for the gap of five years, between 1843 through 1848, when he produced no poems or inscriptions? In that period he was owned by Reverend Landrum and his son B. F. Landrum, both of whom must have known about the poems Dave wrote while at Lewis Miles's pottery. Did the political climate in Edgefield change, or were the Landrums opposed to allowing a literate slave to express himself—or even incise his name—on their vessels? Based on the evidence we can only speculate.

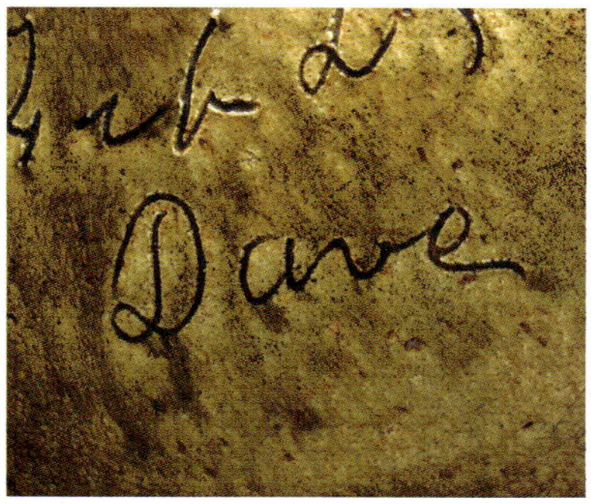

Figure 24 Detail of the inscription on the shoulder of the jug illustrated in fig. 32.

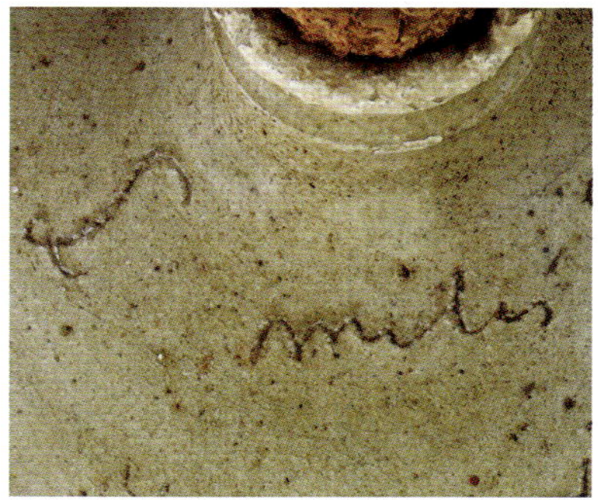

Figure 25 Detail of the signature on the shoulder of the jug illustrated in fig. 15.

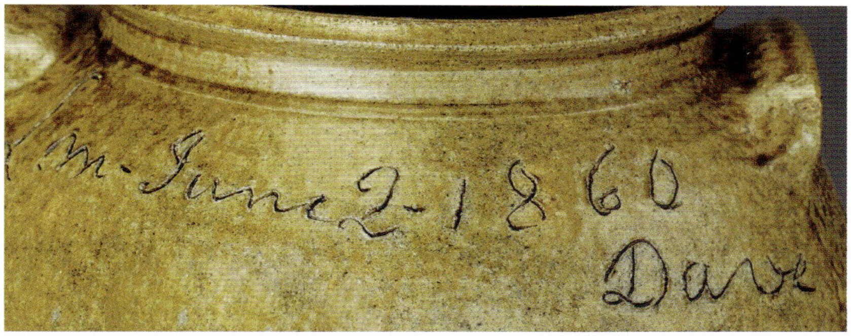

Figure 26 Detail of the inscription, date, and signature on the shoulder of the storage jar illustrated in fig. 21.

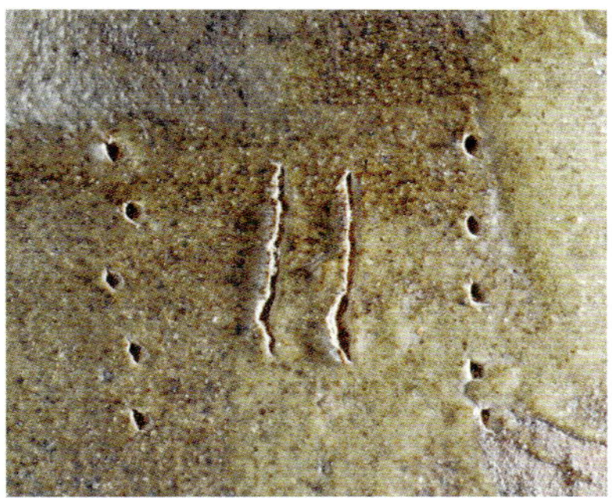

Figure 27 Detail of the two slash marks and ten punctates on the shoulder of the storage jar illustrated in fig. 18.

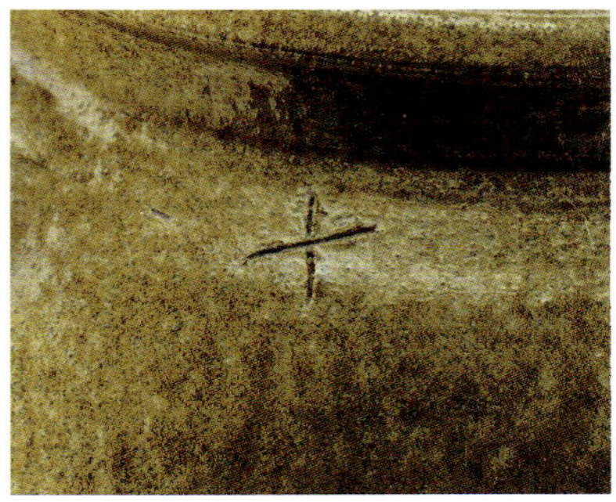

Figure 28 Detail of the X on the shoulder of the reverse side of the storage jar illustrated in fig. 21.

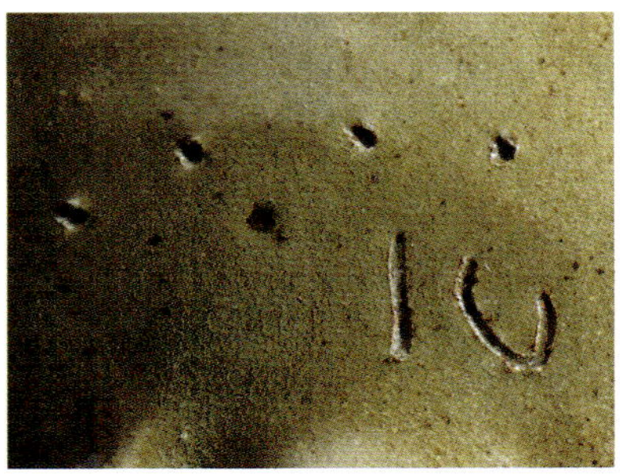

Figure 29 Detail of the four punctates, slash, and U on the shoulder of the jug illustrated in fig. 31.

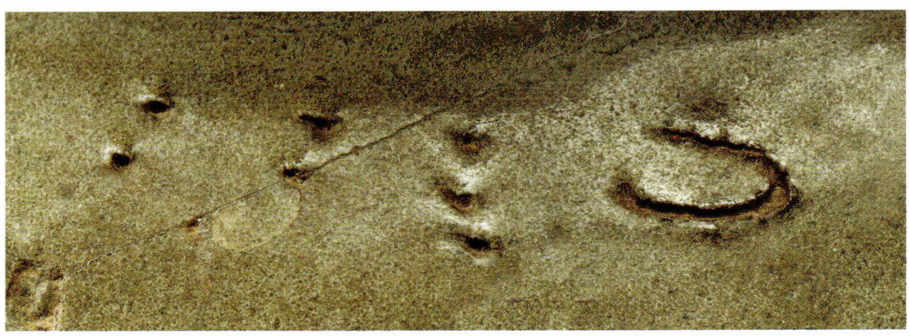

Figure 30 Detail of the backward C and seven punctates on the shoulder of the storage jar illustrated in fig. 22.

Little is known about the political, social, or religious views of Reverend John Landrum and his son, B. F. The reverend served as minister of Horne's Creek Baptist Church—of which his son was also a member—and he was said to "correct the morals and mind the heart of his fellow men."[50] He once "expelled his brother Amos from his church in January 1839 for drunkedness after he had been duly admonished,"[51] an act that gives us insight into the reverend's rigid and perhaps unforgiving personality. At the very least, it would take a person capable of complex rationalizations to be both slave owner and a proper minister.

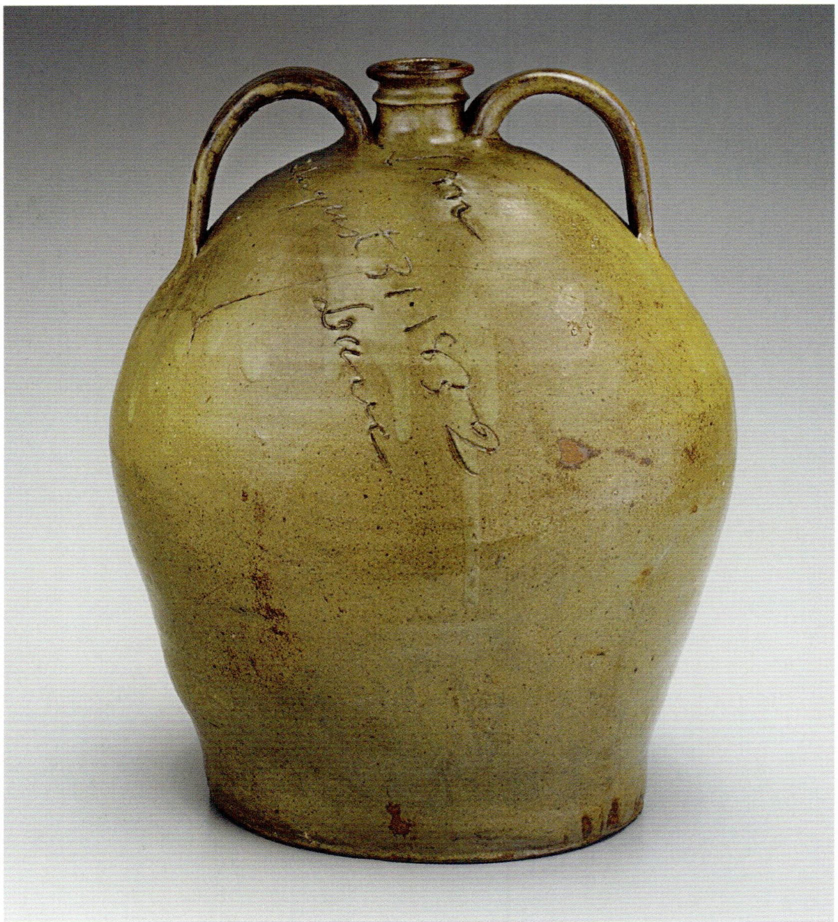

Figure 31 Double-handled jug, Lewis Miles pottery, 1852. Alkaline-glazed stoneware. H. 16". Incised in script: "Lm August 31·1852 / Dave"; on opposite side: slash and U marks with four punctates. (Courtesy, James and Mary Smith.)

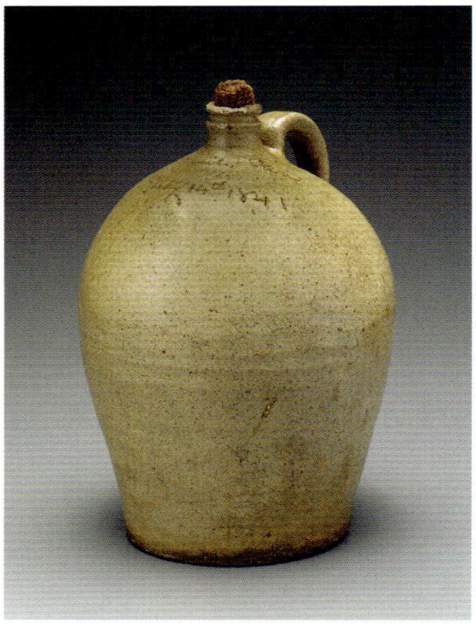

Figure 32 Jug, Lewis Miles pottery, 1841. Alkaline-glazed stoneware. H. 14½". Incised in script on the shoulder: "Feby 14th 1841 L miles Dave" and two slash marks. (Courtesy, James and Mary Smith.)

Reverend Landrum died on December 9, 1846. Excerpts from his laudatory obituary state that he "entered upon his ministerial functions about 1790, and was a zealous and untiring preacher of the gospel for more than fifty years." He was touted for his "simplicity of manners, purity of character, and devotion to religion," as well as for a mind that was "never occupied with amusements, or with anything in its nature, trifling." Reverend Landrum "was a warm friend to education and left no proper opportunity to embrace its advantages to society." The author further described him as "constant in friendship; a kind husband, an affectionate parent and a good citizen." There was no reference to the reverend's stoneware pottery.[52]

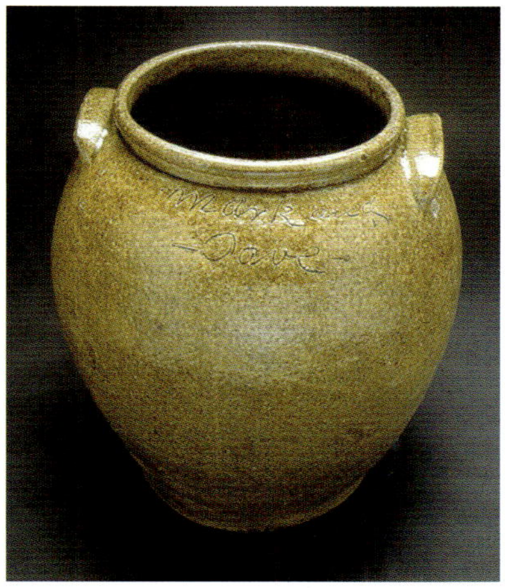

Figure 33 Storage jar, Lewis Miles pottery, 1859. Alkaline-glazed stoneware. H. 15". Incised in script: "Mark and/ –Dave–" and two slash marks; on opposite side: "Lm. March 10·1859." (Courtesy, Museum of American History, Division of Ceramics and Glass, Smithsonian Institution, Washington, D.C.)

Figure 34 Detail of the signature on the shoulder of the storage jar illustrated in fig. 33.

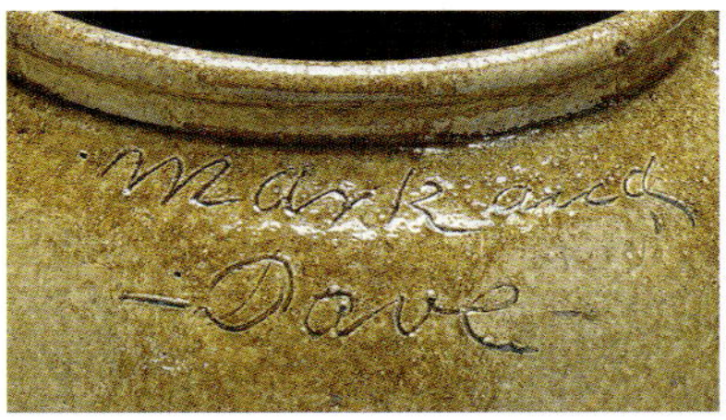

After Reverend Landrum's death, B. F. Landrum and Lewis Miles reportedly cooperated in the management of his pottery but had a "falling out"; by 1850 both are listed as having separate establishments.[53]

There is evidence that B. F. Landrum was authoritarian, like his father, and a harsh slave owner. At the pottery he reputedly whipped and bound a female slave "of turbulent disposition" who had threatened to run away. Shortly after the beating, she hung herself.[54]

In contrast, Dave's relationship to the reverend's brother, Dr. Abner Landrum, appears to have been respectful and affectionate. Five days after the notice of Dr. Landrum's death appeared in the *Edgefield Advertiser* Dave made a jar, dated April 14, 1859, bearing the dedication:

> Over noble Dr. Landrum's head
> May guardian angels visit his bed[55]

Starting in January 1849 Dave was again owned by Lewis Miles and apparently was allowed to do what he had done previously, for there resulted an incredible proliferation of marked vessels. As before, all of the writing was in script, although now only Miles's initials—"Lm"—not his full name, were incised.

From August 1857 through July 1859 eighteen Dave poems are known. Three additional vessels with poems were made between 1860 and 1863, all with Dave's signature. Judging from this volume, it seems certain that Miles supported, even fostered, Dave in his activities. It is intriguing that the provisions of the 1834 law were disregarded by Miles (and, indeed, by Dave himself), that despite monetary incentives it seems no one came forward to inform the local authorities of the breach, and, if anyone had, no action was taken by the local government.

What sort of man was Lewis Miles? Little is known of his personal life. A former Edgefield turner once stated that Miles was known for his charity: "If you gave a dollar he'd give a hundred."[56] Despite being his owner, Lewis Miles allowed Dave to inscribe his poems and to sign his works. Considering the laws and social climate of the time, Miles had the strength of character to risk criticism and even punishment. Were it not for him, the extent of Dave's contributions would have remained largely unknown.[57]

Figure 35 The authors at the Lewis Miles pottery site at Stoney Bluff.

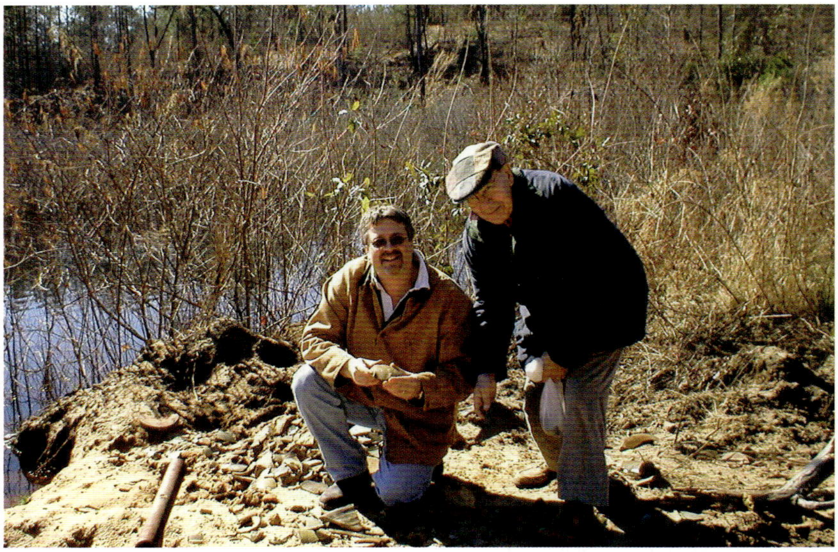

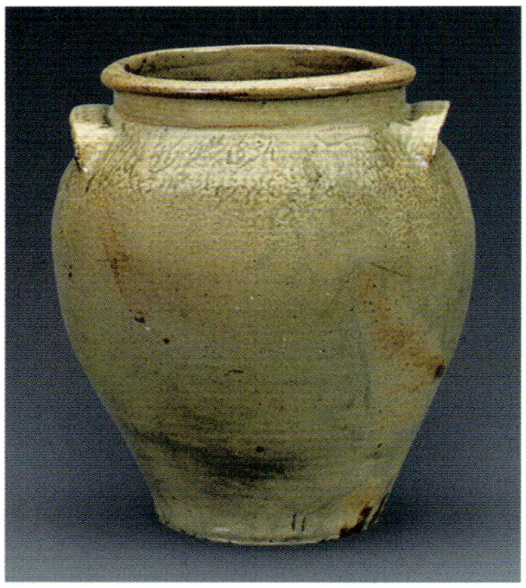

Figure 36 Storage jar, Pottersville pottery, 1834. Alkaline-glazed stoneware. H. 19¼". Incised in script at shoulder: "Put every bit all between / surely this jar will hold 14"; on opposite side: "July 12th 1834" and fourteen punctates; at base: two slash marks. (Courtesy, South Carolina State Museum, Columbia.)

An underlying question that must be asked, then, is why did Miles not free Dave? The answer lies in the laws in South Carolina at the time, which prohibited manumission. Even after 1840 it was all but impossible. Emancipation could be achieved only by appealing to the state legislature, which invariably rejected all claims. The slave system did not allow for any other legal methods by which to obtain, or grant, freedom.[58]

Dave's Poems and Inscriptions

A chronological listing of Dave poems and inscriptions, updating the previous compilation of poems prepared by Jill B. Koverman, is provided in Table 4.[59] It includes thirty poems, nine inscriptions, and one dedication. It is beyond the scope of this paper to give an extensive interpretation of Dave's poems, although those directly related to his masters as well as some provocative words have been discussed.[60]

One can only imagine what Dave's signature and writings meant to him. His literacy in his slave community was a sign of intelligence and some measure of independence. With each succeeding year, Dave gained status and confidence and, with that, his signatures became larger. Every word he wrote and every pot he signed not only represented his own, nonviolent resistance to slavery but also reflected the ambivalence of at least some of his masters to the laws of oppression.

Moreover, with his writings Dave was able to communicate directly with both the slave and white communities. Seeing these vessels surely inspired slaves to become literate and to seek their freedom. Each incised vessel was not an isolated work, because its importance went far beyond.

The full influence of Dave's writings on the white and slave communities of his time is yet to be fully recognized and understood, but his importance is undeniable. He was an innovator in the Southern stoneware tradition, combining the art of his pottery with that of his incised poems—and doing so as a slave who not only could read and write but who used those skills to

TABLE 4. Dave Poems and Inscriptions

6-12-1834	Concatination
7-12-1834	Put every bit all between surely this jar will hold 14
3-29-1836	Horse, mules and hogs – all our cows is in the bogs – there they shall ever stay till the buzzards take them away
4-12-1836	Catination
11-9-1836	A better thing I never saw When I shot off the lions jaw
1-29-1840	Ladys & gentlemens Shoes sell all you can: & nothing you'll loose
2-10-1840	whats better than wishing while we both are at fishing
6-27-1840	Give me silver or; either gold Though they are dangerous; to our soul
7-31-1840	Dave belongs to Mr. Miles wher the oven bakes & the pot biles
8-25-1840	Edgefield District
8-26-1840	Another trick is worst than this Dearest Miss, spare me a kiss
5-16-1843	Not Counted
10-17-1850	Just a mammouth jar...for I not...
5-17-1852	Cash wanted
6-28-1854	Lm says this handle will crack
7-6-1857	for Mr. John Monday
8-16-1857	I wonder where is all my relations Friendship to all – and every nation
8-22-1857	I made this jar for cash – though its called – lucre trash
8-24-1857	A pretty little girl on a virge volca[n]ic mountain, how they burge
1-30-1858	Making this jar: I had all thoughts Lads & gentlemen: never out walks
2-20-1858	A noble jar For lard or tar
3-31-1858	I made this for our Sott It will never – never – rott
4-8-1858	This noble jar will hold 20 fill it with silver then you 'll have plenty
4-12-1858	A very large jar which has four handles pack it full of fresh meat – then light candles
4-21-1858	When you fill this jar with pork or beef Scot will be there to get a peace, — This jar is to Mr Segler who keeps the bar in orange burg For Mr Edwards a Gentleman — who formly kept Mr thos bacons horses
7-29-1858	The sun, moon, and – stars in the west are plenty of – bears

11-3-1858	I saw a leppard & a lions face then I felt the need of – grace
12-6-1858	Nineteen days before Christmas Eve Lots of people after its over How they will grieve
3-10-1859	Mark and Dave
3-25-1859	I made this out 2 number & cross If you do not listen at the bible you'll be lost –
4-14-1859	Over noble Dr. Landrum's head May guardian angels visit his bed
4-18-1859	Hive is eighteen; hundred + fifty nine unto you all I fill in— —cline
5-3-1859	Good for lard or holding fresh meat blest we were, when peter saw the folded sheets
5-13-1859	Made at Stoney Bluff for making lard enuff Dave Baddler
5-13-1859	Great & Noble jar hold sheep goat and bear Dave & Baddler
7-4-1859	The forth of July is surely come to blow the fife = and beat the drum
7-19-1859	Rev. W. A. Lawton
8-7-1860	I saw a leopard & a lions face then I felt, the need of grace
11-9-1860	A noble jar for pork or beef then carry it a round to the Indian chief
5-3-1862	I – made this Jar all of cross If you don't repent, you will be lost

express himself during one of the most oppressive and dangerous periods in American history.

When admiring a Dave pot today, the impulse arises to communicate with it. For many it awakens an intangible spiritual feeling. We want the tactile pleasure of holding it, to grasp it with both hands, to feel its girth and weight, and to rub our hands gently across the vessel, appreciating its texture while admiring its many colors and hues. When the burned and underfired areas, touch marks, and glaze interactions are observed, we reflect on the potting process and the firing, with images of fire and smoke billowing from the kiln chimney. This also brings to mind the historical background and culture of the period. And from this we understand why Dave's utilitarian pots—made for the everyday needs of the households and plantations of rural South Carolina—have been transformed into objects of beauty and art.

ACKNOWLEDGMENTS

The authors are most grateful for the cooperation and assistance of numerous collectors, librarians, and museum curators who shared their information about the dated vessels and history of the period. This paper would not have been possible without the previous research done by the scholars who recognized the significance and importance of Dave and the Southern alkaline-glaze tradition. We also thank Susan Witkowski, Michael Schoenwald, and Deborah A. Goldberg, Ph.D., for their assistance in preparing this paper.

1. Jill Beute Koverman, "Searching for Messages in Clay: What Do We Really Know about the Poetic Potter, Dave?" in *I made this jar . . . : The Life and Works of the Enslaved African-American Potter, Dave,* edited by Jill Beute Koverman, exh. cat. (Columbia: McKissick Museum, University of South Carolina, 1998), pp. 22–25.

2. Ibid., p. 23 n. 10.

3. Ibid., p. 23 n. 12.

4. "A Map of the District of Edgefield, made in conformity to the Resolutions of the Legislature in December 1816, by Thomas Anderson, Dep. Surveyor," South Carolina Department of Archives and History, MB5-6. The original manuscript of this map shows "Landrum's Pottery" approximately one mile northeast of the Edgefield courthouse. When the map was published in 1817 the notation was changed to indicate a "Pottery" at "Landrumsville." This map also shows Reverend John Landrum's pottery.

5. The comment is believed to be a reference to alkaline-glazed stoneware. Robert Mills, *Statistics of South Carolina* (Charleston, S.C.: Hurlbut and Lloyd, 1826), pp. 523–24. Mills further describes the site: "There is another village of sixteen or seventeen houses, and as many families, within a mile and a half of Edgefield court-house, called the Pottery, or Pottersville, but which should be called Landrumville, from its ingenious and scientific founder, Dr. Abner Landrum."

6. Koverman, "Searching for Messages in Clay," p. 23 n. 10.

7. Ibid., p. 23 n. 14.

8. Joe L. Holcomb and Fred E. Holcomb, "South Carolina Potters and Their Wares: The Landrums of Pottersville," *South Carolina Antiquities* 18, nos. 1–2 (1986): 53.

9. "Old Pottersville and Dr. Landrum," *Edgefield Advertiser,* May 11, 1859; South Carolina Library, Columbia, South Carolina.

10. Koverman, "Searching for Messages in Clay," p. 24 n. 16.

11. 1840 Federal Census, Edgefield County, p. 83; South Carolina Department of Archives and History, Columbia, South Carolina.

12. Thomas Anderson's 1816 map of Edgefield District (South Carolina Department of Archives and History, MB 5-6) shows "Rev. Jn. Landrum's Pottery" located approximately twelve miles southwest of the Edgefield courthouse, on Big Horse Creek.

13. Edgefield County Probate Records, Box 56, Package 2312, South Carolina Department of Archives and History, Microfilm Roll ED32.

14. Ibid.

15. Joe L. Holcomb and Fred E. Holcomb, "Archaeological Findings Related to Dave at Edgefield Pottery Sites," in *I made this jar . . . ,* pp. 73–81.

16. Carl Steen, "An Archeological Survey of Production Sites in the Old Edgefield District of South Carolina," unpublished report, Diachronic Research Foundation, Columbia, S.C., December 1994, p. 81.

17. Koverman, "Searching for Messages in Clay," p. 24.

18. Ibid.

19. Joe L. Holcomb and Fred E. Holcomb, "South Carolina Potters and Their Wares: The History of Pottery Manufacture in Edgefield District's Big Horse Creek Section, Part I (ca. 1810–1925)," *South Carolina Antiquities* 21, nos. 1–2 (1989): 26.

20. "Notes made on trip to Seigler's Pottery, near Eureka, S.C., October 4, 1930," Charleston Museum Archives, Edgefield Pottery file, Charleston, S.C. The Seigler Pottery, located near Lewis Miles's Stoney Bluff pottery site, was active from about 1875 to 1898.

The date of Dave's alleged train accident is not known. It could not have happened in the Edgefield village area if it occurred while Dave was working for Drake (1828–1832) because the South Carolina Railroad, starting in Charleston, did not begin service until about 1833, and it bypassed Edgefield when it extended its line to Hamburg (North Augusta). See Cinda K. Baldwin, *Great and Noble Jar: Traditional Stoneware of South Carolina* (Athens: The University of Georgia Press, 1993), p. 46. Perhaps the incident occurred later, or while Dave was working for Lewis Miles, since the Charleston-Augusta railroad passed within a few miles of Miles's pottery (see Isaac Boler, "Map of Edgefield County, S.C.," 1871, obtained from the Old Edgefield District Genealogical Society). To date, however, no report of his accident has been found in the local newspaper, which probably would have mentioned it.

21. Fletcher continued:

> Old Dave used to sit around the place dozing and one day a goat came up and when Dave's head would drop over and jerk back the goat thought Dave was bantering him and that goat just backed off and rammed him, knocking him off his seat. Dave always said that somebody hit him with a plank, and never would believe that a goat did it.

"Notes collected on trip to various potteries. Miss L.M.B. & E.B.C. June 24–26 1930," Charleston Museum Archives, Edgefield Pottery file, Charleston, S.C. "Miss L.M.B." is Laura M. Bragg, director of the Charleston Museum at that time.

22. 1870 Federal Census, Population Schedule, South Carolina, Edgefield County, National Archives M593-1495, p. 433. The other listed members of the household were: Caroline, age 28 (presumed to be Mark's wife); Brister, 10; Pierce, 8; Emma, 6; David, 4; and an infant, 1 month. It is not known whether any of these individuals were related to Dave.

There are two original handwritten versions of this census; one was made for the federal government, the other for the state of South Carolina. On the federal copy, the occupations of both David Drake and Mark Jones appear to be written as "Turner," although the "u" in David Drake's occupation seems to have been altered to form an "a." This is not clarified by the state copy, where overwriting on David Drake's occupation results in an unintelligible word and Mark Jones's occupation is clearly but inaccurately written as "Farmer." The census also has a column which indicates that they could neither read nor write, which is in error because we know that Dave could do both. Census of Shaw's Creek Township, Edgefield County, July 11, 1870; South Carolina Department of Archives and History.

23. 1880 Federal Census, Population Schedule, South Carolina, Edgefield County, Shaw's Creek Township, T9-1228.

24. Howard A. Smith, *Index of Southern Potters* (Mayoden, N.C.: Old America Company, 1986), pp. 84, 161. Two black potters, Mark and Brewster Jones, are listed later on as working at the Seigler Pottery, located thirteen miles from Edgefield. Brewster is perhaps Mark Jones's son, listed in the 1870 Federal Census as Brister. See also "Notes made on trip to Seigler's Pottery," Charleston Museum Archives.

25. Printed in capital letters on the base of a jug dated September 6, 1859: "DADY.GIBS / MAUM HANNAH / STEPHENS / BY TIM." The significance of this inscription is unknown. This jug is in a private collection.

26. Three vessels signed by or attributed to Dave on which are inscribed only the month and year are known: January 1850 (signed "Lm"); June 1855 (signed "Lm"); and October 1855 (signed "Dave" and "Lm").

27. Holcomb and Holcomb, "South Carolina Potters and Their Wares," pp. 52, 57.

28. Holcomb and Holcomb, "Archaeological Findings Related to Dave at Edgefield Pottery Sites," pp. 76–79.

29. Holcomb and Holcomb, "South Carolina Potters and Their Wares," p. 26.

30. The only known exceptions are two vessels—a jug and a jar with a poem—both dated December 6, 1858, and two jars with poems signed by Dave and Baddler (another Edgefield slave) and dated May 13, 1859.

31. A storage jar signed "Dave / Lm" and dated March 31, 1858, has two horizontal loop handles, as opposed to the lug handles typically found on Dave storage jars.

32. Holcomb and Holcomb, "South Carolina Potters and Their Wares," p. 12. See also note 5 above.

33. John A. Burrison, "Alkaline Glaze Stoneware: A Deep South Pottery Tradition," *Southern Folklore Quarterly* 39, no. 4 (December 1975): 382.

34. As Charles Zug explains: "[P]erhaps the simplest form of the alkaline glaze was a mixture of equal parts of clay and wood ashes. The former provided the necessary silica and alumina, while the latter provided additional silica and alumina as well as calcium, sodium, and potassium oxides—the alkaline substances that serve to flux the compound. Pine ash, for example, contains, 24.39 percent silica, 9.71 percent alumina, 39.73 percent lime, 3.77 percent sodium oxide, 8.98 percent potash, and additional oxides of iron, manganese, magnesium, and phosphorous. As this analysis reveals, lime is the principal flux." Charles G. Zug III, *Turners and*

Burners: The Folk Potters of North Carolina (Chapel Hill: University of North Carolina Press, 1986), p. 179.

In "Alkaline Glazes and Groundhog Kilns: Southern Pottery Traditions," *Antiques* (April 1977): 768, Georgeanna H. Greer describes how green hues were achieved:

> The green hues of the glaze were determined primarily by the amount of iron present in the glaze solution and the conditions under which it was fired. In a clear-burning firing 1 to 3 percent iron will produce a creamy tan to straw color, while 5 percent or more will produce brown to black colors. The more common smoky or reducing firings, which chemically rob the iron compounds of oxygen, produce colors in the green range, from a pale blue-green through grassy and olive colors to deep blackish green, again depending upon the amount of iron in the glaze mixture. Streaks of pale sky blue mixed with white and blotches of deep blood red which appear on some of the pottery are also produced by iron under special conditions.

35. Daisy Wade Bridges, *Ash Glaze Traditions in Ancient China and the American South* (Robbins, N.C.: Southern Folk Pottery Collectors Society, 1997), p. 17.

36. Ibid., pp. 13, 14.

37. For an extensive review of the history and different possibilities, see Baldwin, *Great and Noble Jar,* pp. 18, 19. See also Bridges, *Ash Glaze Traditions,* pp. 5–10; Burrison, "Alkaline Glaze Stoneware," pp. 386–88; Zug, *Turners and Burners,* pp. 70–74.

38. Dave's contemporaries in Edgefield—Thomas Chandler, for example—produced wares with a celadon glaze.

39. Orville Vernon Burton, *In My Father's House Are Many Mansions: Family and Community in Edgefield, South Carolina* (Chapel Hill: University of North Carolina Press, 1985), p. 46. This is an excellent, comprehensive study of the social, political, economic, religious, and cultural conditions of whites and slaves in Edgefield, South Carolina, during this period.

40. Janet Duitsman Cornelius, "When I Can Read My Title Clear," in *Literacy, Slavery, and Religion in the Antebellum South* (Columbia: University of South Carolina Press, 1991), p. 37. Anti-literacy laws existed in South Carolina as early as 1740, when the state was under British rule:

> XLV And *whereas,* the having of slaves taught to write, or suffering them to be employed in writing, may be attended with great inconveniences; *Be it therefore enacted* by the authority aforesaid, That all and every person and persons whatsoever, who shall hereafter teach, or cause any slave or slaves to be taught, to write, or shall use or employ any slave as a scribe in any manner or writing whatsoever, hereafter taught to write, every such person and persons, shall, for every such offence, forfeit the sum of one hundred pounds current money.

The 1834 act, legislated nearly one hundred years later, was far more punishing, however, and added corporal punishment for the slaves themselves:

> No. 2639. AN ACT to amend the Laws in relation to Slaves and Free Persons of Color
>
> *Be it enacted,* by the honorable the Senate and House of Representatives, now met and sitting in General Assembly, and by the authority of the same, If any person shall hereafter teach any slave to read or write, or shall aid or assist in teaching any slave to read or write, such person, if a free white person, upon conviction thereof, shall, for each and every offence against this Act, be fined not exceeding one hundred dollars, and imprisoned not more than six months; or if a free person of color, shall be whipped, not exceeding fifty lashes, and fined not exceeding fifty dollars, at the discretion of the court of magistrates and freeholders before which such free person of color is tried; and if a slave, to be whipped at the discretion of the court, not exceeding fifty lashes; to informer to be entitled to one half of the fine, and to be a competent witness. And if any free person of color or slave shall keep any school, or other place of instruction, for teaching any slave or free person of color to read or write, such free person of color or slave shall be liable to the same fine, imprisonment and corporal punishment, as are by this Act imposed and inflicted on free persons of color and slaves for teaching slaves to read or write.

Slavery, Race and the American Legal System 1700–1872, Garland Series 7, edited by Paul Finkelman (New York: Garland, 1988), p. 468. See also John Belton O'Neall, *Negro Law of South Carolina, Collected and Digested by John Belton O'Neall* (Columbia, S.C.: Printed by John G. Bowman, 1848), p. 23.

Fourteen years later, the State Agricultural Society of South Carolina directed that a review of the 1834 act be submitted to the governor and legislature for consideration in the November 1848 session:

> Sec. 42 This Act grew out of a feverish state of excitement produced by the impudent meddling of persons out of the slave States, with their peculiar institutions. That has, however, subsided, and I trust we are now prepared to act the part of wise, humane and fearless masters, and that this law, and all of kindred character, will be repealed. When we reflect, as Christians, how can we justify it, that a slave is not to be permitted to read the Bible? It is in vain to say there is danger in it. The best slaves in the State are those who can and do read the Scriptures. Again, who is it that teach your slaves to read? It generally is done by the children of the owners. Who would tolerate an indictment against his son or daughter for teaching a favorite slave to read? Such laws look to me as rather cowardly. It seems as if we were afraid of our slaves. Such a feeling is unworthy of a Carolina master.

See also *Slavery, Race, and the American Legal System 1700–1872,* p. 413 (Act. no. 670, XLV).

41. See Burton, *In My Father's House.*

42. Cornelius, "When I Can Read My Title Clear," p. 64.

43. Ibid.

44. An unsigned jar from 1834 attributed to Dave has "Concatination" inscribed on it; another, from 1836, bears the word "catination." "Concatenate" and "catenate" both refer to a linking together, in a series or chain. *Merriam-Webster's Collegiate Dictionary,* 10th ed., s.v. "Concatenate," "Catenate." If these words were written as an intended allusion to slavery, it was a very daring act. It is not difficult to believe that it was. The 1834 vessel was recently auctioned at Sotheby's New York (sale cat., May 19, 2005, lot 149).

45. *Edgefield Advertiser,* April 1, 1863, p. 3.

46. Cornelius, "When I Can Read My Title Clear," p. 37.

47. Baldwin, *Great and Noble Jar,* p. 67. See also Smith, *Index of Southern Potters,* pp. 95, 106.

48. Dr. Landrum died on April 3, 1859. His obituary, published in the *Edgefield Advertiser* on April 13, 1859, reads in part: "The *Bulletin* well says of him: 'Mr. Landrum was much respected, and regarded as one of our most useful and enterprising citizens during his public life, notwithstanding bitter partisans were arrayed against him on account of his Union sentiments. He lived, respected by all who knew him; and he died, leaving a record of a well spent life, without a blot or tarnish to mar his good name.'"

49. Cornelius, "When I Can Read My Title Clear," pp. 62, 65, 66.

50. Baldwin, *Great and Noble Jar,* pp. 50, 75; *Edgefield Advertiser,* December 23, 1846.

51. Stephen Ferrell and T. M. Ferrell, *Early Decorated Stoneware of the Edgefield District, South Carolina* (Greenville, S.C.: Greenville Museum of Art, 1976), p. 9.

52. *Edgefield Advertiser,* December 23, 1846.

53. Steen, "An Archaeological Survey," p. 80.

54. Baldwin, *Great and Noble Jar,* p. 75.

55. John A. Burrison, "Dave the Potter and His Place in American Ceramics History," in *Pottery, Poetry, Politics Surrounding the Enslaved African-American Potter, Dave,* symposium, McKissick Museum of the University of South Carolina and the South Carolina Humanities Council, April 25, 1998 (Columbia, S.C.: McKissick Museum, University of South Carolina, 1998), p. 26; *Edgefield Advertiser,* April 13, 1859.

56. "Information given by Mr. G. U. Flesher 7/8/30, Long hand notes taken by Miss Bragg and typed by EAB," Edgefield pottery file, Charleston Museum, Charleston, S.C.

57. "Pottery, Poetry, Politics Surrounding the Enslaved African-American Potter, Dave," symposium handout, McKissick Museum, University of South Carolina, April 25, 1998, Session 2, Question and Answer, p. 64 (first "Q" on page, from A. Goldberg).

58. Burton, *In My Father's House,* pp. 202, 393 nn. 1, 2, and 399 n. 54.

59. Jill Beute Koverman, "Dave's Verse as Social Response," in *I made this jar . . . ,* pp. 90–92.

60. For analyses of interpretations of the poems, see Aaron De Groft, "Eloquent Vessels/Poetics of Power: The Heroic Stoneware of 'Dave the Potter,'" *Winterthur Portfolio* 33, no. 4 (winter 1988): 249–60; and James A. Miller, "Dave the Potter and the Origins of African-American Pottery," in *I made this jar . . . ,* pp. 53–60.

John A. Burrison

Fluid Vessel: Journey of the Jug

▼ ON A SUNNY SATURDAY *in the fall of 1997, Hewell's Pottery in the north Georgia countryside was in the midst of its annual Turning and Burning festival, a homegrown celebration of traditional pottery-making. The guest of honor was Lanier Meaders, arguably America's most famous folk potter, who had made it to the age of eighty. This was his big birthday party, with hundreds of well-wishers on hand. Lanier had kept alive Georgia's old alkaline-glazed stoneware tradition virtually single-handed until others, inspired by his success, picked it up and carried it on, and our meeting many years earlier, in 1968, had inspired me to research and write the first in-depth survey of a Southern state's ceramic heritage.*[1]

The Hewell and Meaders "clay clans" of northeast Georgia (the country's last stronghold, along with North Carolina and Alabama, of Euro-American folk pottery) have maintained ties for nearly a century. William J. Hewell of Gillsville in Hall County did much of his potting at Mossy Creek, fifteen miles to the north, in the Appalachian foothills of White County. It was there, in about 1920, that he introduced the concept of the face jug to Lanier's father, Cheever Meaders. Hewell probably had acquired the concept from his in-laws, the Fergusons, north Georgia's first known makers of jugs with modeled faces, who likely had brought the idea from antebellum South Carolina. When face jugs became Lanier's specialty in the 1970s, earning him a national reputation—and the nickname "Jughead"—his young friend Chester Hewell was prompted to revive his own family's making of them, bringing the local tradition full circle.

The county-fair atmosphere of Lanier's birthday bash was tinged with sadness, for he was gravely ill (and would die four months later). Under the big tent the Hewells had set up for the occasion, I was asked to say a few words over the public address system about Lanier and his importance in American ceramics history, after which I fed him a slice of his birthday cake. When he clutched my sleeve from his wheelchair, indicating he had something to tell me, I leaned down to catch his barely audible whisper: "John, you're Jughead now." In what otherwise might have been taken as an insult he was conferring on me the honor of his nickname, knowing that soon he would have no need for it.

Belying the stereotype of the isolated country craftsman who never ventured beyond his home community, Lanier Meaders served as a U.S. Army paratrooper in World War II, returning to Mossy Creek from his air base in Britain and combat in Germany. The jug with which his career as a potter was to become inextricably bound took much the same route in its journey to America, some three centuries earlier. I have written about jugs in the context of Georgia's ceramics history, but the Turning and Burning festival led me to ponder more broadly, as the newly anointed "Jughead," a clay artifact type that became the quintessential American pot form—even giving rise to an American musical idiom, the jug band—and to attempt to write its life history, much as I had Lanier's.[2]

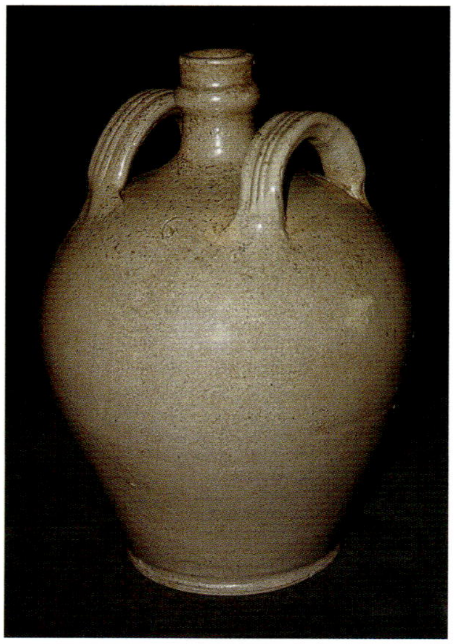

Figure 1 Two-handled vessel, probably American, possibly nineteenth century. Alkaline-glazed stoneware. H. 9½". (Author's collection; unless otherwise noted, all photos by author.) This small mystery vessel, which can be traced only so far as the Chicago antiques trade, shows indications of an early Southern stoneware jug (bulbous form, feldspathic alkaline glaze, foot ledge, collared neck), but the peculiar horizontal loop handles, reeded along their flat tops, depart from the vertical handle that defines a jug.

Semantic Juggling: What's Pot-Bellied, Skinny-Necked, One-Armed, Yet Full of Spirit?

In tracing the travels of this seemingly ubiquitous ceramic form, the first order of business is to clarify just what a jug is,[3] a task complicated by different meanings for the term in American and British English. As George Bernard Shaw wryly put it, "England and America are two countries separated by the same language."[4] In Britain, the word *jug* can mean either a small pitcher (as in a cream jug for tea) or a wide-mouthed drinking vessel, which, in reference to medieval wares, is tall and often baluster-shaped (really an early tankard).[5] In the United States, however, the term has come to mean a narrow-necked (so it can be stoppered), flat-bottomed (free-standing) vessel for keeping and transporting liquids (although certainly one could drink from it), with one or two vertical loop handles for grasping and pouring (figs. 1, 2). In Britain such a vessel is called a *flagon* or *bottle*; what we Americans call a *bottle* normally lacks a handle. Thus, the subject of the eighteenth-century English poem "The Brown Jug" is an ale mug, whereas the American song "The Little Brown Jug" of a century later refers to a narrow-necked container for distilled spirits.[6] It is the American meaning that this essay addresses.

Peregrinating Pots: Old-World Origins and Diffusion

The jug form, as defined above, seems to have arisen in the ancient world as a response to technological advances. As practiced in the Near East by 2500 B.C., the fermentation of date and grape juice into wine and the extraction of oil from olives and aromatic and medicinal plants, along with

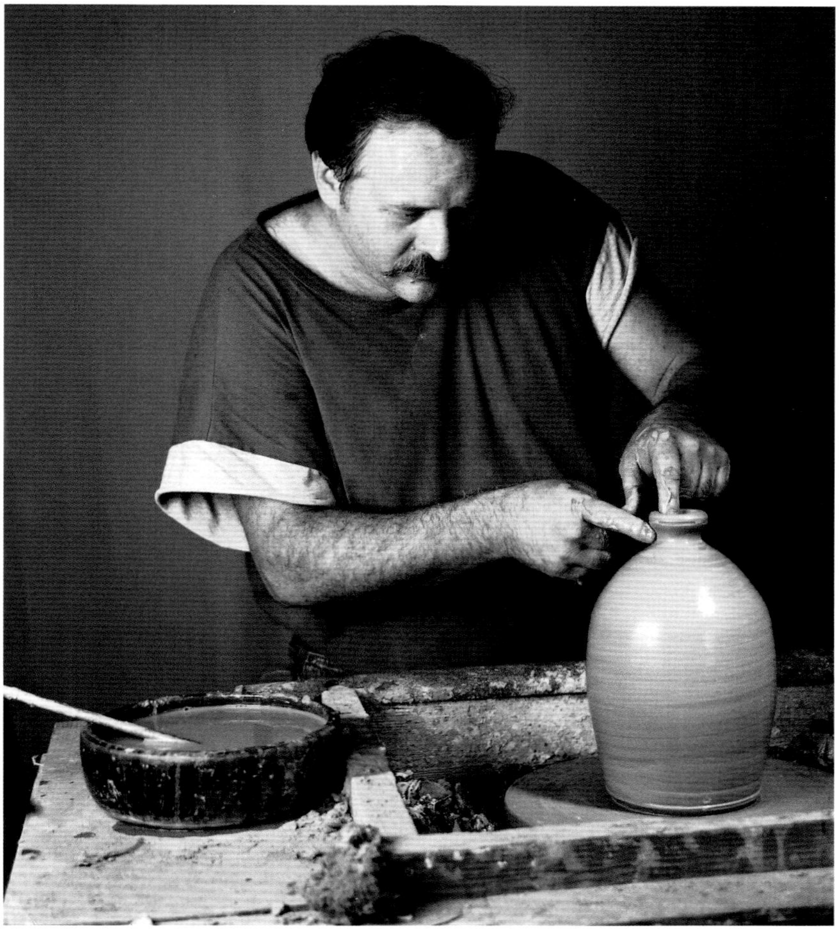

Figure 2 Dwayne Crocker of Gillsville, Georgia, shapes the neck of a jug on his electric wheel, 2001. (Photo, Chris Swanson.) The vertical loop handle that also defines the form will be added once the clay dries a bit.

a growing trade in these liquids, required inexpensive containers to store and transport them. At the same time the potter's wheel, beginning as a low turntable in Mesopotamia and later improved with added flywheel and raised headblock as a kickwheel in Greece and Egypt, offered the speed to support large-scale production and facilitated the continuous drawing in and up of the jug neck.[7] The Far East experienced similar advances, but the loop handle that helps to define the jug never caught on as an addition to the bottle form favored there.

The jug first appeared in Palestine during the second phase of the Early Bronze Age, about 2800 B.C. (fig. 3), then was adopted in Greece (where it was called a *lekythos*), Egypt, and Persia by 1500 B.C., migrating west, with Phoenician trade, to the Etruscans and Romans.[8] The Roman Empire helped to spread the form beyond the Mediterranean into Germany and Britain, where it became part of the Romano-British ceramic repertoire (fig. 4).[9]

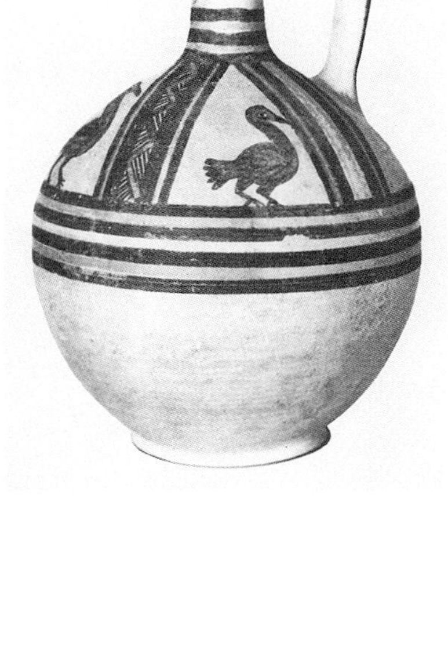

Figure 3 Bichrome wheel-made jug, Gaza, Palestine, ca. 1500 B.C. Earthenware. H. 11⅜". (Courtesy, Israel Antiquities Authority, Jerusalem.)

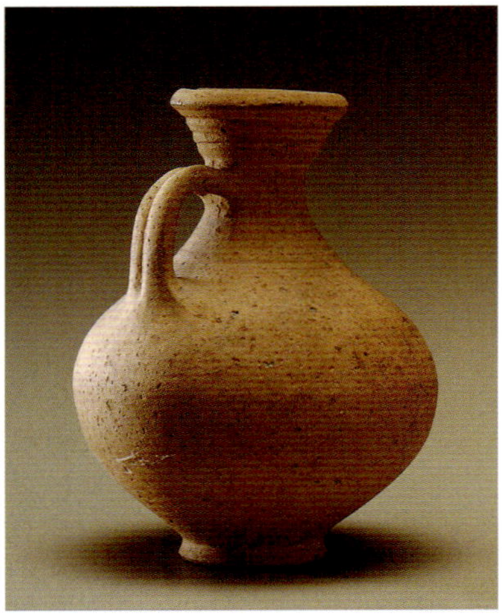

Figure 4 Romano-British jug, first half of second century A.D. Unglazed earthenware. H. 5⅝". (Noël Hume Collection; photo, Gavin Ashworth.) This locally made jug was recovered by a fisherman off the tidal Upchurch Marshes, Kent, England.

Some German earthenware jugs were indeed made in the Middle Ages, although published references to them are scant.[10] The form came into its own there, however, after the emergence of stoneware in the fourteenth century, perhaps to support the expanding trade in Rhenish wine.[11] German stoneware jugs were most notably manifested as the *Bartmannkrug*, or graybeard, with a sprig-molded face mask on the neck and medallion on the belly, thousands of which were shipped to Britain in the sixteenth and seventeenth centuries (figs. 5, 6).[12] Salt-glazed stoneware jug production

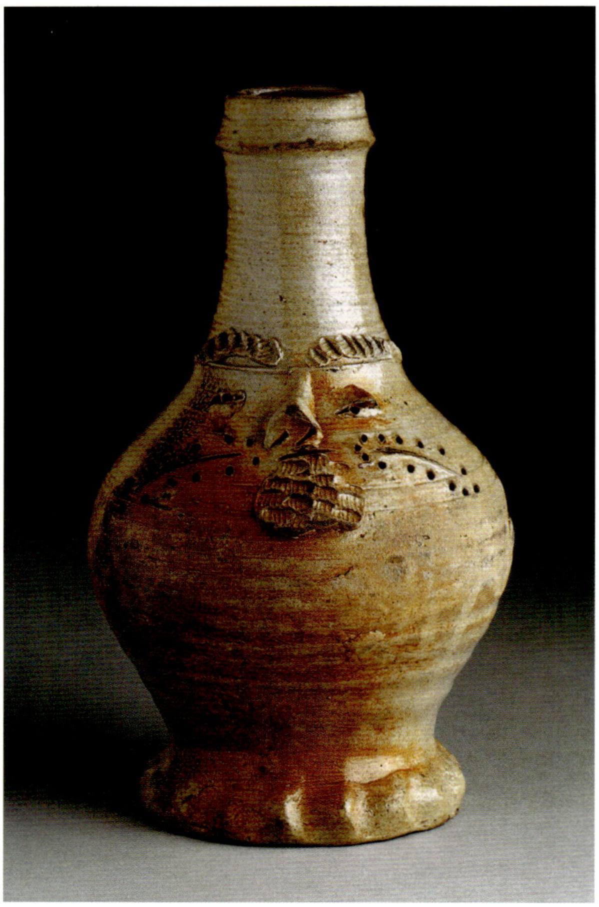

Figure 5 Jug, Raeren or Aachen, Germany, ca. 1475–1525. Stoneware. H. 8⅞". (© The Trustees of The British Museum.) This salt-glazed jug has an applied face, strap handle, and thumbed base.

Figure 6 Graybeards, Frechen, Germany, ca. 1660–1665. Salt-glazed stoneware. (Noël Hume Collection; photo, Gavin Ashworth.) The vessel illustrated on the left was recovered from the Thames at London. The one on the right was found beneath the hearth of a cottage at Stratford St. Mary, Suffolk. Both were used as "witch bottles."

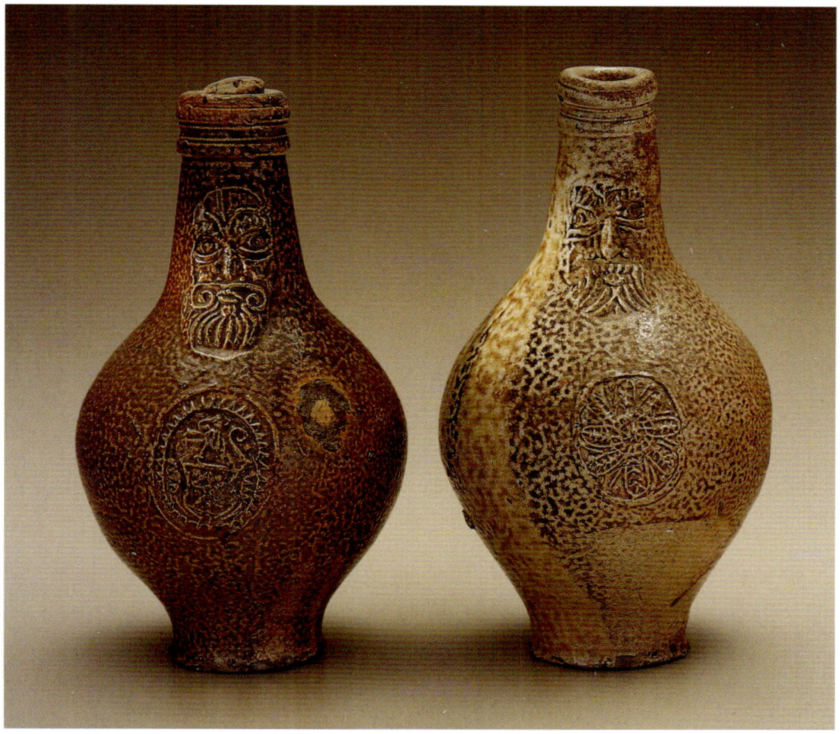

remained robust in Germany through the nineteenth century, and continues in some areas today (fig. 7).[13]

The jug form's link to the potter's wheel is especially evident in the ceramic history of England. When Roman withdrawal and Saxon invasion ushered in the Dark Ages, the wheel fell out of use and shaping technology reverted to hand-building, with open-mouthed forms such as cooking pots

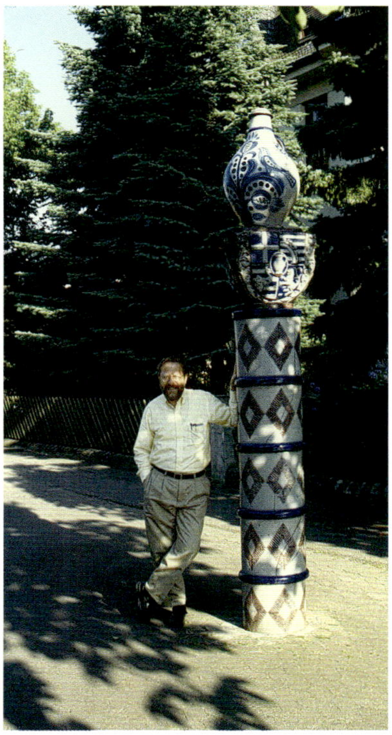

Figure 7 The author, with a modern monument in Höhr-Grenshausen, Germany, featuring a cobalt-decorated jug atop the town arms. (Photo, Linda Carnes-McNaughton.) Such emblems of occupational identity are found throughout the Westerwald's Kannenbäckerland (mug-baking district), where the salt-glazed stoneware tradition is still active.

and cremation urns dominating as in the prehistoric period.[14] Reintroduction of the wheel in the seventh to twelfth centuries by later Saxons and Normans set the stage for the jug's revival in the Middle Ages, although the form was less common than the baluster-shaped drinking vessel known to British ceramics historians as a jug.[15] Beer, the chief alcoholic beverage for ordinary medieval folk, was brewed locally and consumed as fresh as possible, hence there was little need to store and transport it. The upper class drank wine, but England's climate was not conducive to viticulture, and most wine was imported in containers that also were made on the Continent.

Some British earthenware jugs of the Early Modern period were a domestic answer to imported German stoneware, with London delftware potters of the 1620s–1670s producing tin-glazed examples inscribed "sack," "claret," and "Renish [*sic*] wine" as potables became affordable to the growing middle class (fig. 8).[16] Country potters maintained production of

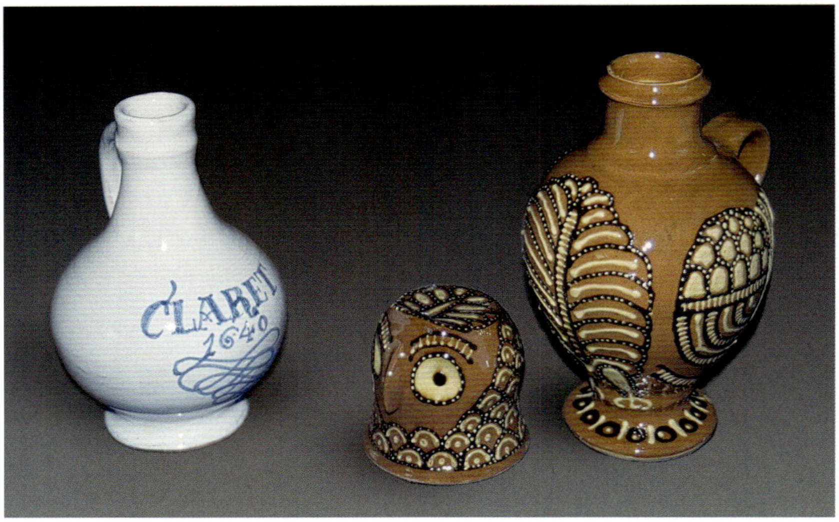

Figure 8 Reproduction jugs, John Hudson, Mirfield, West Yorkshire, late twentieth century. *Left:* Claret jug, after a London delftware example. H. 6¾". *Right:* Slip-trailed owl jug after one attributed to Thomas Toft of Staffordshire, with its top beside it. H. 8¾". (Author's collection.)

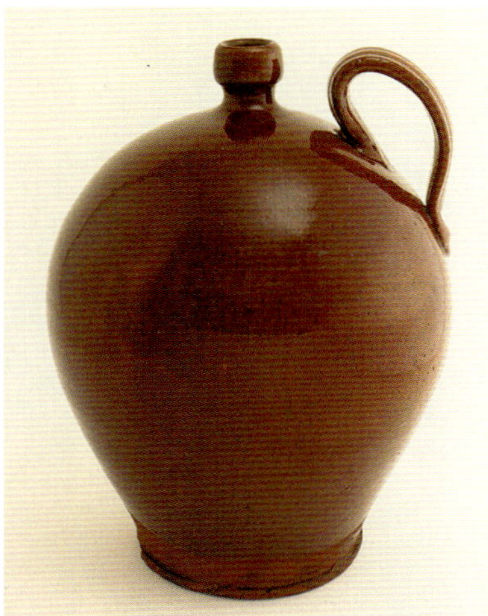

Figure 9 Jug, County Down, Northern Ireland, early to mid-1800s. Lead-glazed earthenware. H. approx. 12". (Courtesy, Museum of Country Life, Castlebar; photo, National Museum of Ireland.) Little has been published on Irish country pottery; this example of the jug form is the only one known to the author.

lead-glazed jugs through the nineteenth century, along with other coarsewares still needed for storing and processing food and drink (fig. 9).[17] The earliest British stoneware (1640s–1670s)—made by German immigrants

Figure 10 Rhenish-style jug, workshop of John Dwight, Fulham, England, last quarter of seventeenth century. Gray salt-glazed stoneware, with varicolored clay and sprig-molded figures. H. 7½". (Willett Collection; © The Trustees of The British Museum.)

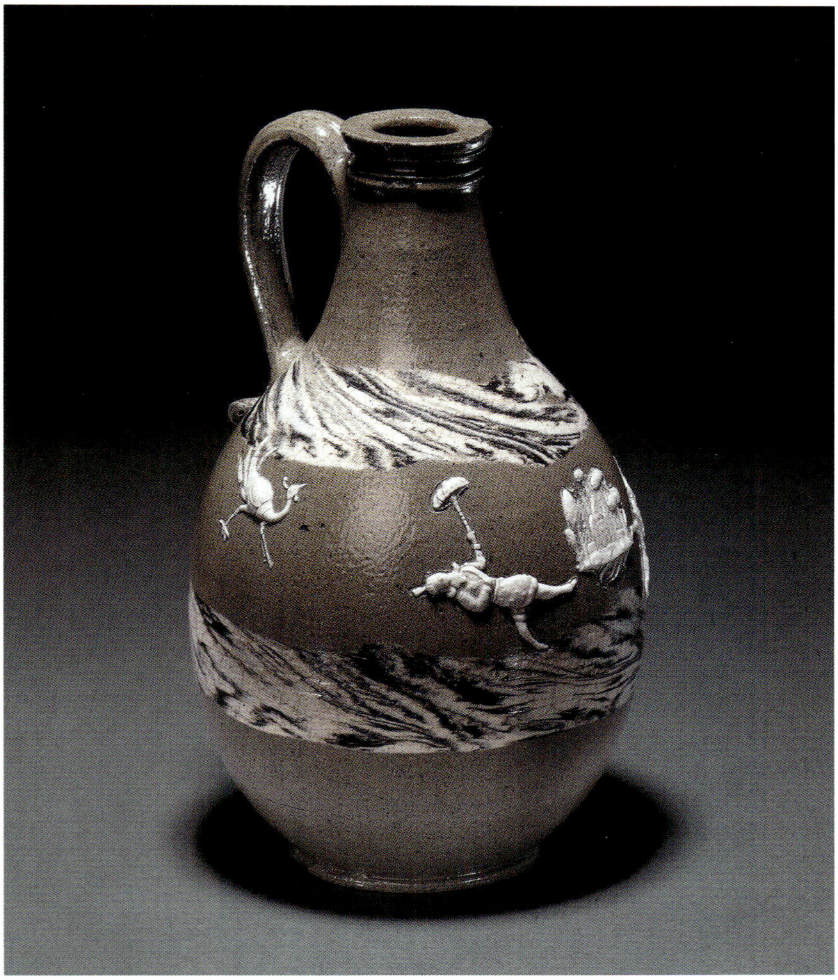

and John Dwight in the London area—included Rhenish-style jugs (fig. 10), whereas salt- and later Bristol-glazed stoneware jugs continued to be made in England and Scotland through the nineteenth century, eventually mass-produced in urban factories (fig. 11).[18]

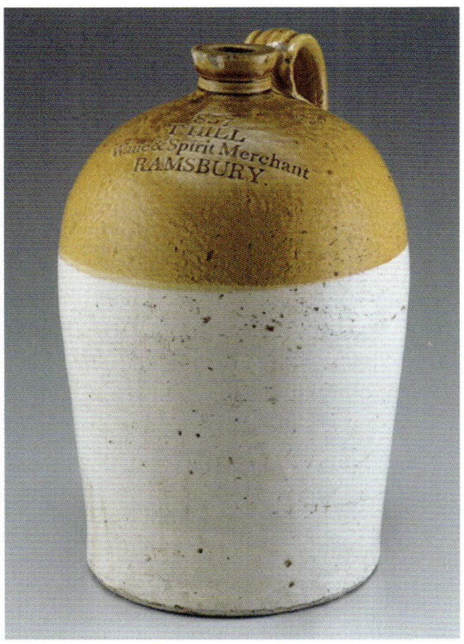

Figure 11 Two-tone jug, Bristol (Avon County), England, late 1800s. Stoneware, with Bristol glaze over partial iron wash. H. 15½". Marks: POWELL + BRISTOL; "857 / T HILL Wine & Spirit Merchant / RAMSBURY." (Photo, Gavin Ashworth.) Developed in 1835, Bristol glaze contains feldspar, whiting (as a flux), and often zinc oxide (as a whitener and opacifier). The jug was made for a merchant at Ramsbury, a village in the adjacent county of Wiltshire. The Powell of the mark probably refers to the firm of William Jr. and Septimus, descendants of William Powell, who codeveloped Bristol glaze. The firm was sold to Charles Price in 1906.

The Jug Comes to America

Documentation for the colonial period, especially the first century of settlement, is too sparse to know what role jugs played in early American life. Again, semantics contribute to the problem, for early written references to jugs would have carried over the English meaning as a wide-mouthed drinking vessel. By the early nineteenth century, however, the American meaning of the word was established (figs. 12, 13).[19]

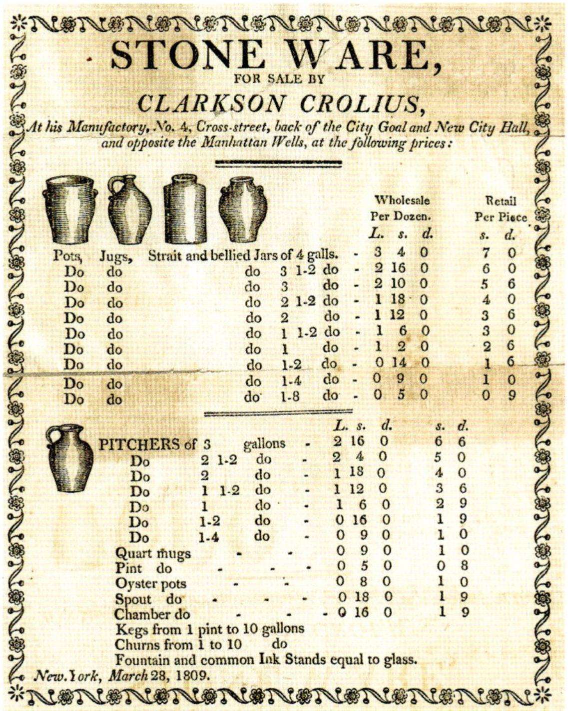

STONE WARE,

FOR SALE BY

CLARKSON CROLIUS,

At his Manufactory, No. 4, Cross-street, back of the City Goal and New City Hall, and opposite the Manhattan Wells, at the following prices:

			Wholesale Per Dozen. L.	s.	d.	Retail Per Piece s.	d.
Pots,	Jugs,	Strait and bellied Jars of 4 galls. -	3	4	0	7	0
Do	do	do 3 1-2 do -	2	16	0	6	0
Do	do	do 3 do -	2	10	0	5	6
Do	do	do 2 1-2 do -	1	18	0	4	0
Do	do	do 2 do -	1	12	0	3	6
Do	do	do 1 1-2 do -	1	6	0	3	0
Do	do	do 1 do -	1	2	0	2	6
Do	do	do 1-2 do -	0	14	0	1	6
Do	do	do 1-4 do -	0	9	0	1	0
Do	do	do 1-8 do -	0	5	0	0	9

	L.	s.	d.	s.	d.
PITCHERS of 3 gallons -	2	16	0	6	6
Do 2 1-2 do -	2	4	0	5	0
Do 2 do -	1	18	0	4	0
Do 1 1-2 do -	1	12	0	3	6
Do 1 do -	1	6	0	2	9
Do 1-2 do -	0	16	0	1	9
Do 1-4 do -	0	9	0	1	0
Quart mugs - -	0	9	0	1	0
Pint do - - -	0	5	0	0	8
Oyster pots - -	0	8	0	1	0
Spout do - - -	0	18	0	1	9
Chamber do - - -	0	16	0	1	9

Kegs from 1 pint to 10 gallons
Churns from 1 to 10 do
Fountain and common Ink Stands equal to glass.

New.York, March 28, 1809.

Figure 12 Price list of Manhattan stoneware potter Clarkson Crolius, 1809. (Courtesy, American Antiquarian Society.) Early illustrations of jugs associated with the word help establish when the American meaning came about.

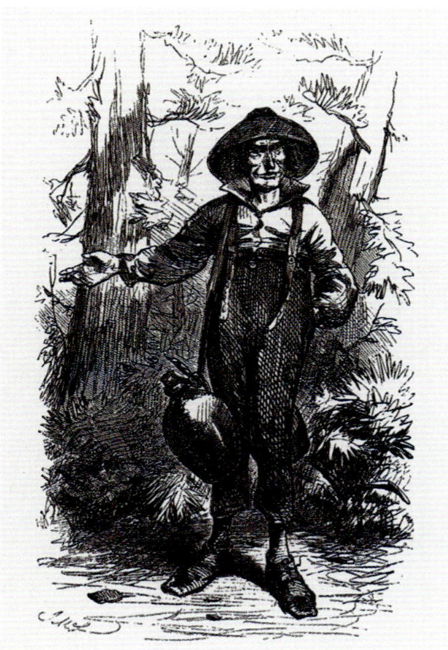

Figure 13 Hardin E. Taliaferro ("Skitt"), *Fisher's River (North Carolina) Scenes and Characters* (New York: Harper and Brothers, 1859), pp. 266–67. This image shows the American notion of a jug.

The foremost Old World influences on Euro-American ceramics were England and Germany, whose immigrant potters transplanted earthenware and stoneware traditions in the seventeenth and eighteenth centuries. England provided precedents for the colonial redware of New England and the Tidewater South, while mid-Atlantic earthenware exhibits a mix of British and German ideas.[20] Influential settlers of these three East Coast regions can be traced to specific areas of Britain, theoretically making it possible to compare early American earthenware jugs with those of key source areas. Thus, for New England one would look to East Anglia, for the mid-Atlantic to northern England and Wales, and for the coastal South to southwest England.[21] Complicating such an approach, however, is the role London played in the settlement of all three American regions, and the difficulty in finding large enough samples of reasonably intact wares from both sides of the Atlantic that were made when influence would have been strongest.

Lura Woodside Watkins, pioneer scholar of New England ceramics, noted that "[j]ugs of redware were . . . not so common in the eighteenth century as in the years that followed."[22] Surviving American earthenware jugs of the eighteenth to early nineteenth centuries are ovoid or turnip-shaped and often have a narrow foot extending around the base. Decoration normally

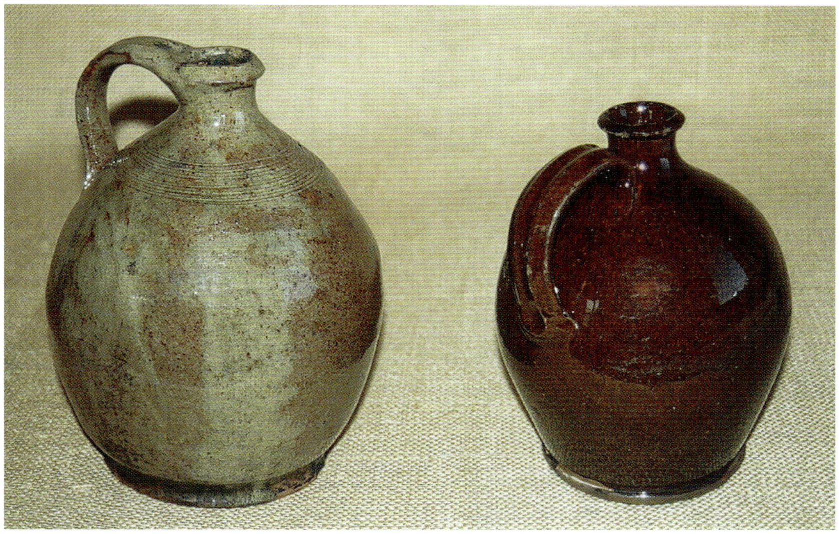

Figure 14 Jugs. Lead-glazed earthenware. *Left:* New England (Massachusetts provenance), ca. 1800. H. 7¾". The lead glaze is accented with copper over shoulder tooling. *Right:* Pennsylvania, early to mid-1800s. H. 6½". Iron or manganese-colored lead glaze and extruded handle. (Author's collection.) Both vessels have a foot-ledge and pronounced lip but are distinguished by characteristic regional handle placements.

was limited to incised circumferal lines (tooling) and use of a metallic oxide (copper or manganese) to color the lead glaze. As for regional distinctions, handle placement is typically lower on mid-Atlantic and Southern examples, with the upper end applied to the jug's shoulder or shoulder-neck juncture, whereas the upper handle terminal on New England examples often was attached to the neck at or near the mouth (fig. 14).[23] This regional handle placement also applies (with exceptions, of course) to American stoneware jugs.

Salt-glazed stoneware technology was introduced to the East Coast in the early eighteenth century from both Germany and England.[24] Many of the oldest surviving American stoneware jugs have a reeded or cordoned (tooled) neck, a feature carried over from German and English stoneware. Those in the Germanic tradition have cobalt blue (or, less common, manganese purple) brushed within an incised design, around handle terminals, and highlighting a stamped maker's mark (fig. 15); a partial iron-oxide dip

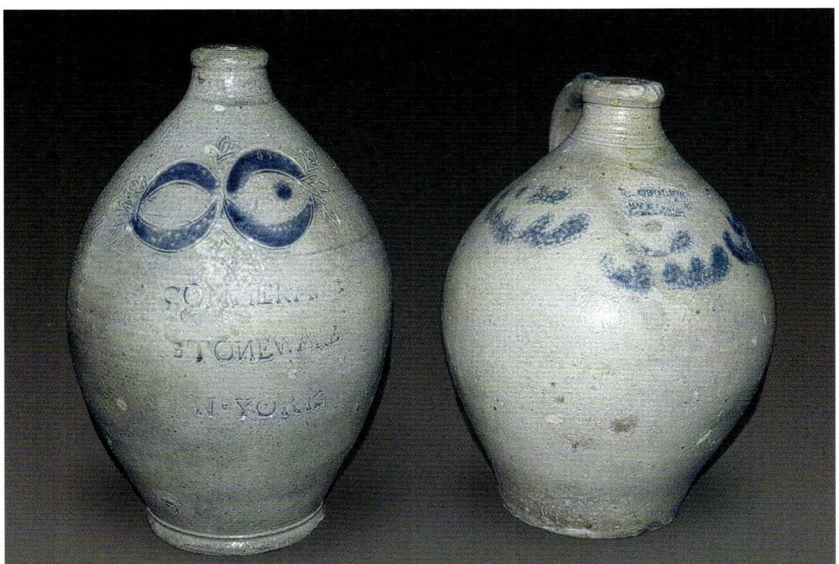

Figure 15 Jugs. Manhattan. Salt-glazed stoneware. *Left:* Thomas Commereau, ca. 1815. H. 11¾". *Right:* Clarkson Crolius Jr., ca. 1840. H. 11". (Author's collection.) Both vessels are in the Germanic tradition, with maker's marks, handle terminals, and floral designs highlighted with cobalt blue.

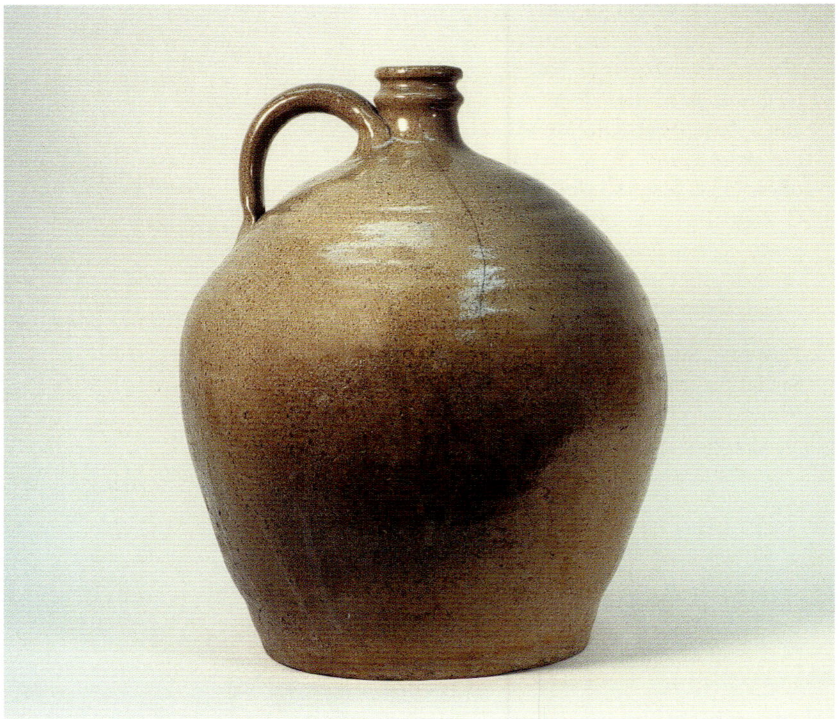

Figure 16 Jug, attributed to Abner Landrum's Pottersville stoneware manufactory, 1821. Alkaline-glazed stoneware. H. 14⅛". Date incised under feldspathic alkaline glaze. (Ex-collection of Tony and Marie Shank; photo, courtesy McKissick Museum, University of South Carolina.) This vessel is the earliest dated Edgefield District, South Carolina, jug. Its double-collared neck became the norm for antebellum Edgefield jugs.

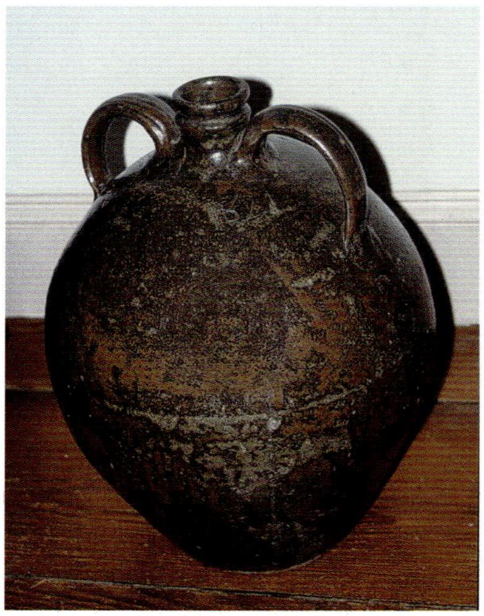

Figure 17 Two-handled jug, attributed to Abraham Massey, Washington County, Georgia, 1821. Alkaline-glazed stoneware. H. 18". Date incised under alkaline glaze. (Courtesy, Collection of Levon and Elmaise Register.) This is the earliest known dated Georgia pot. Associated with Abner and John Landrum back in South Carolina, Massey and Cyrus Cogburn introduced alkaline glazes and the Edgefield-style double-collared neck to middle Georgia.

created a two-tone effect for those in the English tradition.[25] Both salt- and alkaline-glazed antebellum stoneware jugs made in the South often have a pronounced lip; in the Edgefield District of South Carolina a collar was thrown around the middle of the neck as well (figs. 16, 17, 29).[26] These early neck and lip treatments apparently served both as a design element and as an aid in securing a stopper with a cord or wire.

American ceramics historians have noted a general shift in the shape of stoneware jugs from sensuous to severe that began about 1860.[27] Bulbous and ovoid jugs tended to give way to those with straighter sides (figs. 18–20), culminating in the development of a distinctly American type: the cylindrical "shouldered" or "stacker" jug, with a shoulder ledge to support a production collar so that jugs could be stacked in a column to make efficient

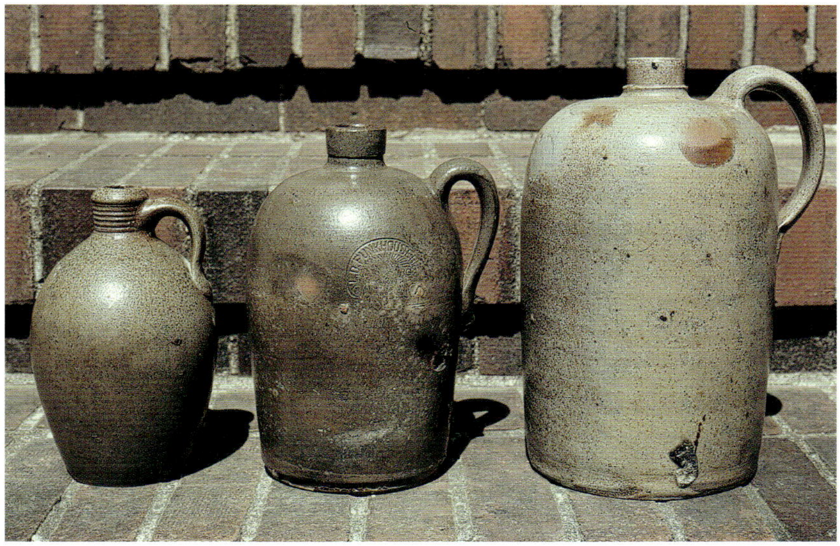

Figure 18 Jugs. Salt-glazed stoneware. *Left:* Nicholas Fox, Chatham County, North Carolina, ca. 1840. *Center:* retailed by Luther Funkhouser, Strasburg, Virginia, 1890s. *Right:* Peter Herrmann, Baltimore, Maryland, 1890s. (Ex-collection of Charles G. Zug III; photo, Charles G. Zug III.) These vessels illustrate the general shift in America from curved to straight sides.

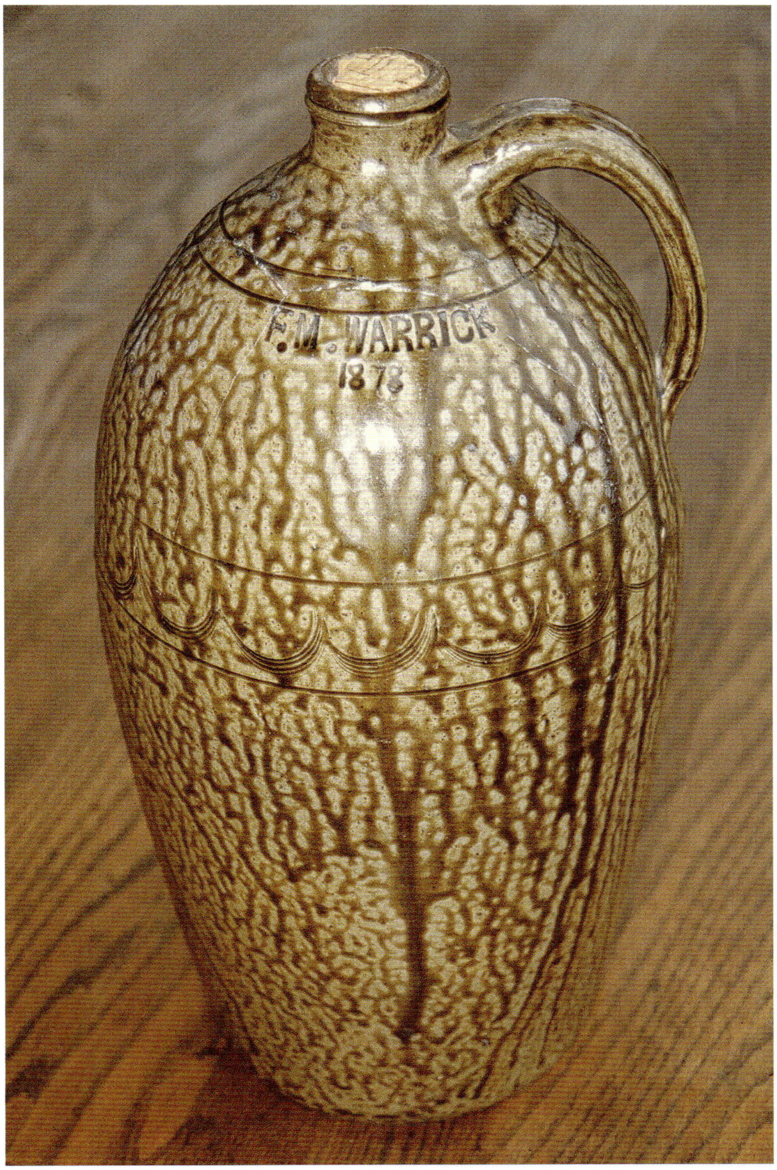

Figure 19 Stoneware jug, F. M. Warrick, Sand Mountain, Alabama, 1878. H. 16". (Courtesy, Collection of Ron Countryman.) This jug bears a rare stamped date and combed decoration under the ash glaze. The moderately curved shape and incised lip are typical for the period.

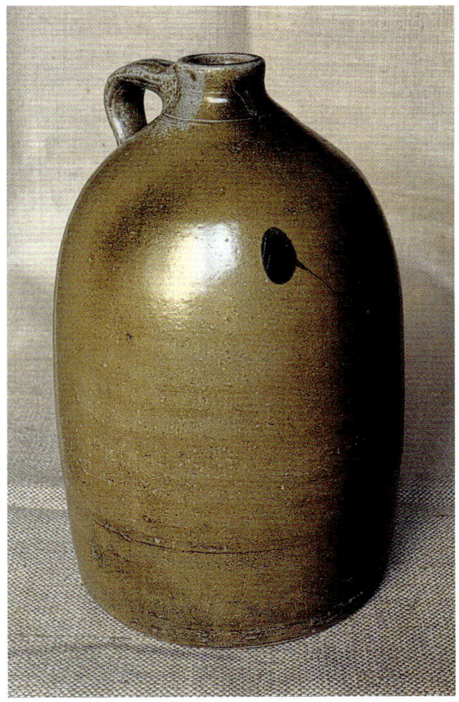

Figure 20 Stoneware jug, Randolph County, Alabama, ca. 1900. Salt-glaze over Albany slip. H. 13½". (Author's collection.) By this time the typical American jug had become straight-sided and lost its lip.

use of kiln height.[28] This shift to less curvaceous forms sacrificed graceful proportions for greater volume, making the clay jug more competitive with factory-made glass and metal containers. The change also may reflect a decline in throwing skills as pottery manufacture became more industrialized, although in conservative, family-run shops like those of the rural South, earlier styles sometimes were maintained alongside later ones. Cheever and Lanier Meaders, for example, made both ovoid and cylindrical jugs in the mid- and late twentieth century (figs. 21, 22). Dating estimates for jugs therefore should not be based on shape alone, as each workshop had its own history of form variation.

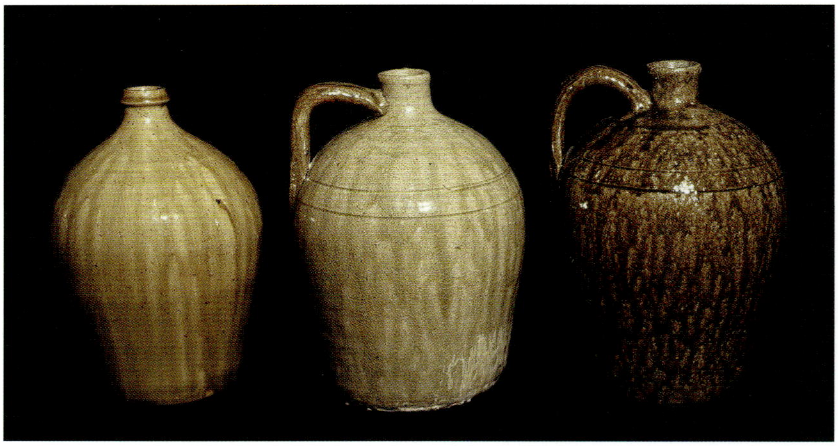

Figure 21 Jugs, White County, Georgia. Alkaline-glazed stoneware. *Left:* David Dorsey, ca. 1850. "Flint" subtype of lime glaze. H. 9¼". (Private collection.) *Center:* Cheever Meaders, 1950s. Flint glaze. H. 9¾". (Author's collection.) *Right:* Lanier Meaders (who maintained his father's accenting lines at shoulder), 1972. Ash glaze. H. 9¾". (Author's collection.) These jugs illustrate continuity in shape at some Southern workshops. Except for their lack of a lip, ovoid Meaders jugs could be mistaken for antebellum ones.

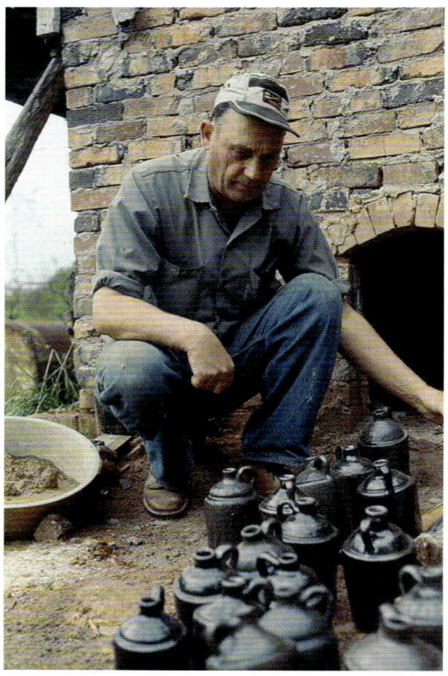

Figure 22 Lanier Meaders, Mossy Creek, White County, Georgia, 1974. Meaders unloads small, ash-glazed "shoulder" jugs from his wood-fueled, rectangular "tunnel" kiln. He and his father, Cheever, both made this cylindrical style of jug alongside the earlier ovoid shape.

Permutations: Form Follows Functions

The generic jugs described so far might have been used to contain any liquid, depending on the needs of the owners. Since the form's origins, however, the emphasis has been on fermented or distilled spirits. In the United States, especially, jugs with the capacity of one quart to two gallons normally were meant for rum or whiskey, sometimes being made to order for distillers (both illicit and licensed). The medieval technology for converting grain to whiskey was brought from Ireland and Scotland in the eighteenth century and adapted to the New World grain, maize, for moonshine and bourbon.[29] As Annie Becham Long, who was born into one middle Georgia "clay clan" and married into another, put it, "Half the people in our part of Crawford County were making jugs and the other half were making liquor to put in 'em."[30] The form's importance in the South is underscored by the frequent occupational identification of potters as "jug makers" in the federal censuses, the nickname "Jug" being used by several potters,[31] and the name "Jugtown" identifying at least six pottery-making

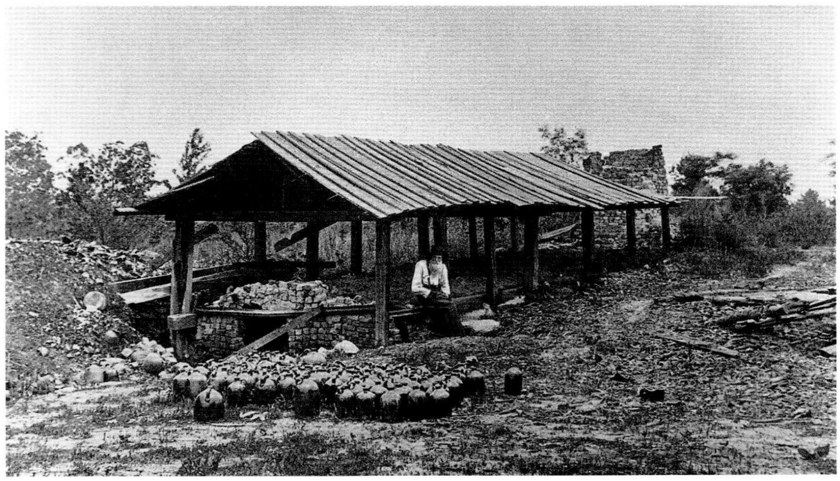

Figure 23 Eddie Averett's Crawford County, Georgia, "groundhog" kiln, 1890s. (Photo, courtesy Georgia Geologic Survey.) The photograph shows whiskey jugs ready for sale to local moonshiners or licensed distilleries in nearby Macon, ranging in shape from squatty "beehive" to ovoid to cylindrical.

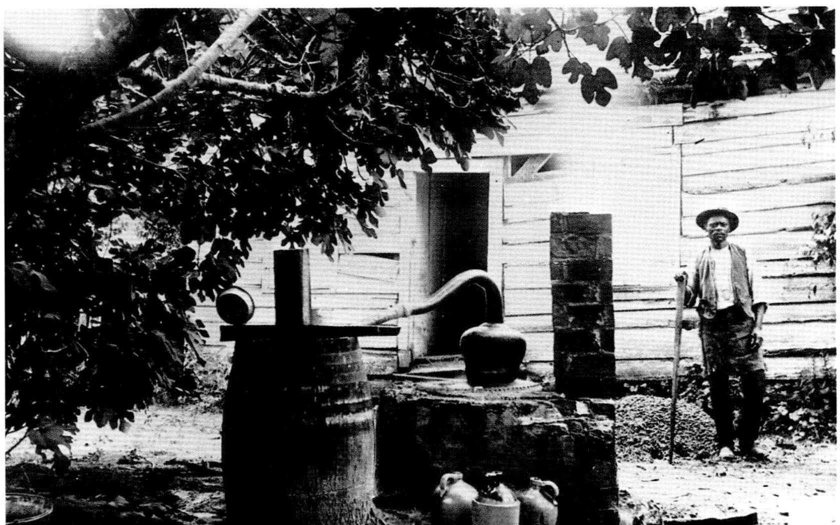

Figure 24 Moonshine still, Savannah, Georgia, ca. 1890. (Photo, William O. Wilson, Vanishing Georgia Photographic Collection, Georgia Division of Archives and History, ge0139-86.) The "shoulder" jug at center, with two-tone Bristol and Albany-slip glazes, is machine-molded; the salt-glazed jugs on either side are handthrown.

communities.[32] Southern potters specializing in whiskey jugs (figs. 23, 24) saw their trade decline in the early twentieth century due to Prohibition and the availability of affordable glass and metal containers. Long before that, though, the basic jug had been modified to serve a number of other, more specialized needs.

Building on the Form: Big Jugs, Squatty Jugs, Multi-Neck Jugs

It required no great leap of imagination for American potters to increase jug size to accommodate other uses, just the skill and strength to throw such "bigwares." A capacity of five gallons was the upper limit of most potters' normal range, but examples as large as twenty gallons are known.[33] Jugs of four and more gallons often had a second loop handle on the other side of the neck, creating a graceful, amphoralike symmetry while making them easier for two people to lift and carry when full. There are northern English precedents for such large, two-handled jugs (figs. 25, 26), but the form could have arisen independently in response to New World conditions.[34]

Figure 25 Cider jug, Halifax, West Yorkshire, England, late 1800s. Lead-glazed earthenware. H. approx. 14¼". (Courtesy, Shibden Hall and Folk Museum, Calderdale Metropolitan Borough Council—Museums and Arts, England.) Leaving the lower wall unglazed is said to have caused the cider to "roll," accelerating fermentation; the hole beneath the neck held a loose stick that allowed gas to escape.

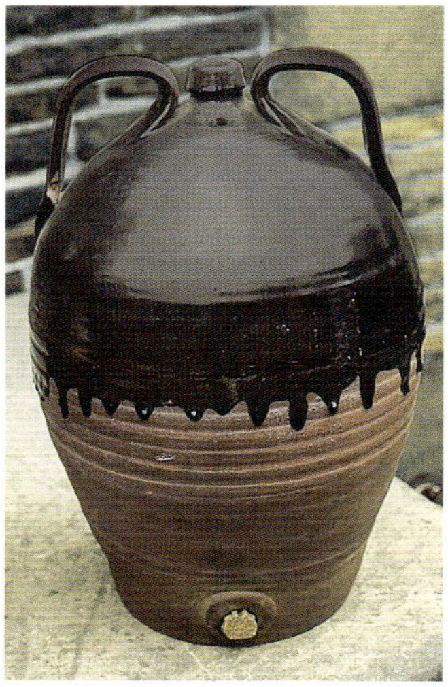

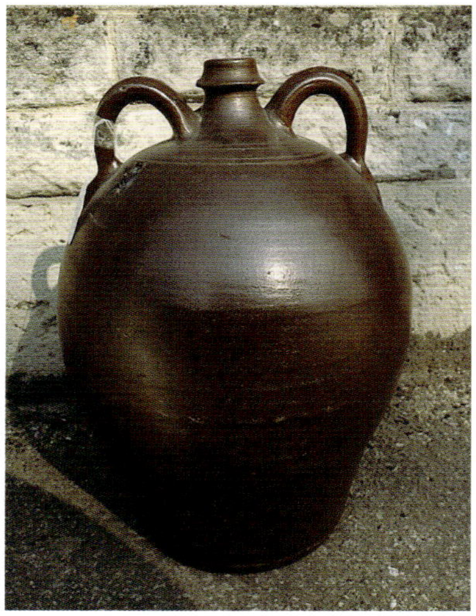

Figure 26 Four-gallon jug, Liverpool area, Merseyside, ca. 1830. Brown salt-glazed stoneware. H. 16½". When found on the East Coast such vessels frequently are misidentified as American. This example was photographed at an antiques shop in England.

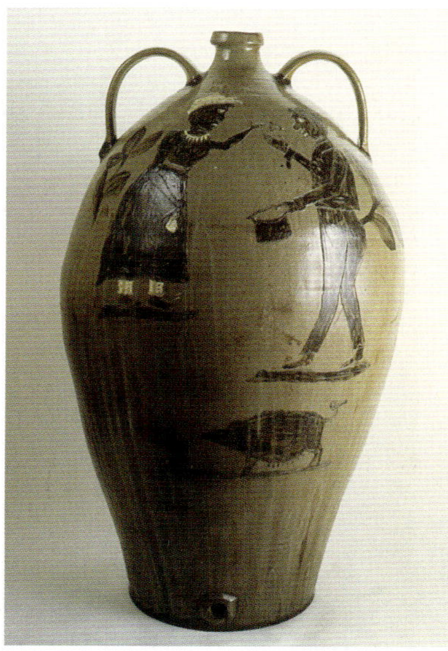

Figure 27 Water cooler, attributed to Thomas Chandler, Phoenix Factory, Edgefield District, South Carolina, ca. 1840. Stoneware. H. 31¼". (Courtesy, High Museum of Art, Purchase in honor of Audrey Shilt, President of the Members Guild, 1996–97, with funds from the Decorative Arts Endowment and Decorative Arts Acquisition Trust; photo, McKissick Museum, University of South Carolina.) The toasting slave couple was brushed under the alkaline glaze in iron-stained and kaolin slips within incised outlines, whereas the sow below (facing a similar pot), the handle terminals, and the leafy vine encircling the other side were painted freehand.

One use for big jugs in the United States was as a watercooler or cistern before indoor plumbing, typically with a spigot hole in the lower wall (figs. 27, 28). Decorated examples might also have been used in taverns as beer, wine, or punch dispensers, since nonporous stoneware would have been more sanitary (easier to clean) than wooden kegs.[35] In the antebellum

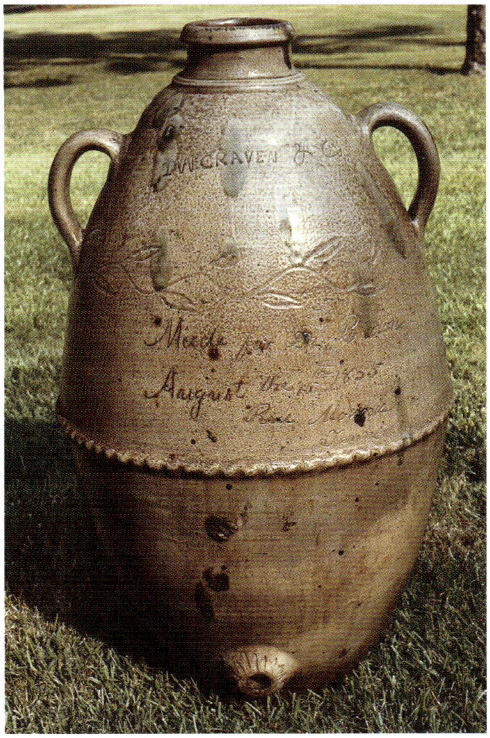

Figure 28 Water cooler, Tinsley Craven, Henderson County, Tennessee, 1855. Salt-glazed stoneware, with incised vine design and fluted joint reinforcement. H. 26⅛". (Ex-collection of Tony and Marie Shank.) With its wide mouth, this piece approaches fountain-type coolers but could still be considered a jug.

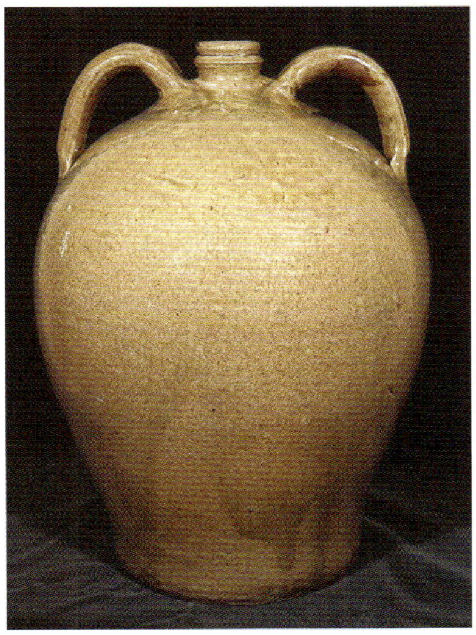

Figure 29 Syrup jug, attributed to Nathaniel Ramey pottery, Edgefield District, South Carolina, late 1830s. Alkaline-glazed stoneware. H. 16½". (Author's collection.) The bulbous shape of this four-gallon example is typical of these vessels.

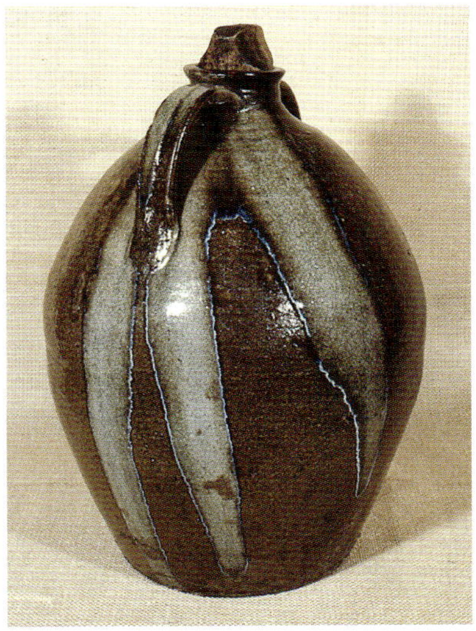

Figure 30 Syrup jug, Catawba Valley, North Carolina, late 1800s. Alkaline-glazed stoneware. H. 16⅜". (Author's collection.) The decorative technique of placing broken bottle glass on handles and rim before firing to melt in contrasting streaks is found mainly on North Carolina stoneware. The neck was widened to facilitate pouring the cane syrup, requiring a carved wooden plug to fit it.

South, sugar, in the form of solid cones, was affordable mainly to the upper class and, as a precious commodity, was locked in an item of furniture known as a sugar chest.[36] Ordinary folk used honey and cane syrup as their chief sweetenings, making syrup jugs of three or more gallons a mainstay of the Southern potter's repertoire (figs. 17, 29, 30).[37] Kept in the kitchen or nearby smokehouse, the syrup would be handy for cooking or table use. Some potters widened the mouth to facilitate pouring the thick liquid, but this required the owner to fabricate a wooden plug, as the usual corncob stopper would not have made a tight fit.

Another approach to adapting the basic form to specialized uses was to compress it, creating a jug with a low center of gravity. In the late nineteenth

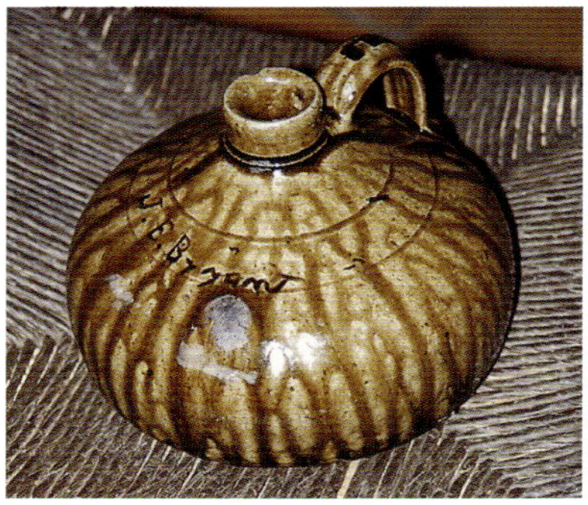

Figure 31 Buggy jug, attributed to Billy Bryant, Crawford County, Georgia, late 1800s. Alkaline-glazed stoneware. H. 5¼". Mark: stamped "BB" on top of handle; presentation inscription "J. E. Bryant." (Courtesy, Museum of Southern Stoneware at River Market Antique Mall, Columbus, Ga.)

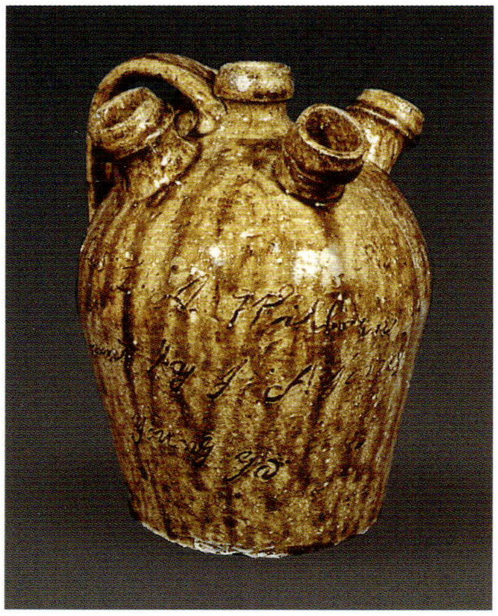

Figure 32 "Flower jug," James Jones, Union County, Georgia, 1870s. Ash-glazed stoneware. H. 9". Marks: signed by the maker; inscribed "Miss L. A. Wilborn." (Author's collection.) A previous owner's oral history claims that the five spouts of this mountain "wedding jug" were meant for the minister, bride, groom, best man, and maid of honor to drink from at the ceremony.

century the Norton Pottery of Bennington, Vermont, produced squatty molasses jugs glazed with Albany slip, adding a well-like drip-catcher lip with a pinched pour-spout.[38] Southern potters made even squatter "buggy jugs" (proportionately, the top half or third of a regular jug), which were designed to keep that special stock of "antifreeze" from tipping over on a bumpy buggy or wagon drive (fig. 31).

Localized to northern Georgia and North Carolina's Catawba Valley, alkaline-glazed "flower jugs" (for displaying cut flowers) elaborated the basic jug form by adding necks (usually four, separately thrown "off the hump") around the central one.[39] Made as presentation pieces—an inscribed Georgia mountain example has an oral history as a "wedding jug" (fig. 32)—in the mid- to late 1800s and revived by folk potters a century later, they seem related to English and Dutch quintals that were based on a vase type introduced to Europe from Persia along with the tulip in the 1600s. Old World examples, however, lack the loop handle that helps to define a jug.

Further Mutations: Water Jugs for the Field

Two types of "harvest jugs" are known in the United States, each with different Old World roots. Although occasionally made in the North, their concentration in the South can be attributed to the region's agrarian economy and warm climate; hours of field work in the heat of the day could produce a powerful thirst. Unlike the jug types discussed above, these designs involved major alteration of the basic form.

The ring jug, built from a hollow ring, is a European form that can be traced to ancient Greece[40] but its inspiration in America most likely was the German stoneware tradition (fig. 33).[41] Its wheel-thrown production is something of a potter's secret: starting with a clay disk, the potter scoops out the center with a rib, leaving a solid ring. A trough is created by pulling up the outer and inner edges, which are then curved toward each other and joined.[42] After cutting the circular tube off the wheel and drying to leather hard, the flat (headblock) side is trimmed to the round and a separately thrown neck is attached over a hole cut in the arch. It is at this stage that the potter has the option of creating a jug or a bottle; if the former, one or two loop handles, and a thrown or solid-block base, are added (fig. 34).

A functional explanation for the donut shape is that it could lie flat for easier transport. The center hole allowed placement on a saddle horn, the

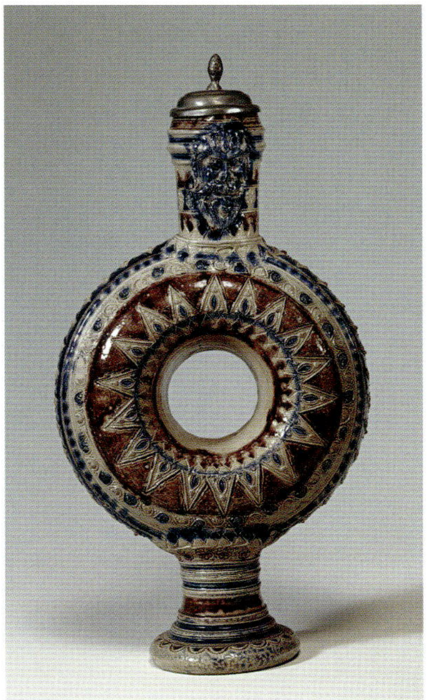

Figure 33 Ring jug with lion mask, Westerwald, Germany, third quarter of seventeenth century. Salt-glazed stoneware with cobalt and manganese. H. 17⅝". (Courtesy, Hetjens-Museum Düsseldorf/Deutsches Keramikmuseum, Düsseldorf, Germany.)

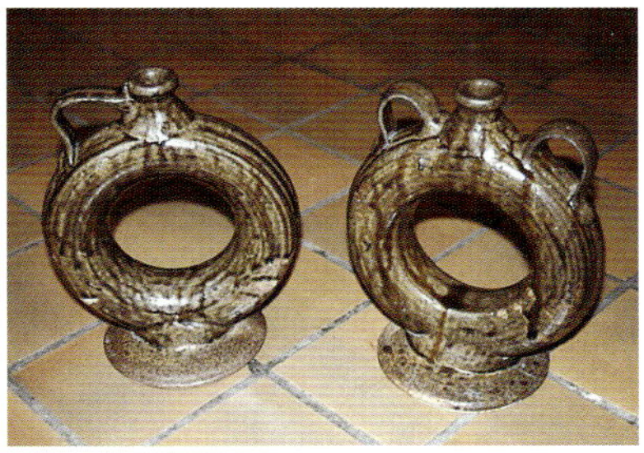

Figure 34 Ring jugs, attributed to Crawford County, Georgia, late 1800s. Alkaline-glazed stoneware. (Courtesy, Museum of Southern Stoneware at River Market Antique Mall, Columbus, Ga.) Although this pair of jugs was made by the same potter, the handles on each appear to be the work of different hands.

hame knob of a plow-animal's collar, or a tree branch; in some cases this opening was large enough for the insertion of an arm, for carrying on the shoulder. Twentieth-century Southern ring jugs and bottles were made as novelties. Georgia-born Javan Brown glazed his "old-time" ring jugs on one side only (that side faced out so that the sun reflected off the shiny glaze, while the porosity of the unglazed side allowed limited evaporation of the water, to keep it cool). Marie Rogers of Jugtown, Georgia, calls them "Confederate jugs," having heard from her husband, Horace (from whom she learned to make them), and his father, Rufus, that they sometimes were used as canteens by Rebel soldiers in the Civil War.[43] Bill Gordy of Cartersville, Georgia, added a mushroom-shaped stopper to his ring jugs, which he coated with his signature "Mountain Gold" glaze.[44]

The "monkey" (an Afro-Caribbean term for thirst) jug is distinguished by a stirrup handle across the top and a tubular, canted (off-center and angled) spout. A second neck (which could have a widened mouth for filling) or airhole would cause the water to stream into the drinker's mouth (fig. 35). The separately thrown necks are located either below the handle terminals

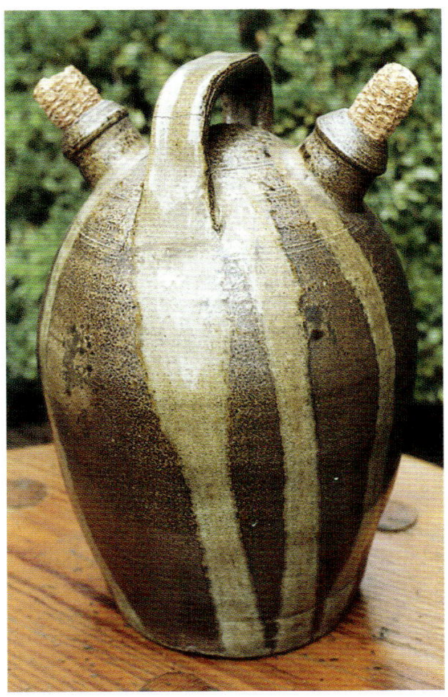

Figure 35 Monkey jug, central North Carolina, ca. 1860. Salt-glazed stoneware, with melted-glass decoration and corncob stoppers. H. 12". (Courtesy, Mary Frances Berrier Collection.) According to oral history, a slave drank from one spout and his master from the other.

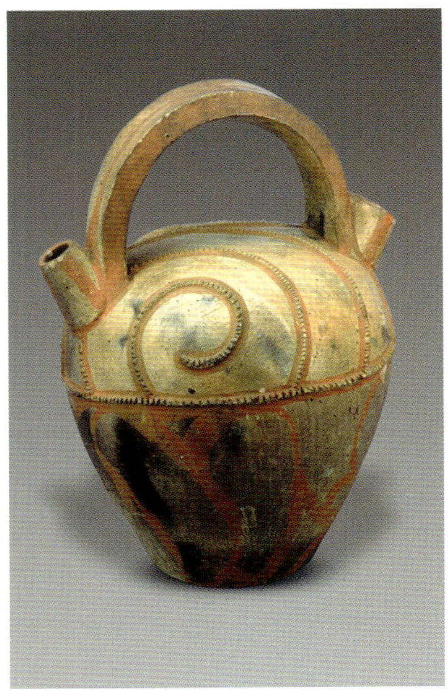

Figure 36 Monkey jug, Ekoi people, Nigeria, West Africa, probably early 1900s. Unglazed earthenware. H. 13⅜". (© The Trustees of The British Museum.)

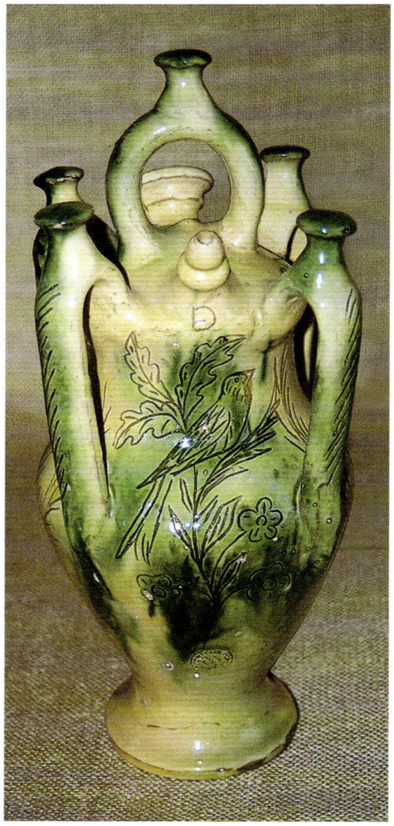

Figure 37 *Botijo,* Felix Perez workshop, Cuenca, Spain, mid-nineteenth century. Lead-glazed earthenware. H. 14". (Author's collection.) Made as a tourist souvenir, the monkey-jug form of this vessel is elaborated with vertical loop handles, button finials, and a sgraffito bird and foliage with copper-green highlighting.

in the plane of the handle or on opposite sides of it. This form, also traceable to ancient Greece, is concentrated in Africa and Mediterranean Europe (figs. 36, 37); in Spain, where it is known as a *botijo,* it is virtually a national emblem, with three museums devoted to its seemingly infinite variety in shape, glaze, and ornamentation.[45] In America, imports from Europe may have inspired Northern monkey jugs; in the South, where the form was made by slave potters, a West African source is probable.[46]

Craft into Art: Making Faces on Jugs

Working in such a malleable medium day in and day out, it is not surprising that potters the world over have seen their clay as a kind of mirror and accepted its challenge to model a human likeness on a pot, pushing utilitarian craft into plastic art. This anthropomorphizing impulse is stronger in some clay-working societies than in others; in England, for example, it surfaced several times, first with Romano-British burial urns, then with medieval "face-on-front jugs," and finally with the "Toby jugs" of late-eighteenth-century Staffordshire, which initiated an industrial tradition of slip-cast character mugs continuing to this day.[47] Evidently, however, England was not the source of America's face-jug traditions.

In reviewing what is known of those traditions, three points should be made by way of introduction. First, although many humanoid vessels made in the United States are indeed jugs, some are pitchers, cups, jars, or bottles. Second, not all depict just faces or heads; some are full figures. Third, Native Americans of the South made "people pots" during the Mississippian era (A.D. 800–1540), too early to have influenced the Southern face jugs discussed here.[48]

As with other adaptations of the jug form in the United States, face vessels—as both a historical and a living tradition—have been concentrated in the South. However, the oldest known Euro-American examples are from the early-nineteenth-century Philadelphia workshop of Henry Remmey Jr. (fig. 38), whose great-grandfather, stoneware potter John Remmey, had immigrated to Manhattan from the Rhineland in the 1730s. John would have been familiar with the graybeard jugs still being made in Germany, and if he made them in New York and passed the concept to his descendants, then Henry's face vessels are Americanizations of the German tradition.[49]

A substantial group of early Southern face vessels was made between 1863 and 1865 by enslaved African-American potters at Colonel Thomas Davies's Palmetto Fire Brick Works at Bath, in the old Edgefield District of west-central South Carolina.[50] Distinguished by bulging eyes and bared teeth of kaolin inset into the stoneware clay body, the iron-based mineral that darkened the alkaline glaze on some, along with the wax resist used to keep the glaze off the white eyes and teeth to maximize contrast, leave

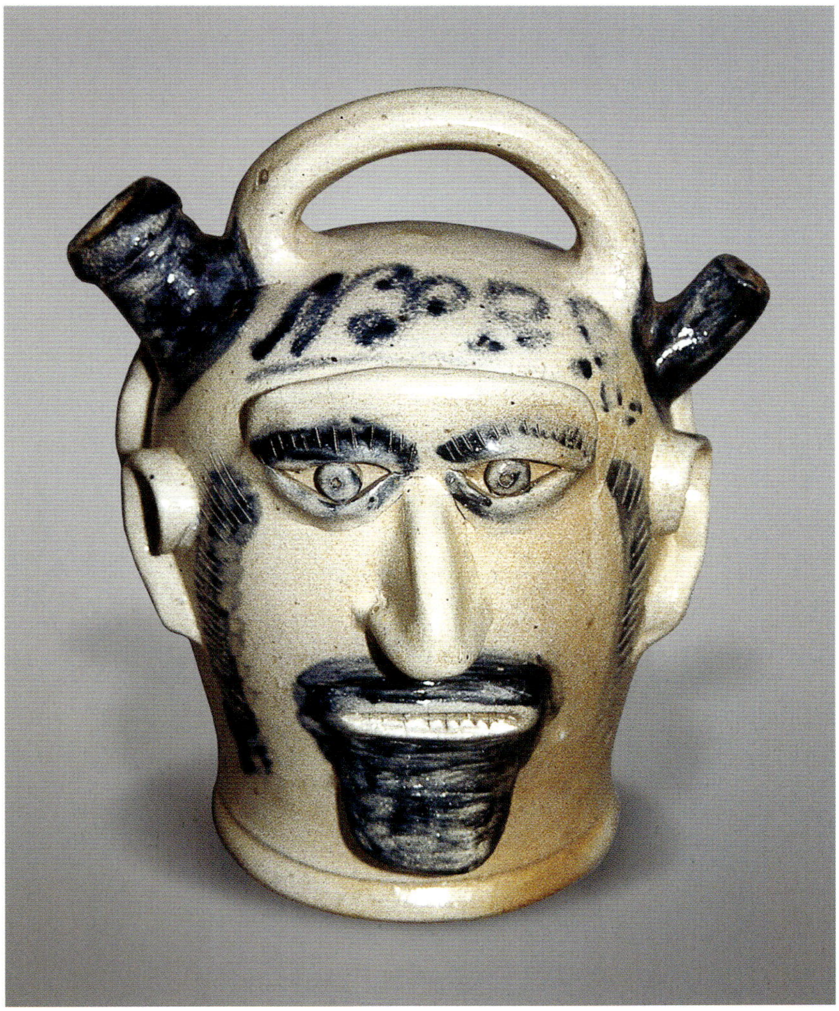

Figure 38 Two-face monkey-form jug, attributed to Henry Remmey Jr. or his son Richard, Philadelphia, Pennsylvania, 1858. Salt-glazed stoneware. H. 11". (Ex-collection of Tony and Marie Shank; photo, Tony Shank.) Highlighted with cobalt blue, this dated jug might represent an Americanization of the German graybeard tradition.

little doubt that they were meant to represent their makers' race (figs. 39, 40). Pioneer ceramics historian Edwin AtLee Barber, after corresponding with Davies, was the first to discuss them in print: "These curious objects . . . possess considerable interest as representing an art of the Southern negroes. . . . The modeling reveals a trace of aboriginal art as formerly prac-

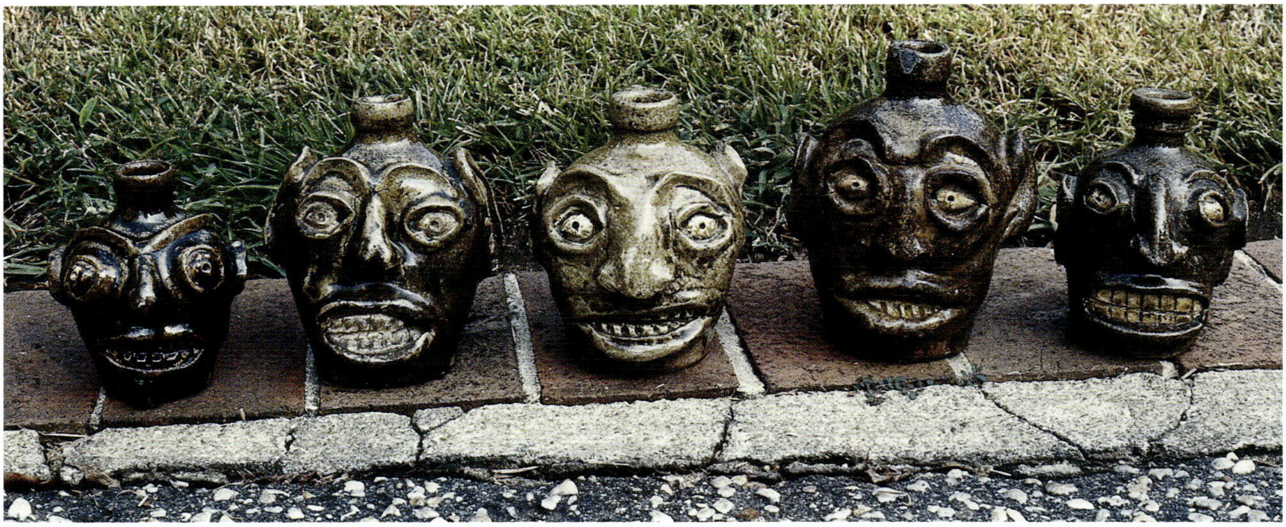

Figure 39 Face jugs, attributed to slave potters at Thomas Davies's Palmetto Fire Brick Works, Bath, South Carolina, early 1860s. Alkaline-glazed stoneware, with inset kaoline eyes and teeth. (Ex-collection of Tony and Marie Shank.)

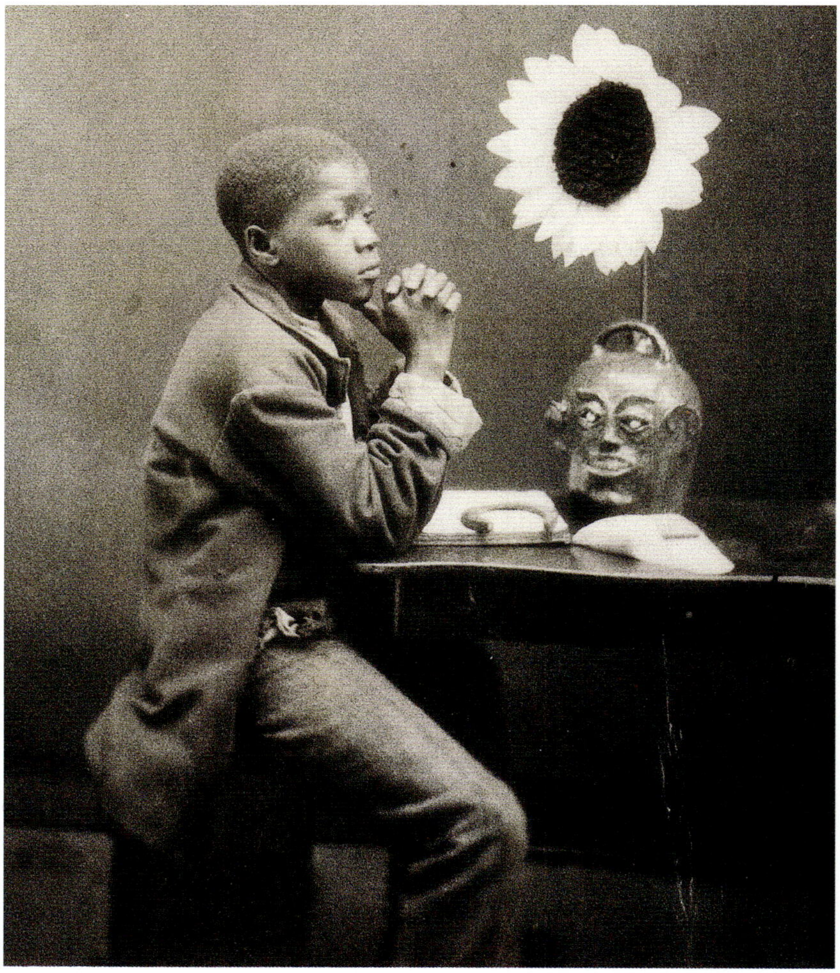

Figure 40 "An Aesthetic Darkey," from a series of stereoscopic cards by photographer J. A. Palmer of Aiken, South Carolina, 1882. (Courtesy, J. Garrison Stradling.) This earliest known published image of an American face jug has a monkey form and probably was made by a local (Edgefield District) African-American potter.

ticed by their ancestors in the Dark Continent."[51] Barber says nothing, however, of the makers' motivations. Sixty years later, Yale University art historian Robert Farris Thompson advanced Barber's suggestion of African origins, arguing that later white face-jug makers such as Cheever Meaders appropriated the "Afro-Carolinian" tradition.[52]

Germany and Africa, then, are two possible sources for American face jugs, with a third possibility that they arose independent of any Old World influence. Can we come any closer to resolving this historical dilemma? Anthropomorphic clay vessels were indeed made in West Africa (the chief source area of the Atlantic slave trade), perhaps early enough for the idea to be brought by slaves. The Yungur of Nigeria, for example, made portrait pots called *wiiso* to honor ancestral spirits at shrines,[53] and the Mambila of Cameroon made similar figural vessels (fig. 41). The angry expressions of some Afro-Carolinian face vessels, which could be interpreted as a nonver-

Figure 41 Figurative jar, Mambila culture, Cameroon, twentieth century. Earthenware with polychrome decoration. H. 20". (Photo, courtesy Douglas Dawson.)

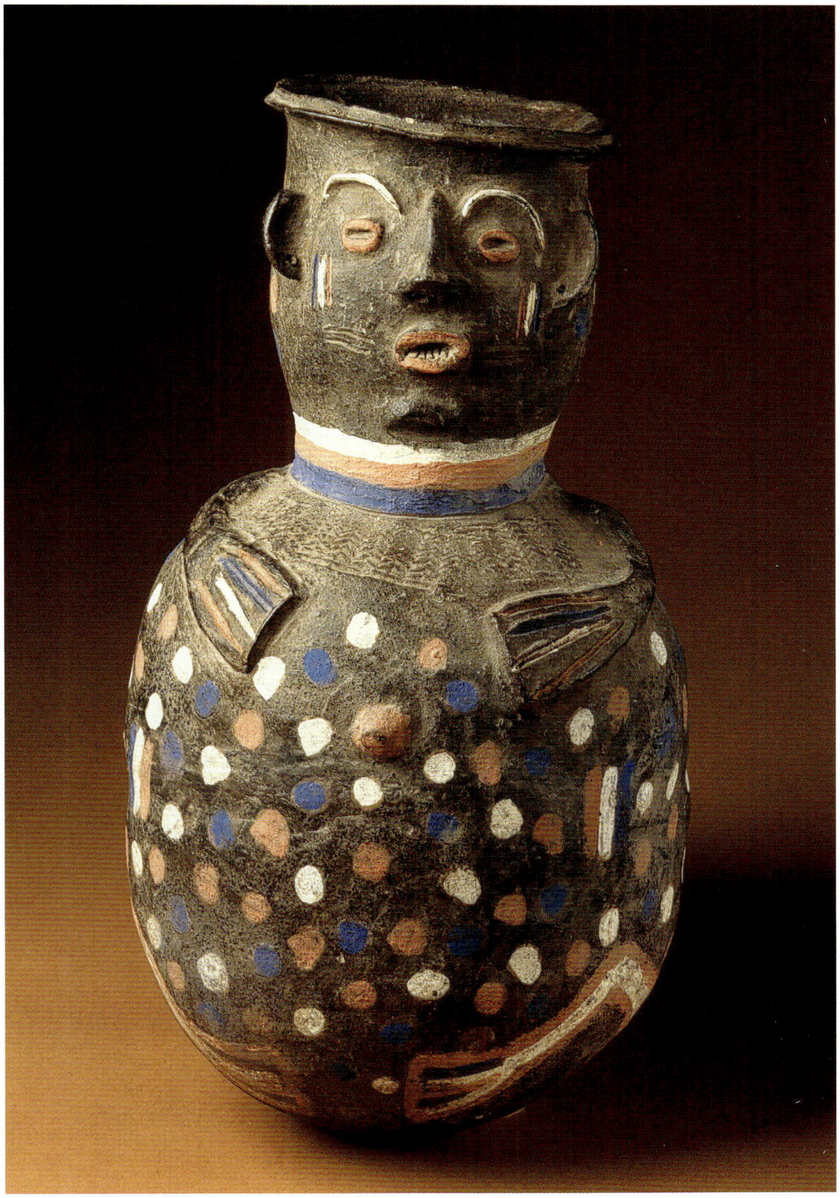

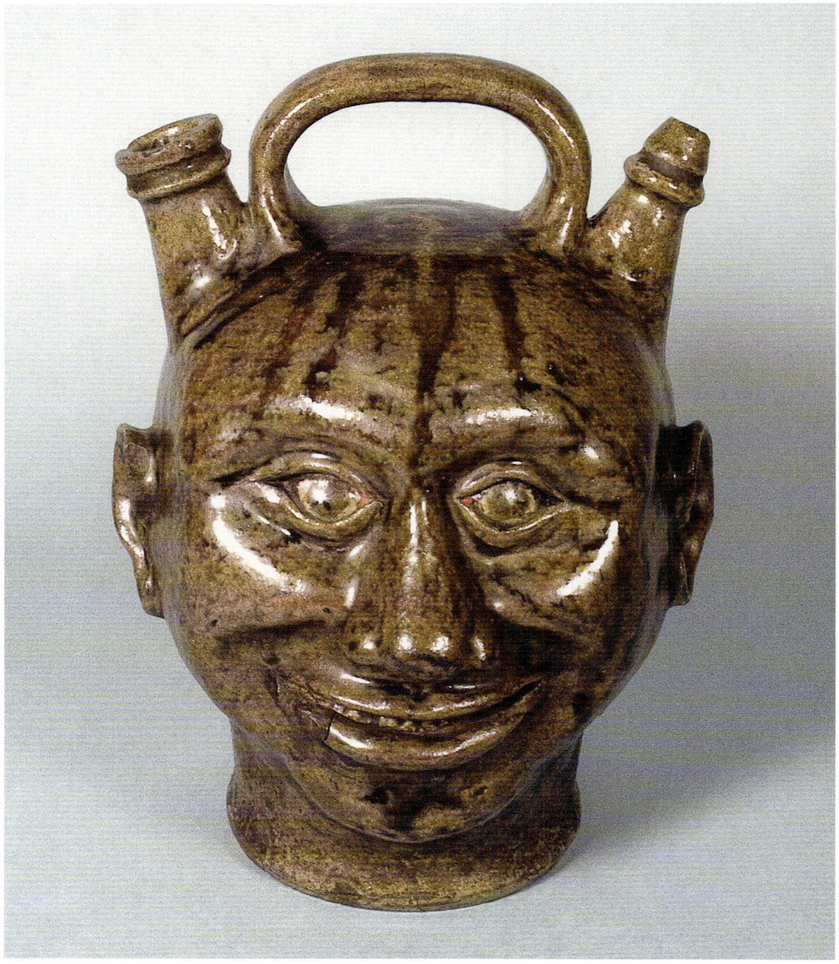

Figure 42 Monkey-form face jug, Edgefield District, South Carolina, ca. 1850. Ash-glazed stoneware. H. 11". Mark: stamped "CHANDLER / MAKER." (Private collection; photo, McKissick Museum, University of South Carolina.) The happy countenance on this jug contrasts with the angry ones of a decade later by slave potters in the same area.

bal protest against enslavement, and a few tantalizing hints that these vessels may have been used in magico-religious or mortuary practices, at least suggest that they had a meaning distinct from face jugs by white potters. This raises the question whether white potters in the South made face vessels as early as the slave-made ones.

In 1995 a piece in a private collection surfaced that addresses this question (fig. 42). An alkaline-glazed, happy-faced jug with the monkey form, it is stamped "CHANDLER / MAKER," the mark of Thomas Chandler, a white potter who worked in Edgefield District from 1838 to 1852.[54] Made about 1850, it preceded the slave-made face vessels from the Davies workshop by over a decade. Before moving to South Carolina, Virginia-born Chandler may have worked as a potter in New York State; it is remotely

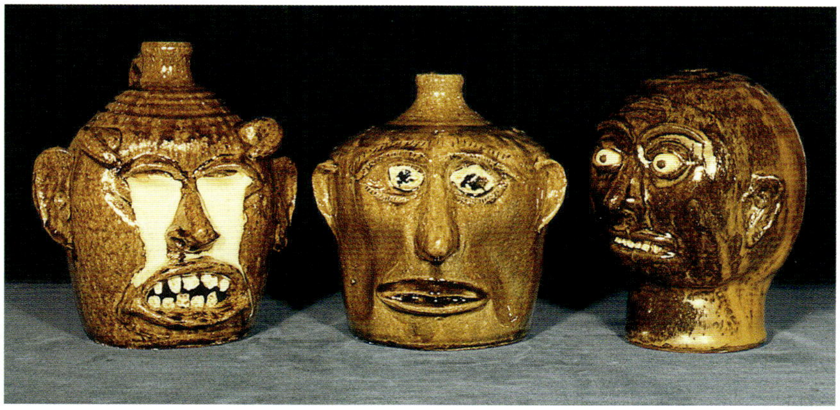

Figure 43 Face jugs and wheel-thrown "wig stand," Lanier Meaders, White County, Georgia, 1968–1974. Alkaline-glazed stoneware. H. (left to right) 9¾", 8¾", 9". (Author's collection). With these items Meaders displayed his sculptural skill, creativity, and sense of humor.

possible that in his Northern sojourn he met and learned of face vessels from one of the Remmeys.[55] Did Chandler then introduce the concept to Edgefield slave potters, or were they working in a separate, perhaps African-based, tradition? We now know that the face jugs of Cheever and Lanier Meaders were part of a continuous Anglo-Southern tradition, but whether ultimately inspired by slave-made examples, as Thompson suggests, might never be learned (fig. 43).

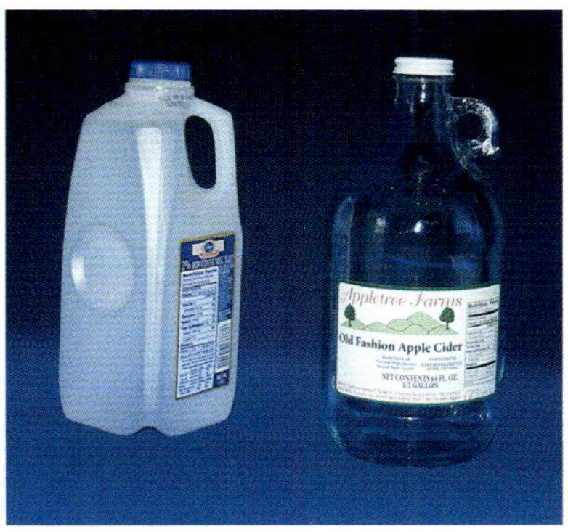

Figure 44 Examples of modern functional jugs, made of glass and plastic rather than clay.

The Jug Today

Jugs are still indispensable for keeping liquids, from bleach to milk and larger volumes of wine. Now, though, they rarely are made of clay; glass and plastic containers are less expensive to produce, and the latter has the further advantage of being less breakable (fig. 44). In recent years, a few whiskey companies have used mold-made clay jugs to reinforce the old-fashioned image of their product, and school-trained studio potters may occasionally throw a jug to demonstrate that they are not embarrassed to make a useful form despite the current trend of pottery as Art.[56]

In the 1980s and 1990s it seemed as though every potter in the South, folk and otherwise, was trying his or her hand at face jugs, which had become something of a regional art icon (for which Lanier Meaders deserves

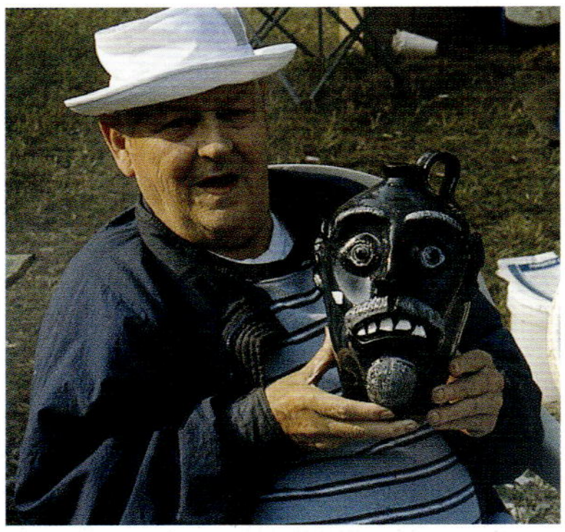

Figure 45 Fifth-generation potter Bobby Ferguson of Gillsville, Georgia, with one of his face jugs in 2004, a year before his death. An ancestor of his may have introduced face jugs to north Georgia from antebellum Edgefield District, South Carolina.

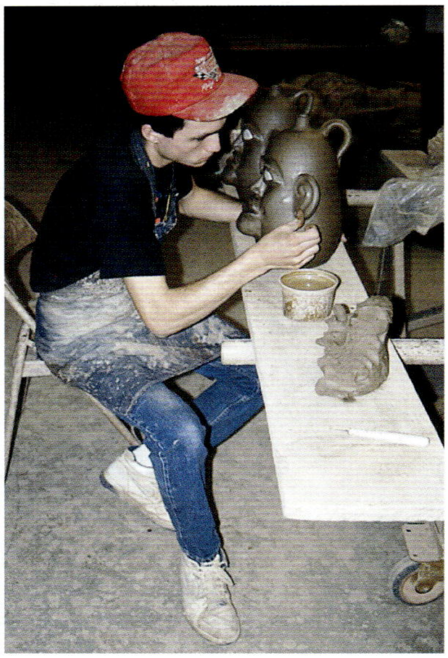

Figure 46 Sixth-generation potter Matthew Hewell of Gillsville, Georgia, sculpts face jugs, 1992. Lanier Meaders's success influenced Matthew's father, Chester, to revive the Hewell face-jug tradition.

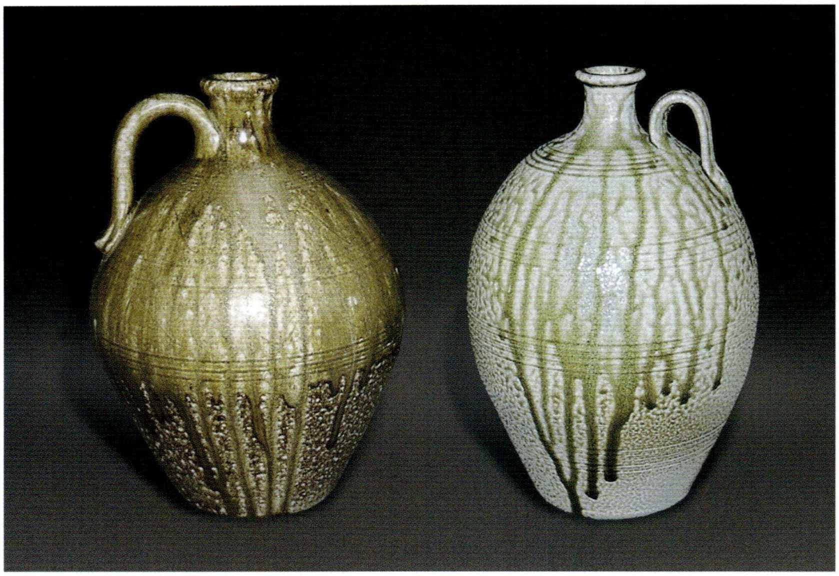

Figure 47 Jugs, late twentieth century. Salt-glazed stoneware. H. 11½". *Left:* David Stuempfle. *Right:* Ben Owen III, Seagrove, North Carolina. (Author's collection.) Both vessels have pronounced lips, tooled decoration, and drips caused by dumping the salt directly over the pots through holes in the kiln arch, although each has a very different handle, and the jug on the left has an ash underglaze. Seagrove is a community of about one hundred shops, a dozen of which are rooted in local tradition.

much of the credit).[57] Today, however, the handcrafting of jugs, with or without faces, is largely the prerogative of a small number of traditionally trained potters, and the domain of a form that once dominated utilitarian American ceramics has shrunk to rural North Carolina, Georgia, and Alabama (figs. 45–47). The market has become one of collectors and home decorators, and the everyday uses of these fluid vessels have all but dried up.

1. Green or brown alkaline glazes used on folk stoneware of the lower South may have been developed about 1810 by Dr. Abner Landrum of Edgefield District, South Carolina, inspired by Jesuit missionary Père d'Entrecolles's account of Chinese high-firing woodash- and lime-based glazes, first published in 1735. The new glazes were then carried westward by Edgefield-trained potters, reaching Texas by 1850. See John A. Burrison, *Brothers in Clay: The Story of Georgia Folk Pottery* (Athens: University of Georgia Press, 1983), pp. 58–62; Daisy Wade Bridges, *Ash Glaze Traditions in Ancient China and the American South* (Robbins, N.C.: Southern Folk

Pottery Collectors Society, 1997); Georgeanna H. Greer, *American Stonewares, The Art and Craft of Utilitarian Potters* (Exton, Pa.: Schiffer Publishing, 1981), pp. 202–10; and Cinda K. Baldwin, *Great and Noble Jar: Traditional Stoneware of South Carolina* (Athens: University of Georgia Press, 1993), pp. 16–20, 144–47.

2. Bengt Olsson, *Memphis Blues and Jug Bands* (London: Studio Vista, 1970); Laurie Wright and Fred Cox, *The Jug Bands of Louisville* (Chigwell, Essex, Eng.: Storyville Publications, 1993).

3. According to the *Oxford English Dictionary,* the word *jug* was in use by 1538 but its etymology is uncertain. Josiah Wedgwood suggested that it came from the pet name for Joan or Judith, but that is inadequate as an explanation.

4. This quote is first attributed to Shaw in "Picturesque Speech and Patter," *Reader's Digest* 41 (November 1942): 100, but I have not been able to trace it to a specific work of his.

5. E.g., R. K. Henrywood, *An Illustrated Guide to British Jugs: From Medieval Times to the Twentieth Century* (Woodbridge, Suffolk, Eng.: Antique Collectors' Club, 1997).

6. "The Brown Jug" was composed in 1761 by Francis Fawkes. It describes the corpse of corpulent toper Toby Filpot (inspired by real-life big drinker and fellow Yorkshireman Henry Elwes, who died that same year) moldering into clay, to be recycled by a potter into "this brown jug, / Now sacred to friendship, to mirth and mild ale." The poem, along with printed images, is thought to have inspired, in turn, Staffordshire's Toby character mug.

"The Little Brown Jug" was composed by Philadelphian Joseph Eastburn Winner in 1869, later becoming a Glenn Miller jazz standard: "Me and my wife live all alone / In a little log hut we call our own; / She loves gin and I love rum, / And don't we have a lot of fun! / Ha, ha, ha, you and me, / Little brown jug, don't I love thee!" The narrator's diction suggests that Winner intended the song as a spoof of Quaker support for temperance.

7. Henry Hodges, *Technology in the Ancient World* (New York: Alfred A. Knopf, 1972), pp. 114–17, 156–57, 179–81, 185–87.

8. Charles A. Burney, *The Ancient Near East* (Ithaca, N.Y.: Cornell University Press, 1977), p. 107; Ruth Amiran, *Ancient Pottery of the Holy Land: From Its Beginnings in the Neolithic Period to the End of the Iron Age* ([New Brunswick, N.J.]: Rutgers University Press, 1970), pp. 58ff.; *World Ceramics,* edited by Robert J. Charleston (London and New York: Paul Hamlyn, 1968), p. 23 (no. 32); Trudy S. Kawami, *Ancient Iranian Ceramics from the Arthur M. Sackler Collections* (New York: Arthur M. Sackler Foundation and Harry N. Abrams, 1992), pp. 94, 110, 217–18; A. D. Lacy, *Greek Pottery in the Bronze Age* (London: Methuen, 1967), pp. 75 (e), 178 (a); James Whitley, *Style and Society in Dark Age Greece: The Changing Face of a Pre-Literate Society 1100–700 B.C.* (Cambridge: Cambridge University Press, 1991), pl. 11; *Meisterwerke Altägyptischer Keramik: 5000 Jahre Kunst und Kunsthandwerk aus Ton und Fayence* (Höhr-Grenzhausen: Rastal-Haus, 1978), pp. 179 (no. 298), 182 (no. 311).

9. John W. Hayes, *Handbook of Mediterranean Roman Pottery* (Norman: University of Oklahoma Press, 1997), pp. 22 (pl. 5 left), 25 (pl. 7), 55 (fig. 20, no. 9), 56 (fig. 21, no. 3), 61 (fig. 25, no. 7), 75 (pl. 28), 93 (pl. 39, top); Val Rigby and Ian Freestone, "Ceramic Changes in Late Iron Age Britain," in *Pottery in the Making: Ceramic Traditions,* edited by Ian Freestone and David Gaimster (Washington, D.C.: Smithsonian Institution Press, 1997), p. 61 (fig. 6); Vivien G. Swan, *Pottery in Roman Britain* (Princes Risborough: Shire Publications, 1975), pp. 38 (pl. 30 center), 43 (no. 21); K. J. Barton, *Pottery in England from 3500 B.C.–A.D. 1730* (Newton Abbot: David and Charles, 1975), p. 96 (nos. 26–27); Ivor Noël Hume, *If These Pots Could Talk: Collecting 2,000 Years of British Household Pottery* (Hanover, N.H.: University Press of New England for the Chipstone Foundation, 2001), pp. 20, 28. Greek and Roman potters also made amphoras, related to jugs but often larger and with two loop handles and an elongated, tapering body.

10. *World Ceramics,* p. 116 (no. 339b); Bernhard Beckmann, "The Main Types of the First Four Production Periods of Siegburg Pottery," in *Medieval Pottery from Excavations,* edited by Vera I. Evison et al. (New York: St. Martin's, 1974), pp. 194, 209–10 (nos. 44–52). The German language adds to semantic confusion: while certain vessel types are given very specific names, what Americans call a jug is referred to variously as a *Krug, Flasche,* or *Pulle*. The latter two terms translate as "bottle," while the more generic *Krug* can mean "jug," "tankard or mug," "jar or crock," and "pitcher."

11. E.g., David Gaimster, *German Stoneware 1200–1900: Archaeology and Cultural History* (London: British Museum Press, 1997), pp. 85ff.; Peter Seewaldt, *Rheinisches Steinzeug* (Trier: Rheinischen Landesmuseums, 1990), pp. 37ff.

12. Called *Bellarmines* by English writers, these anthropomorphized jugs were thought to be caricatures of Cardinal Roberto Bellarmino (1542–1621), an opponent of Protestantism in

central Europe. However, the earliest dated example was made in 1550, while some with hand-modeled faces predate 1500 (Anthony Thwaite, "The Chronology of the Bellarmine Jug," *Connoisseur* 182 [April 1973]: 255–62; Margaret Thomas, *German Stoneware: A Catalogue of the Frank Thomas Collection of German Stoneware* [Woodbridge, Suffolk, Eng.: Antique Collectors' Club, 2003]). Another theory is that the bearded face represents the Wild Man of the Woods, a northern European folklore figure (Gaimster, *German Stoneware,* p. 209). In England, graybeard jugs were recycled as charms to ward off witches, for which see Ralph Merrifield, *The Archaeology of Ritual and Magic* (New York: New Amsterdam Books, 1988), pp. 163–75, and Noël Hume, *If These Pots Could Talk,* pp. 118–26. Some sixteenth-century *Bartmanns* have a wider mouth than jugs as we are defining them herein, and probably were for drinking, not storage (e.g., Gaimster, *German Stoneware,* p. 67 [no. 3.21] and color pl. 12).

13. Wilhelm Elling's *Steinzeug aus Stadtlohn und Vreden* (Vreden: Hamaland-Museum, 1994), for example, documents a tradition of salt-glazed jug-making near the Dutch border, west of Münster, continuous from the seventeenth century to the present.

14. J. N. L. Myres, *Anglo-Saxon Pottery and the Settlement of England* (Oxford: Clarendon Press, 1969); Cathy Haith, "Pottery in Early Anglo-Saxon England," in Freestone and Gaimster, eds., *Pottery in the Making,* pp. 146–51. Globular-bodied, narrow-necked bottles—jugs minus a handle—from Kent burial sites, if English-made, may mark the reintroduction of the potter's wheel from the Continent as early as the seventh century; see Vera I. Evison, "The Asthall Type of Bottle," in Evison et al., eds., *Medieval Pottery from Excavations,* pp. 77–92 and pls. II–IV.

15. Barton, *Pottery in England,* p. 109; Michael R. McCarthy and Catherine M. Brooks, *Medieval Pottery in Britain A.D. 900–1600* (Leicester, Eng.: Leicester University Press, 1988), pp. 231 (no. 667), 248 (no. 782), 261 (no. 861), 278 (no. 983), 279 (no. 985), 383 (no. 1615), 384 (no. 1623), 397 (nos. 1700, 1706).

16. E.g., Noël Hume, *If These Pots Could Talk,* p. 166. For English non-delft earthenware jugs of the 1600s, see R. Coleman-Smith and T. Pearson, *Excavations in the Donyatt [Somerset] Potteries* (Chichester: Phillimore, 1988), pp. 80, 148, 156, showing jugs that are actually called jugs!

17. Peter C. D. Brears, *The English Country Pottery: Its History and Techniques* (Rutland, Vt.: Charles E. Tuttle, 1971), p. 70; Andrew McGarva, *Country Pottery: The Traditional Earthenware of Britain* (London: A and C Black, 2000), pp. 24–26, 29; John Manwaring Baines, *Sussex Pottery* (Brighton: Fisher Publications, 1980), pp. 15, 45. Puzzle and West Country "harvest jugs" were not jugs by our definition but drinking vessels, the former a trick mug, the latter essentially a pitcher.

18. Gaimster, *German Stoneware,* pp. 309–17; Jonathan Horne, "John Dwight, 'The Master Potter' of Fulham," *Antiques* 143 (April 1993): 562–71; Adrian Oswald et al., *English Brown Stoneware 1670–1900* (London: Faber and Faber, 1982); Robin Hildyard, *Browne Muggs: English Brown Stoneware* (London: Victoria and Albert Museum, 1985).

19. One way of determining when the American meaning arose is to find early illustrations of jugs associated with the word, for example in the potters' price lists reproduced in: Harold F. Guilland, *Early American Folk Pottery* (Philadelphia: Chilton, 1971), p. 41 (Clarkson Crolius pottery, Manhattan, 1809); William C. Ketchum Jr., *Early Potters and Potteries of New York State* (New York: Funk and Wagnalls, 1970), p. 179 (John Burger pottery, Rochester, 1857); and Lura Woodside Watkins, *Early New England Potters and Their Wares* (Cambridge, Mass.: Harvard University Press, 1950), pp. 147 (Norton pottery, Bennington, Vermont, 1856), 149 and 151 (Farrar pottery, Fairfax, Vermont, 1840 and 1851); as well as in literary publications such as "Ham Rachel, of Alabama," in Hardin E. Taliaferro ("Skitt"), *Fisher's River (North Carolina) Scenes and Characters* (New York: Harper and Brothers, 1859), pp. 266–67.

20. Diana Stradling and J. Garrison Stradling, in editing *The Art of the Potter* (New York: Main Street/Universe, 1977) from *Antiques,* organized the articles on American redware into two sections, "The Germanic Influence" and "The English Influence."

21. David Hackett Fischer, *Albion's Seed: Four British Folkways in America* (New York: Oxford University Press, 1989); Sarah Peabody Turnbaugh, "Imitation, Innovation, and Permutation: The Americanization of Bay Colony Lead-Glazed Redwares," in *Domestic Pottery of the Northeastern United States, 1625–1850,* edited by Sarah Peabody Turnbaugh (Orlando, Fla.: Academic Press, 1985), pp. 209–28.

22. Watkins, *Early New England Potters and Their Wares,* p. 235.

23. Ibid., figs. 13, 37, 39–41, 73, 84; Guilland, *Early American Folk Pottery,* color pl. [11], pp. 155, 228; Stradling and Stradling, *Art of the Potter,* pp. 43, 47, 51, 57, 65, 66–69; *Regional Aspects of American Folk Pottery,* introduction by William C. Ketchum Jr., exh. cat. (York, Pa.:

Historical Society of York County, 1974), figs. 1, 3, 56, 57, 59, 60; Brian Cullity, *Slipped and Glazed: Regional American Redware* (Hyannis, Mass.: Patriot Press for Heritage Plantation of Sandwich, 1991), pp. 31 (no. 48), 32 (no. 51), 37 (no. 63), 50 (no. 99), color pls. 1, 2, 4, and 6; Arthur E. James, *The Potters and Potteries of Chester County, Pennsylvania* (Exton, Pa.: Schiffer, 1978), p. 88; Jeannette Lasansky, *Central Pennsylvania Redware Pottery 1780–1904* (University Park: Pennsylvania State University Press for Keystone Books, 1979), p. 42; Anthony W. Butera Jr., "'Informed Conjecture': Collecting Long Island Redware," in *Ceramics in America,* edited by Robert Hunter (Hanover, N.H.: University Press of New England for the Chipstone Foundation, 2003), pp. 214, 219; Barbara H. Magid and Bernard K. Means, "In the Philadelphia Style: The Pottery of Henry Piercy," in *Ceramics in America,* edited by Robert Hunter (Hanover, N.H.: University Press of New England for the Chipstone Foundation, 2003), p. 56 (fig. 25); Don Horvath and Richard Duez, "The Potters and Pottery of Morgan's Town, Virginia: The Earthenware Years, Circa 1796–1854," in *Ceramics in America,* edited by Robert Hunter (Hanover, N.H.: University Press of New England for the Chipstone Foundation, 2004), pp. 103 (fig. 5), 113–14 (figs. 28–32); H. E. Comstock, *The Pottery of the Shenandoah Valley Region* (Winston-Salem: Museum of Early Southern Decorative Arts, 1994), pp. 72 (figs. 3.14–.15), 89 (fig. 4.13), 104 (fig. 4.60), 119 (fig. 4.109), 204 (fig. 5.14); John Bivins Jr., *The Moravian Potters in North Carolina* (Chapel Hill: University of North Carolina Press for Old Salem, 1972), p. 124; Charles G. Zug III, *Turners and Burners: The Folk Potters of North Carolina* (Chapel Hill: University of North Carolina Press, 1986), color pl. 3; *Two Centuries of Potters: A Catawba Valley Tradition,* edited by Bill Beam et al. (Lincolnton, N.C.: Lincoln County Historical Association, 1999), front cover and p. 14 (fig. 1). Perhaps due to its emphasis on decorated pieces, there are no jugs in the catalog of the finest public display of Pennsylvania redware, Beatrice B. Garvan's *The Pennsylvania German Collection* (Philadelphia: Philadelphia Museum of Art, 1982).

24. Early American stoneware potters working largely in the German tradition include members of the Crolius and Remmey families, Thomas Commereau, and David Morgan of Manhattan, James Morgan of South Amboy and Old Bridge, New Jersey, and Jonathan Fenton of Boston. Those working more in the English tradition include William Rogers of Yorktown, Virginia, Frederick Carpenter of Charlestown, Massachusetts, and Branch Green of Philadelphia. English-born Anthony Duche of colonial Philadelphia made stoneware in both the German and English styles (Robert L. Giannini III, "Anthony Duche, Sr., Potter and Merchant of Philadelphia," *Antiques* 119 [January 1981]: 198–203).

25. Guilland, *Early American Folk Pottery,* pp. 237, 249, 252, 255–56, 260; Greer, *American Stonewares,* pp. 155–56, 182, 225; Donald Blake Webster, *Decorated Stoneware Pottery of North America* (Rutland, Vt.: Charles E. Tuttle, 1971), pp. 65–66, 98, 160; Stradling and Stradling, *Art of the Potter,* pp. 81–90, 106–7, 114–16, 119–22; James R. Mitchell, "The Potters of Cheesequake, New Jersey," in *Ceramics in America,* Winterthur Conference Report 1972, edited by Ian M. G. Quimby (Charlottesville: University Press of Virginia, 1973), pp. 319–38; Comstock, *Pottery of the Shenandoah Valley Region,* p. 318 (fig. 6.6); Norman F. Barka, "Archaeology of a Colonial Pottery Factory: The Kilns and Ceramics of the 'Poor Potter' of Yorktown," in *Ceramics in America,* edited by Robert Hunter (Hanover, N.H.: University Press of New England for the Chipstone Foundation, 2004), pp. 16, 32; Quincy J. Scarborough Jr., *North Carolina Decorated Stoneware: The Webster School of Folk Potters* (Fayetteville, N.C.: Scarborough Press, 1986).

26. E.g., Baldwin, *Great and Noble Jar,* pp. 35, 37, 39, 40, 45, 53, 100, 147, 150, 153, 154–56, 159, 169, color pls. 2–4, 6, 12; Greer, *American Stonewares,* pp. 170, 173. The double-collared neck sometimes migrated west on jugs made by Edgefield-trained potters (e.g., Baldwin, *Great and Noble Jar,* p. 64, by J. S. Nash of Texas; Burrison, *Brothers in Clay,* p. 123, attributed to Cyrus Cogburn of Georgia).

27. Greer, *American Stonewares,* p. 76; Guilland, *Early American Folk Pottery,* p. 83; Cornelius Osgood, *The Jug and Related Stoneware of Bennington* (Rutland, Vt.: Charles E. Tuttle, 1971), p. 116.

28. A glaring historical inaccuracy of certain films portraying the colonial or frontier era is the inclusion of these late-nineteenth-century shouldered jugs. The earliest examples are two-toned, with salt-glazed wall and Albany slip-glazed shoulder, neck, and interior, as the stacking dish blocked the salt glazing. Later examples, often molded on a jolly machine rather than thrown, either carried on the two-tone effect with Bristol glaze replacing the salt or were fully Bristol glazed. For a sample of such jugs, see Lyndon C. Viel, *The Clay Giants: The Stoneware of Red Wing, Goodhue County, Minnesota,* 3 vols. (Des Moines, Ia.: Wallace-Homestead, 1977–87).

29. John McGuffin, *In Praise of Poteen* (Belfast: Appletree Press, 1978); Joseph Earl Dabney, *Mountain Spirits: A Chronicle of Corn Whiskey from King James' Ulster Plantation to America's Appalachians and the Moonshine Life* (New York: Scribner, 1974).

30. Quoted in Burrison, *Brothers in Clay,* p. 132.

31. Jim Broom in North Carolina and Horace Brown and Norman Smith in Alabama, besides the previously mentioned "Jughead" for Georgia's Lanier Meaders.

32. Shelby County, Alabama; Upson and Pike Counties, Georgia; Catawba, Lincoln, and Buncombe Counties, North Carolina; Greenville County, South Carolina; and White County, Tennessee.

33. E.g., Stradling and Stradling, *Art of the Potter,* p. 94 (twenty gallons, Vermont); Greer, *American Stonewares,* p. 107 (twelve gallons, Ohio). Stoneware "fountain" and keg forms also were used as coolers.

34. Earthenware "cider jars" were a specialty of Halifax-area, West Yorkshire, potters. They typically have a spigot hole in the lower wall; some were lead-glazed only halfway down the outside so the temperature differential would hasten fermentation of the hard cider. The same basic shape in brown salt-glazed stoneware, as made in the Liverpool area, apparently was used in transatlantic trade. Dating from the late eighteenth through the nineteenth century, these large English jugs rarely have been published, but see Graham Wilkinson, *A History of Local Potteries* (Bradford, W. Yorkshire, Eng.: Bradford Art Galleries and Museums, 1981), p. 11 (B); Zug, *Turners and Burners,* p. 31; Stradling and Stradling, *Art of the Potter,* p. 154 (center); John A. Burrison, *Handed On: Folk Crafts in Southern Life* (Atlanta: Atlanta Historical Society, 1993), p. 41 (no. 4).

35. E.g., Stradling and Stradling, *Art of the Potter,* p. 131 (fig. 1, New York); Greer, *American Stonewares,* p. 107 (Ohio); Webster, *Decorated Stoneware Pottery of North America,* pp. 67 and 165 (New York State), 94 (Ohio); Baldwin, *Great and Noble Jar,* pp. 172, 149 (South Carolina).

36. Anne S. McPherson, "'That Article of Household Furniture Peculiar to Earlier Days in the South': Sugar Chests in Middle Tennessee and Central Kentucky, 1800–1835," *Journal of Early Southern Decorative Arts* 23 (winter 1997): 1–65; Robert Hicks and Benjamin H. Caldwell Jr., "A Short History of the Tennessee Sugar Chest," *Antiques* 164 (September 2003): 128–33; *Neat Pieces: The Plain-Style Furniture of Nineteenth Century Georgia,* exh. cat. (Atlanta: Atlanta Historical Society, 1983), p. 126.

37. Greer, *American Stonewares,* pp. 42–43, 178; Baldwin, *Great and Noble Jar,* pp. 37, 45, 49, 150–51, 153, 155–56, 169, color pls. 3–4; Zug, *Turners and Burners,* pp. 46, 48, 77, 87, 307, color pls. 7, 16; Beam, *Two Centuries of Potters: A Catawba Valley Tradition,* pp. 14–15, 33, 36, 39, 75, 80–81, 87, 90; Burrison, *Brothers in Clay,* pp. 23, 65, 123–25, 127, 145, 175, 268.

38. Osgood, *The Jug,* pp. 96, 106.

39. Burrison, *Brothers in Clay,* color pl. 6 and pp. 66 (pl. 35), 116–17, 255; Zug, *Turners and Burners,* p. 368 (fig. 12-14).

40. Lacy, *Greek Pottery in the Bronze Age,* p. 218 (fig. 90d), a Mycenaean example dating to 1200 B.C.

41. See, e.g., Otto Von Falke, *Das Rheinische Steinzeug,* 2 vols. (Osnabrück: Otto Zeller, 1977), 1: 89, and 2: 59–61, 109, for examples dating to the sixteenth and seventeenth centuries.

42. Zug, *Turners and Burners,* pp. 375–77; "Lin Craven Making a Ring Jug," in John A. Burrison, "Folk Pottery," Folklife Section, *New Georgia Encyclopedia* (video clip, available online; www.georgiaencyclopedia.org).

43. Burrison, *Brothers in Clay,* pp. 184–85.

44. Lindsey King Laub, *Evolution of a Potter: Conversations with Bill Gordy* (Cartersville, Ga.: Bartow History Center, 1992), cover, color pl. 10.

45. See Lacy, *Greek Pottery in the Bronze Age,* p. 259 (fig. 102e), for an example of 1800 B.C. Museo del Botijo in Toral de los Guzmanes, León; Museo del Botijo in Villena, Valencia; Museu del Càntir d' Argentona, Cataluña. For recent Spanish jugs, "monkey" and otherwise, see J. Llorens Artigas and J. Corredor-Matheos, *Spanish Folk Ceramics of Today* (New York: Watson-Guptill, 1974).

46. E.g., Guilland, *Early American Folk Pottery,* p. 193 (left, Vermont); Cullity, *Slipped and Glazed,* p. 26 (no. 34, Massachusetts); Lasansky, *Central Pennsylvania Redware Pottery,* p. 43; Greer, *American Stonewares,* pp. 162 and 196 (Ohio). Nigel Barley, *Smashing Pots: Works of Clay from Africa* (Washington, D.C.: Smithsonian Institution Press, 1994), pp. 27, 50, 118, 122; John Michael Vlach, *The Afro-American Tradition in Decorative Arts* (Athens: University of Georgia Press, 1990), pp. 86–90. Monkey jugs were not made in England, with the exception of the

"Sarum kettle" manufactured by Doulton of Lambeth about 1900, based on an African example in the Salisbury Museum.

47. Barton, *Pottery in England,* p. 99; Griselda Lewis, *A Collector's History of English Pottery* (New York: Viking Press, 1969), pp. 15 (fig. 9), 18 (figs. 18, 20, 21), 57 (fig. 106), 123 (fig. 227); McCarthy and Brooks, *Medieval Pottery in Britain,* pp. 229, 269, 337, 367; Jacqueline Pearce and Alan Vince, *Surrey Whitewares* (London: London and Middlesex Archaeological Society, 1988), pp. 24, 78, 127, 129–31; Desmond Eyles, *"Good Sir Toby": The Story of Toby Jugs and Character Jugs through the Ages* (London: Doulton, 1955); Vic Schuler, *Collecting British Toby Jugs . . . 1780 to the Present Day* (London: Francis Joseph, 1997); Noël Hume, *If These Pots Could Talk,* p. 310.

48. Roy S. Dickens Jr., *Of Sky and Earth: Art of the Early Southeastern Indians* (Atlanta, Ga.: High Museum of Art, 1982), pp. 28, 52–53, 58, 82–84; Michael J. O'Brien, *Cat Monsters and Head Pots: The Archaeology of Missouri's Pemiscot Bayou* (Columbia: University of Missouri Press, 1994), pp. xxxiv, xxxix.

49. A face pitcher attributed to Henry Remmey Jr. is dated 1838; a two-faced monkey jug likely by him or his son Richard is dated 1858. Both are salt-glazed stoneware, with cobalt-blue–highlighted facial hair as seen on some German graybeards. For the Remmeys, see Stradling and Stradling, *Art of the Potter,* pp. 114–18; Ketchum, *Early Potters and Potteries of New York State,* pp. 30–32; Susan H. Myers, *Handcraft to Industry: Philadelphia Ceramics in the First Half of the Nineteenth Century* (Washington, D.C.: Smithsonian Institution Press, 1980), pp. 22–25.

50. For discussions of antebellum and later African-American face vessels from South Carolina, see Vlach, *Afro-American Tradition in Decorative Arts,* pp. 81–92, and Baldwin, *Great and Noble Jar,* pp. 79–87, 108–9, color pls. 12–13.

51. Edwin AtLee Barber, *Pottery and Porcelain of the United States* (New York: G. P. Putnam, 1909), p. 466. Stereoscopic cards issued in 1882 by photographer J. A. Palmer of Aiken, South Carolina (in Edgefield District) depict a Negro boy and girl pondering a monkey-form face jug, the earliest known illustrations of a locally made face vessel; titled "An Aesthetic Darkey," the images likely were inspired by W. H. Beard's painting, *The Aesthetic Monkey,* published that year in *Harper's Weekly* (*Making Faces: Southern Face Vessels from 1840–1990,* exh. cat. [Columbia: McKissick Museum, University of South Carolina, 2000], pp. 15–16).

52. Robert Farris Thompson, "African Influence on the Art of the United States," in *Black Studies in the University,* edited by Armstead L. Robinson et al. (New York: Bantam Books, 1969), p. 141.

53. Marla C. Berns, "Pots as People: Yungur Ancestral Portraits," *African Arts* 23 (July 1990): 50–60 and front cover. The Mangbetu of Zaire made juglike figural pots with vertical loop handles, but too late (late 1800s) to have influenced South Carolina slaves (Barley, *Smashing Pots,* pp. 148–49).

54. *Making Faces,* pp. 6–7; Baldwin, *Great and Noble Jar,* pp. 47, 51–54.

55. This scenario is not so far-fetched; a later potter, Charles Decker, began his career at the Remmey Pottery in Philadelphia before establishing northeast Tennessee's Keystone Pottery in 1871, where he and his son William made Remmey-style face jugs (*The Pottery of Charles F. Decker: A Life Well Made* [Jonesborough, Tenn.: Jonesborough/Washington County History Museum, 2004], pp. 52–53, 58).

56. A review of recent issues of *Ceramics Monthly, Studio Potter, Ceramics Art and Perception,* and *American Craft* suggests that the jug is not among the forms privileged by contemporary studio potters.

57. *Making Faces*; William W. Ivey, *North Carolina and Southern Folk Pottery* (Seagrove, N.C.: Museum of North Carolina Traditional Pottery, 1992); Robert C. Lock, *The Traditional Potters of Seagrove, North Carolina* (Greensboro, N.C.: Antiques and Collectibles Press, 1994), pp. 184–90; Barry G. Huffman, *Catawba Clay: Contemporary Southern Face Jug Makers* (Hickory, N.C.: A. W. Huffman, 1997); Michael A. Crocker and W. Newton Crouch Jr., *The Folk Pottery of Cheever, Arie, and Lanier Meaders: A Pictorial Legacy* (Griffin, Ga.: C and C Productions, 1994).

Figure 1 Face jug fragment, Miles Mill, Edgefield, South Carolina, ca. 1867–1872. Alkaline-glazed stoneware. H. 5". (All fragments courtesy Georgia Archaeological Institute; photo, Gavin Ashworth.)

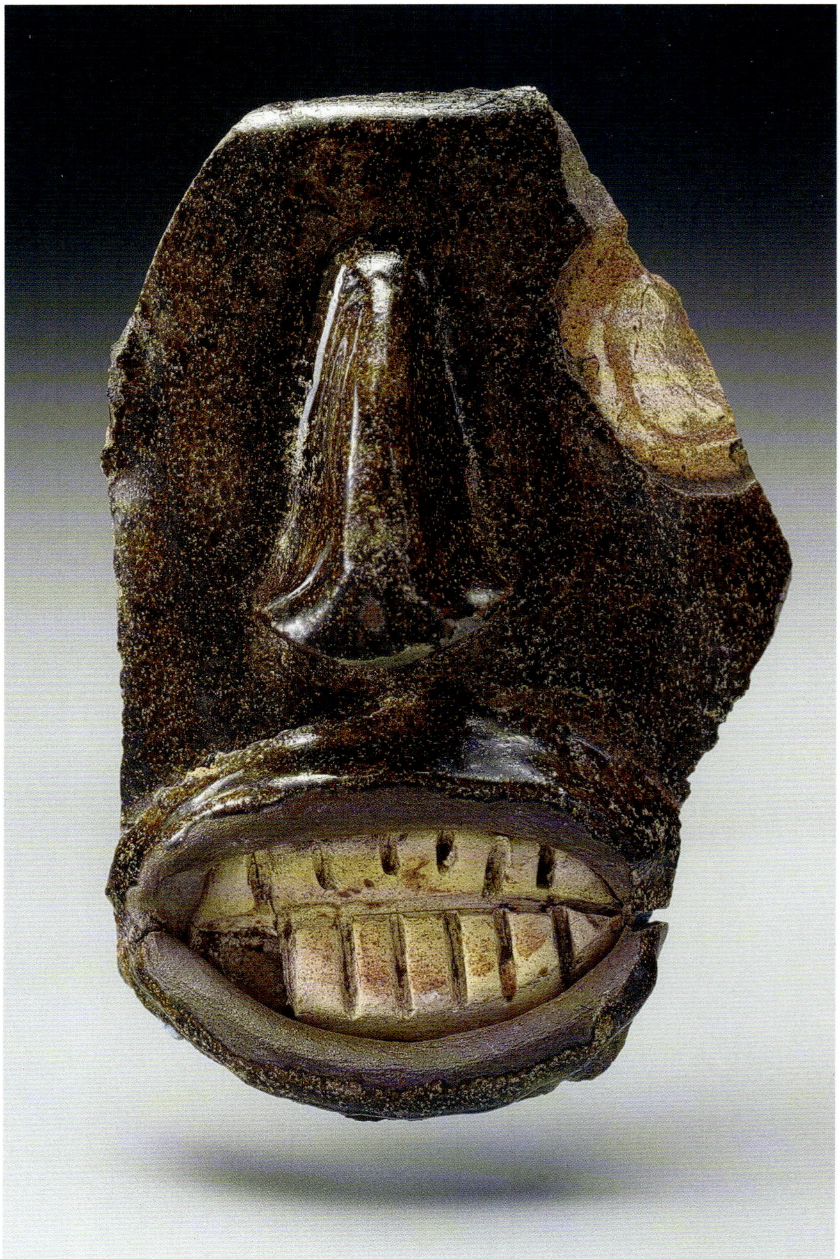

Mark M. Newell
with Peter Lenzo

Making Faces: Archaeological Evidence of African-American Face Jug Production

▼ THE FIRST archaeological evidence of nineteenth-century face jug production within South Carolina's Edgefield District has been announced by the Georgia Archaeological Institute (fig. 1). The production site, located on one of the Miles family potteries near Eureka, sheds light not only on the manufacturing process but also on the origin of some of the vessels in major collections. Study of the recovered fragments proved invaluable, enabling exploration of the technology of face jug production by re-creating several facsimile vessels. This should provide a basis for understanding similar jugs and make it easier to attribute them to specific makers or places.

Origins of Southern Face Vessels

The origin of face vessels made by Southern slaves has been the subject of great debate in recent years.[1] Anthropomorphic representations have appeared on clay pots in virtually all cultures since people first began molding clay. Some researchers claim, perhaps ethnocentrically, that slave-made face jugs got their start with the help of the English Toby jug.[2] The Toby, an eighteenth-century figural vessel most commonly made as a pitcher or decorative piece (fig. 2), was introduced through trade into Africa where it reportedly was adopted and adapted by African artisans. Even a cursory study of African art and religious beliefs, however, would not support such a speculation. Although in 1858 Edgefield saw an influx of slaves from the Congo, the distinct form of the Toby jug, with its traditional tricorn hat and colorful attire, has never been found among the African clay forms of this region. One vessel in the collection of the Charleston Museum, attributed to black potter Jim Lee from Kirksey, South Carolina, is reminiscent of the form—a figural bottle in the shape of a man—but bears no direct morphological connection.[3]

Early American ceramic historian Edwin AtLee Barber interviewed plantation and pottery owner Colonel Davies from Bath, South Carolina. In commenting about the face vessels made circa 1862 by African slave potters, Davies suggested that the slaves had made them "for their own purposes."[4] Even at that early date the strange objects aroused curiosity. Interest in their origin and function has intensified over the years, particularly since the only known vessels had been in private and museum collections (figs. 3–5).

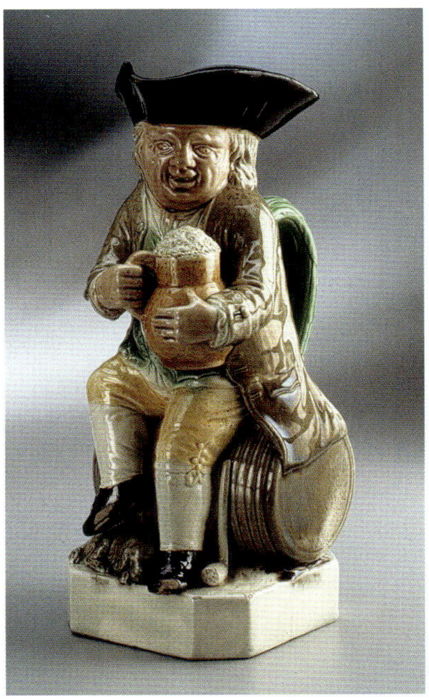

Figure 2 Toby jug, Staffordshire, England, ca. 1785. Pearlware. H. 10". (Chipstone Foundation; photo, Gavin Ashworth.)

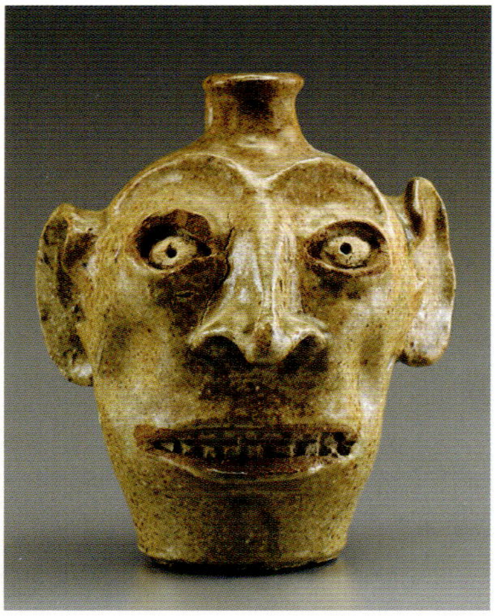

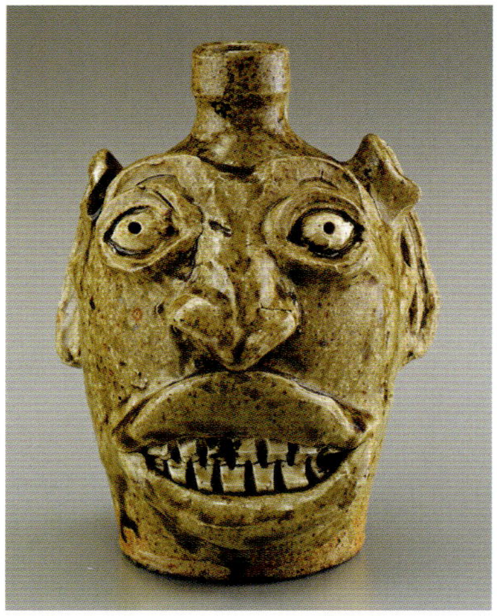

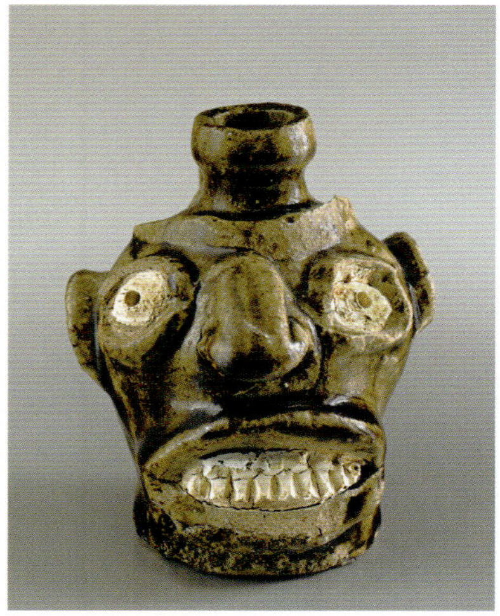

Figure 3 Face jug, Edgefield, South Carolina, ca. 1860–1870. Alkaline-glazed stoneware. H. 5¾". (Private collection; photo, Gavin Ashworth.)

Figure 4 Face jug, Edgefield, South Carolina, ca. 1860–1870. Alkaline-glazed stoneware. H. 6⅜". (Courtesy, James Witkowski; photo, Gavin Ashworth.)

Figure 5 Face jug, Edgefield, South Carolina, ca. 1860–1870. Alkaline-glazed stoneware. H. 4⅞". (Courtesy, Arthur Goldberg; photo, Gavin Ashworth.)

Some of the face jugs, which functioned as water vessels, were called "monkey" jugs—after *monkeyed,* a Southern term for the dehydrating effect of summer heat.[5] The functionality of these jugs does not explain the use of staring eyes and gaping mouths, however, nor does it address why most of the early forms were too small to have any obvious purpose.

Colonel Davies hinted at the connection between the jugs and the "aboriginal art of . . . the Dark Continent," a point further developed by contemporary folklorist John Vlach.[6] Vlach has suggested a direct connection between the face vessels and the *nkisi* figures of central and western Congo, now Zaire.[7]

The African *minkisi* (plural for *nkisi*), also called power figures, are associated with fetishism and ritual magic. Traditionally wooden figures with wide, bright eyes and gaping mouths, they are made up of various components designated for the practice of magic. The *nkisi n'kondi,* or nail figure, for example, is a ritual figurine carved from a light, cypresslike wood (*Canarium schweinfurthii*) believed for this purpose to be sacred (fig. 6). By itself the *nkisi* does not represent a spiritual personality but is a container for one, and it is up to one member of each village, the

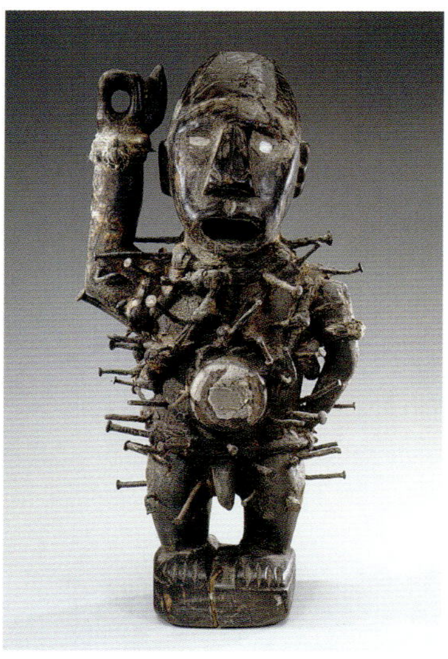

Figure 6 *Nkisi* figure, Zaire, Africa, ca. 1900. *Canarium schweinfurthii.* H. 18⅞". (Courtesy, Georgia Archaeological Institute; photo, Gavin Ashworth.)

nganga n'kondi, to make sure that the capture of the appropriate spirit is accomplished. Taking on the role as expert in such rituals, the *nganga n'kondi* conceals a mixture of minerals, herbs, and other substances inside the *nkisi,* usually within its protruding belly cavity. The aromatic blend lures curious spirits, or spiritual powers, who are entrapped when the cavity is sealed.[8] Spirits are also attracted by the bright whites of *nkisi* eyes, which are either painted with white kaolin clay or inset with glass (fig. 7). The early *nkisi* were made exclusively of kaolin clay, which was

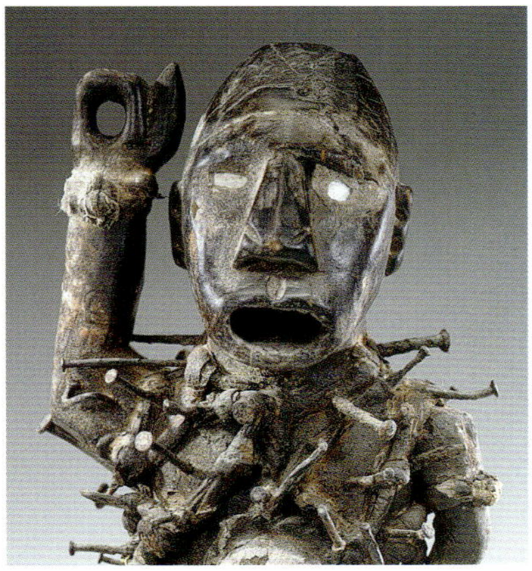

Figure 7 Detail of the *nkisi* figure illustrated in fig. 6.

regarded as a sacred substance.[9] As with other parallels, the functional features of the *nkisi* closely match those of the slave-made face jug, an analogy strengthened by evidence that variations of these rituals are still in existence in America.[10]

It is well accepted that slaves in America practiced their religious beliefs in secrecy. In the tradition of their ancestors, some African Americans, especially in rural areas such as modern-day Edgefield County, still maintain a vigorous belief in "root magic," a fact confirmed by various informants, from local ministers to descendants of slave families.[11]

History of Face Jug Production

The distinctive African-made, green-glazed jug with red clay body and kaolin inserts is thought to date from after 1858, although there is evidence that white potters made face-decorated harvest vessels even earlier. Virginia-born potter Thomas Chandler, for example, made at least one harvest jug at his Shaw Creek site sometime between 1840 and 1857.[12]

The earliest Edgefield face jugs generally have been attributed to potters working in the upper Horse Creek Valley watershed. The Miles Mill pottery is known to have been located in this area, although the actual location of the site discussed in this paper was not known at the time of these attributions. Various extant examples exhibit rich green or brown glazes, high-ridged noses, and eyes and mouths set within separately applied eyelids and lips. The clay bodies fire to a purplish red color, and the eyes and teeth are made of white kaolin clay. These early vessels exhibit two types of teeth construction: sharp and angular, and flat, square teeth etched into a single piece of kaolin.

The tradition of white Southern potters copying the African-style face jug appears to have begun with South Carolina potter Mark Baynham.[13] A recently discovered circa 1900 face jug bearing his stamp ("MARK") was reported in the 2003 issue of *Ceramics in America*.[14] The Baynham vessel appears to have been directly modeled on the slave-made versions: the ears and nose are applied, as are the lips and eyelids, and the eyes and teeth are attached directly onto the vessel's unglazed body. The teeth, which are especially reminiscent of the African-style vessels, and the eyes were left unglazed, implying use of a wax resist (see figs. 31, 32), and an attempt to mimic the African practice. Baynham did not, however, make separate kaolin inserts for each. Like other white potters who followed him, he probably did not consider the detail and extra work necessary, unaware of their possible religious significance.

Discovery of Face Jug Production

Despite the far-ranging interest in these early face jug forms, no archaeological evidence of their manufacture has ever been recorded.[15] An ongoing reassessment by the Georgia Archaeological Institute (GAI) of the Joseph G. Baynham site (38ED221) at Eureka found conclusive proof of early face jug production.[16] During the first few days, GAI encountered evidence that occupation of the Baynham site dated to the 1870s.[17] Furthermore, beneath the Baynham context an earlier pottery was discovered, which had produced wares entirely different from the Baynham wares in body clay, glaze, and style—including a neck style that was known from an existing face jug.

Face jug fragments were first encountered in a test pit selectively located away from the main Baynham context on a wooded slope near a small dam. It was in fact isolated from all other pottery production areas. A two-meter-by-five-meter excavation subsequently revealed a shallow deposit of face jug fragments, bowl fragments, kiln furniture, and kiln debris. Careful testing of the surrounding area indicates that this one spot might have been selected solely for the production of face jugs.

In all, more than two thousand large fragments and small sherds were recovered from the area at depths ranging from 1.18 to 3.93 inches and representing at least seven face jugs (figs. 8–13). The fragments' neck styles, in combination with the Miles Mill face jugs housed in collections, allow us to develop a tentative chronology for their manufacture. The earliest appear to be those with double-collared necks, a style found on the lowest levels of the Sunnybrook site, a mile from the Baynham site on Horse Creek. It is believed that this dates to the earliest Lewis Miles occupation of the site, from the 1850s to circa 1867. Stratigraphic evidence enables us to date with some confidence the flat-lipped neck style (see fig. 10) to about 1867–1872, the John L. Miles occupation of both the Baynham and Sunnybrook sites. This same neck type appears on a Davies pottery vessel dated to circa 1862.[18] The similarity of neck styles is attributed to the movement of the upper Horse Creek slave potters from one local pottery to another. The tubular Bodie-style neck appears to be the last of its kind and is documented on Baynham-made dispensary jugs of the early 1880s.[19]

Figure 8 Face jug fragments, Miles Mill, Edgefield, South Carolina, ca. 1867–1872. Alkaline-glazed stoneware.

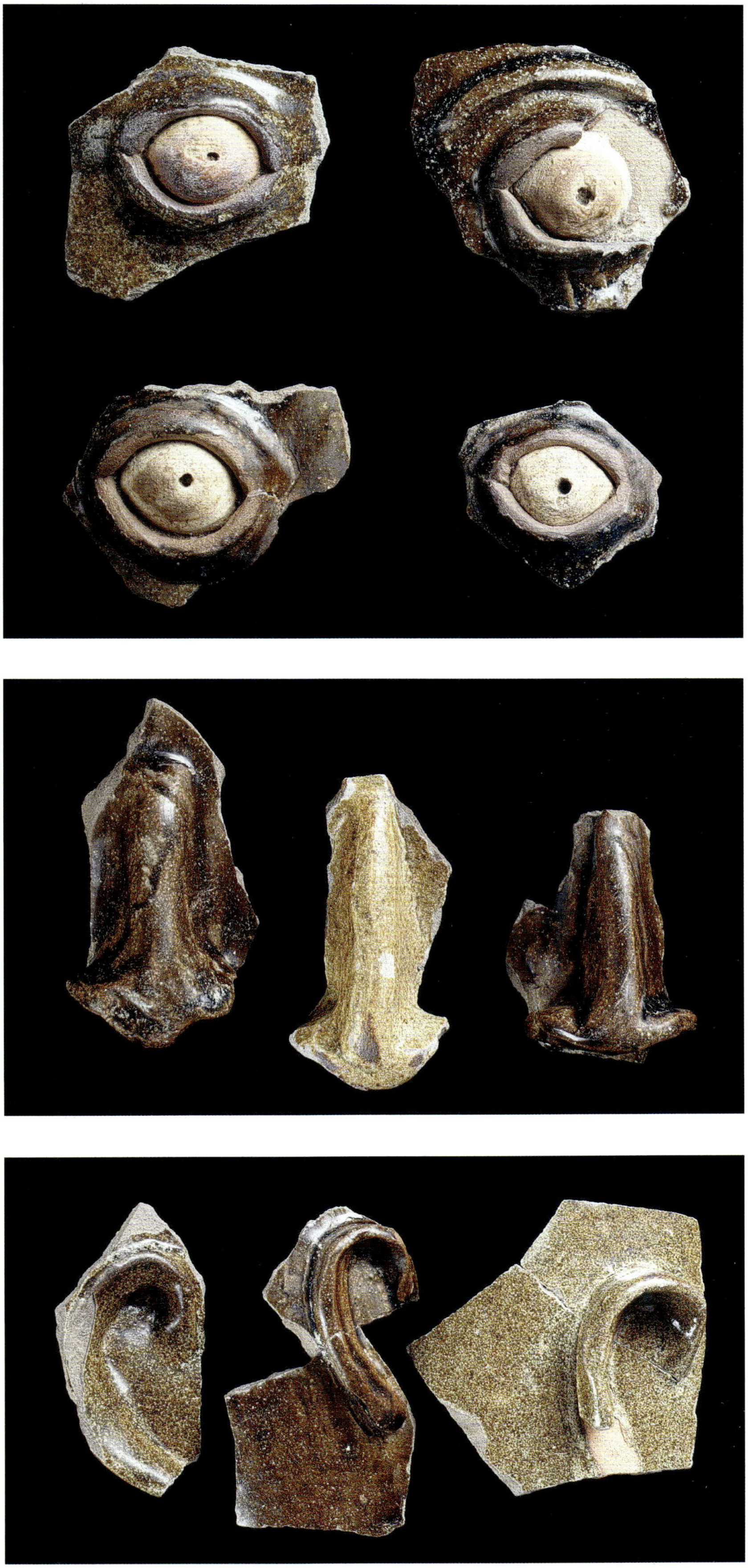

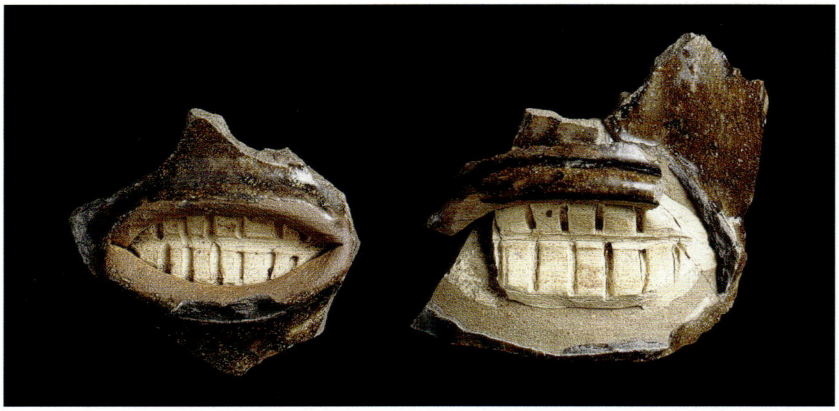

Figure 9 Face jug fragments, Miles Mill, Edgefield, South Carolina, ca. 1867–1872. Alkaline-glazed stoneware.

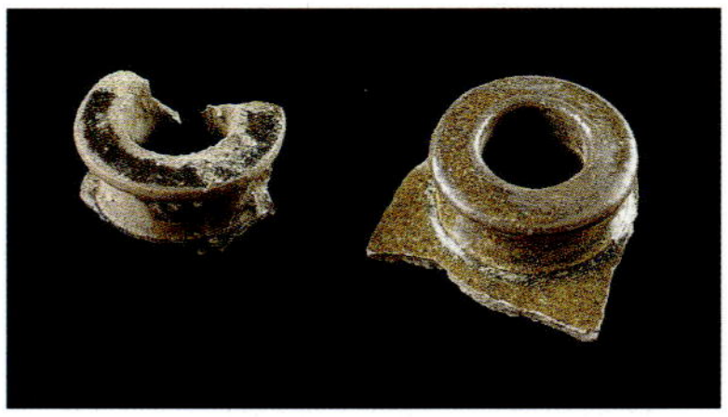

Figure 10 Face jug neck fragments, Miles Mill, Edgefield, South Carolina, ca. 1867–1872. Alkaline-glazed stoneware.

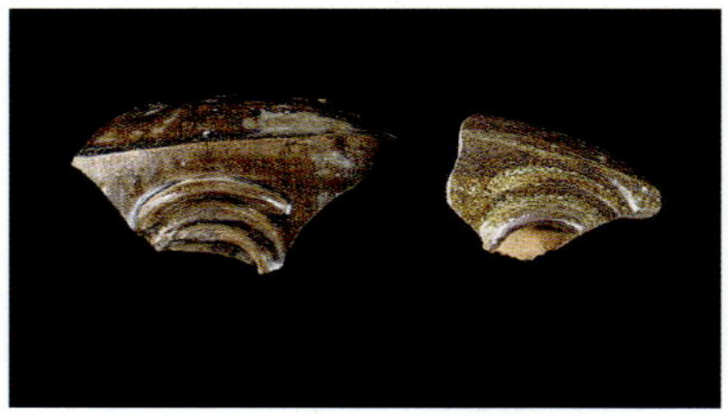

Figure 11 Face jug fragments, Miles Mill, Edgefield, South Carolina, ca. 1867–1872. Alkaline-glazed stoneware.

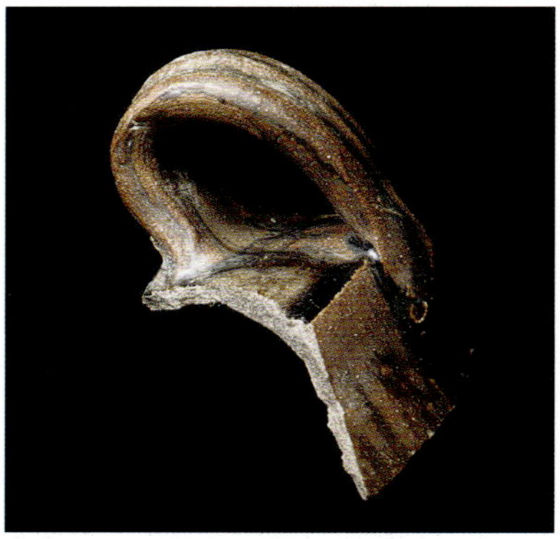

Figure 12 Face jug handle fragment, Miles Mill, Edgefield, South Carolina, ca. 1867–1872. Alkaline-glazed stoneware.

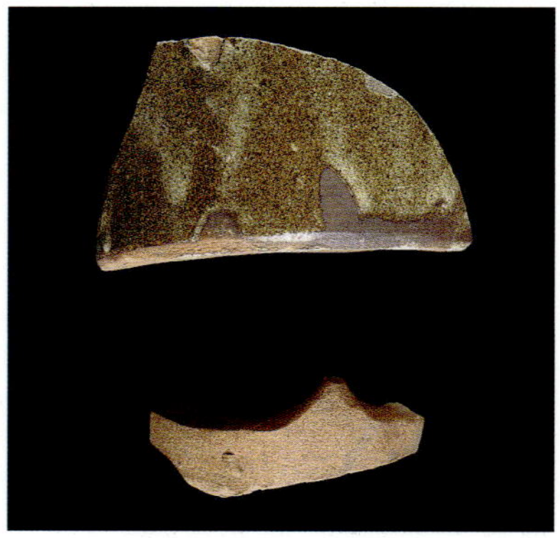

Figure 13 Face jug base fragments, Miles Mill, Edgefield, South Carolina, ca. 1870. Alkaline-glazed stoneware.

Experimental Production of a Face Jug

Following initial development of the construction sequence and the techniques based on analysis of the recovered sherds it was decided that an experimental reconstruction project might shed further light on the process (fig. 14).[20] Peter Lenzo, a potter in Columbia, South Carolina, was enlisted to reproduce the GAI finds. Lenzo, who maintains a studio on Rosewood Avenue, has gained national recognition for his explorative work and interpretation of the American face jug tradition.

The initial task was to select a local clay for the trial pieces, and Bethune clay from a mine east of Columbia was chosen. The clay was used for the last vessels produced by an early-twentieth-century pottery once owned by

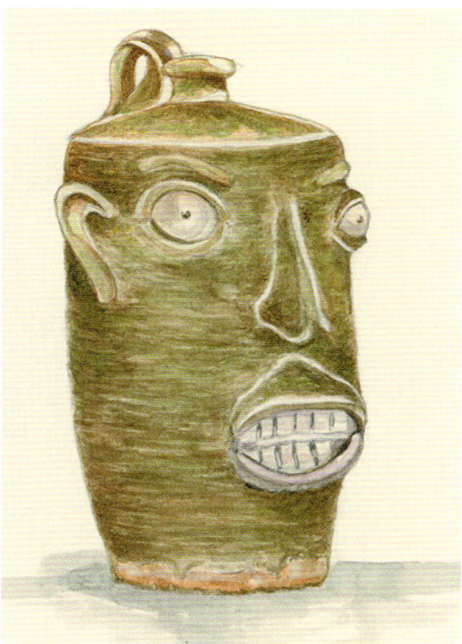

Figure 14 Artist's reconstruction based on the fragments recovered from the John L. Miles site. (Drawing, Christine Madigan.) The best guess for the shape of the face jug came from an intact storage jug found on the Miles site.

Thomas Daugherty, a former partner of Horace Baynham at the Edgefield District pottery at the head of School House Creek.[21] Lenzo used a mixture of Bethune clay and Otis Norris's clay, which Otis digs from around McBee, South Carolina, near the Lynches River. Bethune clay fires to a red color and has a more aggressive tooth (a coarser, stronger body) than Edgefield's Horse Creek clays, but it was deemed acceptable for the project since the latter was unavailable. Lenzo also experimented with a variety of iron-bearing glaze formulas with the hope of reproducing the rich green glaze the original potters applied over the dark red and purplish fired clay body.

After careful study of the sherds, vessel production began with the throwing of a jug on the wheel (fig. 15). In the absence of a readily

Figure 15 In this sequence of photographs, Peter Lenzo forms the jug on the wheel. First the clay is centered and opened, then a cylinder is raised and the shaping of the body is done. After the body contours are defined, the shoulder and neck are closed to create the jug. The initial finishing of the neck and lip begins at this stage.

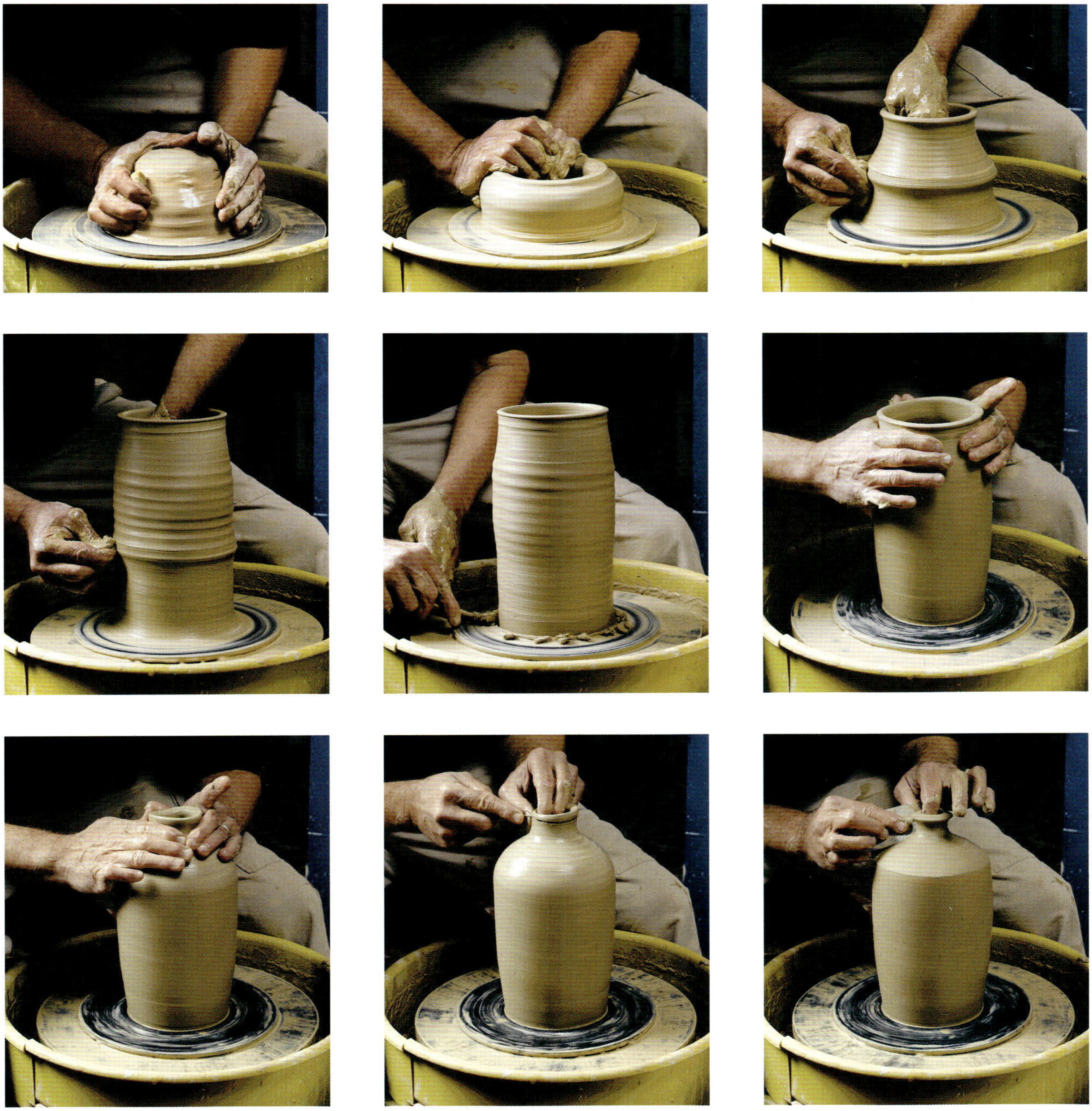

identifiable antique prototype, an intact jug recovered from the site that had similar handle and neck characteristics as the face jug sherds was used as a model. The distinctive broad flat lip at the opening of the face jug is unique in the Edgefield tradition and specific to the John L. Miles pottery (fig. 16). After removing the thrown jug from the wheel (fig. 17), a handle was pulled and attached in the manner suggested by the recovered artifacts (figs. 18, 19).

While the jug was still in a wet, plastic state, work began on the various facial components, beginning with the nose. While there is no direct evidence of sequence, starting with the nose helped set up the proportions and balance of the final face. Created from a cone-shaped section of clay, the nose was attached to the vessel's center and subsequently modeled (fig. 20). It is not known whether these attachments were scored and moistened with water or slip, in keeping with modern hand-building techniques. None of the recovered fragments bears any evidence of scoring. It is clear that considerable force was used to ensure the reliability of the attachment (see fig. 30).

Figure 16 Peter Lenzo finishes the neck and lip of the jug.

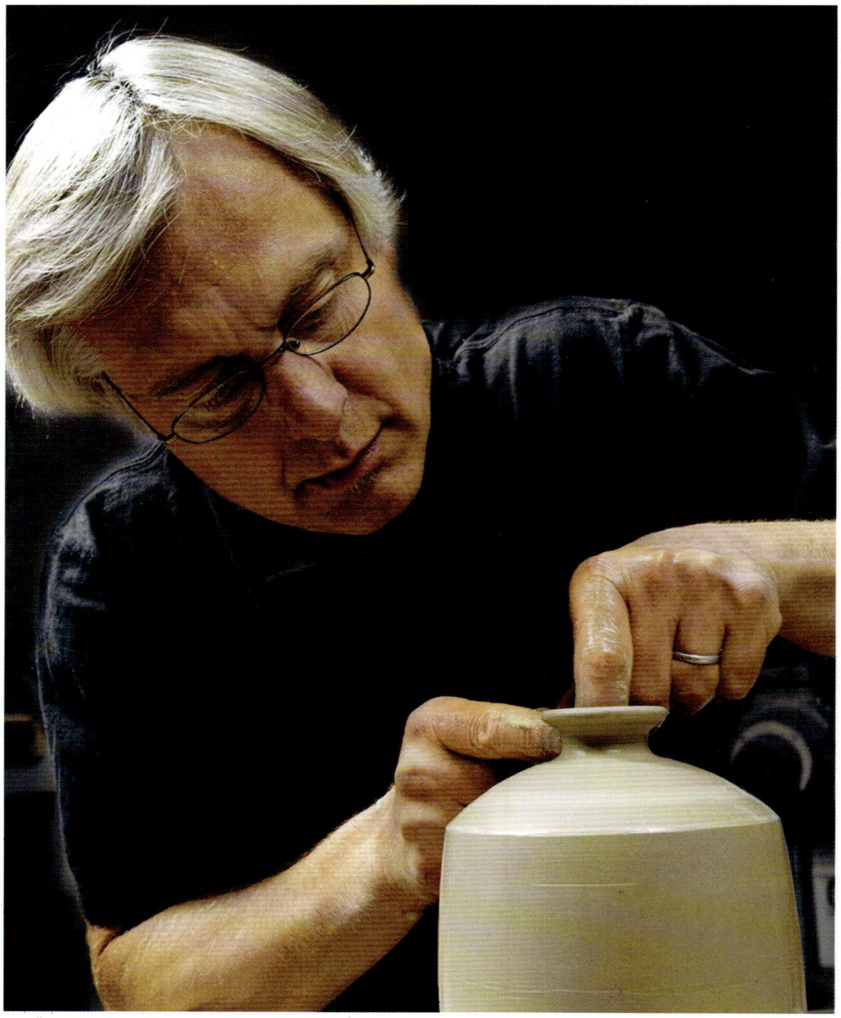

Figure 17 The thrown jug body is cut from the wheel head with a wire tool.

Figure 18 A typical method of forming handles is known as "pulling," whereby the clay is drawn out in successive pulls until the right thickness is achieved. The pulled strip is then cut to size.

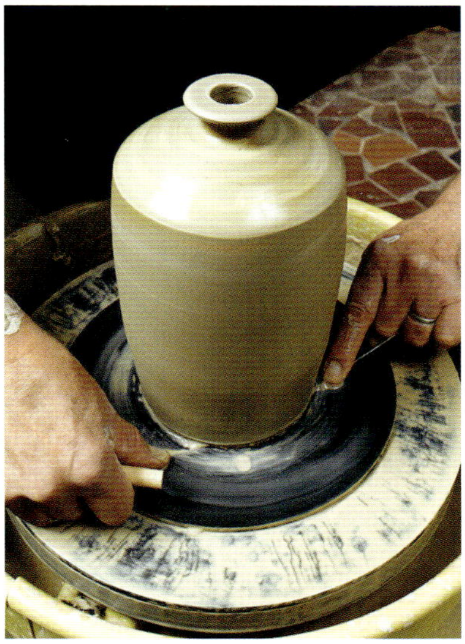

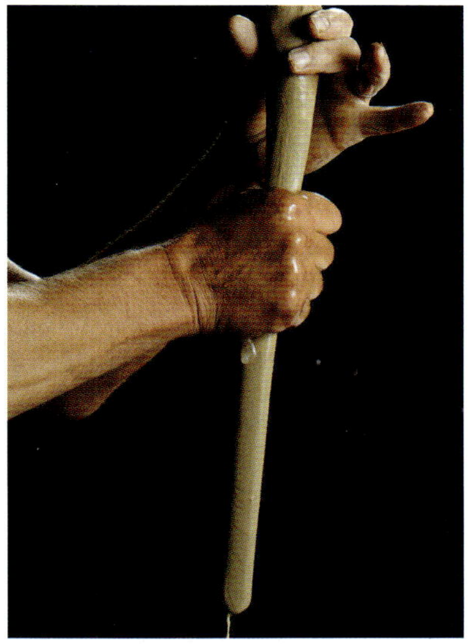

Figure 19 The cut handle is attached first to the shoulder and then to the body.

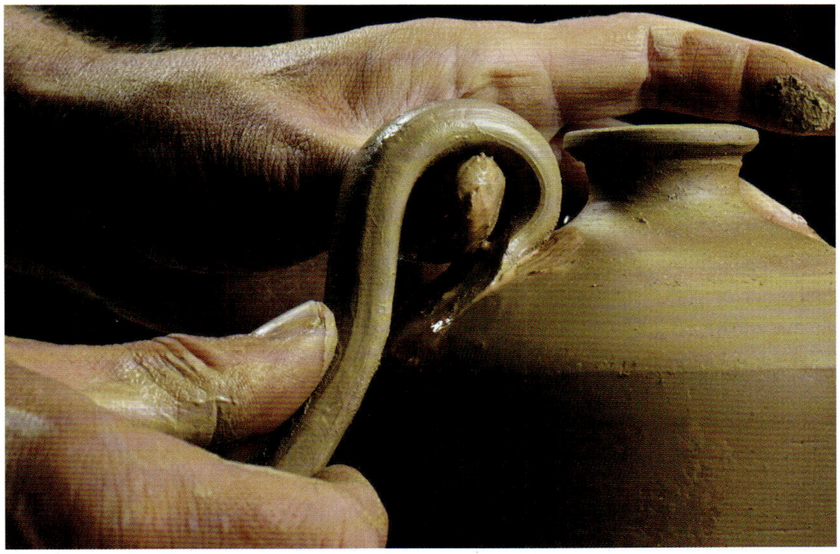

Figure 20 The first facial feature to be placed is the nose, which is attached as a wedge-shaped coil and subsequently modeled by hand.

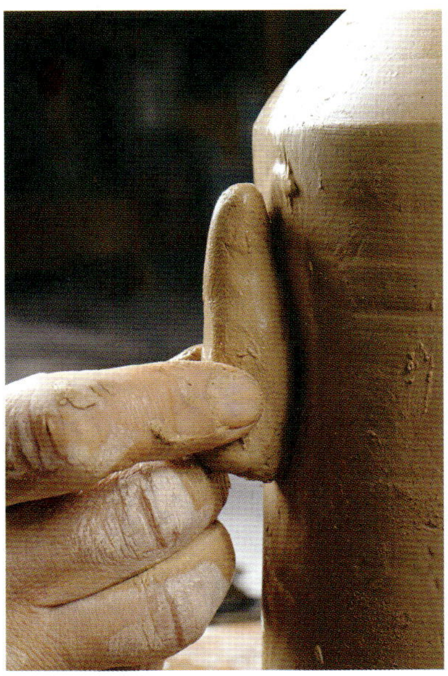

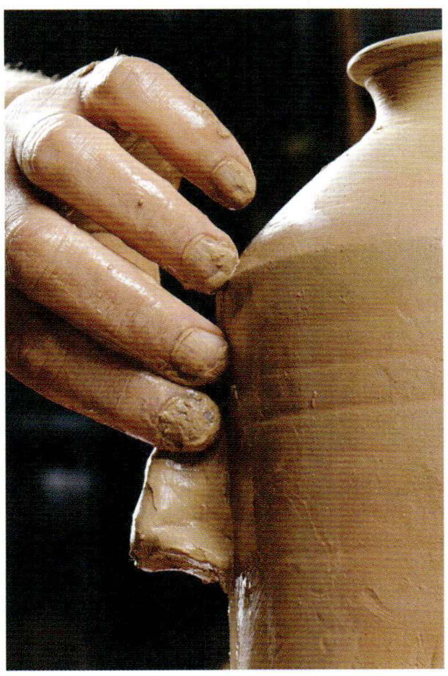

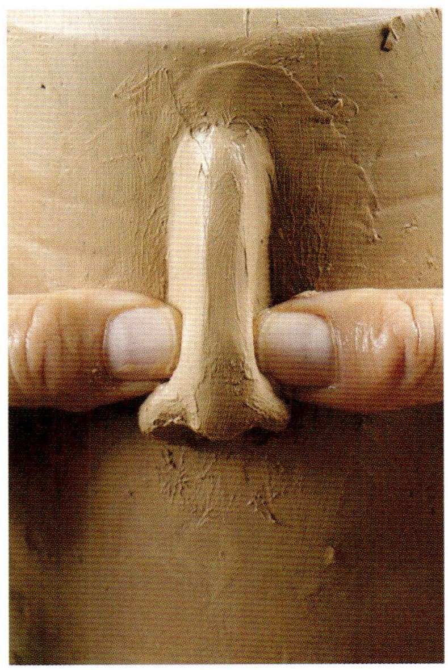

The flat, oval eyes—the vessel's focal point—were achieved by rolling and pressing small balls of kaolin-based white clay in the palm of the hand. Probably mixed with a binding and fluxing agent, each eye was attached to the jug without any application of slip between it and the vessel's surface (fig. 21). Small strips of clay rolled on a flat surface into a coil measuring .12

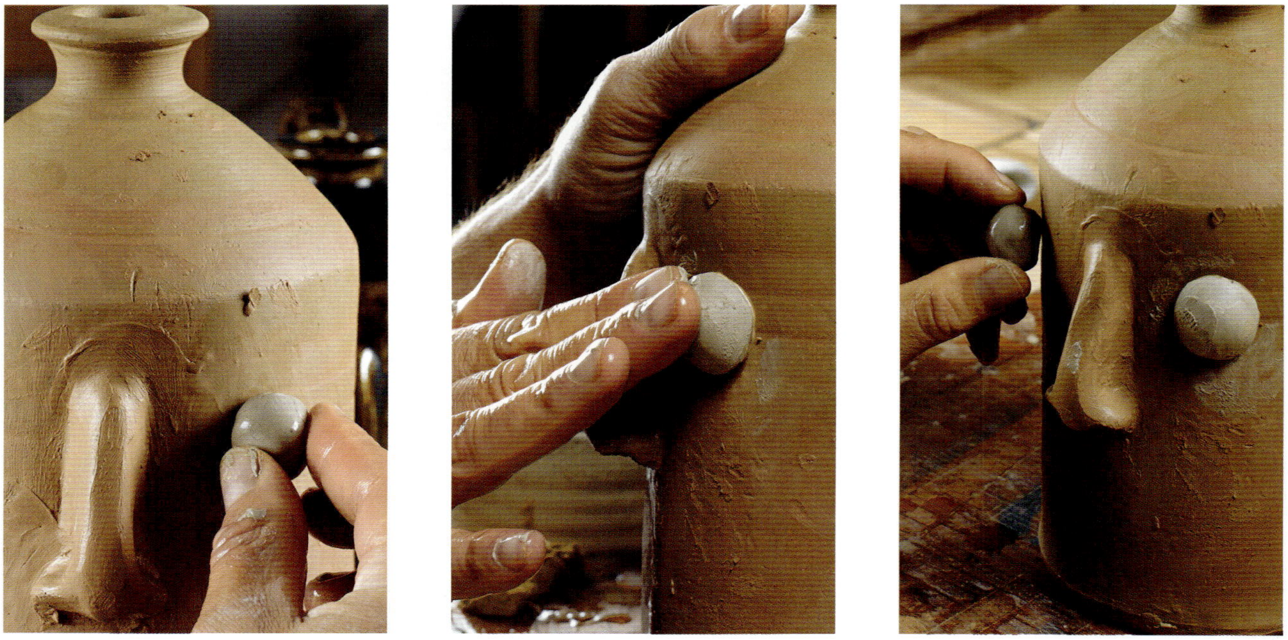

Figure 21 After the nose is formed, two balls of the kaolin that will become the eyes are pressed into position.

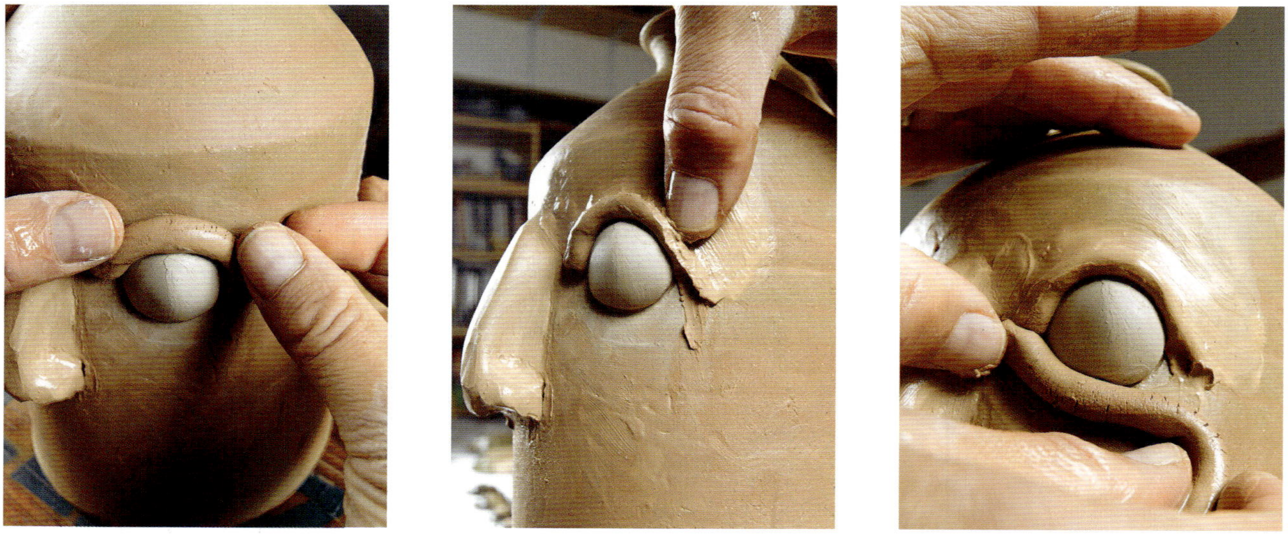

Figure 22 Coils of clay are placed over the top and bottom of the kaolin balls to create the eye sockets. This step is one of the most critical because the shrinkage rate of the two clay bodies differs.

inches in diameter formed the eyelids. Sections of the coil were placed over the eye's top (fig. 22) and bottom edges and then carefully smoothed.

As with the eyes, the teeth were created from a ball of kaolin-based clay that was rolled and flattened into an oval plate and pressed onto the surface of the jug above the vessel base (fig. 23). Since the kaolin clay has a much higher shrinkage rate during firing than the body clay, potters had to use enough overlapping clay for the eyelids and lips to ensure that neither the eyes nor the teeth fell out. Many of the recovered fragments and related

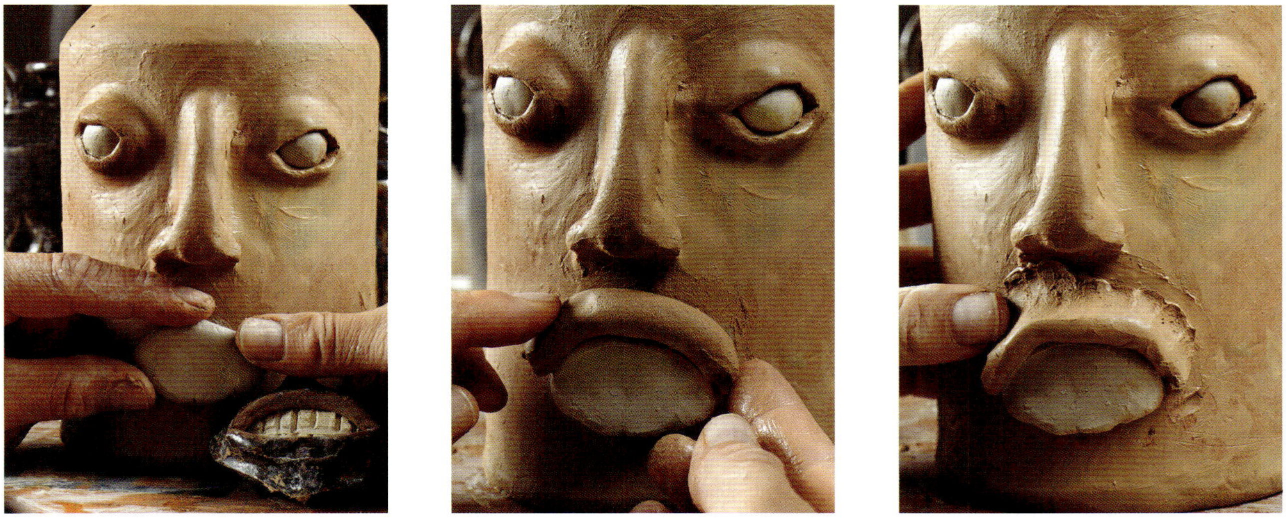

Figure 23 An oval pad of kaolin is prepared and pressed onto the body of the jug. Coils of clay are then placed and modeled to create the lips.

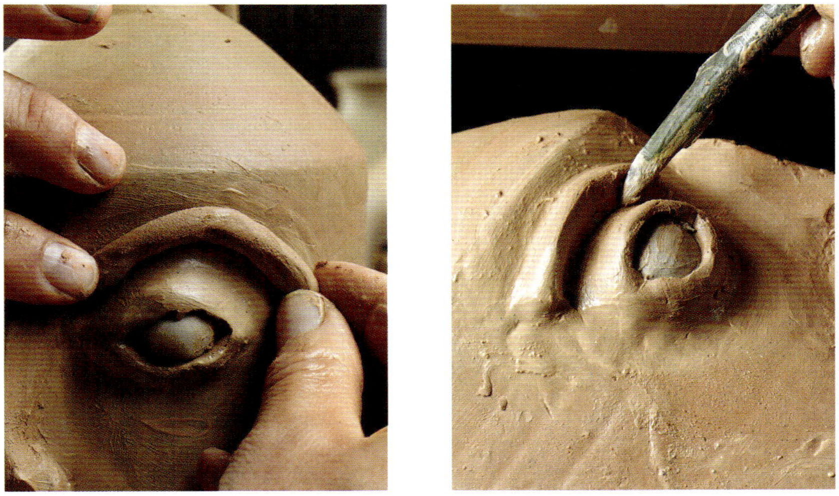

Figure 24 Clay coils are positioned to form the eyebrows, which are modeled with the help of a stick tool.

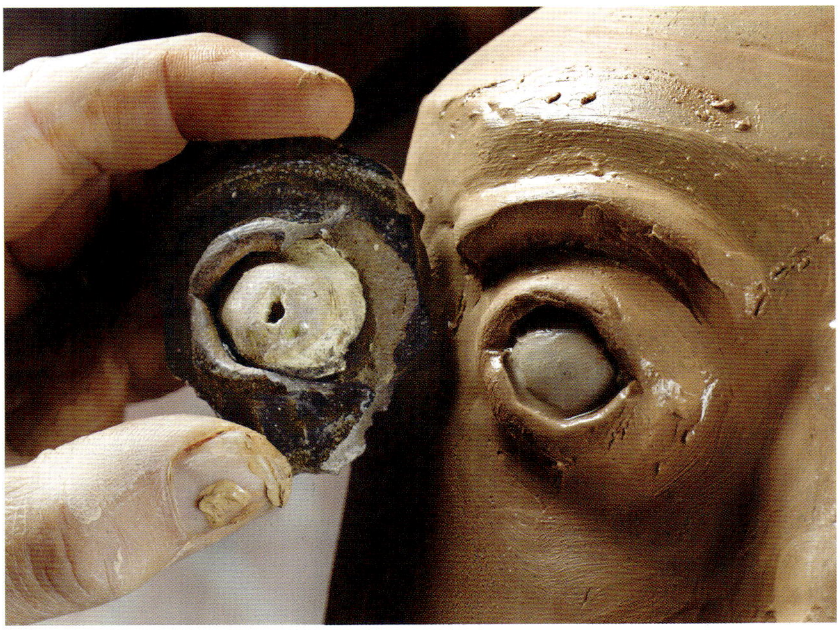

Figure 25 An original eye fragment is compared to Lenzo's prototype.

antique face jugs indicate that the kaolin eyes shrank in the firing and moved unnervingly within their sockets. None of the recovered fragments, however, exhibited loose teeth.

Adding a somewhat quizzical expression to the face, coils of clay were used to fashion eyebrows—modeled with a wooden tool (fig. 24)—and ears (fig. 26).

Figure 26 Clay coils are placed for the ears, which are modeled by hand.

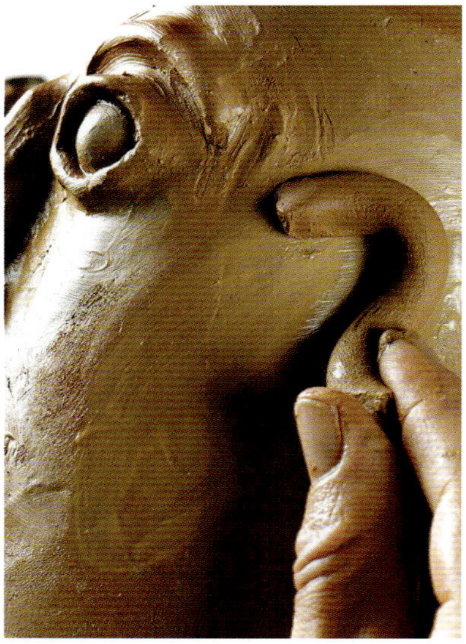

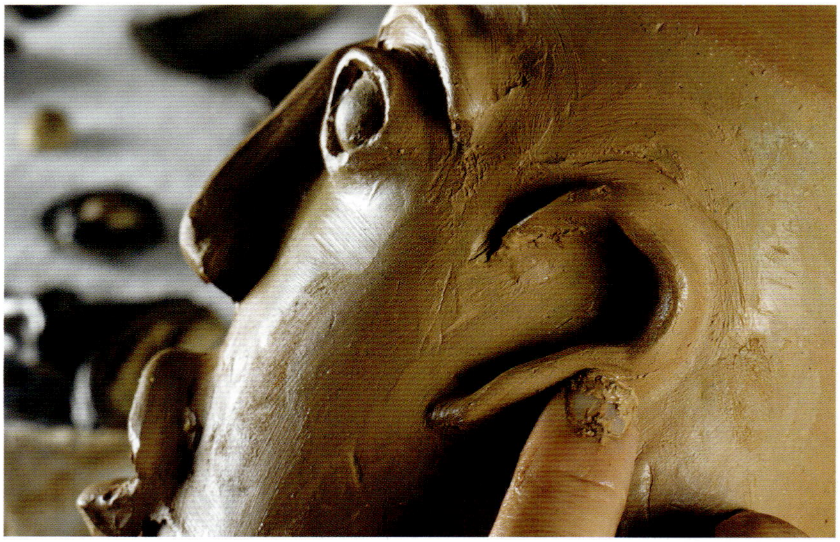

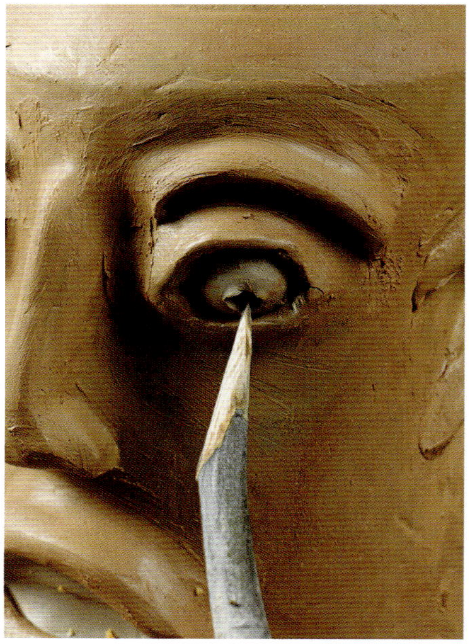

Figure 27 A sharpened stick is used to create the iris. Most of the punctures observed on the fragments were deep and tubular.

Once the facial features had been attached and modeled, attention was turned to the details. To simulate the iris (fig. 27), a sharp stick was used to create a small, triangular hole in each eye. In most of the recovered eyes, the hole was a deep tubular piercing, probably made with an instrument of uniform width. An object that had a worked edge, possibly a wooden rib, scored the clay to make straight, square teeth but only after two coils of clay had formed the lips (fig. 28).

Figure 28 The teeth are etched with a stick. Close examination of the fragments reveals that these marks were made after the lips had been applied.

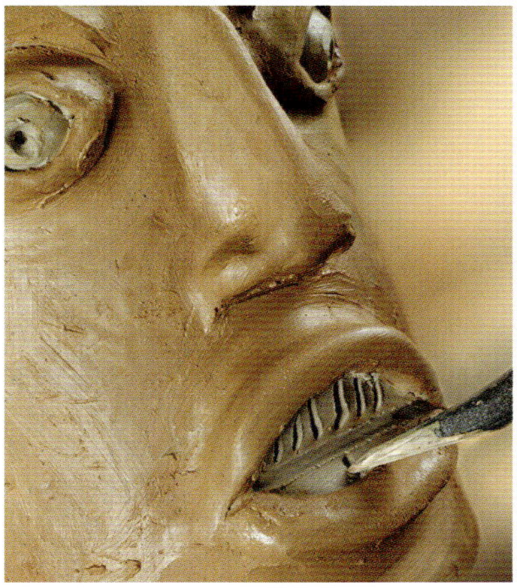

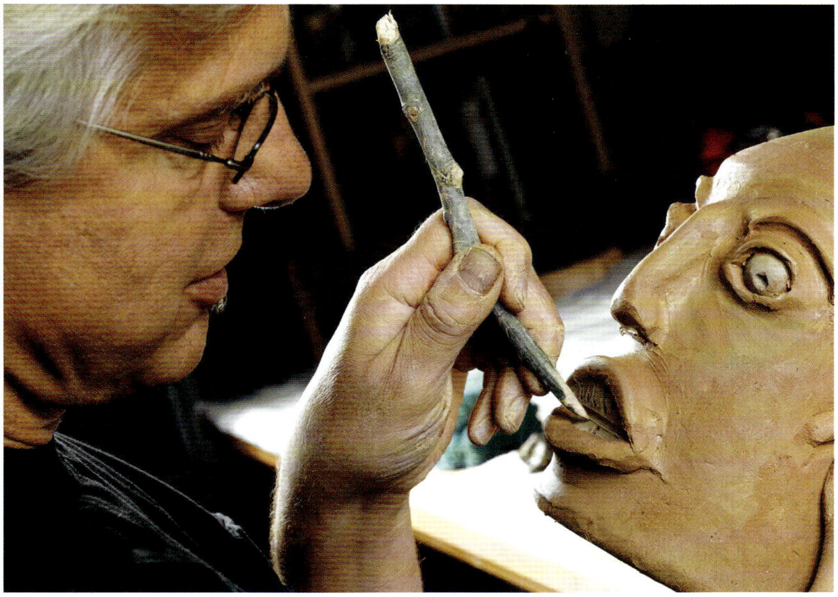

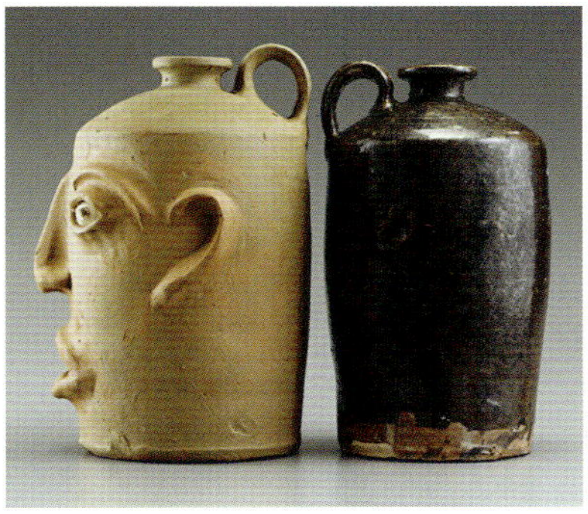

Figure 29 As clay normally shrinks 10 to 15 percent during drying and firing, the unfired prototype face jug appears larger than the intact storage jug recovered from the site.

Figure 30 One of the demonstration jugs was cut in half to show the interior depressions resulting from the pressure applied when attaching the various facial features. These depressions are an exact match for many of the original sherds.

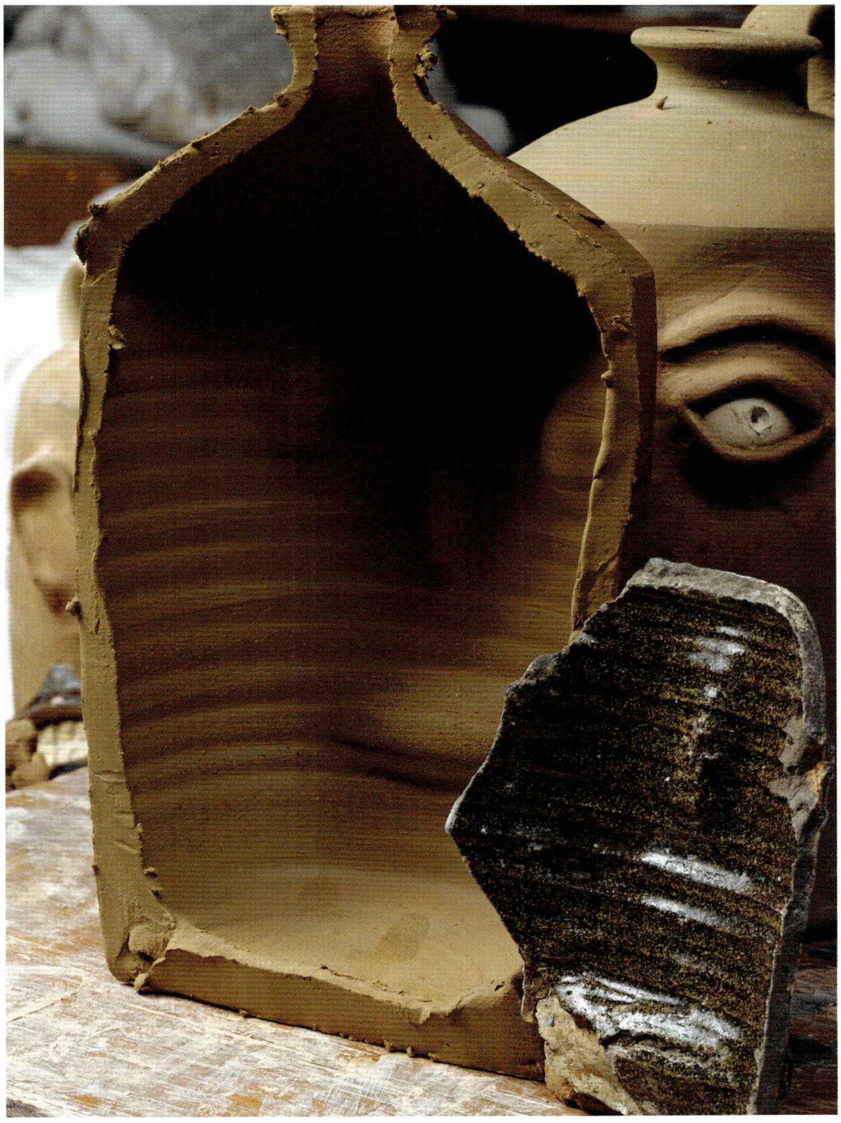

Figure 31 Melted wax is brushed over the kaolin mouth and eyes. This resist, which prevents the glaze from sticking to the kaolin, will burn off during firing leaving the applied areas white and unglazed. Some type of beeswax is thought to have been used by the Edgefield potters.

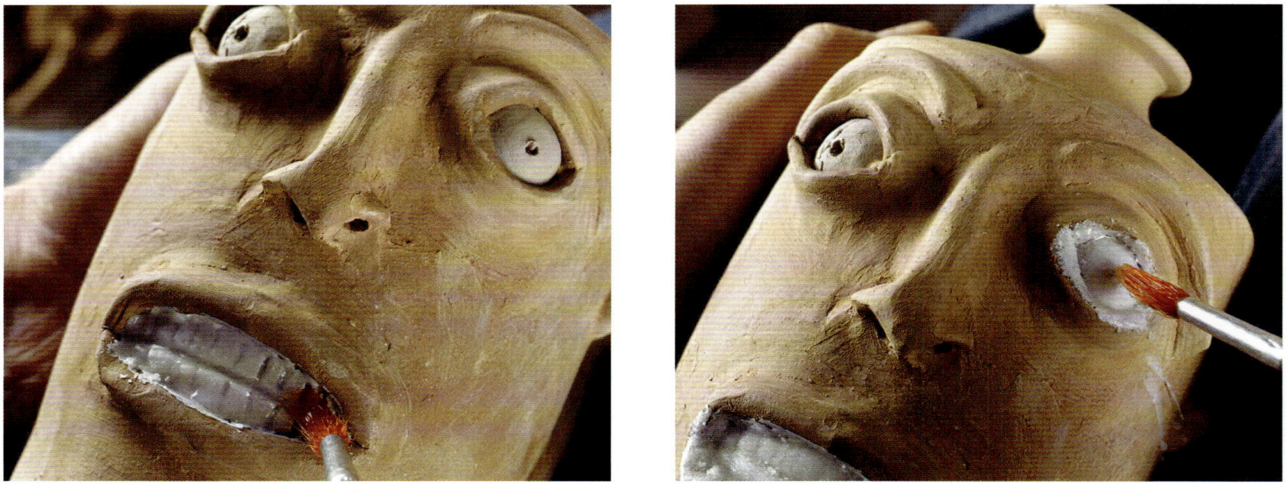

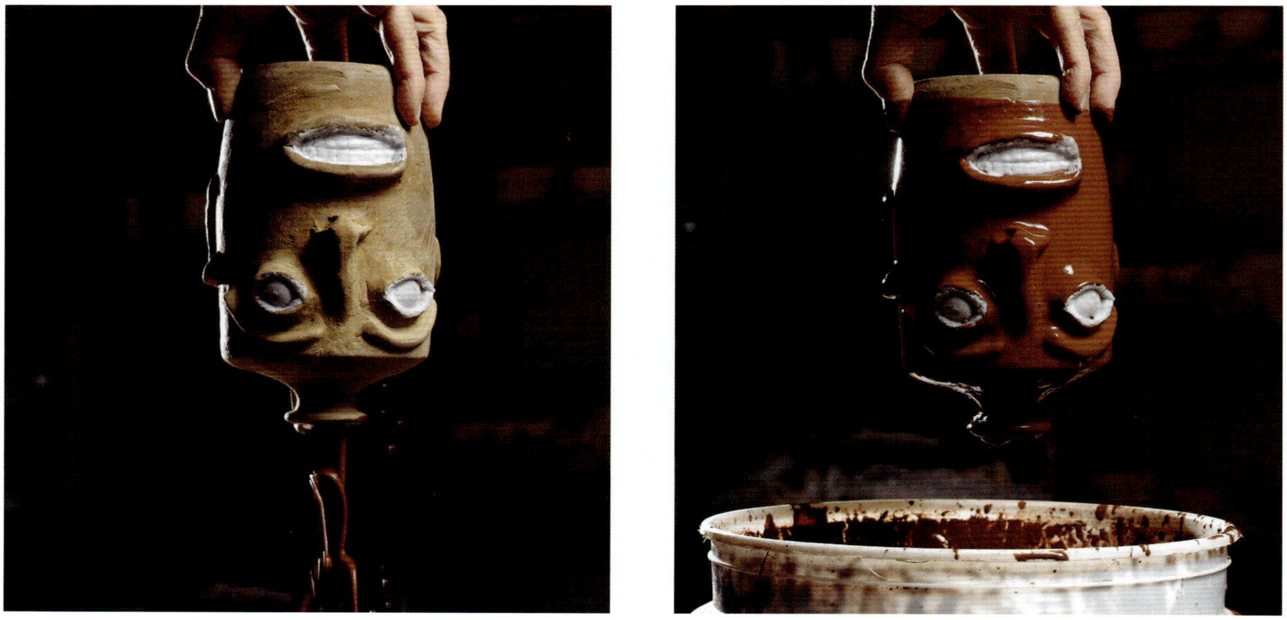

Figure 32 The liquid glaze—a mixture of commercially prepared fine clay, hardwood ash secured from fireplaces and woodstoves, and mined feldspar—is poured into the interior of the jug and allowed to drain. The exterior of the jug is then dipped in the glaze and allowed to air dry. Note how the wax over the eyes and mouth repels the glaze. Many original sherds and vessels exhibit finger marks on the bases of the jugs (see fig. 35). The absence of glaze on the pot's bottom helps keep them from sticking to the kiln during firing.

After the vessel was thoroughly air-dried (fig. 29), preparation for glazing began with the application of wax to the eyes and teeth. This resist technique—the Edgefield potters probably used an application of beeswax—was employed to ensure that the eyes and teeth were left unglazed (fig. 31), and also might have allowed the teeth and eyes to move slightly. The intention, however, is unknown. There is some indication from Edgefield County sources that the effect may have been desirable—that is, the "best" face jugs had eyes that rolled and teeth that chattered.

The final step before firing was coating the interior of the jug and then the exterior in a glaze mixture of a clay slip, wood ash, and possibly some iron (figs. 32, 33).

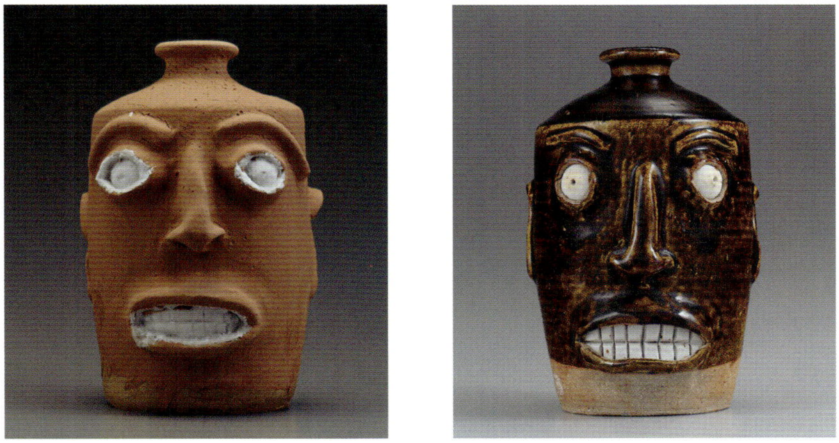

Figure 33 Face jugs, Peter Lenzo, Columbia, South Carolina, 2006. Alkaline-glazed stoneware. H. 9½". At left is the glazed but unfired face jug. At right is a previously fired example of a jug, shown to illustrate the alkaline glaze color.

After the experimental phase of the project had been completed, two antique examples of the face jugs discussed herein came to light. The first was illustrated in the arts section of the *New York Times* during "Antique Week," January 2006 (fig. 34). The jug's overall shape and facial features corresponded very closely with the experimental reconstruction, with the exception of its height. The antique example suggested that some of the face jugs were slightly squatter. A subsequent example confirmed this shape as being more typical of the production on the site (fig. 35).

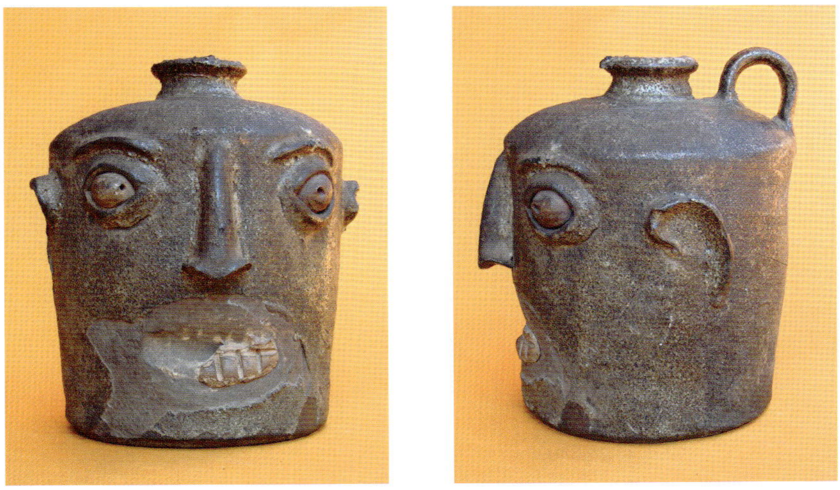

Figure 34 Face jug, Miles Mill, Edgefield, South Carolina, ca. 1867–1872. Alkaline-glazed stoneware. (Courtesy, Charlton Brasher.) Note that the lips and teeth have spalled away on this example at their points of attachment.

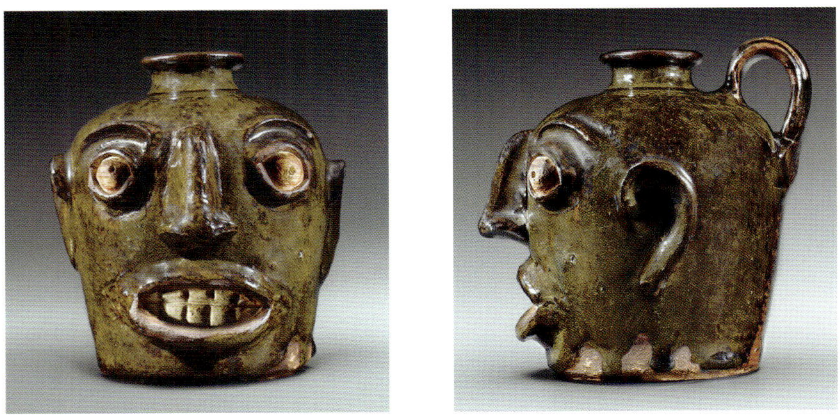

Figure 35 Face jug, Miles Mill, Edgefield, South Carolina, ca. 1867–1872. Alkaline-glazed stoneware. H. 5⁹⁄₁₆". (Private collection; photo, Gavin Ashworth.)

Undoubtedly, more face jug examples residing in antique collections can now be firmly attributed to the Miles Mill production site. A full report on the face jug find is due to be published by GAI late in 2006.

1. See John A. Burrison, *Fluid Vessel: Journey of the Jug,* pp. 92–121 of this volume.

2. Cinda K. Baldwin, *Great and Noble Jar: Traditional Stoneware of South Carolina* (Athens: University of Georgia Press, 1993), p. 81.

3. Ibid., pp. 83–84, figs. 3.11, 3.12.

4. Edwin AtLee Barber, *The Pottery and Porcelain of the United States: An Historical Review of American Ceramic Art from the Earliest Times to the Present Day* (New York: G. P. Putnam, 1909), p. 466.

5. John Michael Vlach, *The Afro-American Tradition in Decorative Arts* (Athens: University of Georgia Press, 1990), pp. 86–90.

6. Barber, *Pottery and Porcelain of the United States.*

7. Vlach, *Afro-American Tradition in Decorative Arts*; Mark M. Newell, "Echoes of Africa," GAI Research Manuscript 433/1966, which covers 1996 iconographic research in Zaire.

8. Robert Farris Thompson, "The Grand Detroit *N'Kondi,*" *Bulletin of the Detroit Institute of the Arts* 56, no. 4 (1978): 207–22.

9. Denba Saccoh, Zaire, personal communication, 1996.

10. Personal communication with an African-American minister in Edgefield, who wishes to remain anonymous.

11. A highly sensitive topic among rural blacks, root magic is a continuing research interest of the Georgia Archaeological Institute. Much of the sensitivity and secrecy are connected directly with the power attributed to the grave sites of important blacks of the past one hundred and fifty years.

12. Baldwin, *Great and Noble Jar,* pp. 47, 51–54.

13. Nancy Baynham, *Baynham: A History of a South Carolinian Family* (Augusta, Ga.: Hamburg Press, 1999).

14. Mark M. Newell, "Making His "MARK," in *Ceramics in America,* edited by Robert Hunter (Hanover, N.H.: University Press of New England for the Chipstone Foundation, 2003), pp. 273–75.

15. Stephen Ferrell, an avocational Edgefield potter-historian, found two face-vessel fragments in the 1980s while raking through a waster pile at Sunnybrook, one of the Lewis Miles sites.

16. George J. Castille, Cinda K. Baldwin, and Carl Steen, *Archaeological Survey of Alkaline-Glazed Pottery Kiln Sites in Old Edgefield District, South Carolina* ([Columbia, S.C.]: McKissick Museum, 1988). Researchers on this site in the 1980s published findings to the effect that Baynham had not been a part of the Edgefield stoneware tradition but that his pottery operation represented a cultural intrusion from Ohio in 1900. GAI's preliminary survey of the site in 1996 also determined that the two sites extended over a far greater area than the earlier archaeologists had supposed—1.5 miles in length as opposed to the 328 feet originally reported. The GAI's reassessment discovered that the boundaries of the site, which previously had been described as encompassed within a few hundred feet, in fact extend beyond a 1.5-by-.5-mile strip.

17. Mark M. Newell, "A Spectacular Find at the Joseph Gregory Baynham Pottery Site," in *Ceramics in America,* edited by Robert Hunter (Hanover, N.H.: University Press of New England for the Chipstone Foundation, 2001), pp. 229–32.

18. *Making Faces: Southern Face Vessels 1840–1990,* exh. cat. ([Columbia, S.C.]: McKissick Museum, 2001), p. 17.

19. Baldwin, *Great and Noble Jar,* p. 103. By 1870 the J. P. Bodie pottery was in operation at Kirksey's Crossroads in the Edgefield District.

20. A grant from the Chipstone Foundation funded the cost of materials and documentation.

21. Baldwin, *Great and Noble Jar,* p. 141.

Diana Edwards

and

Lynne Dakin Hastings

Ceramics at Hampton, Baltimore County, Maryland

▼ THE MOST PROMINENT FAMILY in Baltimore County, Maryland, in the eighteenth and nineteenth centuries was the Ridgely family. By the eighteenth century, it had amassed twenty-four thousand acres of land and built one of the largest and grandest mansions in the country (figs. 1, 2). Hampton Hall, or Hampton Mansion as it is now known, was home to Ridgely families from 1788 to 1948, when it became part of the National Park Service.

The individual styles of each succeeding generation to occupy Hampton are reflected by, among other things, their ceramics. The complete saga requires examination of each generation, but here we concentrate on the period during which most of the ceramics used by the family were purchased, between the last decade of the eighteenth century and about the last quarter of the nineteenth century.

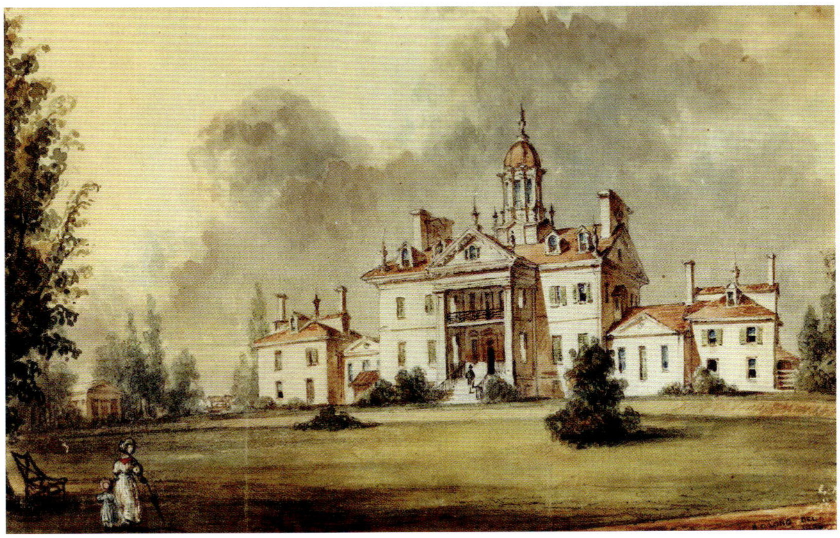

Figure 1 Robery Carey Long Jr., *Hampton Mansion*, 1838. Watercolor on paper. (Private collection; unless otherwise noted, all objects and photos are courtesy of Hampton National Historic Site, National Park Service.)

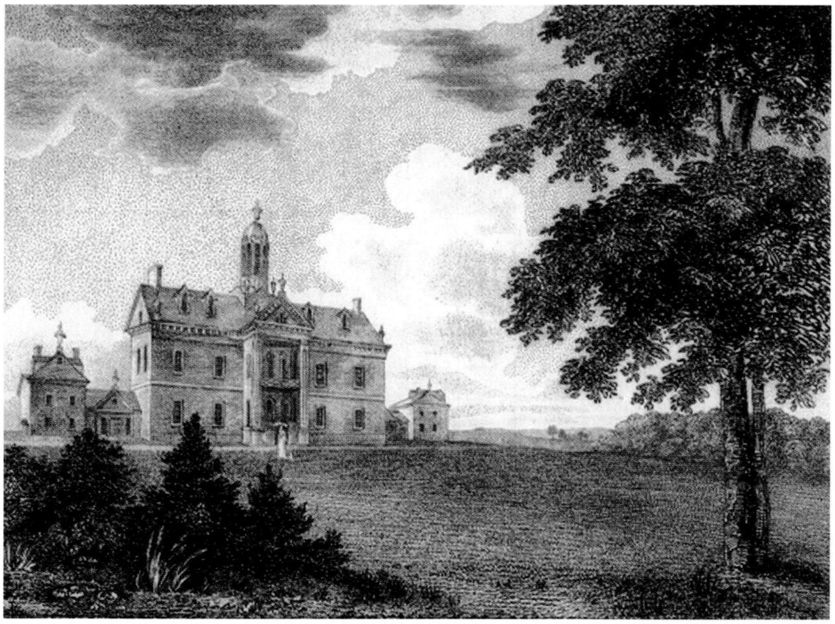

Figure 2 William Russell Birch, "Hampton the Seat of Genl. Chas. Ridgley, Maryland," *The Country Seats of the United States of North America, Springland near Bristol, PA,* 1808. Engraving. 7" x 8½".

The Establishment of an Empire, 1761–1790

The Ridgely family has its roots in the seventeenth century with one Robert Ridgely (1616–1681), a barrister known to have emigrated from England about 1634.[1] By 1666 he was established as a clerk of the Maryland Council and he held other offices in the province thereafter. The early Ridgelys were plantation owners but some—such as Charles the Planter (d. 1705), who died with debts that nearly equaled the value of his estate—are shrouded in obscurity.

It is with Charles's son, Colonel Charles Ridgely (1700–1772), that the history of the family begins to come to life. By 1760 the colonel began moving his interests away from the plantation life of Anne Arundel County, Maryland, to the area north of Baltimore. He purchased one hundred acres situated just north of "Northampton," a tract acquired in 1745, where he established with his two sons, John and Charles, an ironworks put into blast in 1762. The rich local deposits of iron ore combined with tax incentives encouraged by the local government and the British Crown made bar and pig iron in the Chesapeake area a major industry. To help sustain family shares in the iron industry Colonel Ridgely deeded some two thousand acres to his younger son, Captain Charles Ridgely (1733–1790), who effectively undertook the management of the ironworks.

Captain Ridgely's early career had been at sea, possibly on one of his father's ships looking after the family's commercial interests. He later assumed command of the *Baltimore Town,* a ship transporting raw materials such as iron and agricultural products from the colonies to London, where they would be exchanged for finished goods. In 1760 Captain Ridgely married Rebecca Dorsey, the daughter of another wealthy ironmonger in Anne Arundel County. By 1763 he had returned from sea and began diversifying his commercial interests to include, in addition to the Northampton ironworks, a general merchandising business in Baltimore and the operation of several mills, quarries, plantations, and at least one large orchard.

The Revolutionary War and other factors hurt trade with Britain and substantially changed the focus of the iron business, moving it into camp kettles, shot, and cannon production. Essentially a pre-Jeffersonian Democrat, after the war Ridgely espoused the interests of the common man, supported paper money, and campaigned for agrarian reform. He served in the Maryland House of Delegates and was popular with his constituents.

Captain Ridgely's business acumen had made him a very wealthy man, and, as other Marylanders in his position were doing, in 1783 he embarked on building a fine country seat, what would become Hampton Hall. The work on the mansion, completed with the labor of slaves, indentured servants (often Irish), and paid artisans, took seven years. Every aspect of the project was overseen by Captain Ridgely, who was nicknamed "the Builder." The end result was a magnificent five-part, three-story Georgian building made of stuccoed stone with a prominent octagonal cupola rising thirty-four feet above the roof. The cupola, unique in American federal architecture, might have been inspired by the one atop Castle Howard in Yorkshire, England. Unfortunately, Charles Ridgely did not live to enjoy his splendid creation. He died suddenly in 1790, without offspring or a household inventory.

Rebecca Dorsey Ridgely decided not to remain in the large new house, and in 1791 she reached an agreement with her husband's nephew Charles Carnan whereby she would exchange Hampton Hall for a smaller house known as "Auburn." A curious condition in Captain Ridgely's will required all four of his nephews to adopt the surname of Ridgely if they wished to inherit. Thus Charles Carnan became Charles Carnan Ridgely.

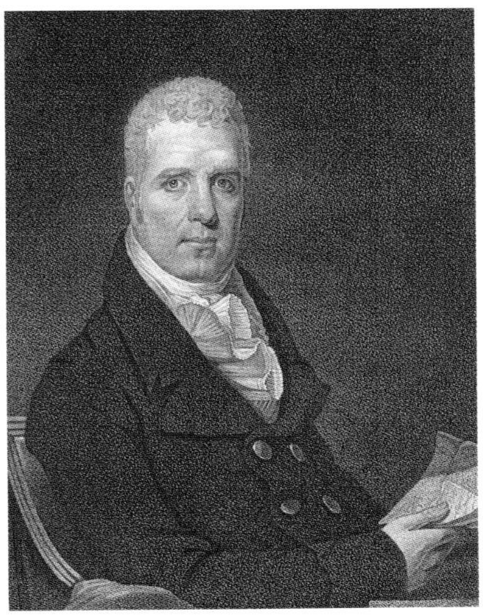

Figure 3 Goodman and Piggot, *Charles Carnan Ridgely (1760–1829)*, ca. 1817–1822. Engraving. 6" x 4¾".

It is with Charles Carnan Ridgely (1760–1829) and the furnishing of Hampton that the picture of grand living in Maryland in the last decade of the eighteenth and the first three decades of the nineteenth century comes into view (fig. 3). Charles Carnan Ridgely was a brigadier general in the state militia by 1796, married Priscilla Dorsey (1762–1814), the younger sister of his aunt Rebecca. Known as General Ridgely, he sat as director on numerous local institutions and was instrumental in establishing the Baltimore and Ohio Railroad. Somewhat surprisingly, considering his uncle's political views, General Ridgely became a leading federalist and promoted the interests of industry and large property ownership. He was elected to the Maryland legislature (1790–1795), the state senate (1796–1800), and served three consecutive terms as governor beginning in 1815. His administration set many precedents, among them establishing free education.

Of particular importance to the image of landed gentry was the presentation of the dining table, and in this Charles Carnan Ridgely excelled. At least by 1818, during his tenure as governor, Ridgely was employing a French chef, who moved between Annapolis, the Baltimore town house, and Hampton Hall in seasonal shifts, but his table and lands were widely known before that. In 1805 the English historian Richard Parkinson wrote in *A Tour of America*: "The General's land is of great note . . . his cattle, sheep and horses & c. of a superior sort. . . . He is very famous for his race-horses. . . . He is a very genteel man, and is said to keep the best table in America. . . . I have often experienced his hospitality."

Although he was given full credit for his gardens and table, perhaps justifiably, one wonders whether his wife had some responsibility for purchasing these objects, listed in the 1829 estate inventory:

4 Pieces of Platto glass	$75.00
18 Platto images	$12.00
3 Platto blocks	$ 3.00
5 Pieces plated ware belonging to Eperne [*sic*]	$5.00
1 Glass Eperne [*sic*]	$15.00[2]

The "images," or figures for the table, were placed on a plateau or a *surtout de table,* a mirror on which the figures appear to be dancing around the table. These small earthenware or porcelain figures often took the form of gods and goddesses (fig. 4). Plateaux typically were seen only in America's elite households and conveyed the classical education of the times. Of George Washington's table, for example, Rosalie Stier Calvert wrote, "Yesterday we dined at the President's House. I have never seen anything as splendid as the table—a superb gilt plateau in the center with gilt baskets filled with artificial flowers. All of the serving dishes were solid silver. . . . The plates were fine French porcelain."[3] In the fall of 1790, when the presidential entourage moved from New York to Philadelphia, Washington's secretary, Tobias Lear, wrote, "The large Images stand on the Side Boards in the Dining Room [of the Philadelphia house] . . . the small Images can be put in the closets with the China."[4] Theophilus Bradbury, a Massachusetts congressman, attending one of the president's Thursday dinners, said, "In the centre [of the table] was a pedestal of Plaster of Paris with [biscuit porcelain] images upon it, and on each end figures, male and female of the same. It was very elegant and used for ornament only."[5] Edward Lloyd IV's table at Wye House, on Maryland's Eastern Shore, danced with images of twenty-nine gods and goddesses, ordered from his London agents in 1792.[6]

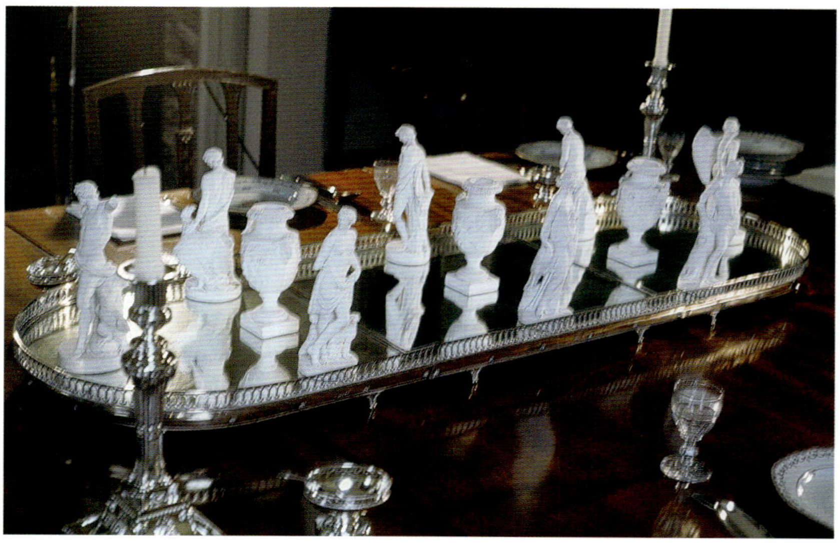

Figure 4 Plateau with figures of gods and goddesses, Niderviller, France, ca. 1792. Porcelain. H. of figures approx. 8¼". (Private collection; photo, Diana Edwards.) Plateaux were mirrored trays on small feet with silver or silver-gilt galleries; larger versions were divided into three or more sections. The one shown here, which included twenty-nine porcelain figures, was ordered by Edward Lloyd IV in 1792. Comprising five parts, it extended to nearly eight feet in length.

On April 22, 1796, Charles Carnan Ridgely purchased "Table Ornaments" totaling three hundred dollars[7] – regrettably, they have not survived. Among the contents of his Baltimore town house on Gay Street were "1 Glass plated Plateau Tray, Plateau Blocks & Plateau."[8]

The 1829 estate inventory of Charles Carnan Ridgely lists eighteen plateau images. There is no way of knowing what the images were made of, but they probably were either French or English porcelain, not plaster of paris or alabaster, as some figures were erroneously described.[9] Those of both Edward Lloyd IV and George Washington were French porcelain, which was readily available in London in the early 1790s, just after the French Revolution.[10]

Other famous tables boasted plateaux and figures to adorn them. Thomas Jefferson used several plateaux while in France and while he was president continued to embellish his tables at Monticello (and presumably in Philadelphia), as well as in Washington. As an arbiter of taste and fashion Jefferson also made purchases for his friends, among them Abigail Adams, for whom he "hunted down French dessert platters of a kind she had her heart set on for her dining table."[11]

Although numerous fine dinner and tea sets from the first quarter of the nineteenth century are in the collection at Hampton, few ceramics survive from the eighteenth century. Those that do include a set of Chinese export famille rose–decorated porcelains, which does not have a history in the family at Hampton (fig. 5), and a pair of porcelain fruit coolers made by the Caughley factory in Shropshire, which does (fig. 6).

Figure 5 Dish, China, ca. 1750–1775. Porcelain. L. 16". NPSHAMP 21464. Although this Chinese export was not part of Hampton's collection during Captain Ridgely's tenure, famille rose Chinese porcelains were popular with wealthy Maryland families, including the family of Charles Carroll the Barrister.

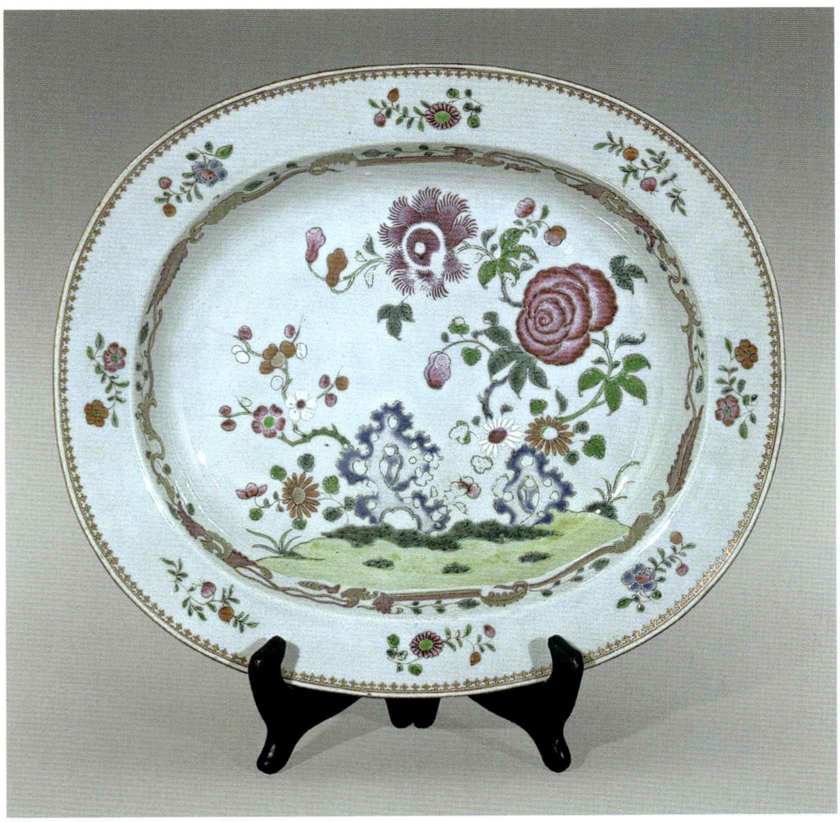

Figure 6 Fruit cooler, Caughley factory, Shropshire, England, ca. 1750–1775. Soft-paste porcelain decorated in the Dresden pattern. H. 9". NPSHAMP 11518. This cooler is one of a pair.

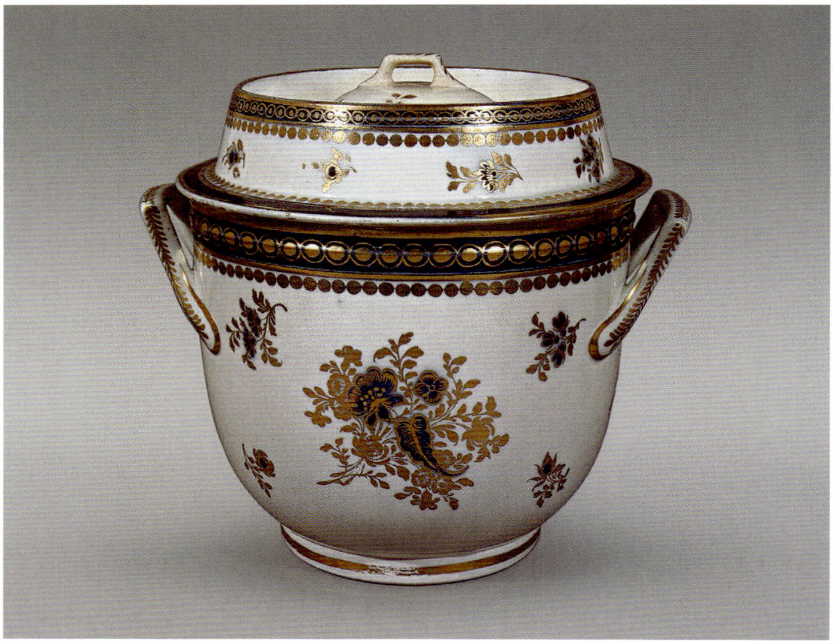

The ceramics purchased by families in colonial America prior to 1790 were primarily Chinese exports or English earthenwares, delft, creamware, and pearlware. Due to either a surfeit of the former or a greater availability of the latter, by 1790 one begins to see a preference for English and French porcelain over Chinese porcelains, which by then were considered old-fashioned. Charles Carnan Ridgely's purchases of English porcelain during his tenure at Hampton shows he was one of the earliest Maryland families to do so. Unfortunately, the brief descriptions in the 1829 estate inventory taken after his death—"1 Set English China (42 pieces) $80.00"[12]—make it impossible to identify individual items. English china sets known to have been at Hampton in that period include ones made by Coalport (figs. 7, 8), Davenport (fig. 9), and two different sets of Worcester (figs. 10, 11) and Spode (fig. 12).

Figure 7 Dessert dish, Coalport factory, Shropshire, England, ca. 1820. Bone china, enamel painted in cobalt blue with gilding. L. 13". NPSHAMP 40353. At the center of each piece from this service is the Ridgely family crest.

Figure 8 Detail of the Coalport factory mark on the reverse of the dessert dish illustrated in fig. 7.

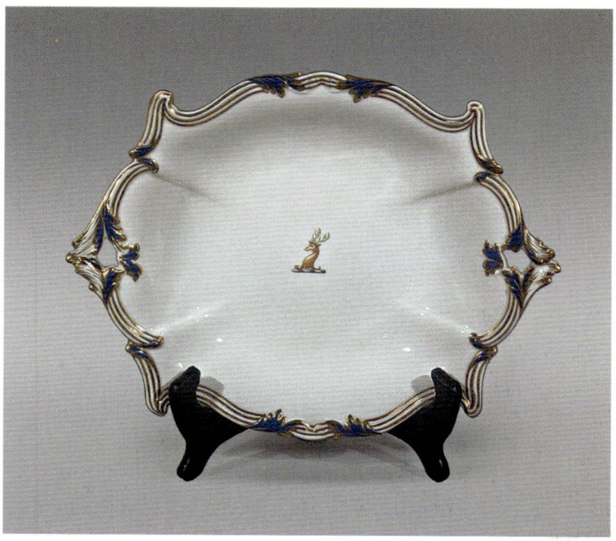

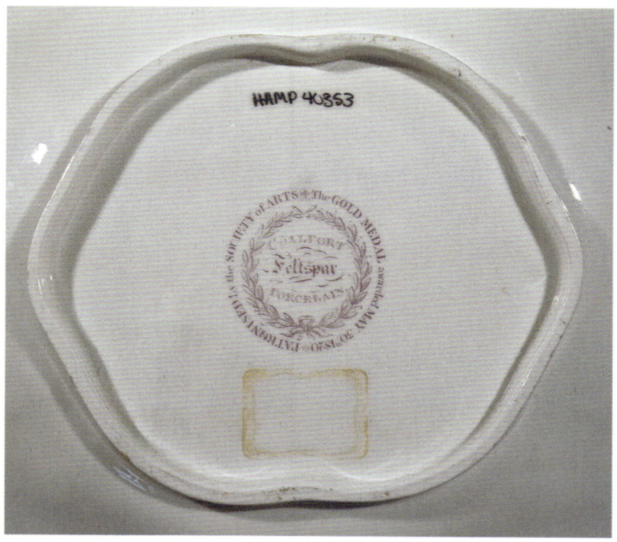

Figure 9 Basket and stand from dessert service, Davenport factory, Staffordshire, England, ca. 1805–1810. Bone china, enamel painted and gilded. H. of basket 6½". L. of stand 10¾". Marks: on both, a red anchor surmounted by "Davenport" in a banner; the pattern number or painter's mark "208." NPSHAMP 1240, 1244.

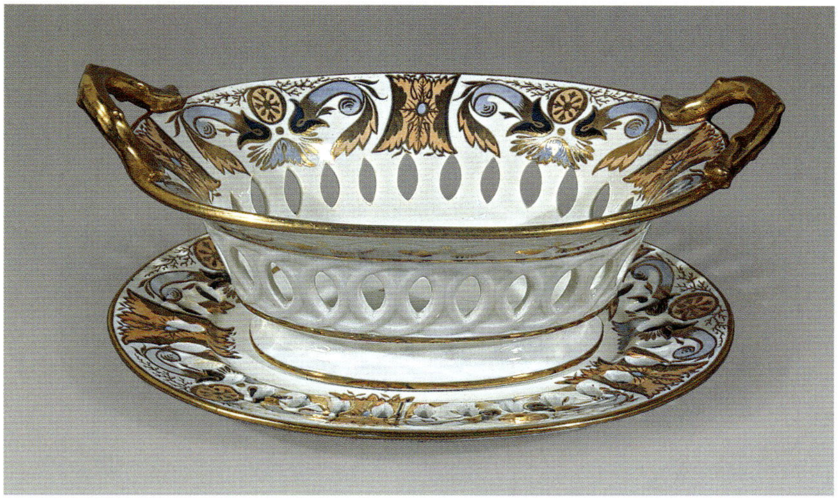

Figure 10 Dish from dessert service, Worcester Porcelain Manufactory (Chamberlain's factory), England, ca. 1810. Soft-paste porcelain, painted with cobalt blue and gilded. L. 12½". Mark: "155" (pattern number). NPSHAMP 10207.

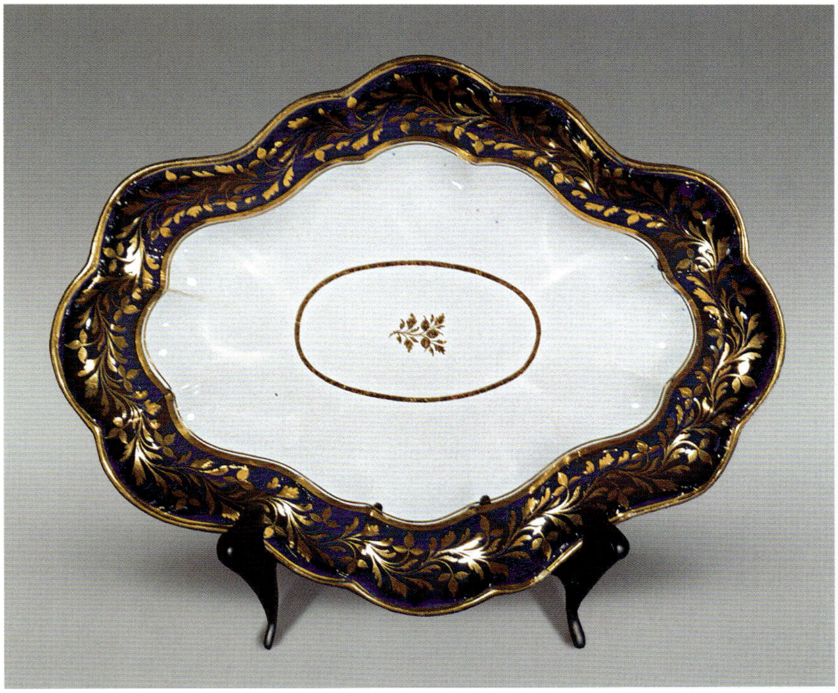

Figure 11 Detail of Chamberlain's Worcester mark on the reverse of a square dessert dish from the same service as the dish illustrated in fig. 10.

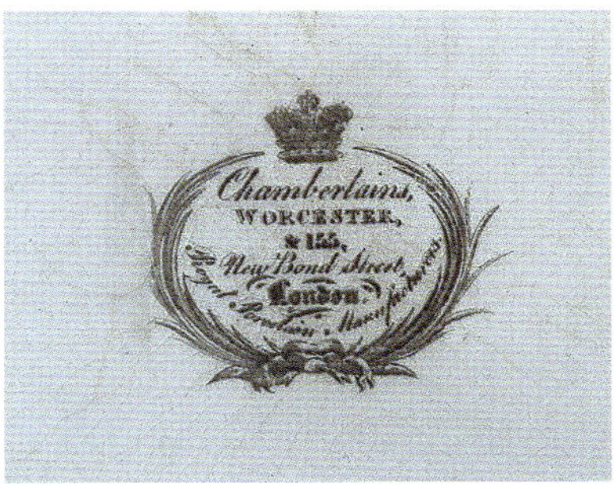

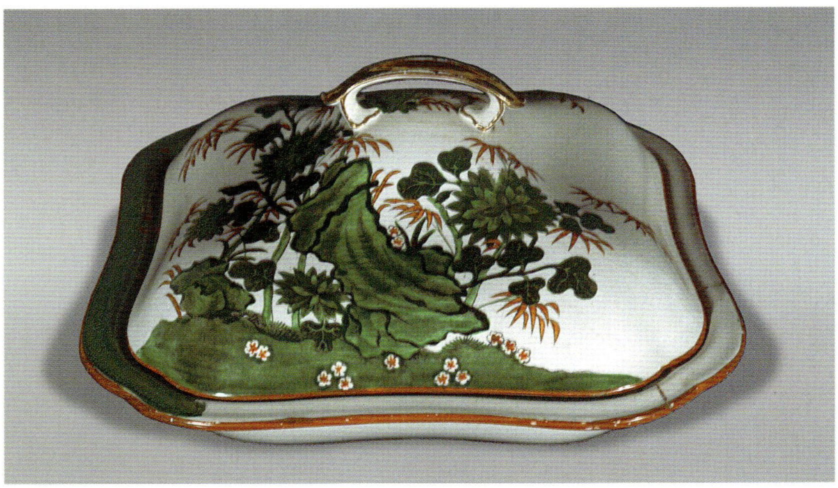

Figure 12 Vegetable dish and cover, Spode, Staffordshire, England, ca. 1808. Stoneware, with green enamels and gilding. H. (with handle) 6", L. 8¾". Mark: "1666" (pattern number). NPSHAMP 9003a, b.

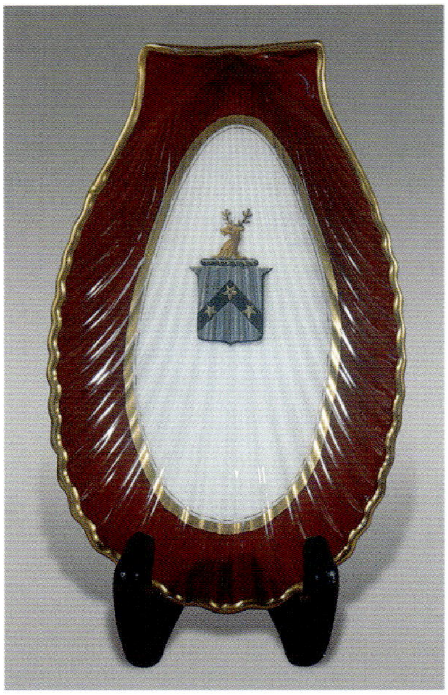

Figure 13 Serving dish from a dinner service, decorated by Feuillet, Boyer, Blot, and Hébert, Paris, ca. 1825. French porcelain, with enamel and gilding. L. 9⅛". Mark: "Feuillet." NPSHAMP 4158. At the center of each piece in the set is the Ridgely family coat of arms.

Another ceramic entry in the 1829 estate inventory was "1 Set French China (68 pieces)," also valued at eighty dollars. This is a rare instance in which such a generic inventory description can reasonably be assigned to particular objects, in this case the bold set of "Feuillet" porcelains, made in Paris and decorated with the Ridgely coat of arms, that are in the dining room at Hampton (fig. 13). In cupboards holding literally dozens of dinner services, tea sets, and accessory pieces, however, one must use other means to distinguish the purchases of Charles Carnan Ridgely from those of the dynamic Eliza Eichelberger Ridgely (1803–1867) of the next generation.

Listed as part of the 1829 estate inventory were "16 Blom Monge moulds," or *blanc mange* molds, which are no longer in the collection, as well as a "Black Tea Pot," of the basalt so popular at the time. One set of 297 pieces of "Blue Canton"—clearly the service the family used informally—was valued at two hundred and fifty dollars (fig. 14).

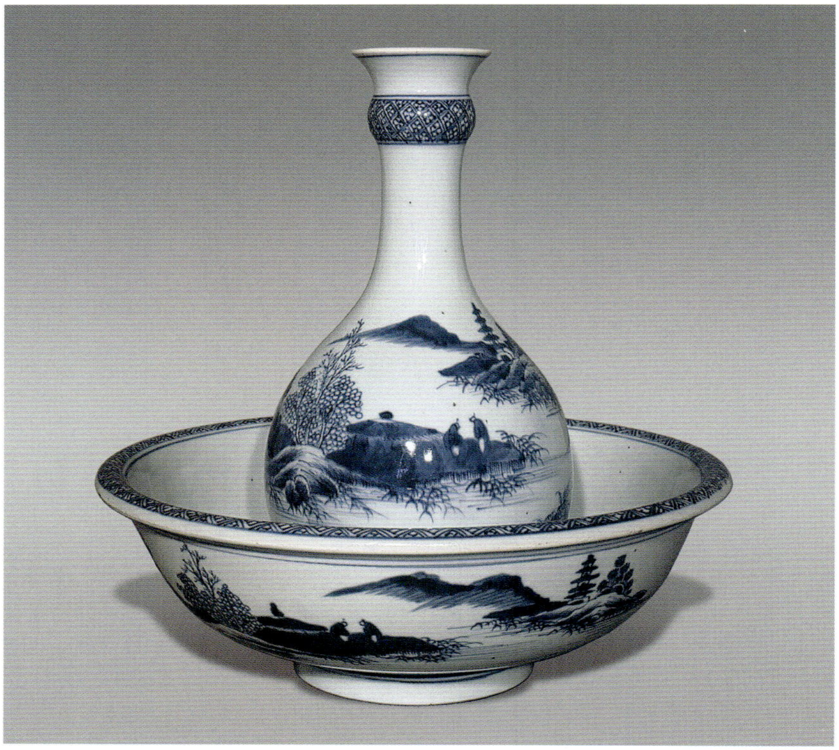

Figure 14 Ewer and basin, China, ca. 1750. Porcelain, with underglaze blue painting. H. of bottle 10½"; D. of basin 12". NPSHAMP 4646, 4647. These Chinese exports were not part of the Ridgely estate but a museum purchase appropriate to furnishing the master bedchamber to circa 1790.

The Gay Street house sale after General Ridgely's death included another set of Canton china, a sixty-eight-piece dessert set of English china, and another dessert set in French porcelain, none of which can be identified with any certainty. "2 White & Brown Pitchers," also among the Gay Street inventory, were probably similar to the one purchased for Hampton by the museum staff in order to restore the inventoried objects (fig. 15). The circa 1810 pearlware cachepot with "moonlight luster" decoration (fig. 16) is in the style of Charles Carnan Ridgely's purchases.

By the early part of the nineteenth century French porcelain was very fashionable in Maryland, and most of the wealthy families—the Ridgelys among them—possessed at least some examples of it. Somewhat surpris-

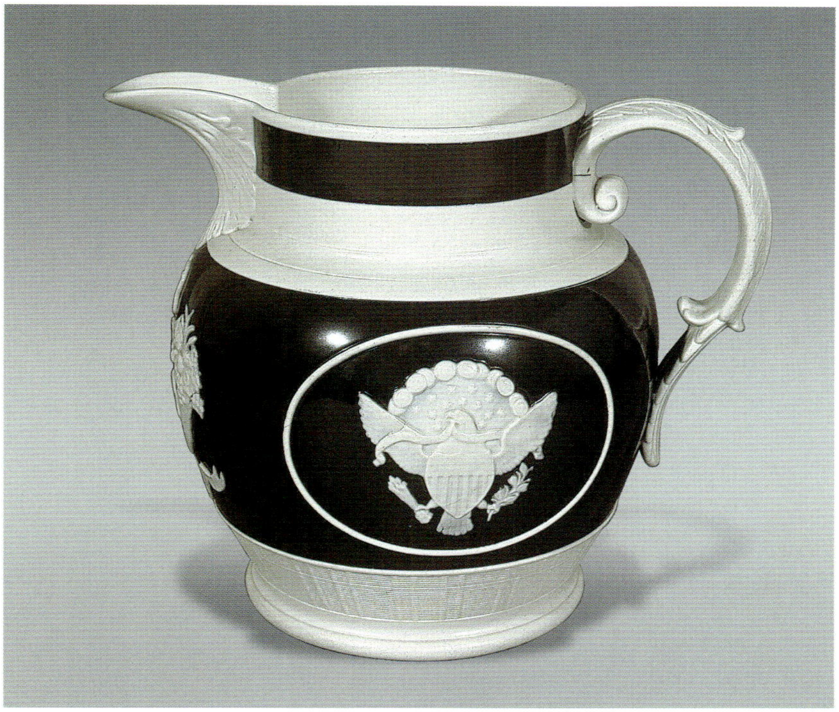

Figure 15 Pitcher, England, ca. 1805–1815. White stoneware, with brown dipped surface decoration. H. 6¼". NPSHAMP 4178. Pitchers made for the American market were produced by several Staffordshire and Yorkshire factories and are almost always unmarked. This example bearing sprig-relief American motifs was a museum purchase, relating to sherds found archaeologically at Hampton and in 1829 estate references.

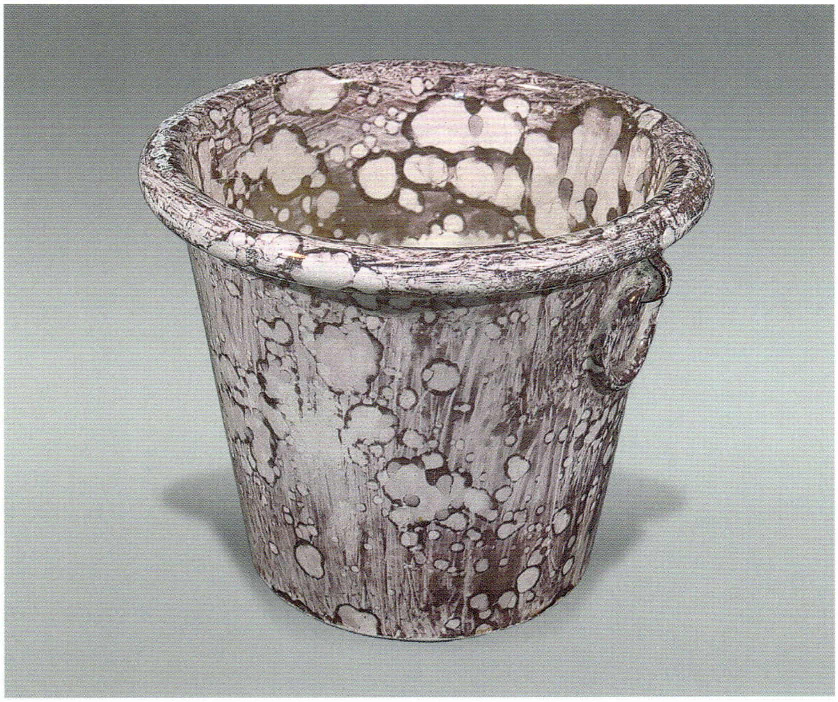

Figure 16 Cachepot, England, ca. 1810. Pearlware, with "moonlight luster" decoration. H. 5". NPSHAMP 10217. This cachepot, one of a pair acquired by the park in 1982, is typical of Charles Carnan Ridgely's purchases in this period and helps to interpret his significant horticultural interest. Luster-decorated earthenwares are rarely attributable to a specific manufacturer.

ing, considering it was on the heels of the French Revolution, Philadelphia newspapers carried advertising for French porcelain in the 1790s. Pasquier and Company, at 91 South 2nd Street, for example, advertised "French China" in the *Aurora and General Advertiser* on February 18, 1796: "Received from different factories in France. . . . Table sets, dessert sets, tea sets . . . & flower pots of different shapes . . . china figures, hyacinths with stands. . . ."

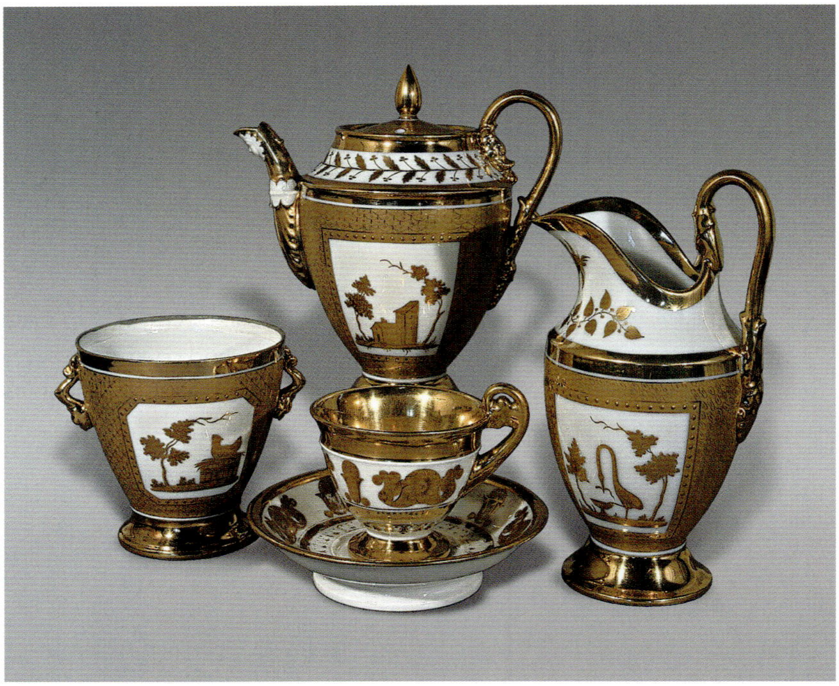

Figure 17 Partial tea/coffee set, decorated by Feuillet, Boyer, Blot, and Hébert, Paris, ca. 1825. Porcelain, with elaborate gilding. H. of tea/coffee pot 8⅜". NPSHAMP 21413, 21419, 21424, 21425, 21426.

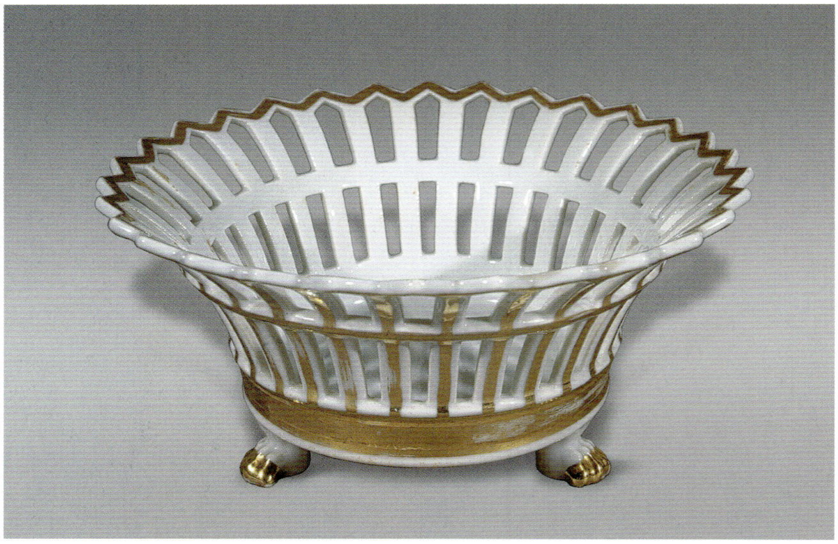

Figure 18 Fruit basket, Paris, ca. 1825. Porcelain, with gilding. H. 7", D. 9⅜". NPSHAMP 4773.

In 1806 and 1807 Rosalie Stier Calvert expressed her desire to have both dinner and tea services in French porcelain, pointing out that it was used at the White House and at the English ambassador's residence.[13] Stephen Decatur's estate inventory of 1820 included twenty dozen pieces of "Fine French China," valued at eighty dollars, and ten dozen French china plates, valued at one hundred and twenty dollars.[14] The 1827 estate inventory of John Eager Howard (1752–1827) listed French "blue and gold" porcelain as well as one hundred and fourteen pieces of "French China gilt insides."[15]

One wonders how much influence the marriage of Elizabeth ("Betsy") Patterson, the beautiful daughter of Bank of Baltimore president William Patterson, to Napoleon Bonaparte's younger brother Jérôme had on the

French porcelain market in Maryland. Betsy met and married Jérôme Bonaparte in 1803, during the Baltimore "season." Incensed that his brother contracted a marriage without his permission, Napoleon insisted that Jérôme return to France or forfeit the crown of Westphalia. In 1805 they sailed for France, but Betsy was denied entry. Jérôme made the decision to obey his brother and she continued on to England, where she gave birth to their son, Jerome. Their marriage was annulled in 1806, and by the end of 1807 Jérôme had remarried and was occupying the throne of Westphalia. Betsy returned to Baltimore, where she was known as Madame Bonaparte and continued to live in the style of French royalty. Her numerous sets of French porcelain, many of which were purchased after her marriage was annulled, are now in the Maryland Historical Society.[16]

Perhaps the most beautiful example of fine French porcelain in the Hampton collection is the ornately gilded tea/coffee set decorated by the Paris workshop of Feuillet, Boyer, Blot, and Hébert (1817–1834) (fig. 17). The Feuillet workshop, among the most prestigious of the Paris porcelain decorating studios, also provided Betsy Patterson Bonaparte with some very handsome pieces. Other Paris porcelain at Hampton is a partial dessert set including the fruit basket illustrated in figure 18.

Charles Carnan Ridgely purchased a large dinner service and numerous pieces of sprig floral porcelains (figs. 19–21)—some French and English but

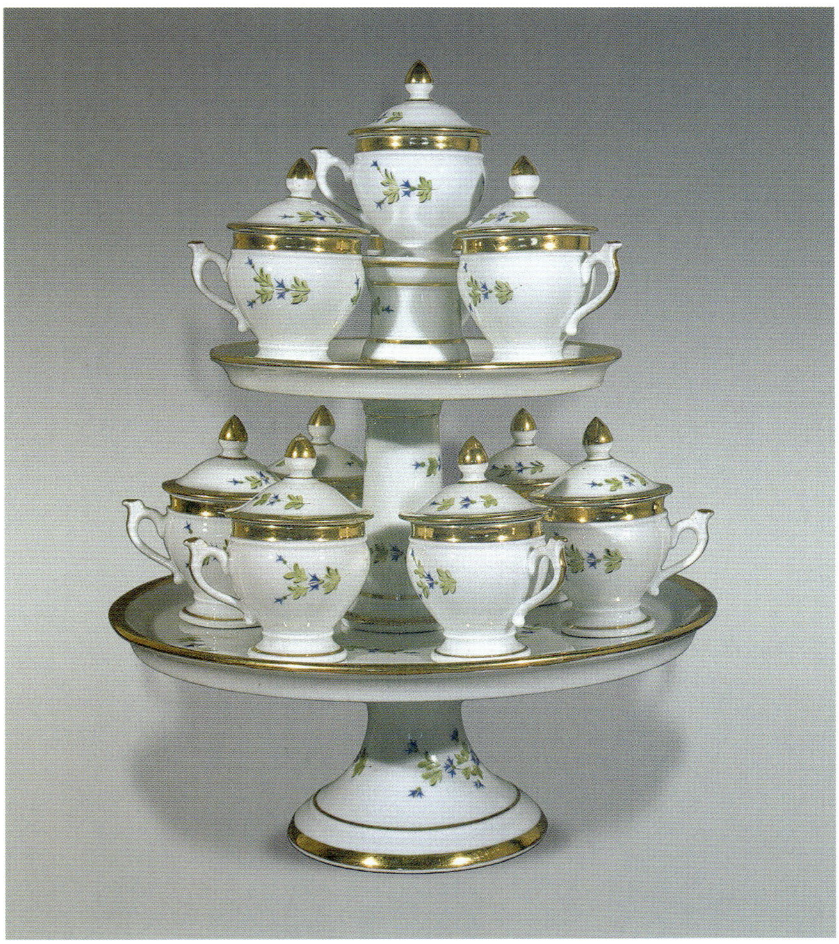

Figure 19 Stand with *pots de crèmes,* Paris, ca. 1810–1830. Porcelain, with enamel decoration. H. of stand 15". H. of each *pot de crème* 4⅜". The stand comprises five pieces held together by a metal rod. NPSHAMP 41998a–e, 41979–41989.

Figure 20 Covered butter dish on stand, France, ca. 1810–1830. Porcelain, with sprig or cornflower decoration. L. 12". NPSHAMP 42005.

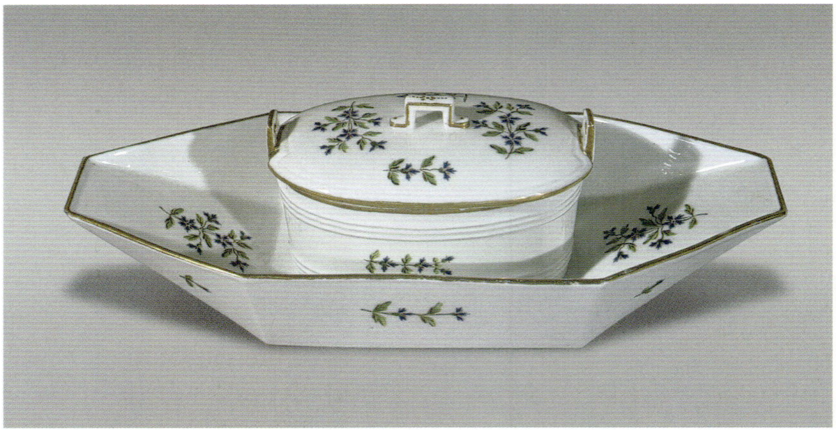

Figure 21 Teapot and lid, probably Worcester, Flight, and Barr, or Caughley (Shropshire), England, ca. 1790. Soft-paste porcelain, with sprig or cornflower decoration. H. 7½". (Courtesy, private collection).

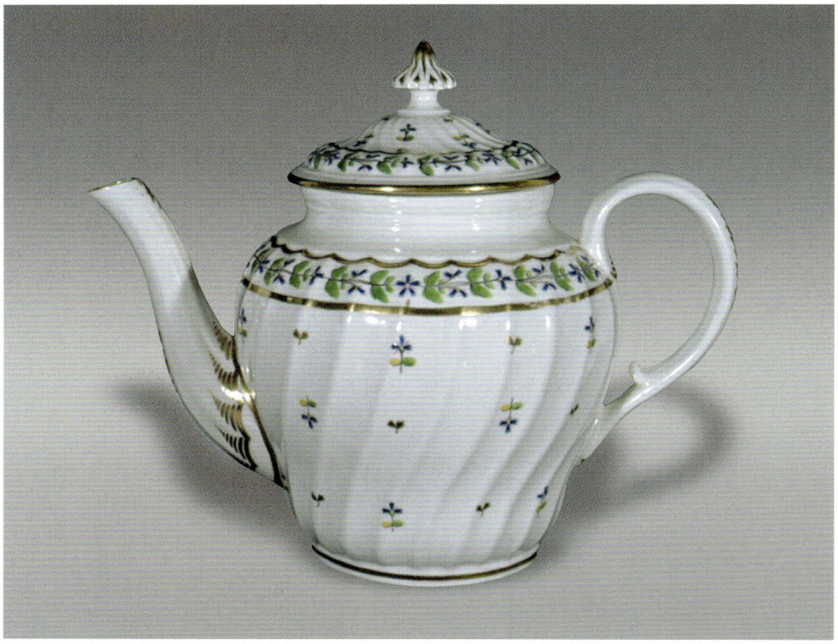

Figure 22 Plate, cup and saucer, Haviland and Grouvelle, Paris, ca 1834. *Plate:* Hard-paste porcelain. D. 8½". Mark: "H & C." *Cup:* Hard-paste porcelain. D. 2⅛". Mark: "H & C." *Saucer:* Paris porcelain. D. 5". Mark: in script, "Halley." NPSHAMP 9483, 9443, 9701. A group of teawares from two different periods was purchased by Eliza Ridgely, perhaps in 1834, as replacement pieces for an earlier service.

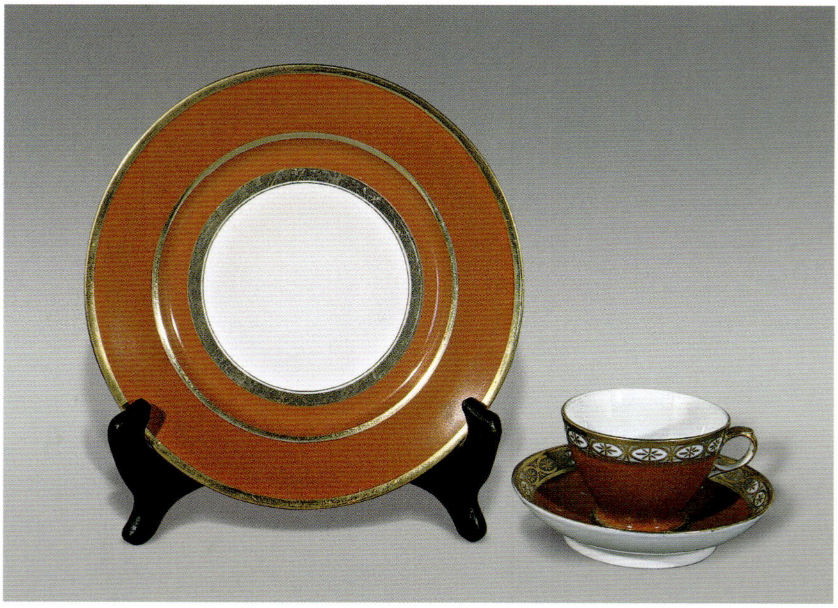

most of them French. Although some remain at Hampton, the majority has been given to the Maryland Historical Society or has been otherwise dispersed. The sprig porcelains with cornflower enamel decoration were very popular with prominent Americans such as Benjamin Franklin, Christopher Gore, Thomas Jefferson, and John Tayloe III.[17] The duc d'Angoulême's Paris manufacture was commonly associated with the cornflower sprig motif, although many factories used it to decorate their porcelains.

Charles Carnan Ridgely was still acquiring furnishings for Hampton as late as 1828, a year before he died, as he purchased several costly items at the estate sale of his close friend John Eager Howard.[18] Ridgely's will, dated April 28, 1828, contained a provision to free more than three hundred slaves listed as part of the estate at the time of his death, a decision that would affect the fortunes of the estate in years to come. After his death, the Ridgely family's great empire was divided among the general's eleven surviving children. The eldest, Charles Carnan Ridgely Jr., had been groomed to take the reins of the Hampton estate, but he had died in a riding accident in 1819 at age thirty-six. His brother John Carnan Ridgely (1790–1867) assumed ownership of the mansion and approximately four thousand surrounding acres, a fraction of the estate's original twenty-four thousand.

Unlike his brother Charles, John Carnan Ridgely was unprepared to be master at Hampton, nor was his life marked by the ambition and desire for public service characteristic of the mansion's two previous masters. His first wife, Prudence Gough Carroll of the Mount Clare Carrolls, died young, as did their six infant children. His second wife, Eliza Eichelberger Ridgely (1803–1867), was a wealthy Baltimore woman whose family, despite the name, had no apparent connection to the Ridgely family into which she married. Eliza, who by all accounts was beautiful, charming, and accomplished, had attended boarding school in Philadelphia at Miss Lyman's Institution. In 1818 the Philadelphia portraitist Thomas Sully painted her playing the harp; coincidentally, he also painted likenesses of both her future husband, John Ridgely (in 1841), and her father-in-law, Charles Carnan Ridgely (in 1820). It is said she captivated the marquis de Lafayette on his visit to Baltimore in 1824 with her playing of the harp, and they corresponded thereafter. When John and Eliza visited France in 1833–1834, they were guests at La Grange, Lafayette's home.

John and Eliza had five children, two of whom survived to adulthood.[19] Eliza's daughter's diaries from the 1840s reveal the seasonal migration between Hampton and the house in town, the tutoring of the children, and illnesses that resulted in death for two of the children. (The diaries do not mention how the third child died.) The Ridgelys were at Hampton every year from approximately April to October, returning to Baltimore for the Christmas holidays and house parties. The house in town during this generation was at 25 Hanover Street, near the harbor.[20]

John and Eliza Ridgely traveled widely, taking several extended tours of Europe: 1833–1834, 1846–1848, 1852, and 1859. It was on tours to Austria, Belgium, England, France, Germany, Italy, Malta, and Turkey that Eliza bought many decorative objects to adorn Hampton. It was a popular time for wealthy Americans to travel abroad collecting objects of culture and refinement. Philip Hone, mayor of New York in 1825 and a dedicated diarist (1828–1851), declared, "All the world is going to Europe."[21]

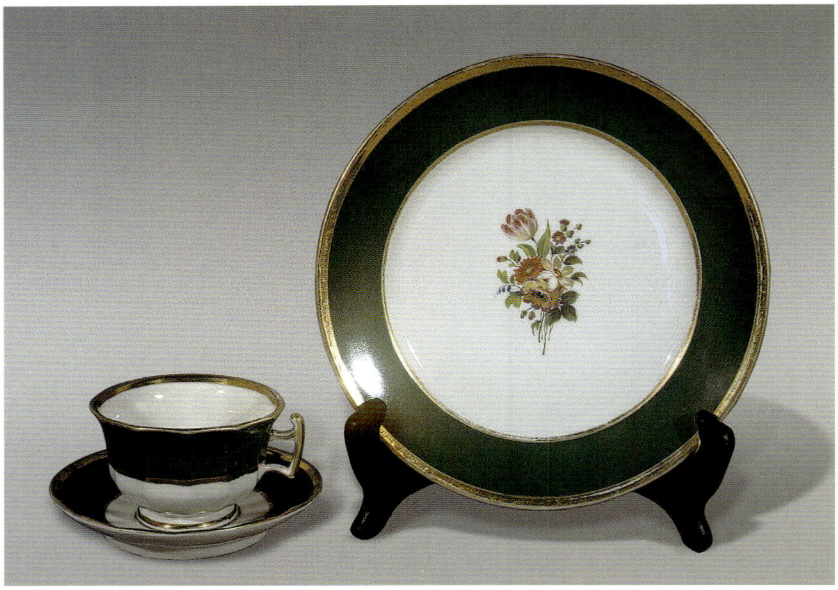

Figure 23 Plate, cup, and saucer, decorated by Feuillet, Boyer, Blot, and Hébert, Paris, ca. 1830–1834. Porcelain, with gilding. D. of plate 8¾"; H. of cup 2½"; D. of saucer 5½". Mark: "Feuillet." NPSHAMP 4036–4077. This set, with a green band and floral center, was selected by Eliza Ridgely on a trip to Paris in October 1834.

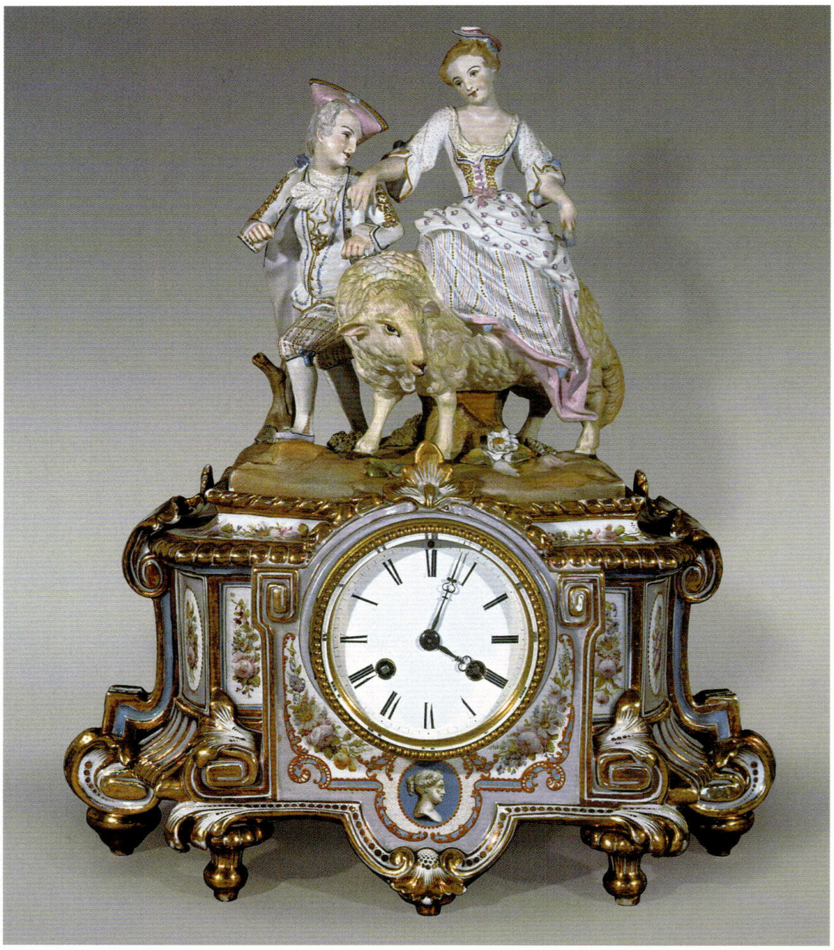

Figure 24 Mantel clock, France, ca. 1840–1850. Porcelain base with enamel painting, surmounted by biscuit porcelain figures. H. 17¾". NPSHAMP 4280a, b. Eliza Ridgely appears to have purchased this clock on one of her trips to Europe. Photos of the music room from the late nineteenth century show the clock on the mantel, where it is today.

Figure 25 Pair of vases, China, ca. 1837–1840. Porcelain, decorated in the Rose Canton style. H. 48¾". NPSHAMP 1025, 1026. A Caroline Butler of New York brought a smaller pair of vases of this variety back from her trip to China in 1837 (David Sanctuary Howard, *New York and the China Trade* [New York: New-York Historical Society, 1984], p. 122).

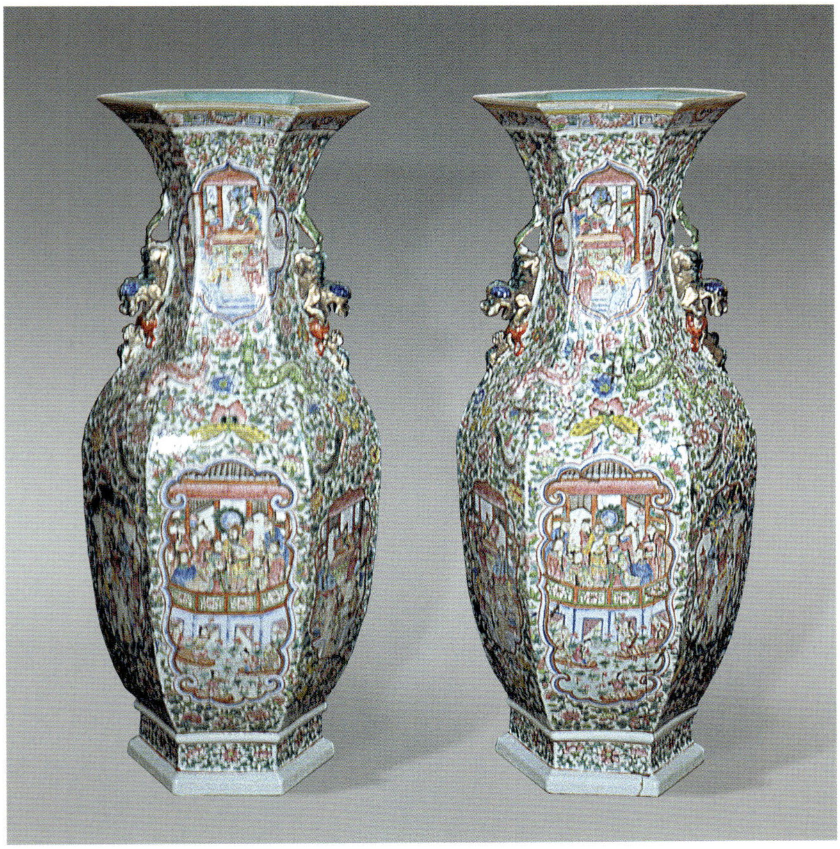

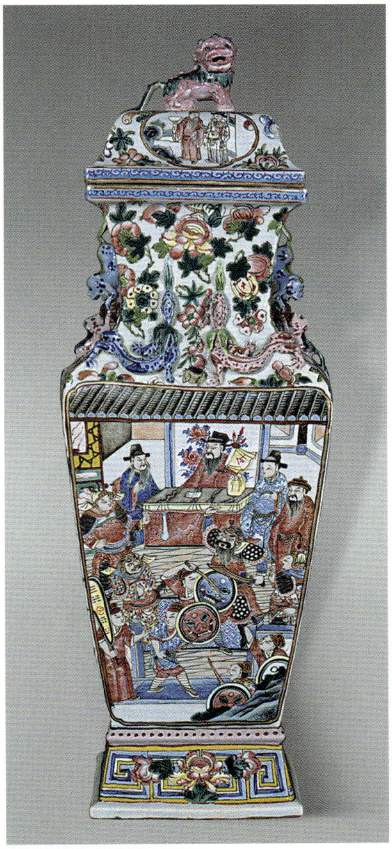

Figure 26 Vase, China, ca. 1840–1850. Porcelain, decorated in the Rose Canton style. H. 20". NPSHAMP 1172. This large vase, with elaborate applied decoration and a cover surmounted by a foo dog finial, is one of a pair.

John and Eliza bought two sets of French porcelain (figs. 22, 23), although it was Eliza who steered the grand ceramic acquisitions—the large bowls and the numerous, impressive Chinese vases, for example, which apparently were purchased abroad. Accounts record diverse purchases including French porcelains, English silverware and carpets, Turkish mirrors, mantel clocks (fig. 24), Venetian glass, Bavarian tumblers, and numerous engravings of European views.[22] In Paris Eliza purchased two leopard skin rugs, which she placed in the Great Hall, where they resided jaw to jaw for the next one hundred years. Nineteenth-century Rose Canton jars standing four feet high (fig. 25), a large pair of Rose Canton vases with attached lids (fig. 26), and a pair of large vases with a black ground (fig. 27), all currently in Hampton's music room, were choices made by John and Eliza when they were traveling overseas. Other large bowls dating to the late Qing period (figs. 28, 29) are impressive monuments to Eliza's superior decorative style. Even the chamber pots and dressing table bottles—in gilded and enameled Chinese porcelain—were fashion statements (figs. 30, 31). A large Rose Canton punch bowl (fig. 32) on view in the hall, though a later gift to Hampton, reflects Eliza Ridgely's taste. A pair of French

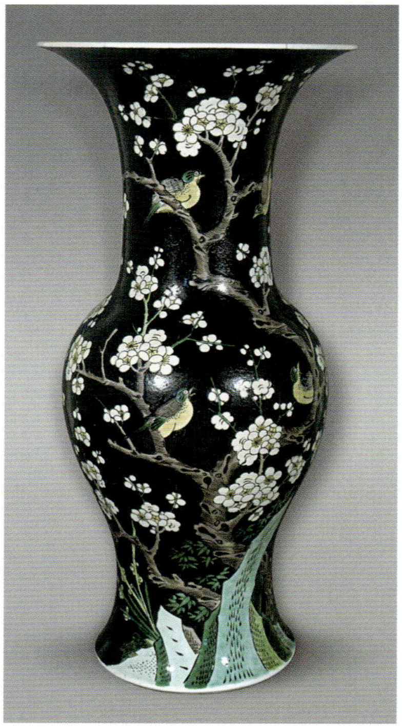

Figure 27 Vase, China, ca. 1860–1870. Porcelain. H. 17⅞". NPSHAMP 10291. One of a pair, this large vase is decorated in a black ground with white prunus blossoms and translucent enamels in the famille verte style. The shape is known as *yen yen* and the vase is like one Warren Delano, a China trade merchant, brought back to New York in 1870. (Howard, *New York and the China Trade,* p. 134.)

Figure 28 Bowl, China, ca. 1785–1800. Porcelain. D. 15½". NPSHAMP 21445. This late Qing dynasty bowl is decorated with a green enamel floral border and a cartouche, painted in sepia, of a landscape.

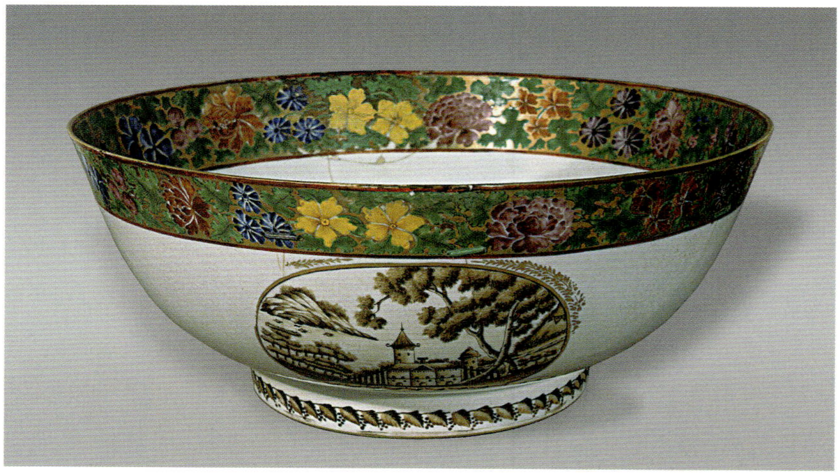

Figure 29 Bowl, China, ca. 1800–1820. Porcelain, with enamel and elaborate gilding. D. 12 1/16". NPSHAMP 17629.

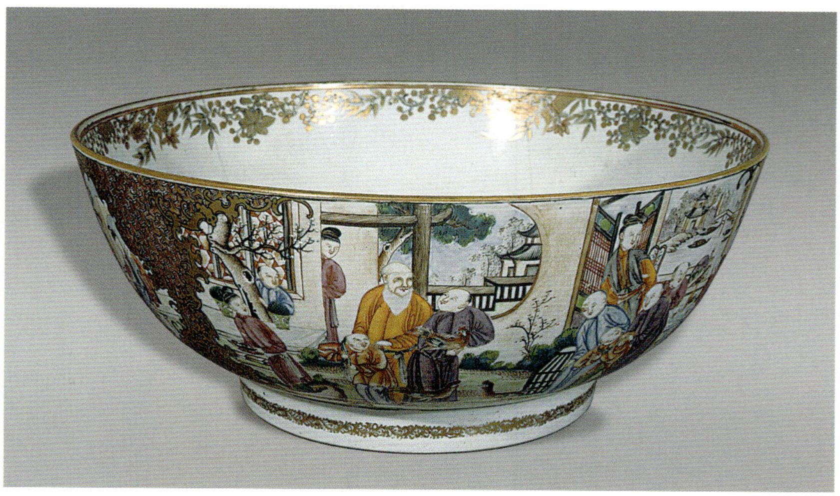

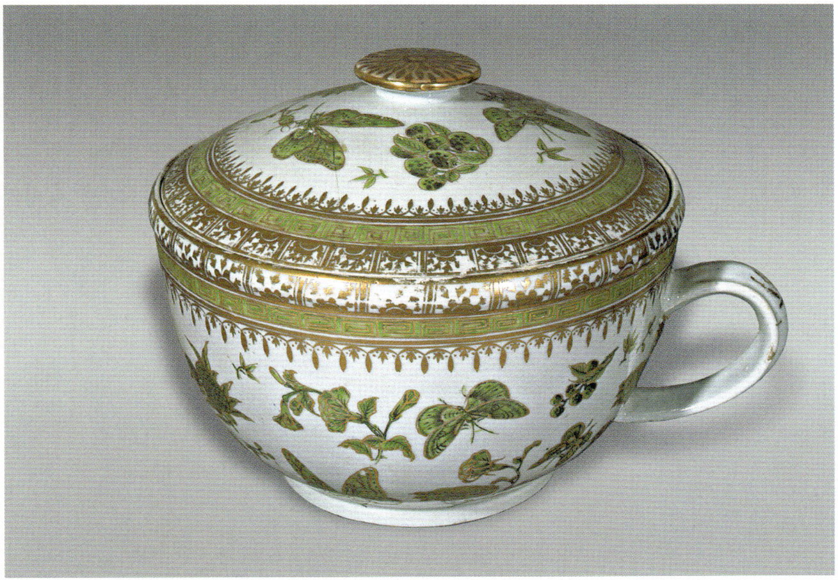

Figure 30 Chamber pot and cover, China, ca. 1825–1835. Porcelain, with green enamel and gilding. H. 8¼". NPSHAMP 10236.

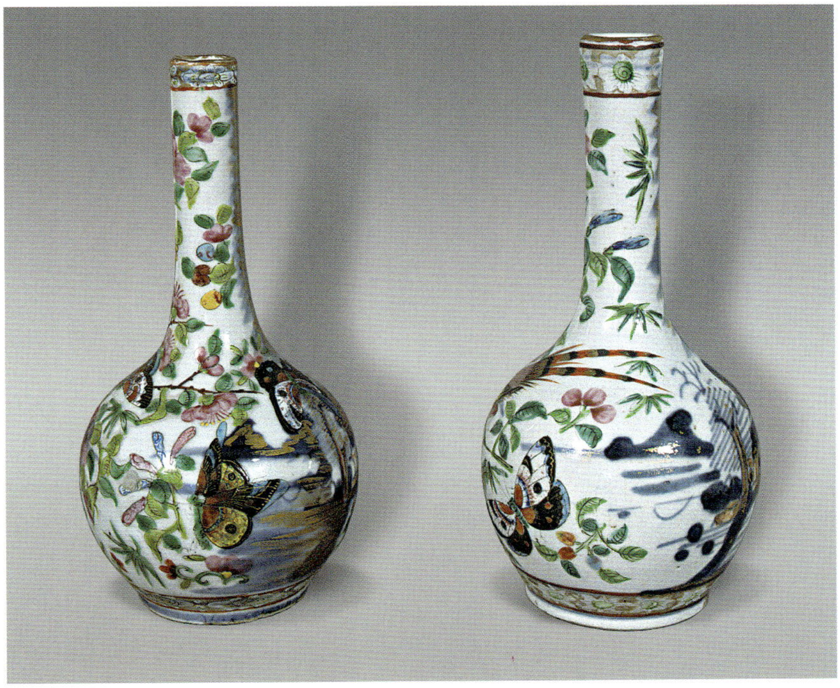

Figure 31 Bottles, China, ca. 1835–1850. Porcelain. H. 6½". NPSHAMP 17630, 17631. These bottles are enamel decorated with the black butterfly motif in the Rose Canton style.

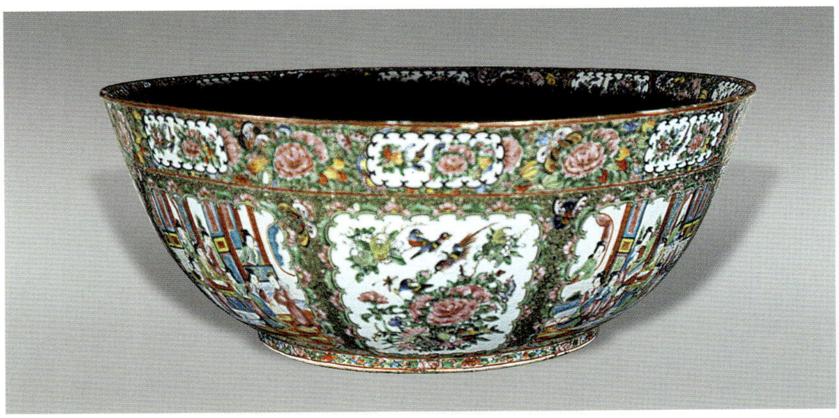

Figure 32 Punch bowl, China, ca. 1850–1870. Porcelain. D. 25". NPSHAMP 4770. This large bowl, decorated in the Rose Canton style, was not part of the Ridgely family possessions but a later gift. It is, however, very much in the taste of the grand purchases Eliza and John Ridgely made in the mid-nineteenth century.

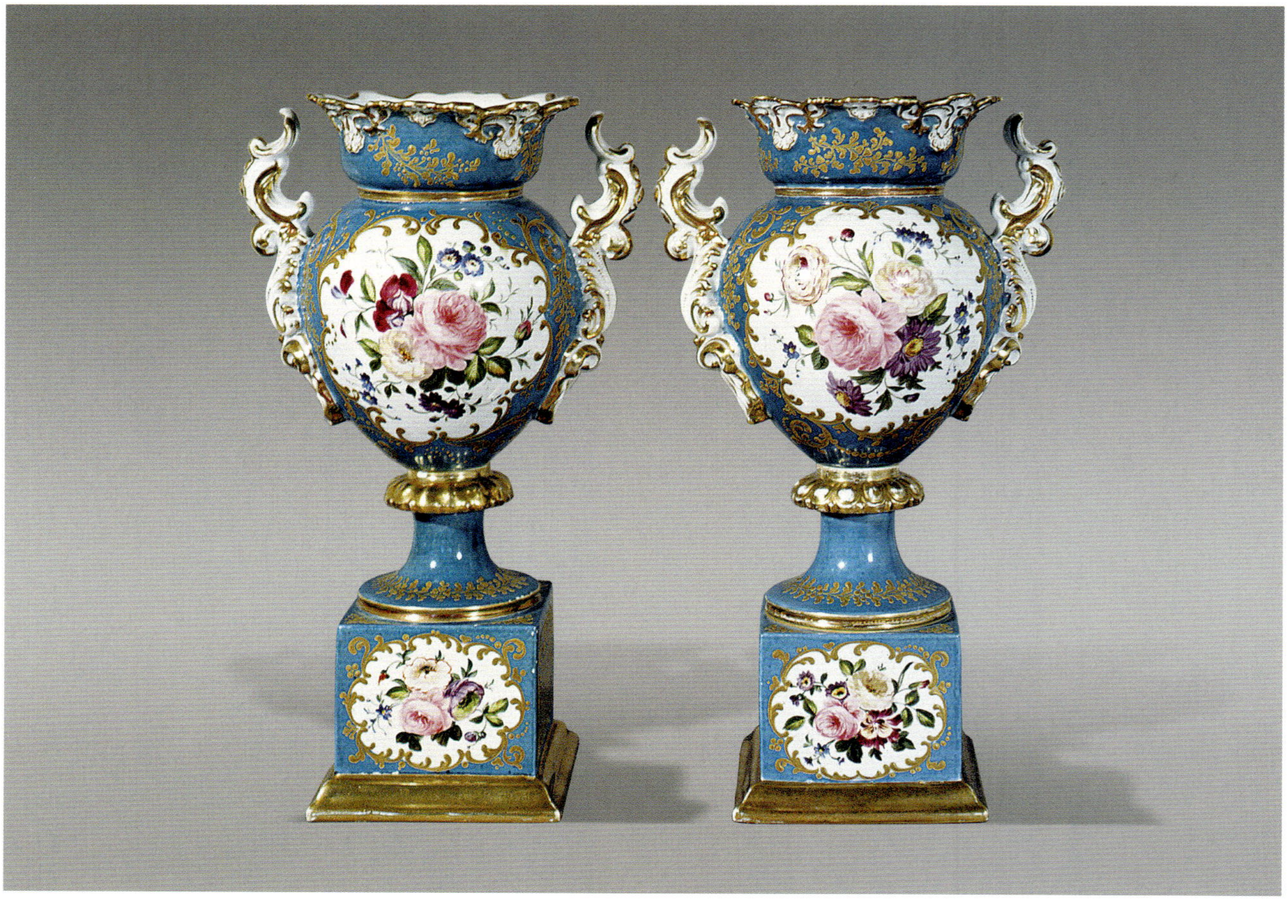

Figure 33 Pair of vases, France, ca. 1850. Porcelain. H. 12¾". NPSHAMP 10246, 10247. According to late-nineteenth-century photos, these vases, decorated with *bleu céleste* ground and reserve panels painted with flowers, sat on the mantel of the music room, where they still reside.

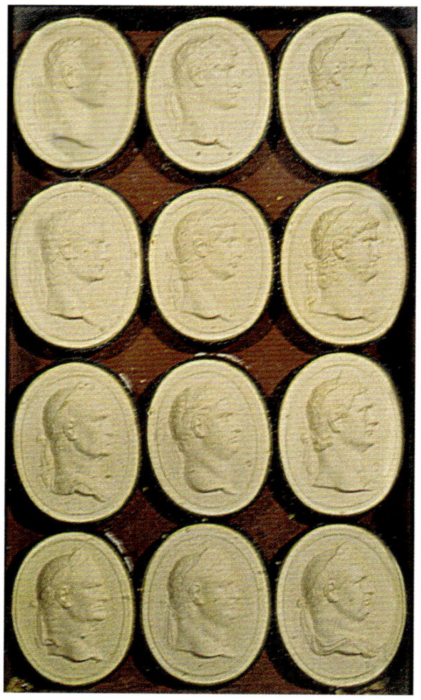

Figure 34 Box with medallions of the heads of Roman emperors, ca. 1778–1800. Wood (box) and plaster of paris (medallions). 12" x 8". NPSHAMP 22538. This box of portrait medallions, part of a set, was probably purchased by John and Eliza Ridgely on one of their European tours.

[illegible]

Domiziano

[illegible] Bartolomeo Paoletti, e Figli, dimoranti di Studio in Piazza di Spagna

— num.° 49 —

Figure 35 Handwritten list of the names of the emperors depicted on the medallions illustrated in 34. The list is affixed to the back of the box.

porcelain vases on the mantel in the music room (fig. 33) might have been acquired when the Ridgelys were in Paris. A set of plaster cast relief models of the heads of the Roman emperors (figs. 34, 35), which date to the last quarter of the eighteenth century, could have been purchased by John and Eliza when they were abroad, or even by Charles Carnan Ridgely.

A large English willow pattern cheese wheel stand (fig. 36) dating to the mid-nineteenth century speaks to a lifestyle at Hampton that would, after the Civil War, be eclipsed by reversals of fortune. By 1850 the Hampton estate had moved from an industrial to an agrarian economy. Fortunes had

Figure 36 Cheese wheel stand, possibly Staffordshire, England, ca. 1850–1860. Earthenware. L. 17". NPSHAMP 10234. Transfer-printed with an underglaze blue willow pattern, this stand seems especially appropriate to find at Hampton, which had its own dairy dating from the eighteenth century.

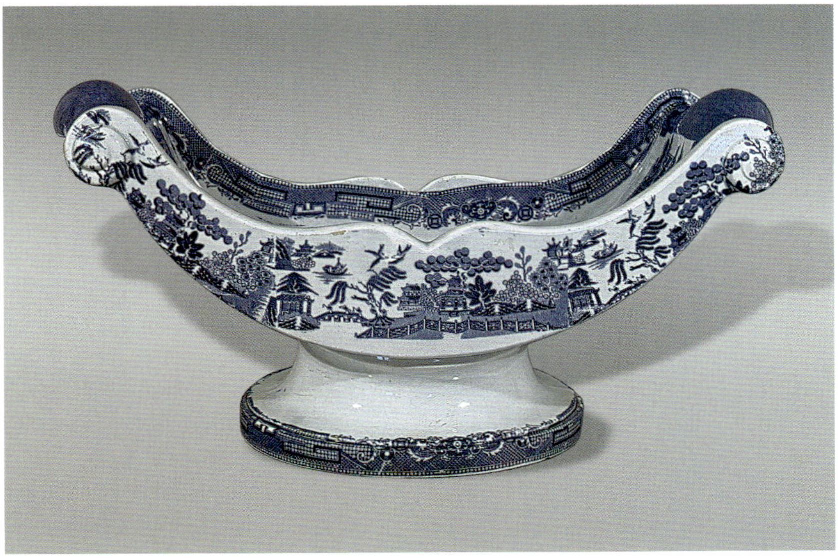

Figure 37 Jug, Minton, Staffordshire, England, ca. 1870. Earthenware, with turquoise scale ground and twisted lizard handle in gilt. H. 11". Marks: incised pattern number "260"; impressed "9A." NPSHAMP 17713. These Japonisme-style vases were the height of fashion at the period.

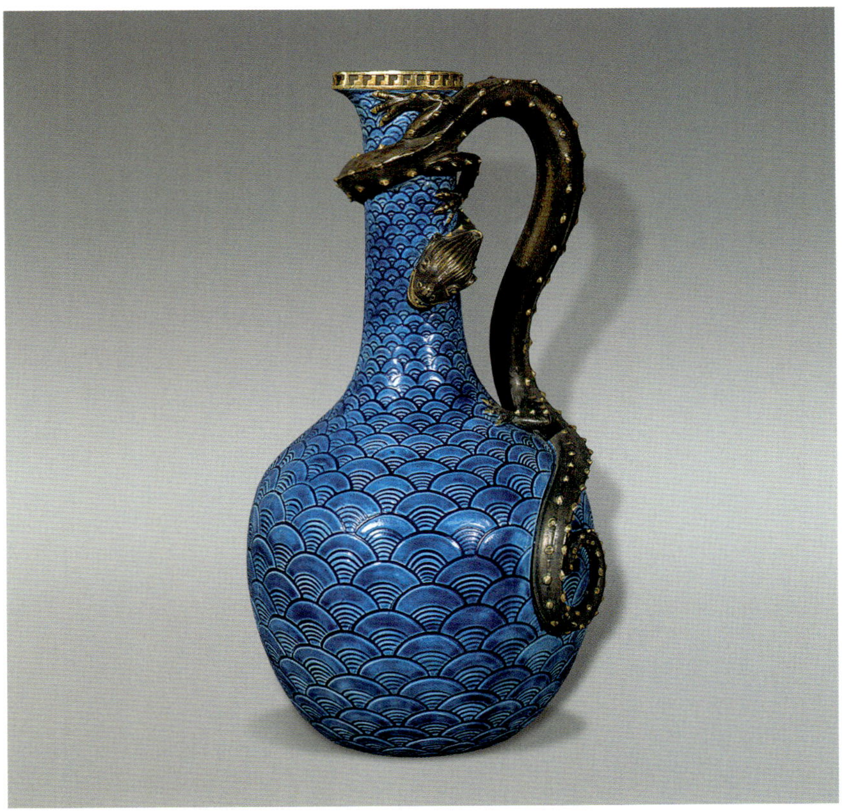

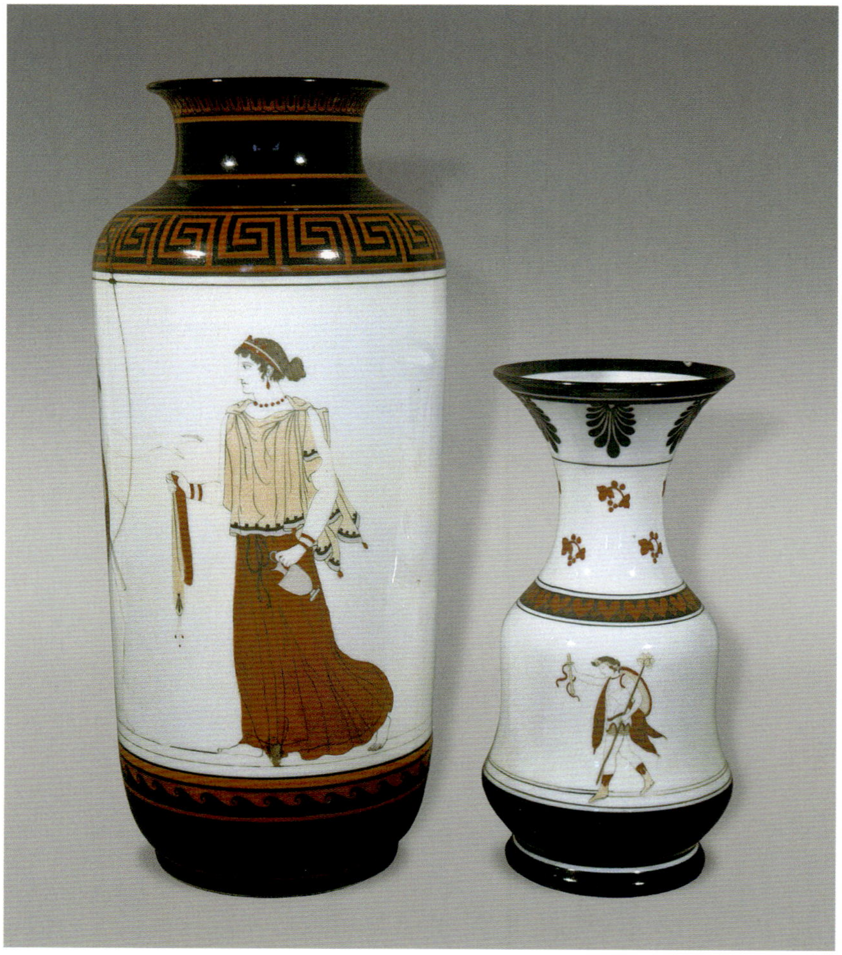

Figure 38 Two vases, England, ca. 1870. Earthenware, painted in Greek revival style. H. 13⅞" and 9½". NPSHAMP 10210, 21459.

further declined due to the need to hire servants after the will of Charles Carnan Ridgely freed the slaves in 1829. John and Eliza had to engage servants as well as buy and/or rent a new slave population. Eliza's fortune assured that Hampton was still a showplace during her lifetime.

The great era at Hampton ended with the death of Eliza Ridgely in 1867. A few examples of contemporary art pottery exist from later in the century during the tenure of Hampton's fourth master, John and Eliza Ridgely's son Charles Ridgely: a turquoise ground lizard jug made by the Minton factory (fig. 37) and two vases in Greek revival style (fig. 38), both from about 1870. The family continued at Hampton until the mid-twentieth century.

Hampton National Historic Site, including the mansion, home farm with dairy, formal gardens, and twenty-three dependency buildings is open to the public, with furnished period rooms that bring this important national landmark back to life, addressing the most significant periods of the estate's history, from Captain Ridgely's dynamic creation of a Maryland showplace in the eighteenth century to the denouement of his American dream after World War II.

1. Lynne Dakin Hastings, *Hampton National Historic Site* (Historic Hampton, Inc. in cooperation with the National Park Service, U.S. Department of the Interior, 1986), p. 3.

2. Estate Inventory of Charles Ridgely of Hampton, 1829, Maryland Hall of Records, Annapolis; copy in the archives of the Hampton National Historic Site. See also Lynne Dakin Hastings, *"The Best Table in America": Furnishing the Dining Room at Hampton (1810–1829),* Historic Furnishings Report (National Park Service, U.S. Department of the Interior, 1994), pp. 172–80.

3. Letter from Rosalie Stier Calvert to her father, H. J. Steir, dated March 13, 1819. Margaret Law Callcott, *Mistress of Riversdale: The Plantation Letters of Rosalie Stier Calvert, 1795–1821* (Baltimore, Md.: Johns Hopkins University Press, 1991), pp. 343–44.

4. Susan Gray Detweiler, *George Washington's Chinaware* (New York: Harry N. Abrams, 1982), p. 112.

5. Ibid.

6. For unknown reasons, two orders for figures were placed on the same day to two different London agents, Messrs. Thomas Eden and Co., and Oxley, Hancock and Co. The order to Thomas Eden and Co. reads: "Fashionable Ornamental Decoration to set of a Dining or Supper Table, that will accommodate 20 People with a Slides of the Images . . . ," as quoted in Diana Edwards, *Taste and Table: A Century of Ceramics in Early Maryland* (Baltimore, Md.: Johns Hopkins University Press, 2003), p. 13.

7. Ridgely Papers, MS 692, Maryland Historical Society, Baltimore.

8. These items sold for thirty-five dollars on October 1, 1829, at the sale following the death of General Ridgely. Also listed for sale were thirty-seven pieces of Liverpool ware, probably transfer-printed creamware. Archives, Hampton National Historic Site.

9. Edward Lloyd IV and George Washington both owned a number of biscuit porcelain figures made at Niderviller and Angoulême; these could have been described erroneously as plaster of paris or alabaster.

10. The order to Edward Lloyd's London agents specified that the plateau images not exceed one hundred guineas. (A guinea was equal to one pound and one shilling, or twenty-one shillings.) The 1796 inventory of Edward Lloyd IV listed the value of the plateau and "29 Alabaster Images" at seventy-five pounds. See Diana Edwards, "Gods and Goddesses: A Brief Note about a Set of Niderviller Figures in a Maryland Family," *French Porcelain Society Journal* 2 (2005): 115–19. In the 1790s prices were still being quoted in either sterling or dollars, but a pound traded at approximately $4.50; www.globalfinancialdata.com/index.

11. David G. McCullough, *John Adams* (New York: Simon and Schuster, 2001), p. 347.

12. The quantities listed in various inventories are unreliable because upon the death of each master the contents of the house were divided among family members; some pieces were subsequently returned. Moreover, Hampton was entailed to the eldest son of each generation and many items were assumed to be part of those transfers.

13. Callcott, *Mistress of Riversdale,* pp. 138, 142, 157–58, 343–44.

14. Probate Court Records, District of Columbia Record Group 21, National Archives, Washington, D.C.

15. MFH5476.H851F9, Maryland Historical Society, Baltimore. Howard, one of George Washington's officers during the Revolution and hero of Cowpens (1781), also served as governor of Maryland (1788–1791) and as a United States senator; in 1816 he was an unsuccessful candidate for vice president. Two of his sons married two of Ridgely's daughters. The total value of Howard's real estate at the time of his death was more than one million dollars.

16. Diana Edwards, "English Aristocrats in Maryland Society," *American Ceramic Circle Journal* 7 (1989): 91.

17. Hastings, *"Best Table in America,"* p. 146.

18. John Eager Howard Estate Sale (1828), MS 2450, Maryland Historical Society, Baltimore.

19. The focus of this article does not include John and Eliza's children, as by their generation the purchases of household furnishings declined with the family fortunes.

20. Lynne Dakin Hastings, introduction to Charles E. Peterson, *Notes on Hampton Mansion,* 2nd ed., rev. (College Park, Md.: National Trust for Historic Preservation Library Collection of the University of Maryland Libraries, 2000), p. xiv.

21. Wendy A. Cooper, *Classical Taste in America, 1800–1840* (New York: Abbeville Press, 1993), p. 27.

22. Hastings, *"Best Table in America,"* p. xiv.

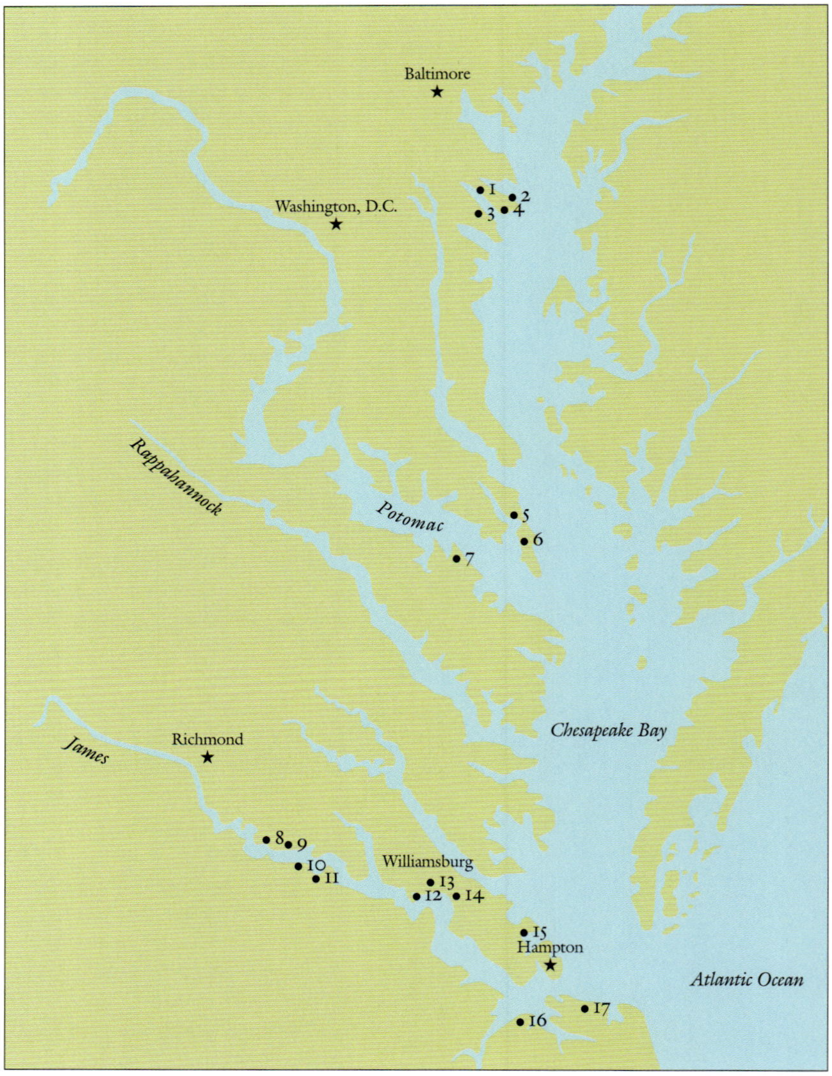

Figure 1 Map of the Chesapeake region showing the location of the principal archaeological sites discussed in the text. (Artwork, Nichole Drgan.)

1. Swan Cove
2. Homewood's Lot
3. Burle's Town Land
4. Broadneck
5. Old Chapel Field
6. St. Mary's City (Pope's Fort, St. Johns)
7. Nomini Plantation
8. Shirley Plantation
9. Walter Aston
10. Jordan's Point
11. Flowerdew Hundred
12. Jamestown
13. College Landing
14. Kingsmill Tenement – Pettus
15. Bennett Farm
16. Harborview (Nansemond Fort)
17. Chesopean

Al Luckenbach and Taft Kiser

Seventeenth-Century Tobacco Pipe Manufacturing in the Chesapeake Region: A Preliminary Delineation of Makers and Their Styles

▼ IN ADDITION TO their obvious functional utility, locally made seventeenth-century tobacco pipes have been recognized as the earliest surviving examples of folk art in the Chesapeake region.[1] The clay tobacco pipe, especially in its "terracotta" versions, is arguably the most studied colonial artifact type in the New World. Despite the fact that numerous authors have addressed the subject over the past several decades—particularly in regard to their social significance—the pipes have remained largely unclassified as to specific makers or geographic points of origin.[2]

This paper asserts that European colonists adopted an existing Native American tradition, and as the industry expanded it followed a model similar to that of early postmedieval country potters in England. Typical English country potters, like a local tobacco pipe maker, produced quantities intended for a limited regional market. Examples of long-distance distributions are rare, but significant when they occur.

These local manufacturers are often recognizable by surprisingly distinctive tobacco pipes, a fact no doubt related to the strong artistic component of their production. In this paper we will name and characterize the products of a number of idiosyncratic terracotta pipe manufacturers in the Chesapeake region and discuss their known distributions.

The discovery of the first known terracotta pipe kiln site at Swan Cove in Providence, Maryland, has provided a baseline against which to interpret existing pipe data from the region.[3] Many previous assumptions now seem questionable at best, and new avenues of interpretation are developing.

The Study of Chesapeake Terracotta Pipes

For several years, and with varying degrees of formality, the authors have surveyed archaeological collections of terracotta pipes from Maryland and Virginia (fig. 1). The idea of this study initially revolved somewhat around the concept that Emanuel Drue's products (whose production site at Swan Cove had just been discovered) would be found in collections from distant sites. This proved not to be the case. In fact, it soon became apparent that terracotta pipe assemblages from different areas of the region actually had very little typological overlap.

With two notable exceptions, terracotta pipes did not travel far from their source of production. Relying on a limited catchment area for consumers, production amounts were necessarily small, although calculating estimates is not possible with currently available information.

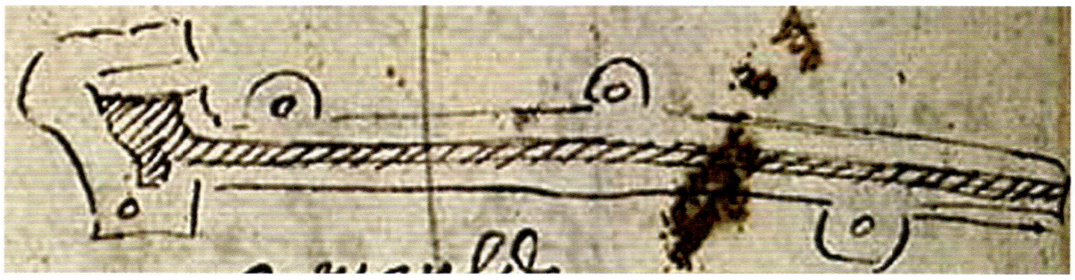

Figure 2 Drawing of pipe mold, 1688. From Randle Holme's *Academy of Armory,* shelfmark HARLEY.5955; FJ/245. (Courtesy, The British Library.)

While pipes are relatively easy to make, some researchers have overestimated their simplicity, postulating production in home fireplaces. An occasional "expedient pipe" might have been made in this fashion, but such limited production certainly would not have much impact on the archaeological record.[4] In reality, even a small-scale pipe maker like Emanuel Drue used a fairly complex arrangement of kiln, kiln furniture, clay processing, and decorating tools (fig. 2). We can assume that other makers discussed here worked on the same order of magnitude.

Terracotta pipes in the Chesapeake region were functional objects, made to smoke tobacco, but the plasticity of the medium and ease of decoration allowed for considerable decorative and stylistic variation. Apparent geographic isolation from other producers heightened this proclivity for stylistic diversity, resulting in variations in basic pipe shape and, most important, a palette on which producers could indulge in decorative embellishments. Daniel Mouer has termed the results "the most intriguing surviving examples of folk art of the early Chesapeake," and further notes that the "care and effort which attended the creation of these artifacts attests to their symbolic importance."[5] Such a perspective is fully supported here.

We contend that the regional variations of pipe makers (and perhaps their apprentices or later generations of related makers) are readily recognizable, and that a basic classification scheme will facilitate scholarly communication regarding these objects. While many previous works, including those in which we were involved, have used properly scientific classificatory terminology (such as "Type 5.08"), we feel that a return to the style of type nomenclature stressing geographic or stylistic characteristics would ultimately make for easier scientific discourse.[6]

Emanuel Drue

Situated on the banks of a relic cove off Mill Creek in Anne Arundel County, Maryland, the Swan Cove site (18AN934) was occupied by planter and tobacco pipe maker Emanuel Drue from the early 1650s until his death in 1669. Drue used state-of-the-art European production techniques in the manufacture of tobacco pipes made of native clays. The pipes and kiln debris recovered from Swan Cove (fig. 3) characterize the nature of a seventeenth-century industry that has long been sought in the Chesapeake.[7]

In addition to the unique assemblage of production-related artifacts, one of the most significant discoveries made at this site was that Drue produced two distinct pipe forms, a Native American–derived "Chesapeake" style angular elbow pipe and a European "belly bowl" (fig. 4), and each seems to have had its own decorative grammar. The elbow form (referred to as

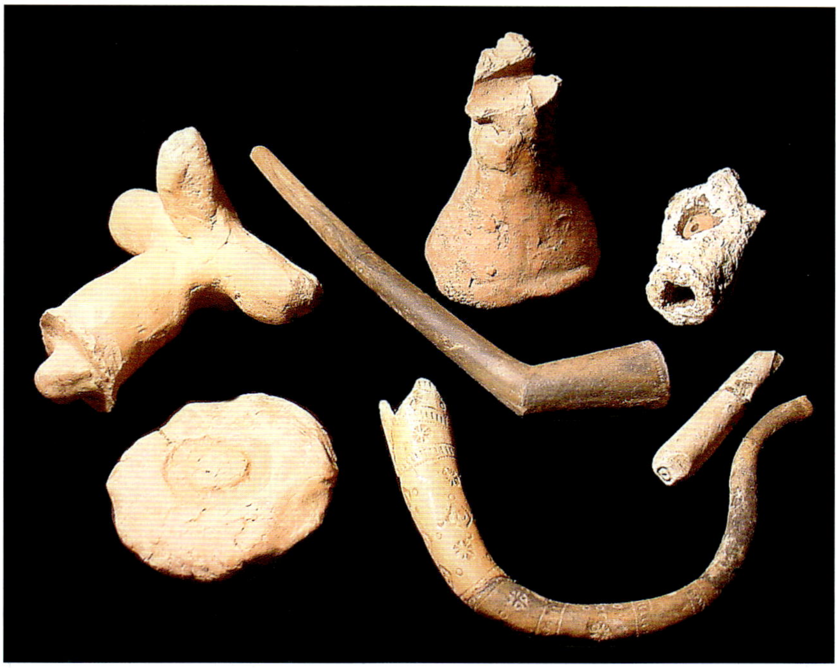

Figure 3 Kiln furniture and pipes, Emanuel Drue, Maryland, ca. 1660. Earthenware. (Courtesy, Anne Arundel County, Maryland; photo, Al Luckenbach.) Recovered from the Swan Cove site, these artifacts include cross-pipe props, decorating tool, bun, and pipes.

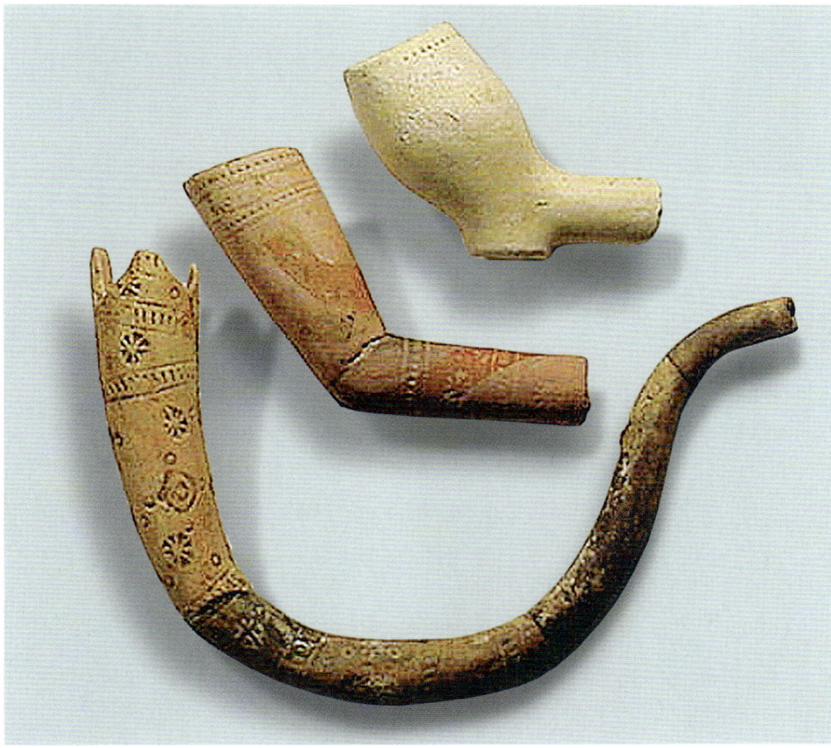

Figure 4 Tobacco pipes, Emanuel Drue, Maryland, ca. 1660. Earthenware. *Top:* Drue Type B, belly bowl. *Center:* Drue Type A, angular elbow. *Bottom:* presentation "Crumm Horn" pipe. (Courtesy, Anne Arundel County, Maryland; photo, Al Luckenbach.)

Drue Type A) was often decorated more exuberantly—with stamps and rouletted bowls and stems—in keeping with other products from the region. In stark contrast, with one exception the belly-bowl form (Drue Type B) was decorated in the same austere manner as similar European products—simply by rouletting the rim.

Another significant discovery was that Emanuel Drue produced a remarkable range of colored products—not because of the vagaries of manufacture, but because he selected clays specifically for their color variation. This fact was dramatically illustrated by hunks of raw clays discovered in intact trash pits that included gray, pink, and green varieties as well as a marbleized white/rose. These "gourmet" clays were used to produce both agatized pipes and pipes with slip-trailed surface decorations.

Perhaps equally noteworthy was the presence of large quantities of raw white clay, amply reflected in pure white examples of both types of Drue products. The presence of pure white Drue belly bowls with a standard simple roulette decoration at the rim has ominous implications for archaeologists trying to differentiate imported tobacco pipes from local ones.

The possibility that individuals might be producing unique expedient pipes has already been mentioned. Before such a statement can be supported, however, the possibility that they are "presentation pipes" needs to be considered. At least two examples have been recovered showing that Emanuel Drue produced elaborately decorated pipes of unique form (see fig. 3). Since one of these was recovered at the home site of his neighbor Robert Burle, it is assumed that it was specially made as a gift.[8]

The information gained from the excavation of Drue's production site has greatly expanded our knowledge of how local tobacco pipe manufacturers operated as well as the range of their products, which, of course, has implications for interpreting the data available on other producers.

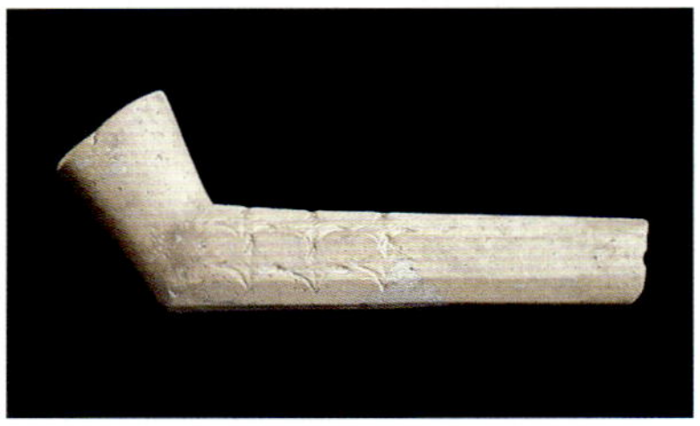

Figure 5 Tobacco pipe, Robert Cotton, Virginia, ca. 1608. Earthenware. (APVA, Jamestown Rediscovery; photo, Beverly A. Straube.) This decorated elbow pipe is from the Jamestown excavations.

Robert Cotton

In 1608 a "Robert Cotton tobacco pipe maker" arrived in Jamestown.[9] Although Cotton then disappears from the documentary record, the Jamestown Rediscovery researchers have recovered a number of distinctive, local mold-made pipes with small bowls, chamfered stems, and European stamp decorations that appear to be his products (fig. 5). The stamps are apparently a match for some used in London between 1580 and 1610, and it is assumed that Cotton came from that city.[10] Obviously, he takes pride of place as the earliest European producer of pipes in the Chesapeake.

Cotton's products, like those of most of the other makers discussed herein, apparently did not travel far from their source. In fact, to date none has been found outside the Jamestown settlement. Although fragments of what might be a muffle have been recovered, the exact location of Cotton's kiln—and the wealth of data it might contain—await rediscovery.

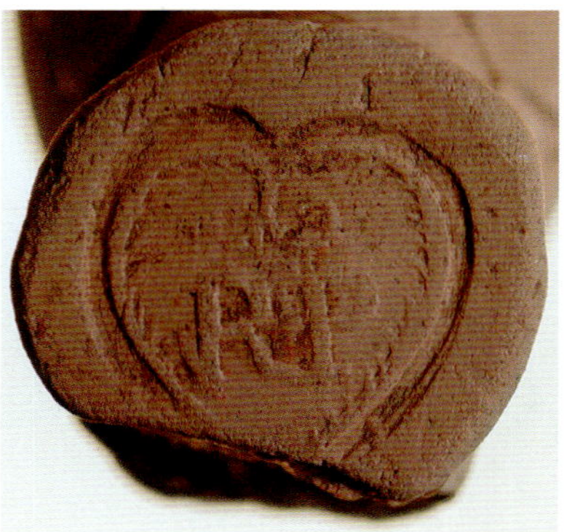

Figure 6 Tobacco pipe, Richard Pimmer, Virginia, ca. 1659. Earthenware. D. of spur .67"; H. of bowl 1.69". Mark: "RP" on pipe spur. (Courtesy, Virginia Department of Historic Resources; photo, Taft Kiser.) This pipe was found at the Nansemond Fort site (44SK192).

Richard Pimmer

In 2001 two researchers working independently for the Lost Towns Project simultaneously produced information that allowed a third Chesapeake pipe maker (after Drue and Cotton) to be identified. In October of that year Taft Kiser, who was surveying and photographing Virginia collections of terracotta pipes, sent Al Luckenbach a picture of a pipe whose heel was marked with a clear *RP* (fig. 6). Examples of the pipe had been recovered from Calvert County, Maryland, among other locations, but given the large numbers originating from Nansemond Fort (44SK192) in southside Virginia, Kiser clearly was closer to the production source.[11]

At the same time, John Coombs was conducting documentary research for the project, scouring the records of the southside Virginia area for evidence relating to Emanuel Drue. Although unsuccessful in his primary goal, Coombs did discover important documents relating to two pipe makers, John Prescott and Richard Pimmer. Pimmer is recorded as having obtained materials to make "forty gross of pipes" as well as securing an indentured servant, Silvanus Gilpin, who was a pipe maker. Both references date from 1659 and place Pimmer in the region of Nansemond Fort. There seems little reason to doubt that the RP pipes are Richard Pimmer products.

Bookbinder

Some of the most distinctive clay tobacco pipes in the Chesapeake region are characterized by the use of marbleized clays and the most elaborate and detailed decorative treatments of any known maker (fig. 7). These marbled

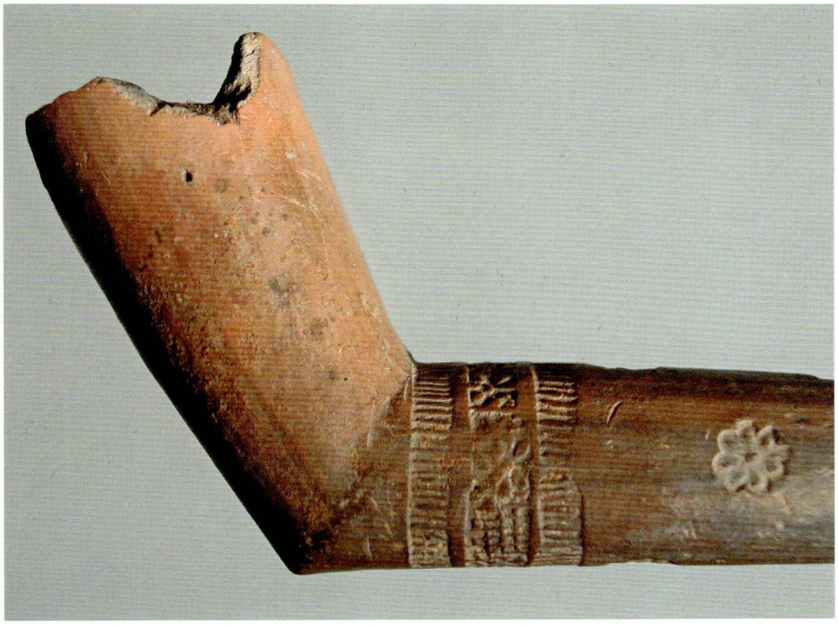

Figure 7 Tobacco pipe, Bookbinder, Virginia, ca. 1640. Earthenware. D. of stem .47"; D. of bore .15". (Courtesy, Virginia Department of Historic Resources; photo, Taft Kiser.) This angular elbow pipe is from the Chesopean site (44VB48).

and stamped pipes have attracted attention from archaeologists for decades, and when Taft Kiser began collecting the distribution data in 1985 he applied the name "Bookbinder," because the stamps and marbleizing resembled decorative work on book covers. William H. Ruppert, a Williamsburg leather worker, suggests the tools used to decorate the pipes might originally have been used to apply gilt to leather purses.

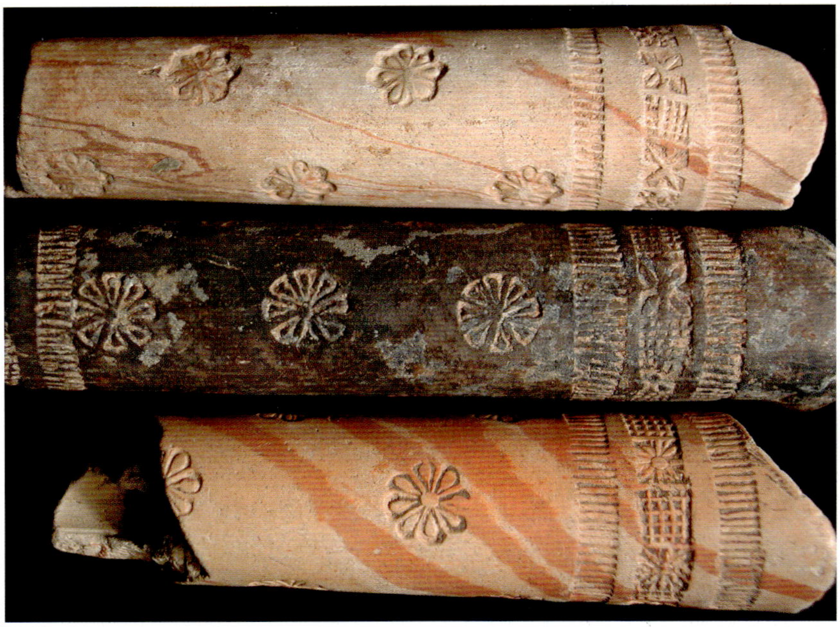

Figure 8 Tobacco pipe stems, Bookbinder, Virginia, ca. 1640. Earthenware. D. of pipe stem (average) .47"; D. of bore .14". (Courtesy, Virginia Department of Historic Resources; photo, Taft Kiser.) These three stems, found at the Chesopean site (44VB48), show the strict decorative grammar used by Bookbinder.

Figure 9 Tobacco pipe, Bookbinder, Virginia, ca. 1640. Earthenware. D. of stem .47"; D. of bore .14". (Courtesy, Virginia Department of Historic Resources; photo, Taft Kiser.) This close-up of a pipe from the Chesopean site (44VB48) shows Bookbinder's detailed "grill work" decoration.

These intricate stamps occur on a single-elbow bowl form in a highly systematized and rigid grammar, the consistency of which, although executed with different tools on slightly variant forms, suggests a generational "Bookbinder school" of pipe making (figs. 8, 9).

The Bookbinder manufactory appears to have been in the vicinity of the Chesopean site (44VB48) in Virginia Beach, Virginia. Unlike nearly every other producer discussed in this article, however, Bookbinder's products have been recovered from strikingly disparate localities.

In Virginia the greatest numbers came from the Chesopean site, but examples have also been recovered from Shirley Plantation (44CC15), Pettus (44JC33), Kingsmill Tenement (44JC39), Wills Cove (44SK56), Nomini Plantation (44WM12), Nansemond Fort (44SK192), Walter Aston

(44CC178), Bennett Farm (44YO68), site 44JC568, College Landing (44JC952), and Jamestown.[12] In Maryland, Bookbinder products have been found at St. Mary's City (Pope's Fort and St. Johns),[13] in Old Chapel Field,[14] and at two sites in the Providence settlement in Anne Arundel County (Burle's Town Land and Homewood's Lot).[15] Bookbinder pipes have even been recovered at Renews, Newfoundland, fourteen hundred miles from their source.[16] These data indicate that these products were distributed under a notably different system from that used by the other makers discussed here.

With one exception, Bookbinder and associated Bookbinder-school pipes are the only local ones known to be decorated with elaborately cut rectangular wheel stamps. The exception is discussed in the following section.

Broadneck

The "Broadneck," another distinctive pipe form, is found in large numbers at the Broadneck site (18AN818) in Anne Arundel County, Maryland, constituting 70 percent of the assemblage (fig. 10). At Pope's Fort in St. Mary's City this type was referred to as "a mold made variety with a double chevron motif."[17]

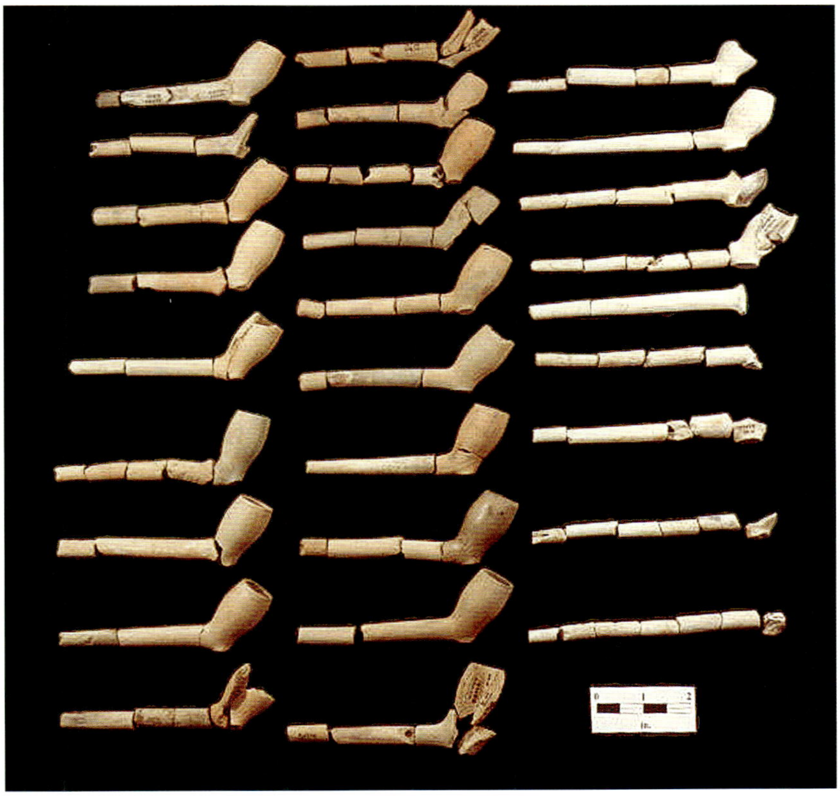

Figure 10 Tobacco pipes, Broadneck maker, Maryland, ca. 1650. Earthenware. (Courtesy, Anne Arundel County; photo, Al Luckenbach.) This complete assemblage of pipes from the Broadneck site (18AN818) shows the predominance of the Broadneck style.

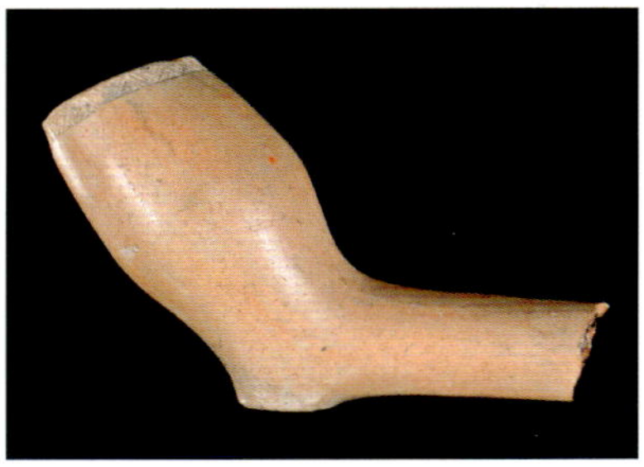

Figure 11 Tobacco pipe, Broadneck maker, Virginia, ca. 1650. Earthenware. H. of bowl 1.61"; D. of bore .12". (Courtesy, Virginia Department of Historic Resources; photo, Taft Kiser.) This Broadneck bowl was found at the Wills Cove site (44SK56).

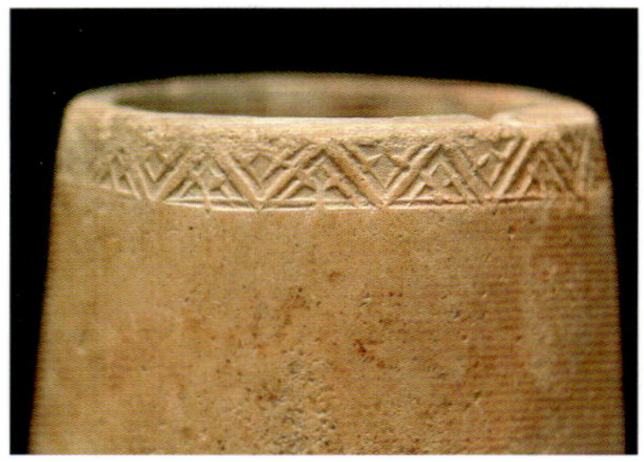

Figure 12 Tobacco pipe, Broadneck maker, Virginia, ca. 1650. Earthenware. D. of rim .75". (Courtesy, Virginia Department of Historic Resources; photo, Taft Kiser.) Broadneck rim decoration adorns this pipe from the Wills Cove site (44SK56).

The Broadneck maker used a rectangular wheel stamp with chevrons as rim decoration on otherwise undecorated pipes. The pipes are a weak oval-heeled form with thick walls and a profile that is distinctive enough to be readily recognized (figs. 11, 12). Marbleized clays were never used.

Broadneck pipes have been recovered in large numbers from the Chesopean site and Martin's Hundred in Virginia; Pope's Fort in southern Maryland; and the Broadneck site, Burle's Town Land, and Homewood's Lot in Providence, Maryland.[18] This is a fairly large distribution for a Chesapeake maker, nearly as large as that of Bookbinder. Drawing on an analogy from Emanuel Drue, this wide distribution suggests that the Broadneck form might, in fact, be a Type B belly-bowl form produced by the Bookbinder school, which otherwise is known only for angular elbow styles.

Dramatic new evidence bearing on such speculations was discovered in 2004 during the Jamestown Rediscovery excavations. Beverly Straube, lab director for the project, noticed a unique example of a Broadneck pipe that has initials on the heel (fig. 13). The initials appear to be *ID*—with the *I* barred, presumably meaning *JD*—although a minority opinion believes the second letter could be a distorted *P*. If the initials are in fact *JD,* then perhaps a member of the Drue family made the pipe. Given the numerous characteristics shared by pipes made by Emanuel Drue and Bookbinder (marbleizing, slip trailed decoration, and white clays, among others) the possibility that Broadneck forms are Bookbinder Type B belly-bowl forms and produced by a Drue means that Bookbinder might have been a Drue; thus the shared traits would have a familial explanation.

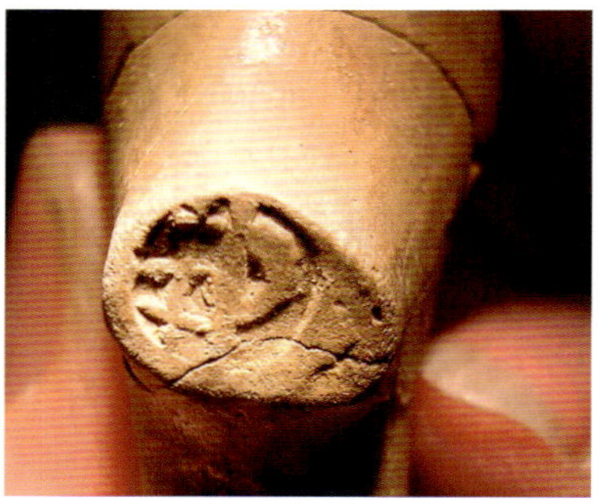

Figure 13 Tobacco pipe, possibly J. Drue, ca. 1645. Earthenware. Mark: "JD." (Courtesy, APVA, Jamestown Rediscovery; photo, Beverly A. Straube.) This Broadneck-style pipe was recovered at the Jamestown excavations.

The theory that the initials are *JP* leads to other interesting possibilities—most notably, that Broadneck pipes might be the products of John Prescott, the pipe maker mentioned previously.

Star Maker

"Star Maker" produced and decorated terracotta elbow pipes with slightly bulbous bowls. The distinctive stars and paneled areas, which often contain initials, were achieved using rouletting and stamps (sometimes with white infilling) (figs. 14–16). At least three different molds were used in the production of these pipes. There is also some indication that apprentices might have been involved.

Star Maker worked either at the Walter Aston site (44CC178) in Charles City County, Virginia, or within a relatively short distance of it. Current data suggests that these pipes were distributed within a limited geographic

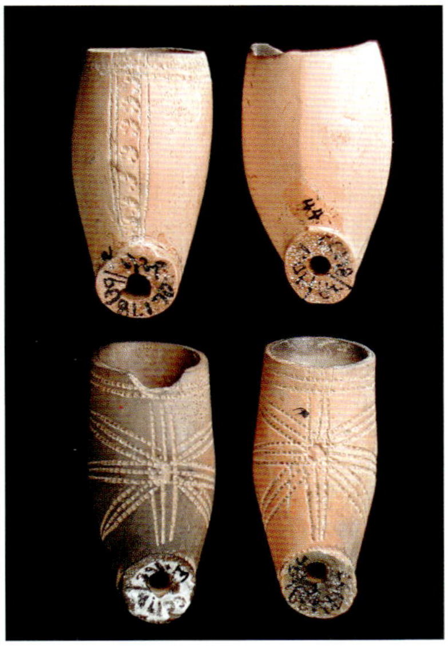

Figure 14 Tobacco pipes, Star Maker, Virginia, ca. 1640. Earthenware. H. of bowls (average) 1.65"; D. of bores .11–.12". (Courtesy, Virginia Department of Historic Resources; photo, Taft Kiser.) The Star Maker's decorative grammar is represented on these four pipes from the Walter Aston site (44CC178).

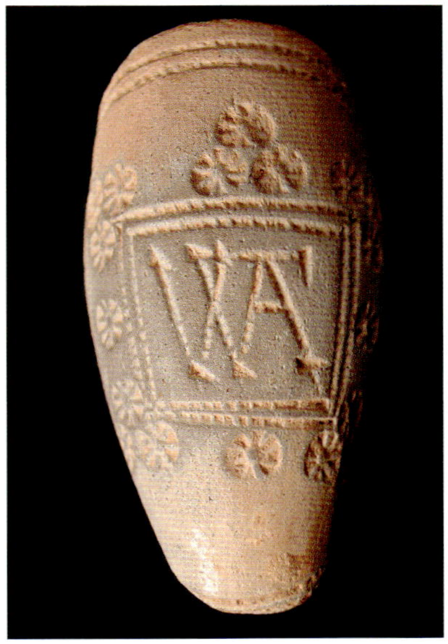

Figure 15 Tobacco pipe, Star Maker school, Virginia, ca. 1640. Earthenware. H. of bowl 1.5"; D. of bore .12". Mark: "WA" (Walter Aston?). (Courtesy, Virginia Department of Historic Resources; photo, Taft Kiser.) This Star Maker pipe was found at the Walter Aston site (44CC178).

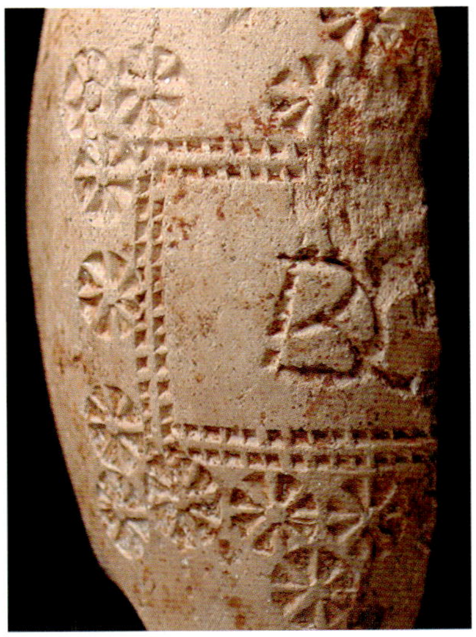

Figure 16 Tobacco pipe, Star Maker school, Virginia, ca. 1640. Earthenware. H. of bowl 1.26". Mark: "DK" (David Kirke?). (Courtesy, Virginia Department of Historic Resources; photo, Taft Kiser.) This example is from the Walter Aston site (44CC178).

area, with one notable exception. A decorated bowl with the initials *DK* was found at the house site of Sir David Kirke, in Ferryland, Newfoundland (fig. 17). The similarity of its style and stamps suggests that it was made by the same individual who produced the Star Maker–style bowl initialed *WA* that was found at the Walter Aston site (see fig. 15). Production at this location seems to have taken place between about 1635 and 1660; there is evidence to suggest that some of these molds moved to Flowerdew Hundred site 44PG77 after 1660.[19]

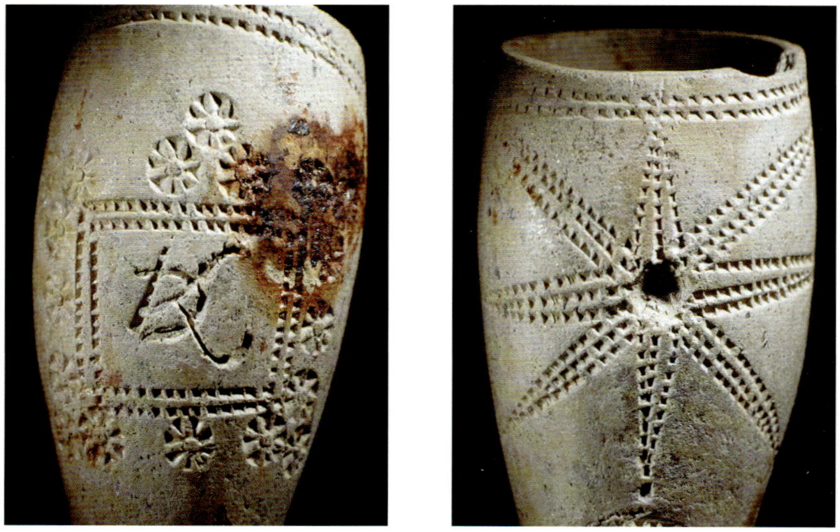

Figure 17 Tobacco pipe, Star Maker school, Virginia, ca. 1640. Earthenware. Mark: "DK" (David Kirke?). (Courtesy, Colony of Avalon Foundation; photo, Jim Tuck and Barry Gaulton.)

This Star Maker pipe was found in Newfoundland. At left is the decorated mark that appears on one side of the bowl; at right is the bowl's reverse side, showing a star decoration.

A total of five distinct decorative grammars can be classified:

Star front and back. An eight-ray star on the front and back adorns the bowl, with the vertical rays open rather than pointed—the primary signature of this individual. The center of each star is marked with a circular stamp. Two rows of rouletting around the rim complete the grammar.

Star back with front panel. This decorative grammar includes a star on the back and two identical rows of rouletting around the rim. The front of the bowl has a rectangular panel that can contain initials, symbols, or stamped rosettes. Outside each corner of the panel is a cluster of stamped rosettes. Recorded initials are: *EK, TF, IT, EA*[?] and *EM, IG, R*[?]*B, FB, E-, H-, T-*. Apprentice bowls have *HE, AH*, and *WA* (the latter presumably Walter Aston).

Star back with undecorated front. The bowl has a star on the back and two rows of rouletting around the rim. The front is undecorated.

Back with two parallel vertical lines and stamps. On the back of the bowl is a vertical band consisting of two or more lines of parallel rouletting, with overlying rosette stamps. Two additional rows of rouletting are around the rim. There is no other decoration. Based on the form of the bowl and the character of the decoration, pipes with these elements are believed to be Star Maker products.

Bowls with no decoration. The bowl is molded and fired without any surface decoration, but based on the bowl form and the character of the decoration it is believed to be a Star Maker product. At least nineteen of these bowls are found in the Flowerdew Hundred Plantation archaeological collections, and these molds may have moved to that plantation after Walter Aston's death, in 1656.

Nomini Maker

The Nomini Maker produced tall, slender elbow pipes in the vicinity of the Nomini Plantation (44WM12) in Westmoreland County, Virginia. The decorative grammar primarily involved the use of rouletting with occasional punctate and stamps and the use of white infill in the decoration (figs. 18–20).

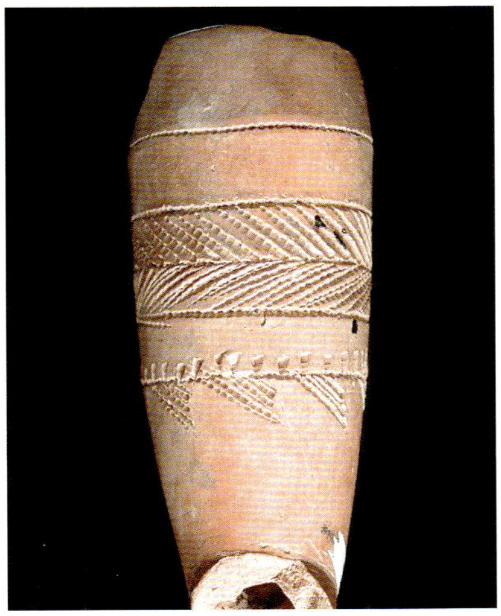

Figure 18 Tobacco pipe, Nomini Maker, Virginia, ca. 1660. Earthenware. H. of bowl 2.17"; D. of bore .17". (Courtesy, Virginia Department of Historic Resources; photo, Taft Kiser.) This version of a Nomini pipe was found at the Nomini Plantation (44WM12).

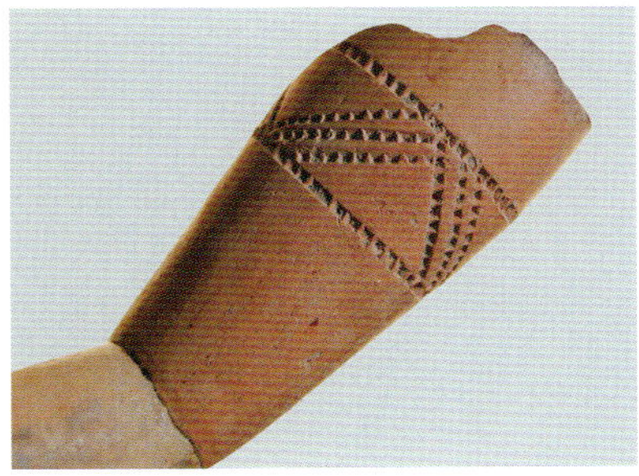

Figure 19 Tobacco pipe, Nomini Maker, Virginia, ca. 1660. Earthenware. H. of bowl 1.93"; D. of bore .15". (Courtesy, Virginia Department of Historic Resources; photo, Taft Kiser.)

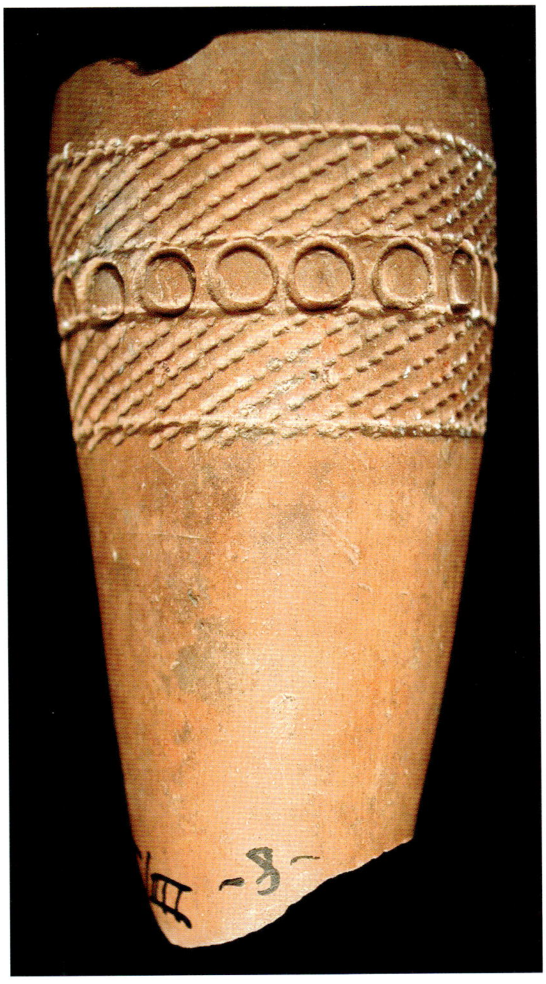

Figure 20 Tobacco pipe, Nomini Maker, Virginia, ca. 1660. Earthenware. H. of bowl 1.85". (Courtesy, Virginia Department of Historic Resources; photo, Taft Kiser.)

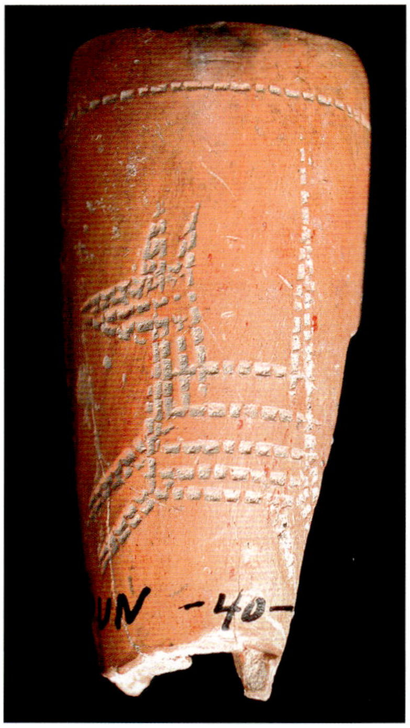

Figure 21 Tobacco pipe, Nomini Maker, Virginia, ca. 1660. Earthenware. H. of bowl 2.17". (Courtesy, Virginia Department of Historic Resources; photo, Taft Kiser.) Illustrated here is an example of a Running Deer pipe.

Based on the style and decoration of the bowls, the Nomini Maker is believed to have been a Native American Algonquin who adapted a new pipe style from the "Running Deer" grammar (fig. 21). This is supported by a 1659 deed reference from Nomini that refers to "near the side of an Indian field commonly known as the Pipemaker's field. . . ."[20]

Banded bowls have a rouletted band around the center of the bowl—in some cases simply two parallel lines, in others the two parallel lines enclose a wider band of diagonal rouletting. The decoration tends to go up from the central band, leaving the lower half of the bowl relatively undecorated, and may consist of individual rouletted bands, double bands, or double bands enclosing diagonals. A circular stamp is sometimes used—in one case forming a band, in another hanging from a band—and marks the ends of a four-pointed cross or star. The rouletted bands can also enclose hanging triangles and, in at least two cases, what appears to be an *H*. One bowl is decorated with incised or impressed cordons running in bands down the bowl.

Freehand bowls have a single band of rouletting around the rim and rouletted symbols. The symbols are typically in a pair: one faces the smoker, the other is on the bowl's front. One symbol resembles a flat-bottomed *V* or a pair of spiked horns. Other symbols probably represent plants. Running deer are also used.

This collection contains a large amount of waste material that has yielded clues to the maker's techniques. Apparently the pipes were rolled by hand, then shaped without using molds.

The overwhelming dominance of Nomini Maker among the local pipes at Nomini Plantation indicates that an individual, or related individuals producing a single style, worked at the site for an extended period of time.

When death or departure ended the work of the maker—which Mitchell places prior to the introduction of the Morgan Jones pottery, in about 1677—the site continued to be occupied, although the dearth of other local products indicates that the occupants returned to those made in Europe. This suggests that the end date for this particular local pipe tradition falls in the late 1670s. Although this corresponds with the terminal dates for other makers discussed here, Native American makers continued to make decorated pipes into the eighteenth century.[21]

The Carver

Pipes that have been finished with a knife have been recovered at a number of sites. Some of these pipes, which often have a faceted appearance, are probably attempts to decorate home-produced expedient bowls, but there were clearly individuals making this style for local trade. Carving was labor-intensive, and production numbers would have been low compared to individuals working with molds.

One of these makers, "The Carver," was working in the Jamestown area from about 1630 to 1645.[22] The Carver produced a square bowl beveled into an octagon, stopping the rear flats at the rounded stem while carrying the front facets under the bowl to a raised heel (fig. 22). In the final deco-

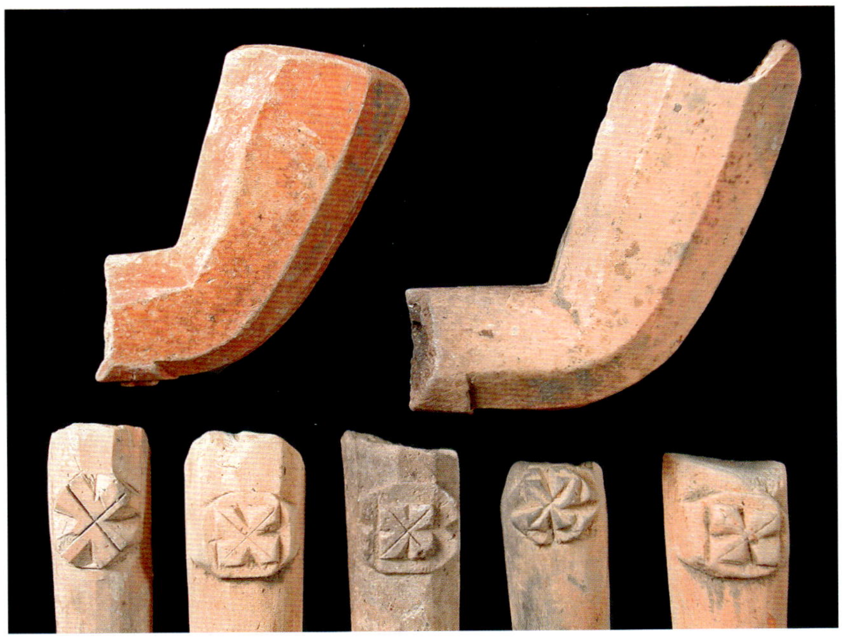

Figure 22 Tobacco pipes, The Carver, Virginia, ca. 1640. Earthenware. H. of bowl (average) 1.65"; D. of bore (average) .14". (Courtesy, Virginia Department of Historic Resources; photo, Taft Kiser.) This assemblage of bowls and stems is from the Kingsmill Tenement (44JC39).

rative stage the heel was incised with an X and elaborated into a raised cruciform pattern with clipped ends. The Carver was exceptionally skilled, creating simple yet distinctive bowls with balanced symmetry and flowing lines.

The Pope's Fort Maker

St. Mary's City in southern Maryland has clear evidence of a resident pipe maker for at least a brief period in the 1640s. Although no kiln debris has been found, wasters of hand-formed pipes have been discovered among

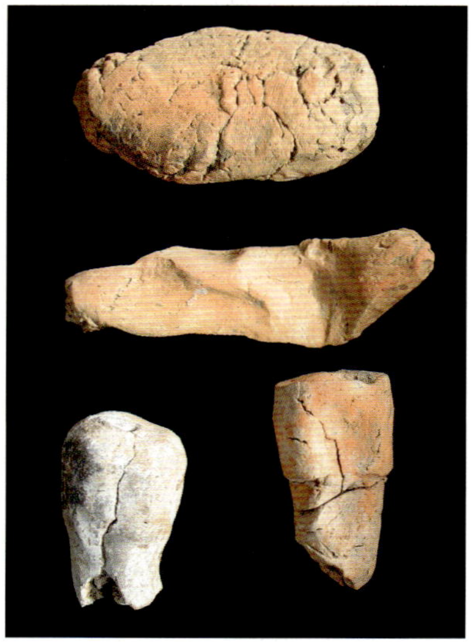

Figure 23 Pipe-making waste, Pope's Fort Maker, Maryland, ca. 1646. Earthenware. (Courtesy, Historic St. Mary's City, Maryland; photo, Silas Hurry.) This waste is from the moat at Pope's Fort.

the artifacts deposited in the fort ditch constructed by forces loyal to Parliament in 1645–1646 (fig. 23).[23] This trash may have been deposited about 1647/8 by William Bennett's forces, who regained control of St. Mary's City for the Calvert family at that time.

The Pope's Fort Maker represents a fascinating and important example of the blending of pipe making traditions. These curved pipes with collars clearly represent a "Susquehanna" pipe shape, common among Iroquoian groups of the Northeast. They are, however, decorated with punctuate Running Deer designs directly related to Native American Algonquin forms. This amalgam is without parallel, and especially intriguing given the strong likelihood that the maker—or, perhaps, maker and decorator—were Caucasian Puritans, as there is no evidence of Native Americans participating in this expedition.

Conclusions

Identifying nine distinct manufacturers is important and satisfying, but numerous products exist that clearly are not those of any of the makers discussed here. Most are either unique examples (which might, in fact, be expedient or presentation pipes) or are found in very small numbers in known archaeological collections.

One major type of terracotta pipe—the Running Deer (figs. 21, 24)—has not been discussed here except in passing. This variety is perhaps the most discussed in the archaeological literature, given the central role it played in theories relating to African iconography. The position affirmed here, and supported by a majority of researchers, is that these are clearly derived from a Native American Algonquin tradition. It is believed that the production of this stylistic tradition encompassed many makers and thus constitutes a distinct product category from the more idiosyncratic types examined in this essay. However, both the Nomini Maker and the Pope's Fort Maker clearly demonstrate a direct influence from the Running Deer style; the Nomini Maker is likely a creolized Native American.

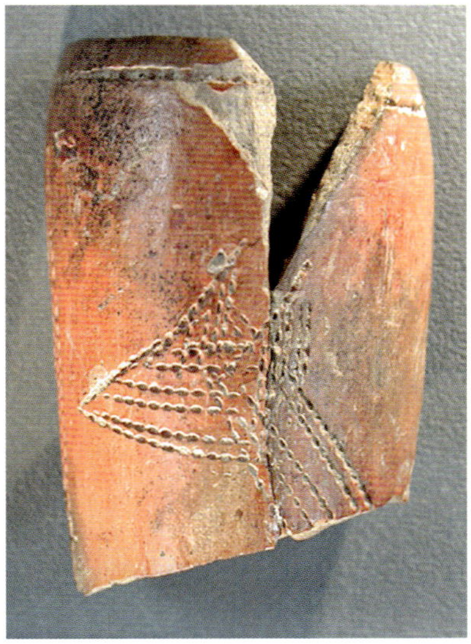

Figure 24 Tobacco pipe, Maryland, ca. 1660. Earthenware. (Courtesy, Anne Arundel County; photo, Shawn Sharpe.) Traditional Running Deer pipe from the Homewood's Lot site, Anne Arundel County, Maryland.

The main points of this paper are clearly demonstrable. The first—that the potential for individualistic expression in the manufacture of clay tobacco pipes allowed for products with considerable variation in shape and decoration—is obvious. More novel is the contention that many of these variations are clearly attributable to individual makers (or their "schools"), at least three of whom can now be named. The remainder have been given nominals capable of facilitating the exchange of future scientific discourse.

One of the unexpected outgrowths of this study was the realization that most of the pipes created by these makers were distributed in a limited geographic area. For all the evidence of elaborate kiln techniques discovered at Emanuel Drue's pottery, his products—with one exception—served only the immediate vicinity of the Providence settlement. It appears that this was also true of six of the nine makers we have described.

Three of the makers had a much larger reach. Broadneck and Pimmer pipes are found over a large area of the Chesapeake region. Bookbinder achieves this distribution and more, having been carried into Newfoundland. The instance of the pipe of the Star Maker school also reaching Newfoundland seems predicated more on the novelty of the initialed bowl than on regular distribution.

In addition to distributional trends, one of the more important discoveries was of Emanuel Drue's large production of two distinct pipe types, a belly bowl and an angular elbow pipe. If other pipe makers were following a similar practice this would greatly compound attributional difficulties. The possibility that Bookbinder's angular elbows were being produced alongside either Broadneck belly bowls or RP belly bowls would, for example, reduce the numbers of makers who were operating beyond a local catchment, since all three enjoyed wider regional distribution.

Another important physical aspect of Drue's products is the presence of white belly bowls with a rouletted rim, which are virtually indistinguishable from English products. How widespread such a practice might have been is currently unknown. Bookbinder produced off-white products, but their shape and decoration would not be confused with imports. The possibility therefore exists that some of the local versus imported pipe attributions made in numerous archaeological studies over the years might have unsuspected errors.

It is clear that many more tobacco pipe makers existed in the seventeenth-century Chesapeake region than have been identified; nevertheless, the ability to name three specific makers and to recognize their products is an important new phase in the history of such studies. It is hoped that future research will add significantly to the archaeological database.

1. See L. Daniel Mouer, "Chesapeake Creoles: The Creation of Folk Culture in Colonial Virginia," in *The Archaeology of Seventeenth Century Virginia,* edited by Theodore J. Reinhart and Dennis J. Pogue (Richmond: Archaeological Society of Virginia, 1993), pp. 105–66.

2. See, for example, Matthew C. Emerson, "Decorated Clay Pipes from the Chesapeake" (Ph.D. diss., University of California, Berkeley, 1988); idem, "Decorated Clay Pipes from the Chesapeake: An African Connection," in *Historical Archaeology of the Chesapeake,* edited by Paul A. Shackel and Barbara J. Little (Washington, D.C.: Smithsonian Institution Press, 1994), pp. 45–46; Mouer, "Chesapeake Creoles"; Fraser Neiman and Julia King, "Who Smoked Chesapeake Pipes?" (paper presented at Society for Historical Archaeology Conference, Salt Lake City, 1999); see also L. Daniel Mouer, Mary Ellen N. Hodges, Stephen R. Potter, Susan L. Henry Renaud, Ivor Noël Hume, Dennis J. Pogue, Martha W. McCartney, and Thomas E. Davidson, "Colonoware Pottery, Chesapeake Pipes, and 'Uncritical Assumptions,'" in *"I, too, am America": Archaeological Studies of African-American Life,* edited by Theresa A. Singleton (Charlottesville: University of Virginia Press, 1999), pp. 82–115.

3. Al Luckenbach, "The Swan Cove Kiln: Chesapeake Tobacco Pipe Production, Circa 1650–1669," in *Ceramics in America,* edited by Robert Hunter (Hanover, N.H.: University Press of New England for the Chipstone Foundation, 2004), pp. 1–14; Al Luckenbach and C. Jane Cox, "Tobacco-Pipe Manufacturing in Early Maryland: The Swan Cove Site (ca. 1660–1669)," in *The Clay Tobacco-Pipe in Anne Arundel County, Maryland (1650–1730),* edited by Al Luckenbach, C. Jane Cox, and John Kille (Annapolis: Anne Arundel County Trust for Preservation, 2002), pp. 46–63.

4. Emerson, "Decorated Clay Pipes," p. 36.

5. Mouer, "Chesapeake Creoles," p. 129.

6. See, e.g., David Gadsby and Shawn Sharpe, "The Lost Towns Project Tobacco Pipe Classification Scheme," in *The Clay Tobacco-Pipe in Anne Arundel County, Maryland (1650–1730),* edited by Al Luckenbach, C. Jane Cox, and John Kille (Annapolis: Anne Arundel County Trust for Preservation, 2002), pp. 4–10.

7. For expanded descriptions, see Luckenbach and Cox, "Tobacco-Pipe Manufacturing in Early Maryland," and Luckenbach, "Swan Cove Kiln."

8. See Al Luckenbach, *Providence 1649: The History and Archaeology of Anne Arundel County, Maryland's First European Settlement* (Annapolis: Maryland State Archives; Crownsville: Maryland Historical Trust, 1995).

9. William M. Kelso and Beverly A. Straube, *Jamestown Rediscovery, 1994–2004* (Richmond: Association for the Preservation of Virginia Antiquities, 2004), pp. 163–66.

10. Beverly A. Straube, personal communication, 2005.

11. Julia A. King, personal communication, 2002.

12. Many of these citations are based on collection analyses by Taft Kiser. For Nomini, see also Vivienne Mitchell, "Decorated Brown Clay Pipebowls from Nominy Plantation: Progress Report," *Quarterly Bulletin of the Archeological Society of Virginia* 31, no. 2 (December 1976): 83; idem, "The History of Nominy Plantation with Emphasis on the Clay Tobacco Pipes," in *Historic Clay Tobacco Pipe Studies,* edited by Byron Sudbury, 3 vols. (1980–), vol. 2 (1983), pp. 1–38; for Jamestown, see John L. Cotter, *Archaeological Excavations at Jamestown Virginia* (Archeological Society of Virginia Special Publication 4, 1958).

13. Henry M. Miller, "Tobacco Pipes from Pope's Fort, St. Mary's City, Maryland: An English Civil War Site on the American Frontier," in *The Archaeology of the Clay Tobacco Pipe,* vol. 12, *Chesapeake Bay,* edited by Peter Davey and Dennis Pogue, British Archaeological Reports British Series, 566 (Oxford: B.A.R., 1991), pp. 73–88; Silas D. Hurry and Robert W. Keeler, "A

Descriptive Analysis of the White Clay Tobacco Pipes from the St. John's Site in St. Mary's City, Maryland," in ibid., pp. 37–72.

14. Christopher I. Sperling and Laura J. Galke, *Phase II Archaeological Investigations of 18ST233 and 18ST329 Aboard Webster Field Annex, Naval Air Station, Patuxent River, St. Mary's County, Maryland,* report prepared for the Department of Public Works, Naval Air Station, Patuxent River, St. Mary's County, Maryland.

15. Luckenbach and Cox, "Tobacco-Pipe Manufacturing in Early Maryland," pp. 46–63.

16. Stephen Mills, "Seventeenth-Century Life in Renews, Newfoundland: Archaeological Analysis of an English West Country Planter's House" (master's thesis, Memorial University of Newfoundland, 2000), p. 159, fig. 13a.

17. Silas D. Hurry and Henry M. Miller, "The Varieties and Origins of Chesapeake Red Clay Tobacco Pipes: A Perspective from the Potomac Shore" (paper presented at the Society for Historical Archaeology Conference, Kingston, Jamaica, 1992).

18. See Ivor Noël Hume and Audrey Noël Hume, *The Archaeology of Martin's Hundred* (Philadelphia: University of Pennsylvania; Williamsburg, Va.: Colonial Williamsburg Foundation, 2001); Miller, "Tobacco Pipes from Pope's Fort," pp. 73–88. For Providence sites, see *The Clay Tobacco-Pipe in Anne Arundel County, Maryland (1650–1730),* edited by Al Luckenbach, C. Jane Cox, and John Kille (Annapolis, Md.: Anne Arundel County Trust for Preservation, Inc., 2002), pp. 46–63.

19. For additional information and images, see www.flowerdew.org/pipe.html.

20. Mitchell, "Decorated Brown Clay Pipebowls from Nominy Plantation," p. 83; see also idem, "History of Nominy Plantation," pp. 1–38.

21. Dane T. Magoon, "'Chesapeake' Pipes and Uncritical Assumptions: A View from Northeastern North Carolina," *North Carolina Archaeology* 49 (1999): 107–27.

22. Alain Charles Outlaw, *Governor's Land: Archaeology of Early Seventeenth-Century Virginia Settlements* (Charlottesville: University of Virginia Press, 1990), pp. 78–79, esp. fig. 49.

23. See Henry M. Miller, *Discovering Maryland's First City: A Summary Report on the 1981–1984 Archaeological Excavations in St. Mary's City, Maryland,* St. Mary's City Archaeology Series, no. 2 (St. Mary's City, Md.: St. Mary's City Commission, 1986), p. 47; Miller, "Tobacco Pipes from Pope's Fort," pp. 73–88; Hurry and Miller, "The Varieties and Origins of Chesapeake Red Clay Tobacco Pipes."

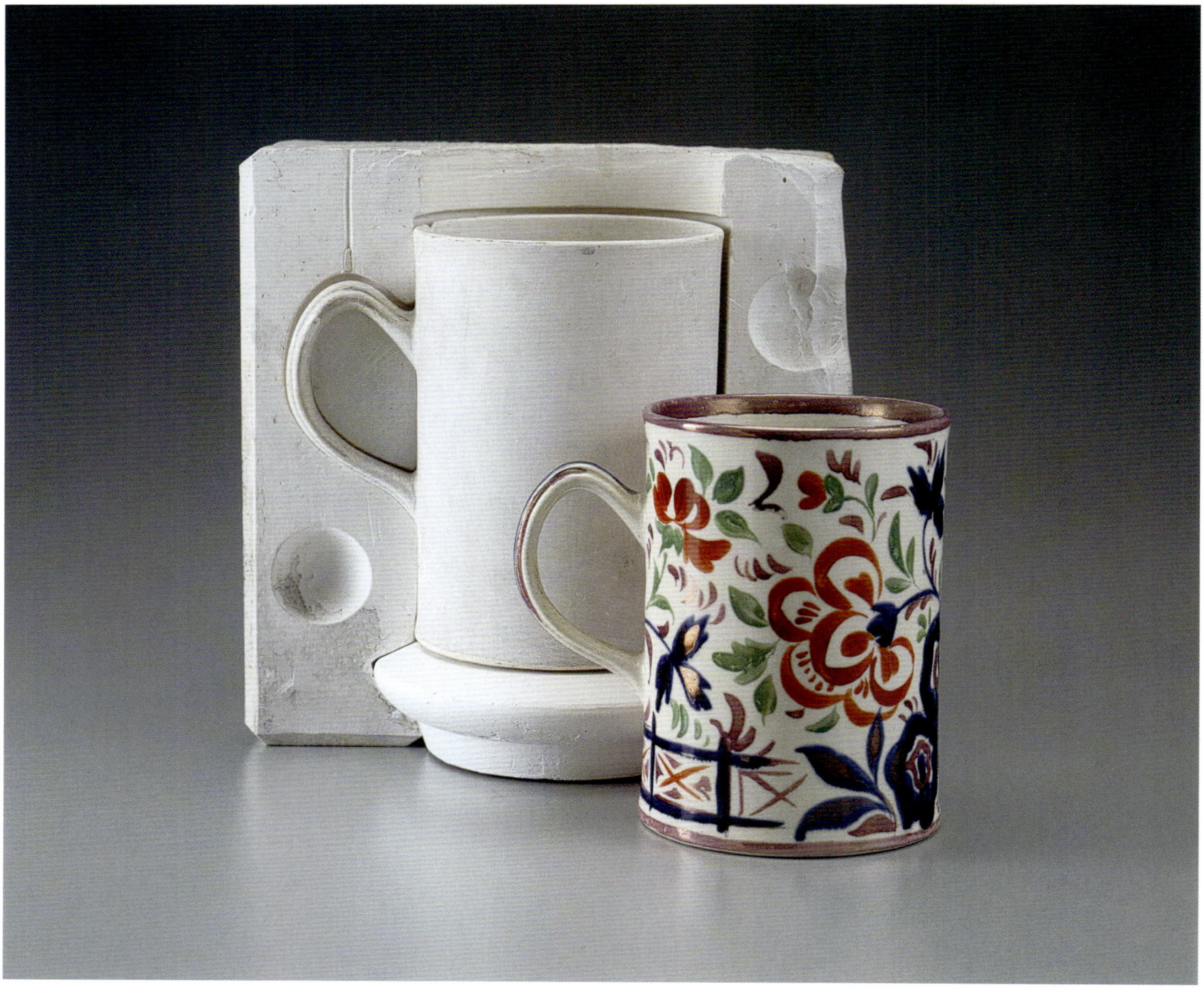

Figure 1 Mold, bisque mug, and finished mug, 1950–1970. Plaster of paris and whiteware. H. of bisque mug 4¾"; H. of finished 4½". (Courtesy, White Hall Restaurant Collection; unless otherwise noted, all photos by Gavin Ashworth.)

It cannot be determined whether this mug, with its pure cylindrical form and loop handle, is copied from an eighteenth-century piece. There is no mug at Williamsburg of this exact form and size, nor is there any record of Thorley borrowing one to copy. It is most likely that Thorley designed this form based on his knowledge of eighteenth-century shapes and proportions. An eighteenth-century vessel of this shape would not have been cast in a mold, as this was, but rather thrown on a wheel, the handle either shaped by hand or pressed in a small, separate mold and then attached. The decoration on this mug is reminiscent of the ware he designed or adapted while he was designer at Allerton Ltd., Longton, England, from about 1924 until 1927, when he immigrated to the United States. (See Austin, "Thorley, Part 1," pp. 179–81, figs. 31–34.) Thorley used the pink luster again during his latter days of production.

Unless otherwise noted, the illustrated items in this article were made by Palin Thorley, Williamsburg, Virginia, and are in the author's collection.

John C. Austin

J. Palin Thorley (1892–1987), Potter and Designer: Part 2, Williamsburg

▼ JOSEPH PALIN THORLEY was born in Staffordshire in 1892. He apprenticed at Wedgwood and designed for several major commercial potteries before becoming involved with Colonial Williamsburg. The earlier phase of his career was published in the 2005 issue of *Ceramics in America*.[1] In Williamsburg not only did he design the wares, as he had done before, but he took on the time-consuming task of producing them as well, a distraction that increasingly interfered with his ability to create new forms, his forte.

The transition from the larger potteries to Williamsburg took many years. Thorley began designing for Colonial Williamsburg in 1937, when he was still living and working in East Liverpool, Ohio. He did not move to the city of Williamsburg until 1949, and even then he continued to design for at least one other pottery, Hall China Company in East Liverpool, until about 1959.

In 1937 a friend suggested that Thorley visit the newly created Colonial Williamsburg and arranged for him to meet with several important people in town, among them the president of the College of William and Mary and the president and several vice presidents of Colonial Williamsburg. When he was shown some archaeological fragments of a silver luster jug that had been excavated in the historic area, Thorley boasted, "Oh, I can reproduce that"—which, of course, he could—and thus began a new episode in his life.

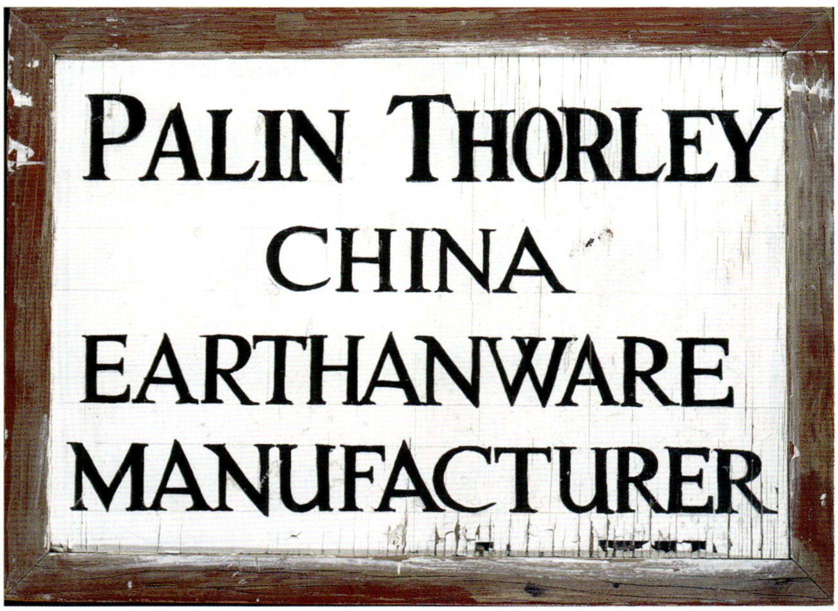

Figure 2 Secondary sign that hung outside Thorley's studio-shop, 1950s–1970s. The primary, much more elaborate sign that hung near the road illustrated a potter at his wheel. (See Austin, "Thorley, Part 1," fig. 1.)

In 1941 Thorley was made a licensed manufacturer for Colonial Williamsburg's Craft House, which oversees the sale of reproductions of antique objects in the collections of Colonial Williamsburg, although it was designed to be educational as well as a moneymaking outlet. The objects sold at the Craft House were produced by licensed manufacturers who were not employees of Williamsburg and who sold their products to many other outlets as well. The objects were selected, and finished samples approved, by the Craft Advisory Committee (now the Product Review Board), which is made up of employees from different facets of the foundation—curators, architects, archaeologists, and business administrators. The committee was concerned only with those products sold at the Craft House. Other Williamsburg outlets, such as the craft shops in the historic area and the gift shops in the Williamsburg hotels, were separate responsibilities. Thorley remained a licensed manufacturer until the fall of 1956, when his license was terminated. The termination probably was the result of his inability to keep up with orders, although he also had a reputation for being difficult to work with and unrealistic in his demands.

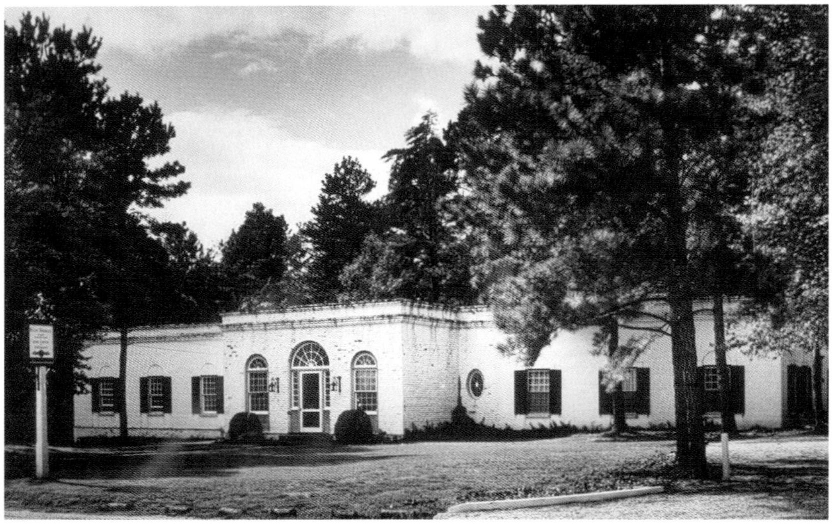

Figure 3 White Hall, as it appeared in the British trade magazine *Pottery and Glass,* November 1950, with the caption: "Studio, laboratory and offices at Williamsburg, Virginia USA, built 1945–1950 by Joseph Palin Thorley for research on commercial ceramics processes and the reproduction of wares of historic interest to Williamsburg, under special licence from the authorities."

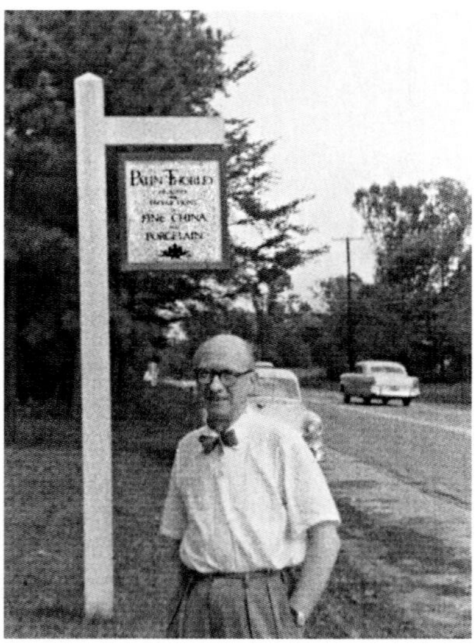

Figure 4 Palin Thorley standing beneath the sign that hung in front of White Hall until about 1956, when it was replaced with one illustrating a potter at his wheel.

In 1946 Thorley purchased the Van Rensselaer Museum building on Jamestown Road, which at one time had exhibited a collection of antique firearms. His move to the area was supported by the administrators of Colonial Williamsburg, who provided a loan for the down payment. Thorley enlarged the building and converted it to a studio, where he designed and made his pottery. He continued to produce items in Ohio until 1949, when he opened the studio, but in the interim he was in Williamsburg frequently enough to rent at nearby Ludwell Apartments.

The central, projected area of the studio building is where Thorley's wife, Edith, managed a small gift shop, selling wares her husband made in Williamsburg as well as those designed by him and produced by other potteries. Behind the shop was Thorley's studio, office, and storage areas, as well as the couple's living quarters. As they planned to buy or build a large house similar to the one they had left in Ohio, their residence in the

Figure 5 Interior of Thorley's studio, 1964 or 1965, showing barrels, probably filled with slip, and stacks of plaster of paris molds.

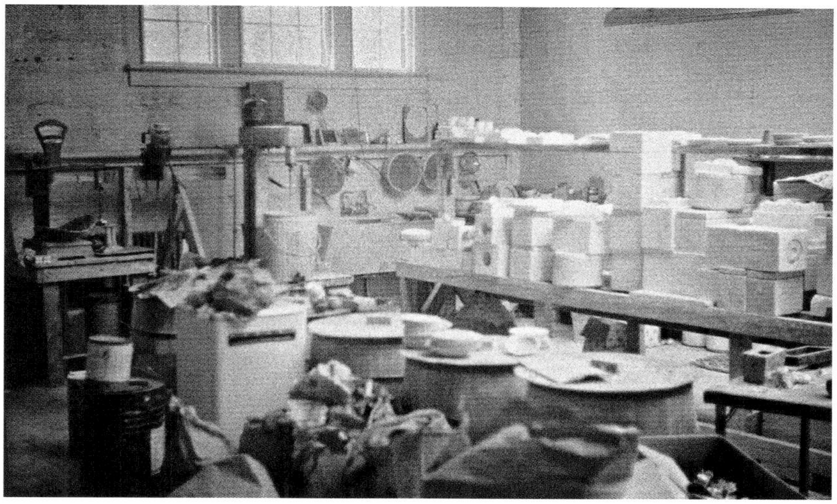

Figure 6 Gloss (glazing) kiln, 1970, with "square bricks" placed on a shelf inside.

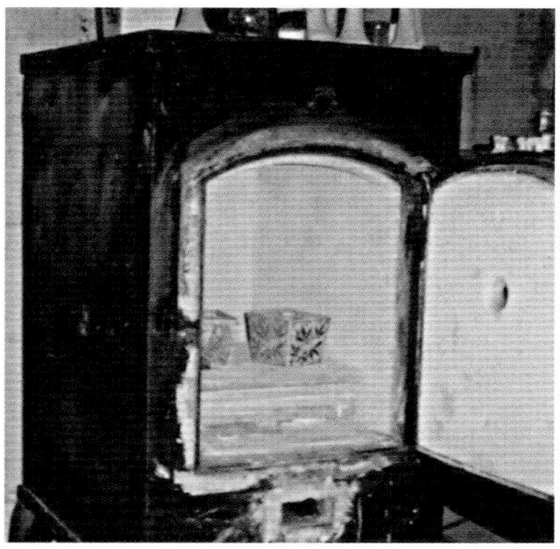

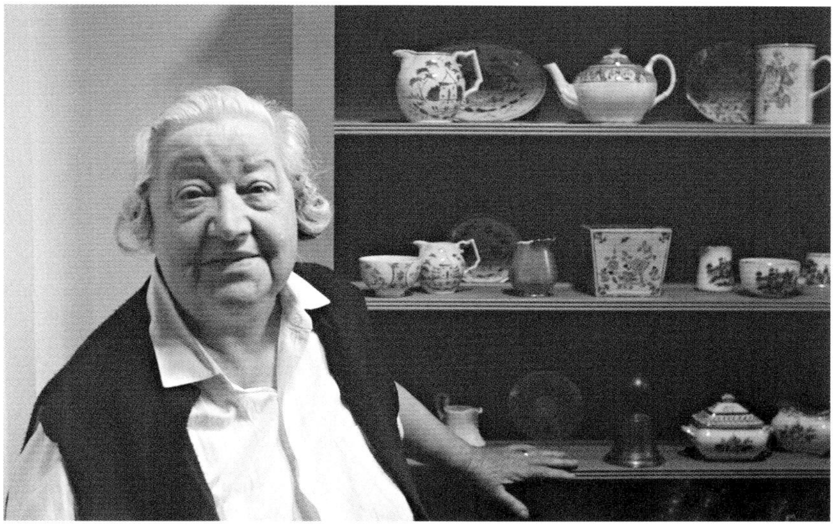

Figure 7 Edith Thorley in her shop, 1964 or 1965, and the interior of the shop, probably 1960s. Both pictures show forms developed for the Craft House that Thorley continued to make, examples made or decorated by Thorley after he stopped producing for Colonial Williamsburg, and unidentified objects that perhaps were designed by him earlier and still in production by the East Liverpool (Ohio) factories.

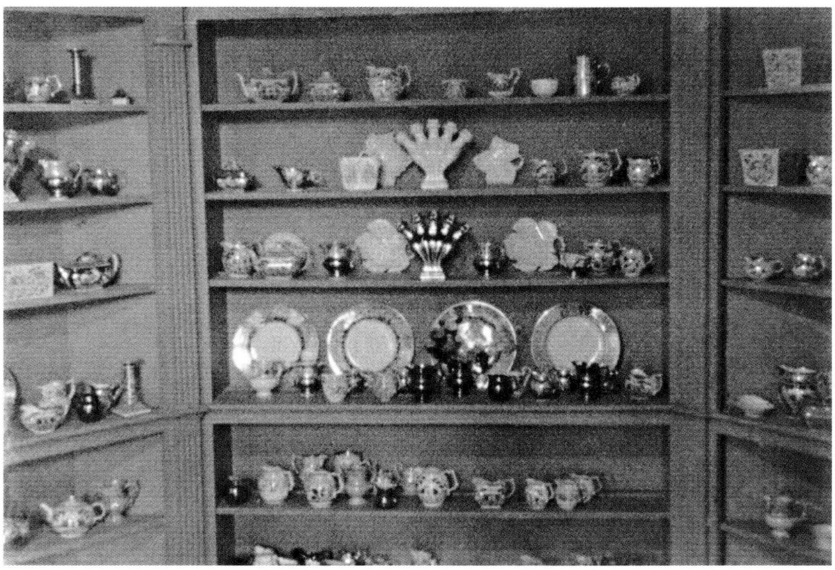

museum building was intended to be temporary. Nevertheless, Thorley referred to the building as "White Hall" (or "Whitehall" or "WhiteHall"), though the name never appeared on the several signs that hung in front. It now is home to the White Hall Restaurant, which has many of Thorley's ceramic objects and molds on display.

An undated but post-1950 inventory in the Thorley Papers lists more than six thousand pieces of ceramic in clay, biscuit, or finished state. The inventory includes:

Kilns, #4906 Denver	
High temp—with Carborundum	
filling & kiln furniture	10,000
Kiln #403 Denver	1,000
small Decorating	200
medium Decorating	250
"Needs repairs	100
Machines Mills, etc	
Mechanics—Motor	
5 motor grinding wheels	500
2 jigger and Jolleys complete	1,500
Throwing wheels	300
mill one de-airing Pug	1,200
8 lining wheels	400
Molds & casting bench	250
Slip clay mixer & equipment	1,000
Glaze fritts—oxides	

Thorley had long since become an American citizen, but his love for England and the royal family was steadfast. One of the highlights of his life occurred in 1946, when Elizabeth, the Queen Mother, visited Williamsburg and bought Craft House reproductions to take home as gifts, among them Thorley's heart-shaped inkwell and the jardiniere, or "square brick." Thorley wrote his friend Joseph Thompson at Hall China Company, "We had a very wonderful week with Her Majesty the Queen Mother here. Her Majesty purchased some of my productions and made a few highly complimentary comments. I am going to name these items 'A Queen's Choice.'" Among the Thorley Papers are clippings about the queen mother's purchases of his objects in newspapers from Ohio, Virginia, and Staffordshire, England. All the articles repeat the popular but erroneous story that twenty years earlier Queen Mary had placed an order for Thorley's Chatelaine figure (see Austin, "Thorley, Part 1," fig. 28).

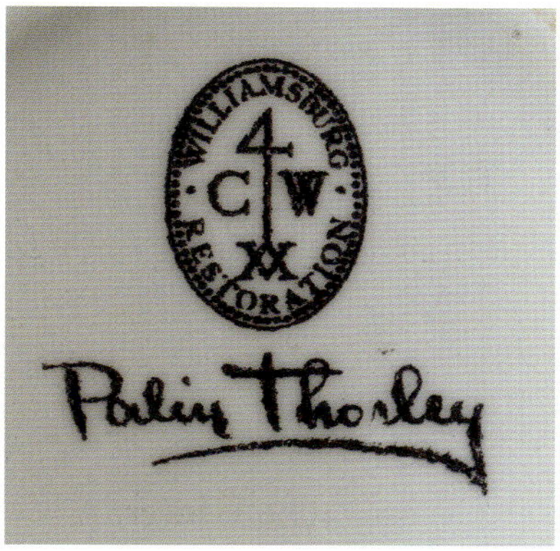

Figure 8 Printed mark used by Thorley on wares produced for the Craft House, Colonial Williamsburg's merchandise outlet, between 1941 and 1956. All of Thorley's forms that he produced for the Craft House—whether sold through this outlet or directly to other customers—bear this mark. The oval emblem is based on seventeenth- and eighteenth-century British shippers' marks.

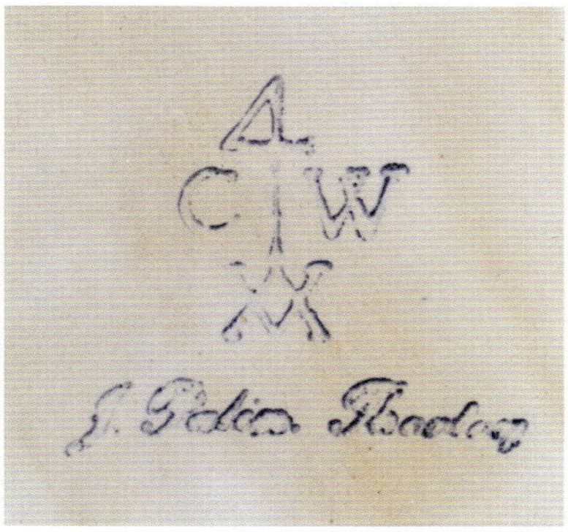

Figure 9 Rare variation of the Craft House/Thorley mark. This seems to appear only on the "GR" plate, which was produced between 1942 and 1945 (see fig. 52).

Figure 10 Thorley's printed mark was registered in 1956. In the application he states that he used the mark as early as 1955; therefore, pieces with this mark must have been made between 1955 and the 1970s, when Thorley ceased producing. Objects with this mark and the oval Colonial Williamsburg mark are rare and were made in 1955 or 1956.

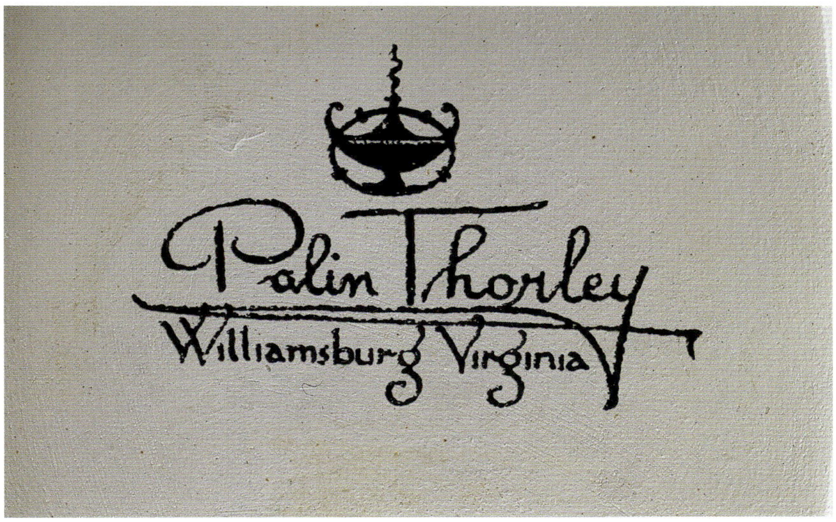

Figure 11 Examples of handwritten marks possibly used from 1949, when Thorley moved to Williamsburg, until the 1970s, when he stopped producing ceramics. His monogram, illustrated on the lower part with the red ground, is sometimes the only mark on a piece and dates the same as his full signature.

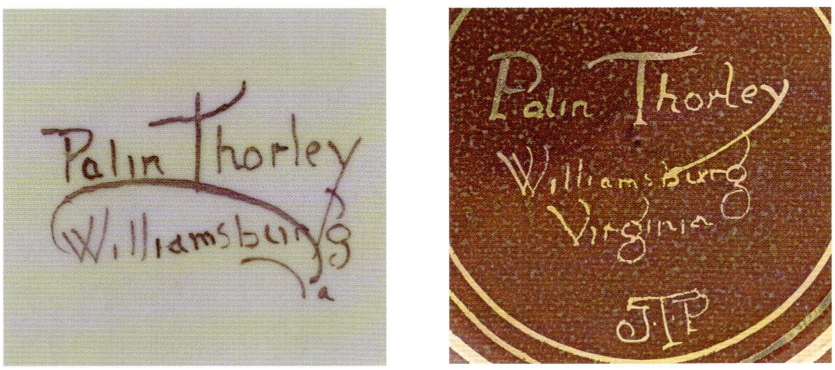

Colonial Williamsburg Craft House Products

Although Thorley made other objects for Colonial Williamsburg, only five of his forms are illustrated in the Craft House catalogs. The 1942 catalog illustrated only the luster jug, and only in the largest size. Included on the war years price list (1942–1945) but not shown in the catalog were jugs (in the two larger sizes), the jardiniere, and a "Lambeth Delft Plate (blue and white). . . ."[2] The posy holder was added to the 1949 catalog, and in 1951, the inkstand. The catalog illustrated in figure 12 is undated but probably was first printed in 1953 or 1954 and included just two sizes of jugs (largest

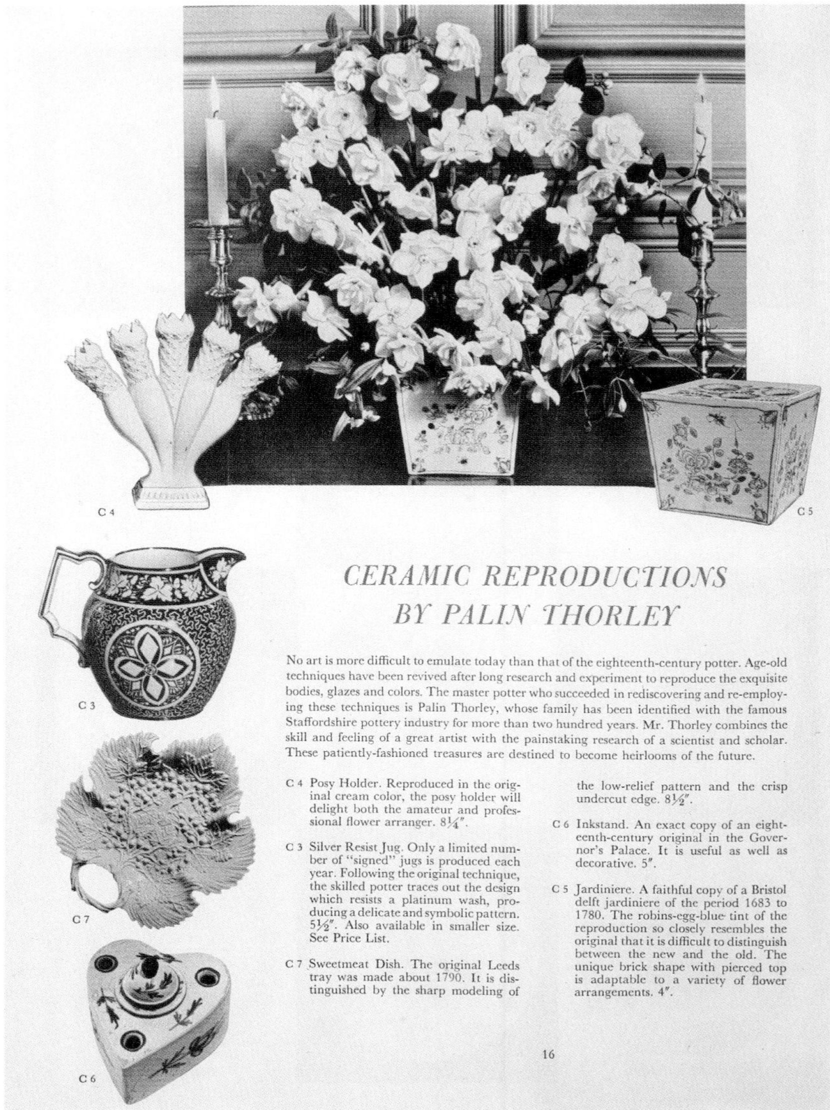

C 4

C 5

C 3

C 7

C 6

CERAMIC REPRODUCTIONS BY PALIN THORLEY

No art is more difficult to emulate today than that of the eighteenth-century potter. Age-old techniques have been revived after long research and experiment to reproduce the exquisite bodies, glazes and colors. The master potter who succeeded in rediscovering and re-employing these techniques is Palin Thorley, whose family has been identified with the famous Staffordshire pottery industry for more than two hundred years. Mr. Thorley combines the skill and feeling of a great artist with the painstaking research of a scientist and scholar. These patiently-fashioned treasures are destined to become heirlooms of the future.

C 4 Posy Holder. Reproduced in the original cream color, the posy holder will delight both the amateur and professional flower arranger. 8¼".

C 3 Silver Resist Jug. Only a limited number of "signed" jugs is produced each year. Following the original technique, the skilled potter traces out the design which resists a platinum wash, producing a delicate and symbolic pattern. 5½". Also available in smaller size. See Price List.

C 7 Sweetmeat Dish. The original Leeds tray was made about 1790. It is distinguished by the sharp modeling of the low-relief pattern and the crisp undercut edge. 8½".

C 6 Inkstand. An exact copy of an eighteenth-century original in the Governor's Palace. It is useful as well as decorative. 5".

C 5 Jardiniere. A faithful copy of a Bristol delft jardiniere of the period 1683 to 1780. The robins-egg-blue tint of the reproduction so closely resembles the original that it is difficult to distinguish between the new and the old. The unique brick shape with pierced top is adaptable to a variety of flower arrangements. 4".

16

Figure 12 Page from Colonial Williamsburg's *Craft House Catalog*, probably 1953 or 1954.

and middle) plus the leaf-shaped sweetmeat dish. The price lists for the spring and fall of 1955 and the spring of 1956 accompanied the same catalog; on the 1956 list, however, the jardiniere was listed as "Not available." A new catalog, published later in 1956, offered no Thorley ware (by then the Craft House had severed relations with the potter), although it did include a posy holder and a jardiniere similar to his but manufactured by other potteries. Thorley had been working on new forms for the Craft House, but none was submitted to the Craft Advisory Committee after the sweetmeat dish.

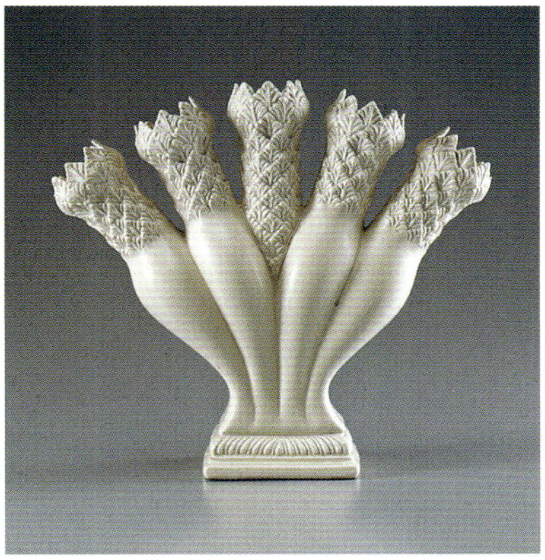

Figure 14 Posy holder. Earthenware. H. 8". Approved for reproduction in 1946, last produced for the Craft House in 1956.

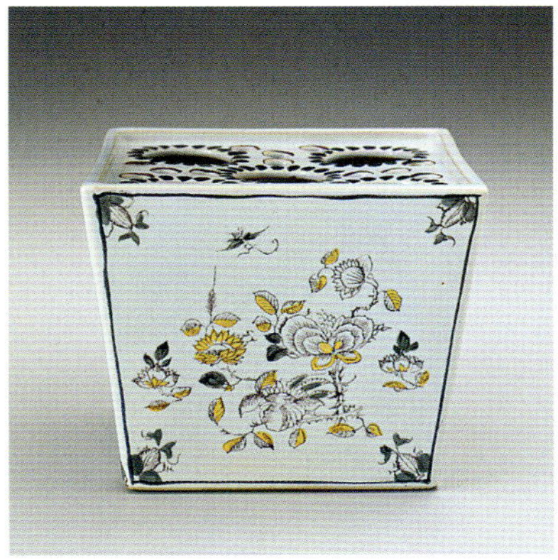

Figure 15 Jardiniere. Earthenware. H. 4". Approved for reproduction in 1942; last produced for the Craft House in 1955.

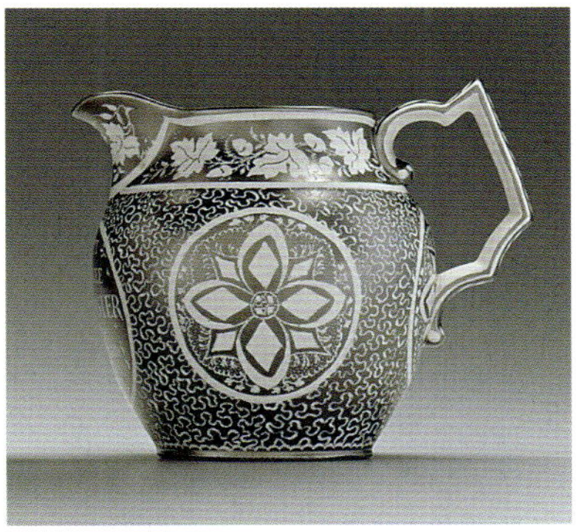

Figure 13 Silver resist jug, 1938–1956. Earthenware. H. 5½". The dates indicate the year the piece was approved for reproduction and the last year it was produced for the Craft House. If a Thorley-made object bears the Williamsburg mark, it was made between those dates.

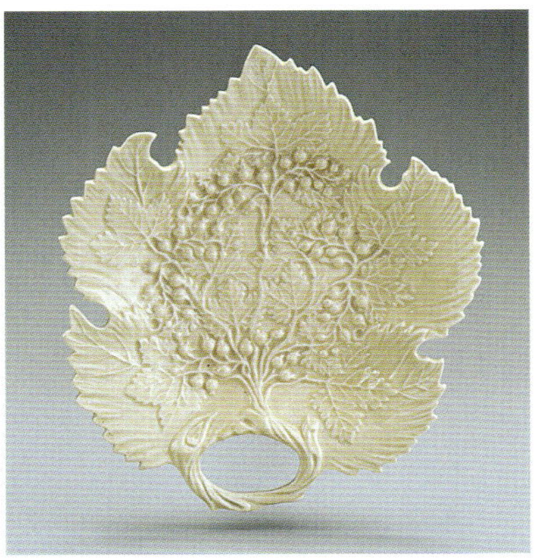

Figure 16 Sweetmeat dish. Earthenware. L. 8". Approved for reproduction in 1952; last produced for the Craft House in 1956.

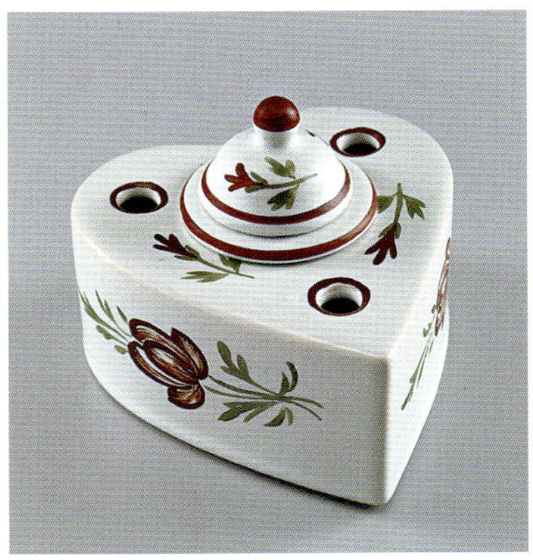

Figure 17 Inkstand. Earthenware. H. 4½". Approved for reproduction in 1951; last produced for the Craft House in 1956.

Thorley Jugs

The silver luster jugs Thorley produced for Colonial Williamsburg differ from his other Craft House reproductions in two ways. First, he did not initially create the form—called a Thorley Jug—for Colonial Williamsburg. It was already in production by Hall China Company with simple gold asterisk-like decoration. Second, it was not based on an antique example in the collections of Colonel Williamsburg but rather on archaeological fragments excavated in the historic city. Perhaps Thorley had trouble producing the smallest jug, for apparently it was not offered for sale after 1945.

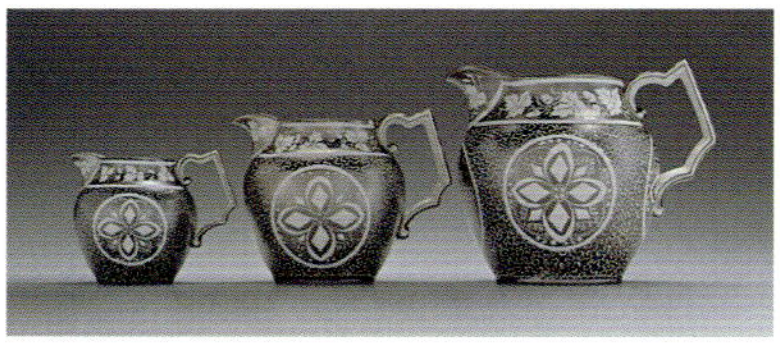

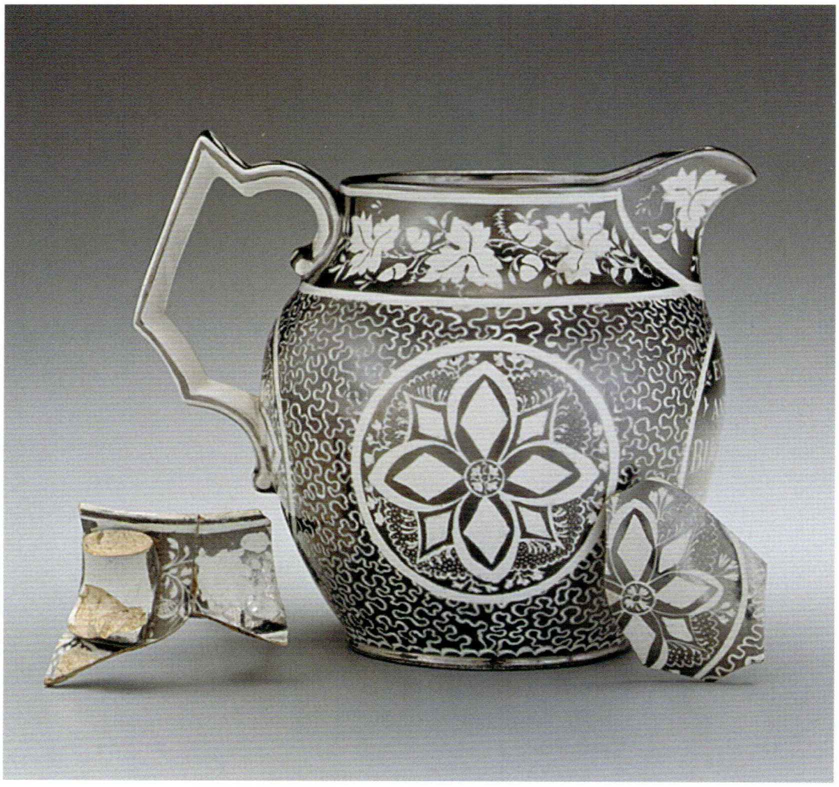

Figure 18 Thorley Jugs, 1945–1956. Earthenware. *Left to right:* H. 3", 4½", and 5½". These jugs were produced for the Craft House, and all bear the stamped "CW" logo and Thorley's name.

Figure 19 Silver luster jug and archaeological sherds from a similar, early-nineteenth-century jug excavated from the Donigan site, in Williamsburg (Windmill lot). Earthenware. H. of jug 5½".

Figure 20 Thorley/Craft House mark on the jug illustrated in fig. 19. (Courtesy, Collins Collection.)

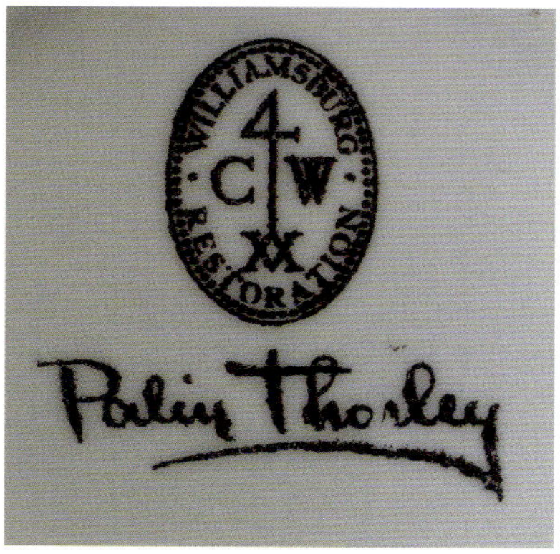

In the late 1950s and 1960s Thorley often used the Thorley Jug form for special-order or commemorative pieces and typically marked them with his name, either stamped or handwritten. Inscriptions on the bottom are rarer than those on the side. On examples with the Craft House silver luster decoration, one of the rosettes was replaced by the inscription. The date on the jug does not necessarily indicate the date the piece was made or decorated. For example, on a set of three made for the Hunt family of Richmond and dated, respectively, 1957, 1960, and 1961, the first was made in 1961 but commemorates a wedding that took place in 1957; the second was made in 1961 and commemorates the birth of the first son, in 1960; and the third, made in 1962 and utilizing a bird motif rather than the Craft House design, commemorates the birth of the second son, in 1961, and is known through family pictures of the three. A thank-you note that was written in January 1961 after the receipt of the first two is in the Thorley Papers.

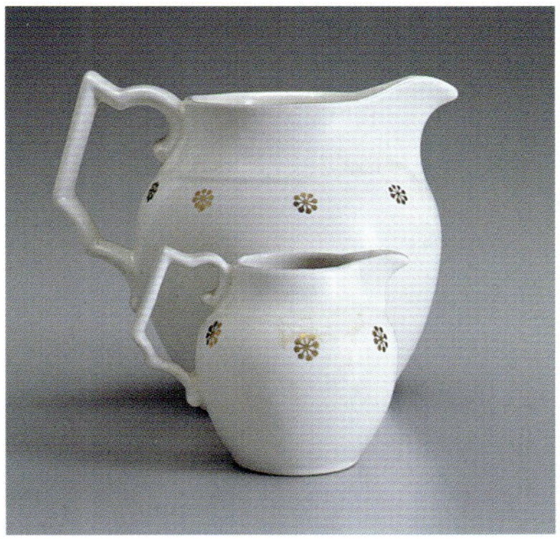

Figure 21 For comparison, the largest and the smallest of Hall China Co.'s Thorley Jugs. Earthenware H. 5½", 3¼".

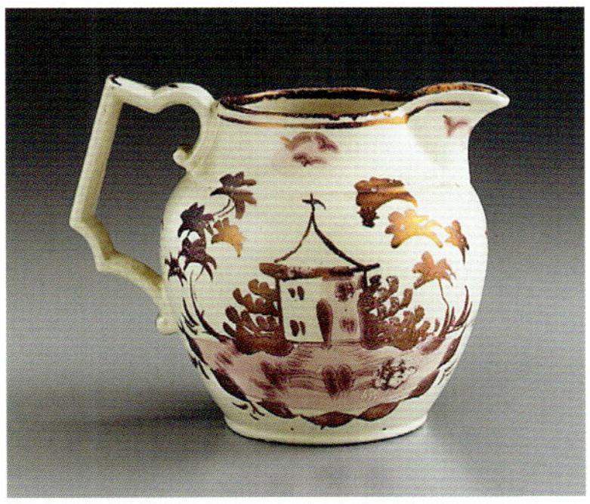

Figure 22 Thorley Jug, with tower decoration. Earthenware. H. 4½". Thorley was producing objects for other outlets while he was making ware for the Craft House. A 1954 order from the Williamsburg Lodge gift shop included jugs in eight different patterns, among them "Pink Luster tower pattern." The motif on this unmarked pink luster jug was probably inspired by his work on eighteenth-century English delft.

Figure 23 Inscription on the bottom of the jug illustrated in fig. 24. The inscription is used infrequently; the numbers in blue suggest that the jug had been used earlier for experimental purposes.

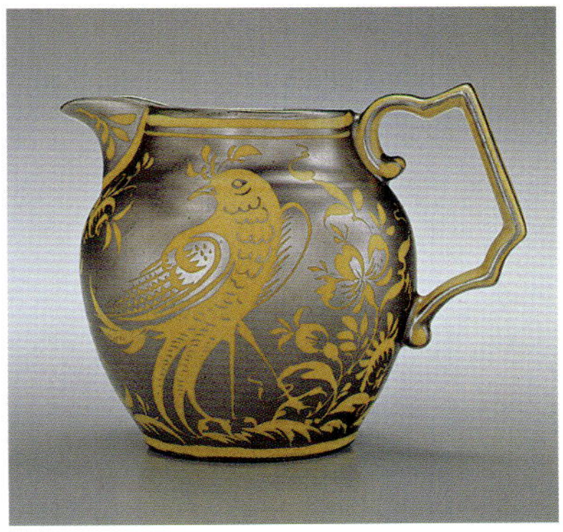

Figure 24 Silver and yellow luster jug, ca. 1960. Earthenware. H. 3¼". (Courtesy, Collins Collection.) The bird pattern is not common for Thorley commemorative wares. This jug is one of two inscribed for the Buckmaster family.

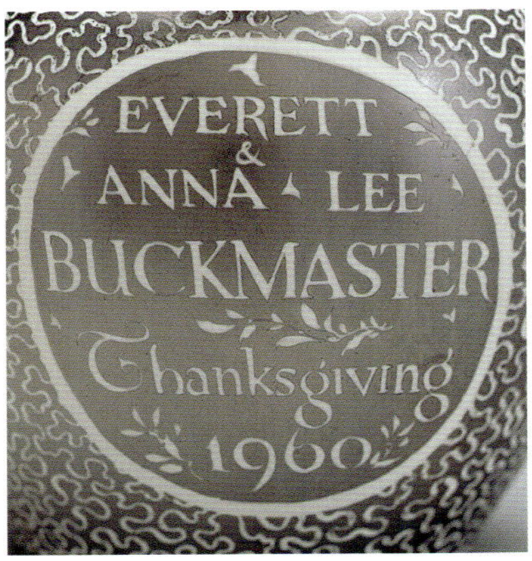

Figure 25 Inscription on a silver luster jug. (Courtesy, Collins Collection.) Inscriptions typically were placed on the side, like the rosette on the jugs illustrated in figs. 18 and 19.

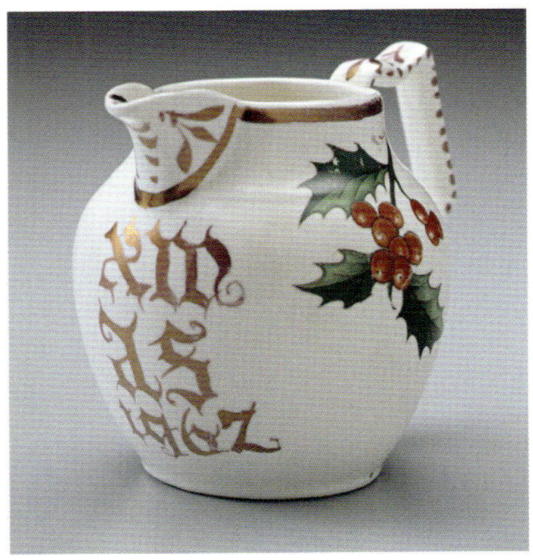

Figure 26 Jug, dated 1967. Whiteware. H. 3¼". (Courtesy, Clark Taggart Collection.) Probably decorated as a Christmas present and never sold or given, this small Thorley jug was in Thorley's possession in 1986. The inscription, "XM/AS/1967," is written in Old English script, popular in the 1920s and 1930s but used by Thorley on many experimental pieces and also found in drawings among his papers.

Jardinieres (Square Bricks)

The jardiniere, often referred to as a "square brick" because of its similarity to the more common brick-shaped flower containers, is, with its flat sides, a perfect form on which to apply decoration. Using the reproduction's mold, Thorley created a myriad of designs. He drew or painted many of his ideas first on paper and, if satisfied, applied them to the undecorated ware.

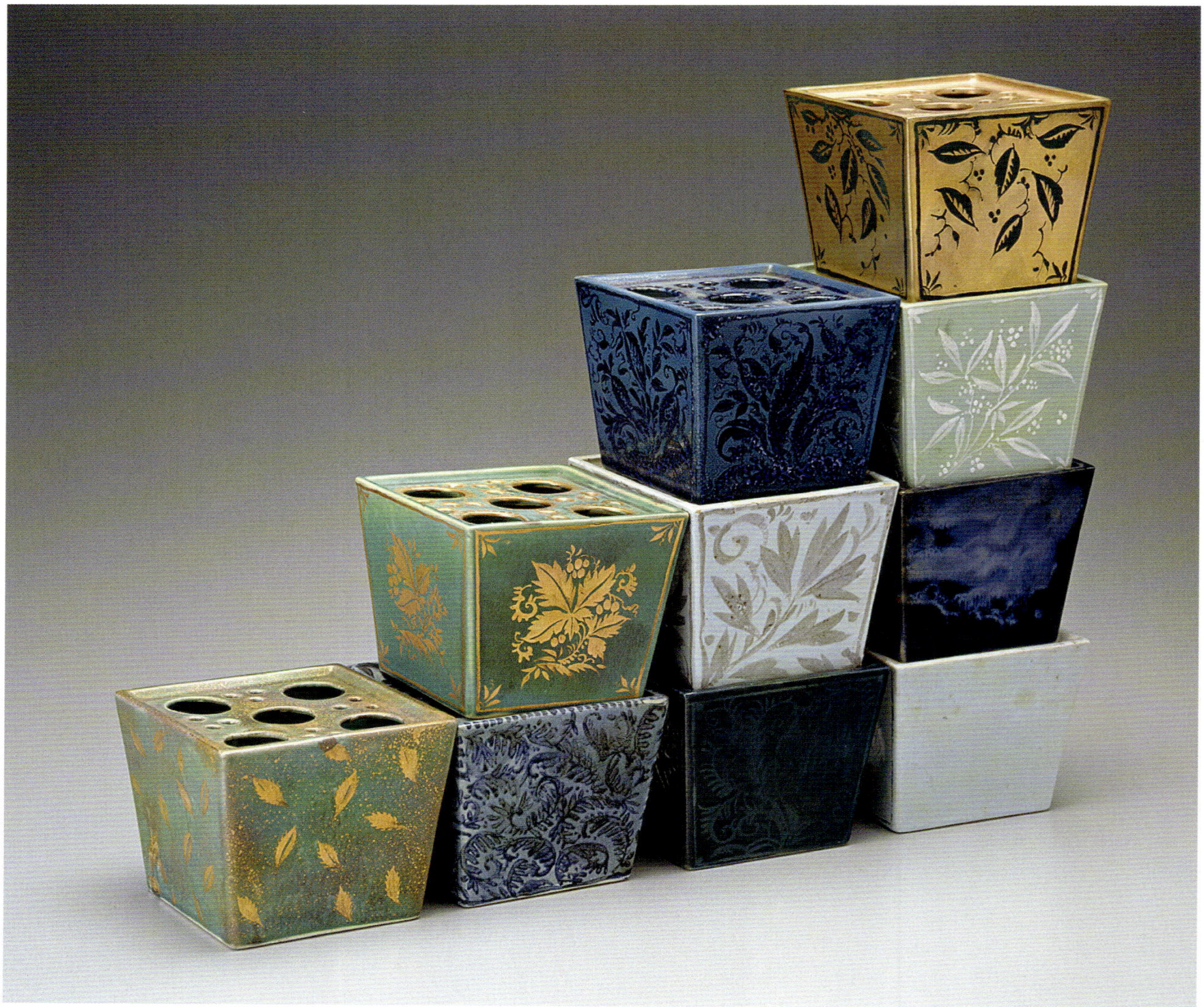

Figure 27 Jardinieres, or square bricks, 1946–1970. Earthenware. H. of each 4". This stack of jardinieres shows the variety of colors and designs used by Thorley.

Figure 28 Jardiniere, 1942–1956. Whiteware. H. 4". This brick is a well-executed, finished example. It does not bear a mark, as most production pieces would, but "#1" appears on a paper label affixed to the bottom. It is possible that this, like the five-finger posy holder illustrated in fig. 41, is the sample that was presented to and approved by the Craft Advisory Committee of the Craft House. If so, its date of manufacture would be 1942; if not, between 1942 and 1956.

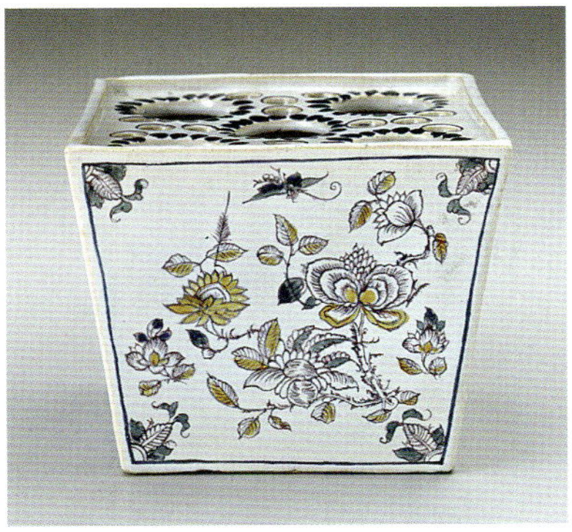

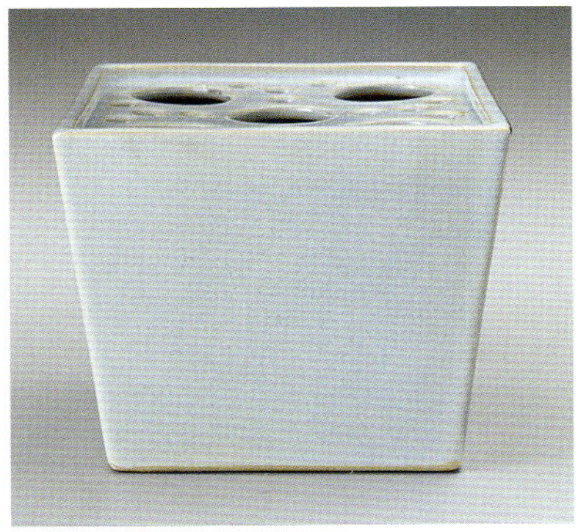

Figure 29 Jardiniere, 1942–1970. Earthenware. H. 4". This example illustrates an undecorated surface to which decoration could be added.

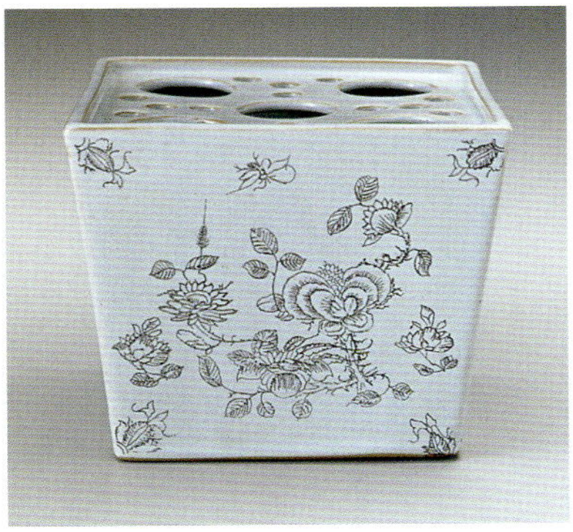

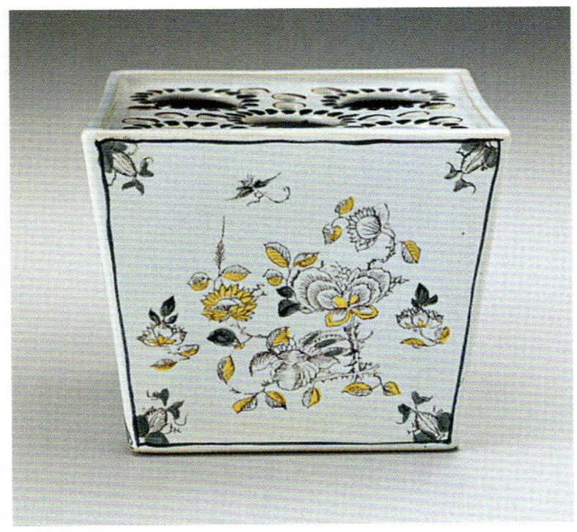

Figure 30 Jardiniere, 1942–1970. Earthenware. H. 4". Even though the eighteenth-century example from which this was copied was decorated by hand, the basic design of a reproduction was silk-screened and applied in monochrome—here, in manganese purple—and the piece was fired. Were the decoration painted entirely by hand, it would be impossible to produce the quantity necessary and still have the decoration as close to the original as required.

Figure 31 Jardiniere, 1942–1956. Earthenware. H. 4". (Courtesy, Collins Collection.) The yellow and blue decoration was painted by hand and the piece refired. As this bears the printed Craft House mark illustrated in fig. 8, it can be dated between 1942 and 1956.

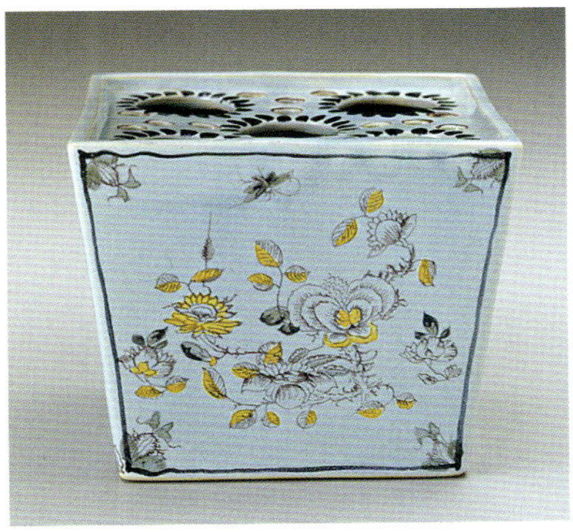

Figure 32 Jardiniere, after 1956. Earthenware. H. 4". Mark: Thorley stamp (illustrated in fig. 10). (Courtesy, Collins Collection.) This is a rare example of a reproduction made after 1956, when Thorley stopped producing for Colonial Williamsburg, as it bears Thorley's printed mark, not the Colonial Williamsburg marks probably made shortly after that date.

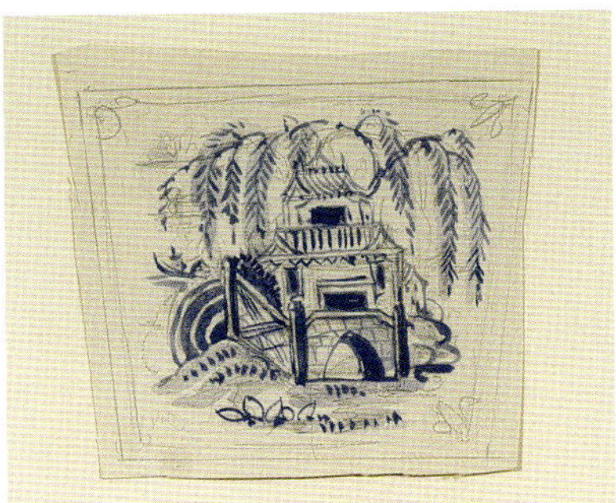

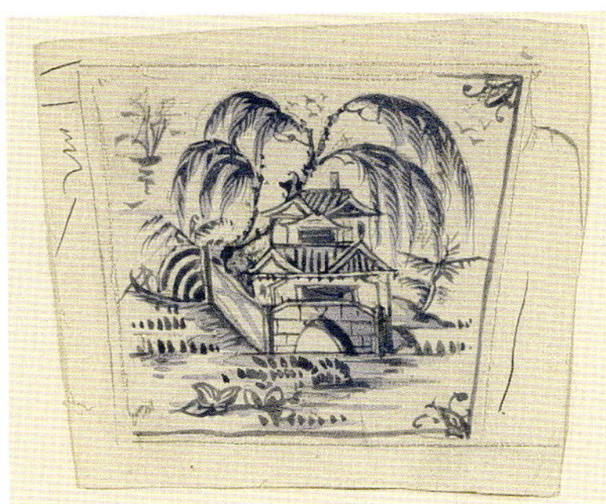

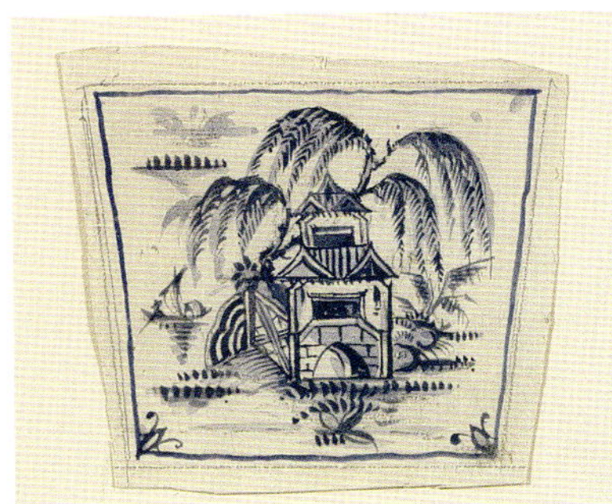

Figure 33 Drawings for jardinieres. Watercolor on paper. 7" x 16".

Flower Bricks

Even though these brick-shaped containers are sometimes confused with inkwells, it is generally believed that vessels of this form were used as flower containers in the middle of the eighteenth century. Colonial Williamsburg has a large collection of antique examples, two of which Thorley borrowed to reproduce. Even though there is no evidence that he ever produced any for the Craft House, he did make bricks in two sizes that in shape were somewhat related to the ones he borrowed but not in the decoration. As many drawings for bricks with eighteenth-century-style decoration have been found among Thorley's papers, it is likely that he had intended to produce examples relating to the originals. His surviving examples tend to have experimental glazes or are decorated with decals more reminiscent of eighteenth- or nineteenth-century porcelain.

Figure 34 Flower bricks. Whiteware. *Rear:* 1955–1970. Mottled glaze. L. 7½". Mark: Thorley stamp (illustrated in fig. 10). *Front:* 1942–1970. Foliated decoration. L. 5". Mark written in black and experiment mark in white.

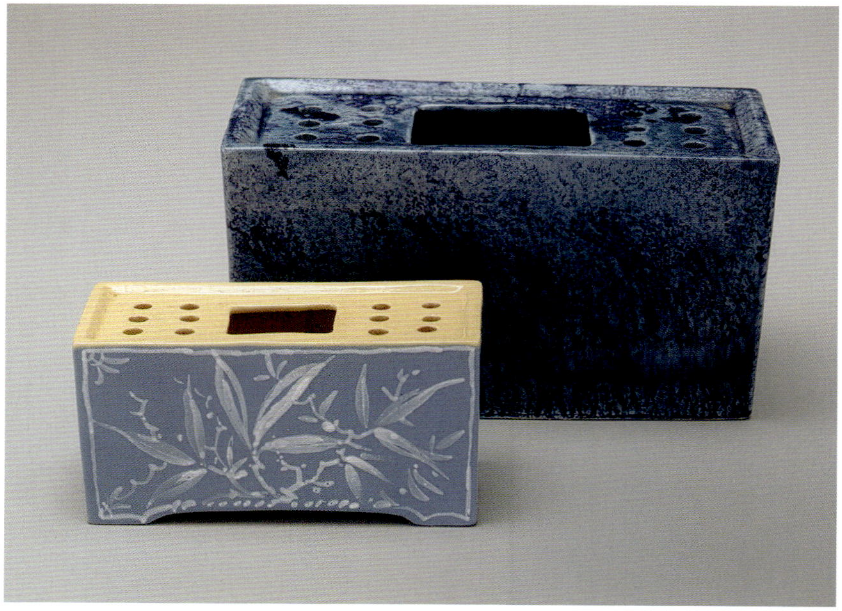

Figure 35 Flower brick and jug, 1955–1970. Earthenware. Mark: Thorley stamp (illustrated in fig. 10). *Brick:* L. 7⅛". *Jug:* H. 5½". (Courtesy, Collins Collection.) Hand-painted with applied decal decoration.

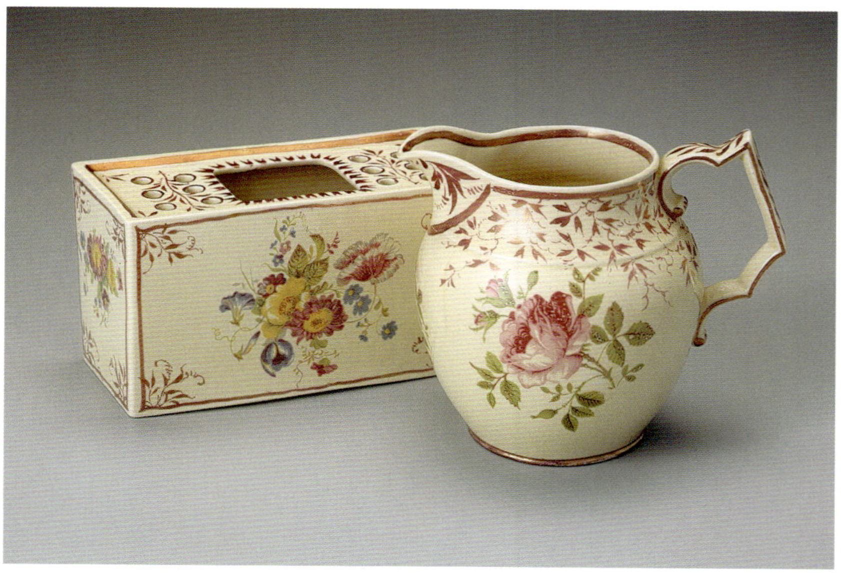

Figures 36 Drawings for flower bricks in the style of the eighteenth century, ca. 1954. Watercolor on paper. L. of images approx. 7". Probably intended to be used to decorate examples sold by Colonial Williamsburg. In 1954 Thorley presented to the Craft Advisory Committee hand-painted designs "in the spirit of" eighteenth-century style and technique, to be used on jardinieres. Even though one design was approved, none appears to have been produced and presented to the committee for approval.

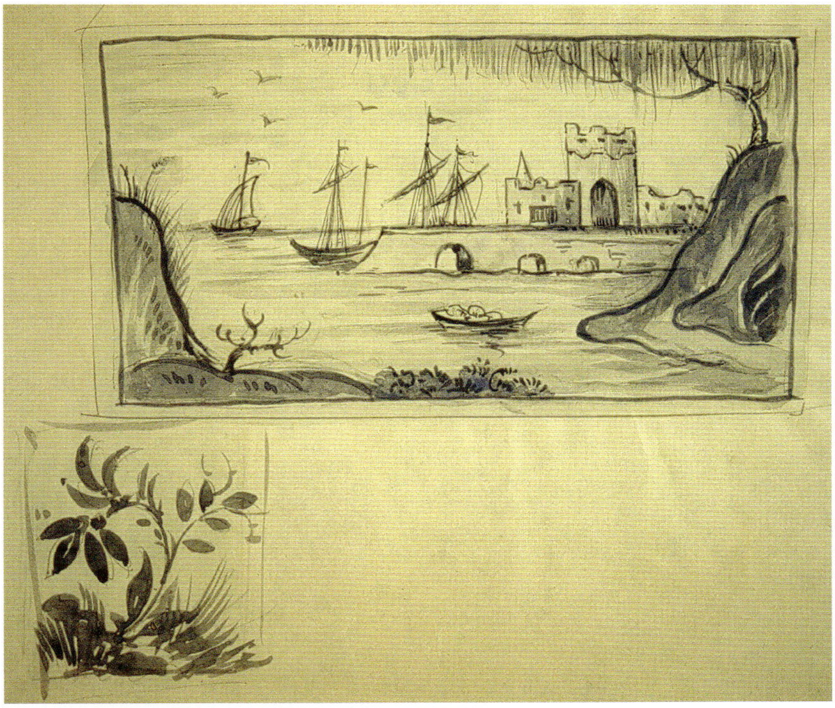

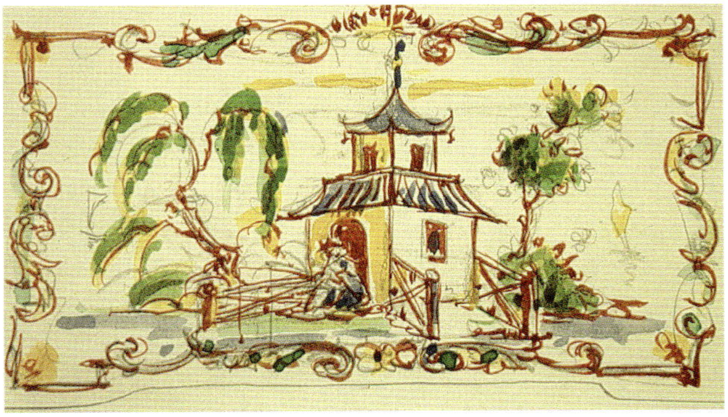

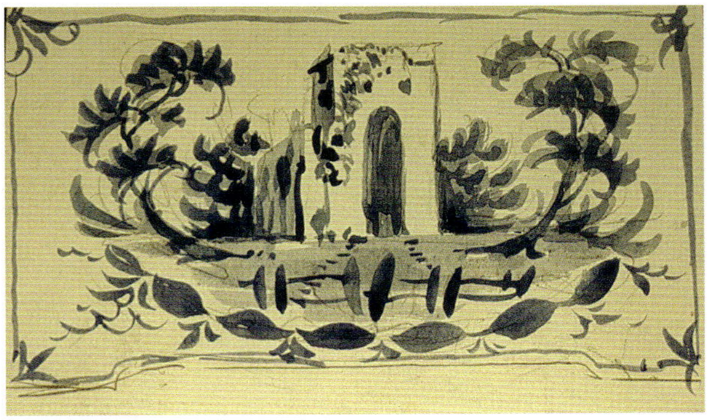

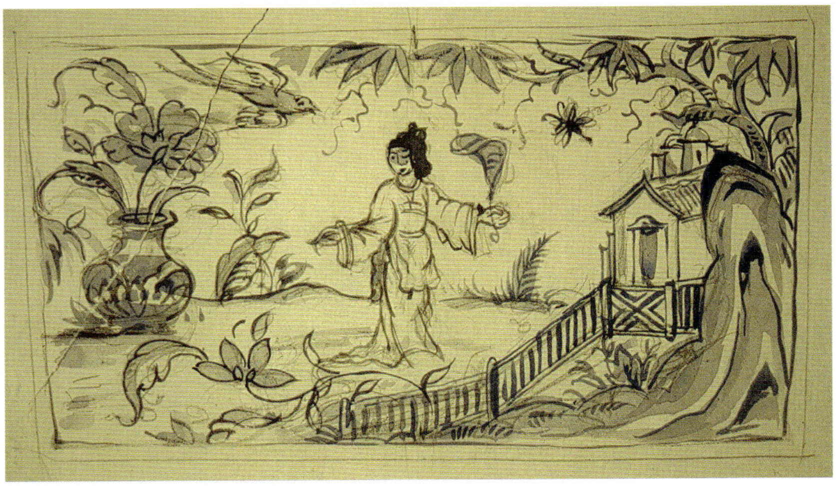

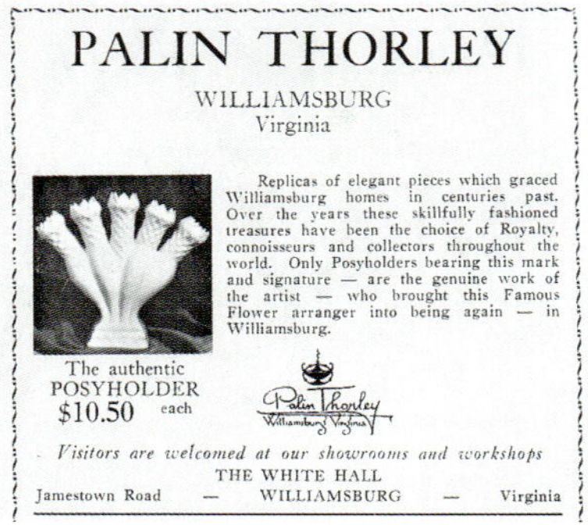

Figure 37 Palin Thorley's most well-known reproduction was used to illustrate this advertisement for White Hall, the shop in Thorley's pottery, which appeared in the 1959 *Historic Virginia Week in Williamsburg* catalog. At this time Thorley was still producing the form for his own sales; Wedgwood was producing it for Colonial Williamsburg.

Posy Holders

Thorley was the first in the twentieth century to reproduce the five-finger vase, or posy holder, popular in the fourth quarter of the eighteenth century. His 1946 reproduction of a creamware example in the collection of Colonial Williamsburg was so successful that he was unable to keep up with the orders from the Craft House, let alone the many requests from gift shops and department stores throughout the country. Colonial Williamsburg flower arrangements, as published by Louise B. Fisher in her books on the subject,[3] became the way to do colonial bouquets, and the posy holder became, to many, synonymous with Colonial Williamsburg and its potter.

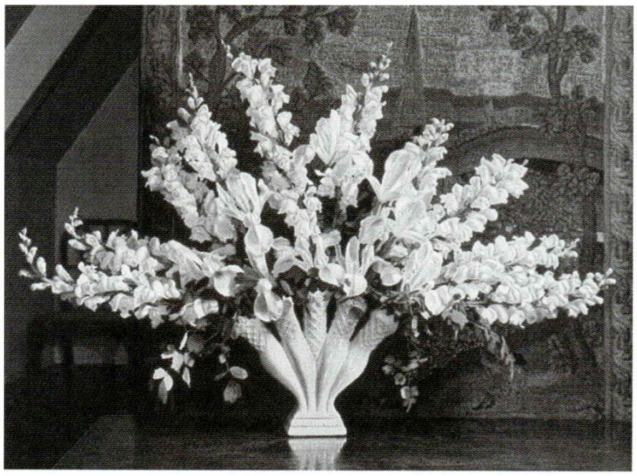

Figure 38 An illustration showing a Palin Thorley posy holder in use, in Louise Bang Fisher, *An Eighteenth-Century Garland: The Flower and Fruit Arrangements of Colonial Williamsburg* (Williamsburg, Va.: Colonial Williamsburg, 1951), sixth page insert between pp. 26 and 27.

The eighteenth-century examples made in Leeds and Staffordshire in England were of the warm creamy color of the reproduction and left without surface decoration. Late in the eighteenth and into the nineteenth century bright enamel colors were added to variations of the form that relate to those illustrated in figures 42 and 43. Luster-decorated Thorley examples can be seen in figures 81 and 86.

Thorley continued to produce this vase after he and the Craft House parted ways, the only change being to remove Colonial Williamsburg's icon from the mark. Colonial Williamsburg, however, retained Wedgwood to make the identical form to sell at the Craft House, which infuriated Thorley, who believed this created confusion about the maker and resulted in Wedgwood receiving credit for better-quality examples that he himself had made.

The five-finger vase remained a popular form for the rest of the century, made in Europe and the Orient as well as by Thorley, who produced at least three versions of it.

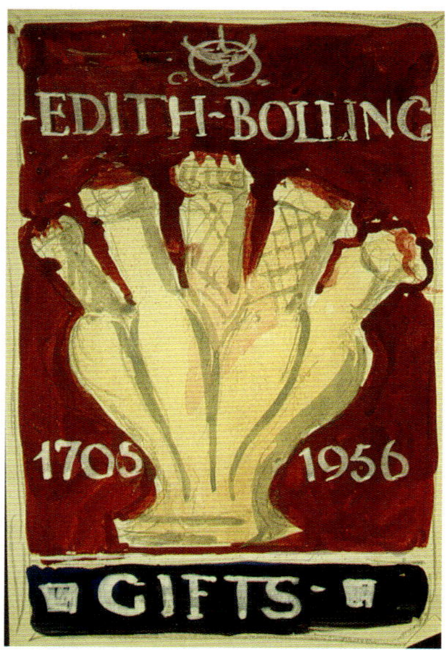

Figure 39 Shop sign design, 1956. Paint on paper. Approx. 18" x 24". (Swem Library, College of William and Mary.) In 1956 Thorley considered naming the shop for his wife, Edith, who managed it. He designed a sign incorporating the posy holder, which he eventually replaced with the "potter at his wheel" sign for which he is best known (see Austin, "Thorley, Part 1," fig. 1).

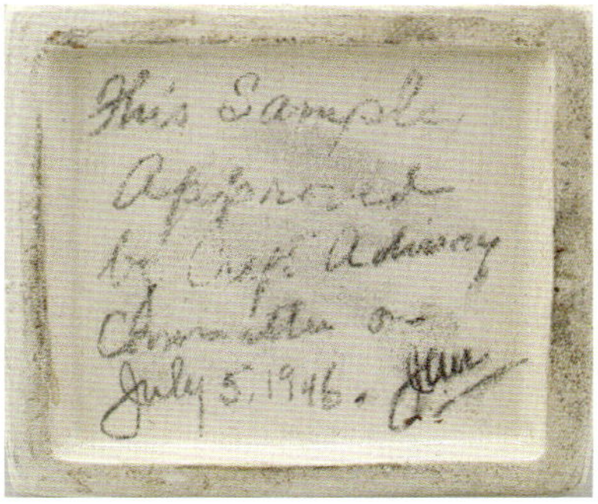

Figure 40 Base of manufacturer's model for the posy holder illustrated in fig. 41. Inscribed in pencil on the base: "This Sample Approved by Craft Advisory Committee on July 5, 1946. JAU." The initials *JAU* refer to John ("Jack") A. Upshir, who was in charge of the Craft House during Thorley's tenure.

It was customary for the Craft House to require the licensed manufacturers to present to the Craft Advisory Committee two examples of each object to be produced. When a product was approved, one example remained with the committee and the other stayed with the manufacturer. This helped to ensure that the merchandise did not vary from the accepted model. Few of these accepted models have survived.

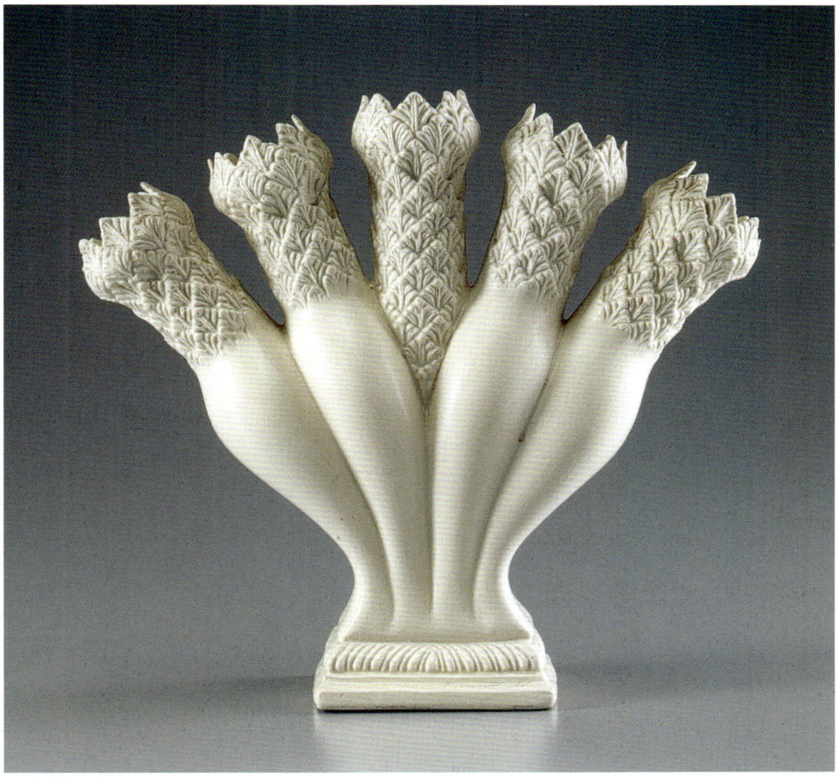

Figure 41 Posy holder, 1946. Creamware. H. 8". This is the original model produced by Thorley after an eighteenth-century example in Colonial Williamsburg's collection. This was used as a guide by Thorley for the hundreds more he made during his productive life.

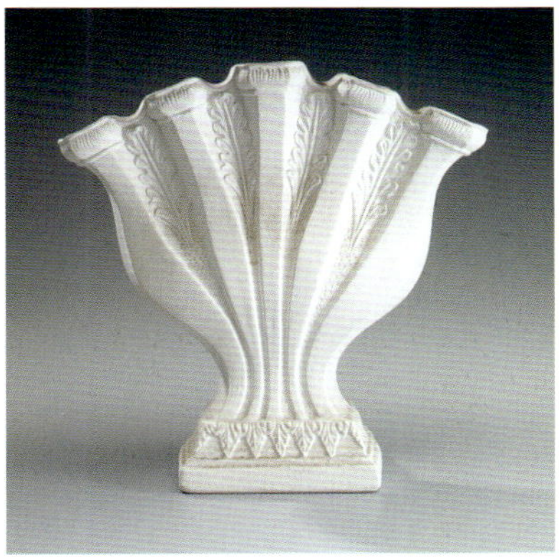

Figure 42 Posy holder model, 1955–1970. Earthenware. H. 7½". Mark: Thorley stamp (illustrated in fig. 10). Fired once and never glazed, this vase is one half inch taller than it would have been had it been glazed and refired. This model is very much like colorful early-nineteenth-century examples. Although never produced for the Craft House, it might have been copied from a related antique example in Colonial Williamsburg's collection. The Thorley Papers show that one was returned to Colonial Williamsburg in the 1950s.

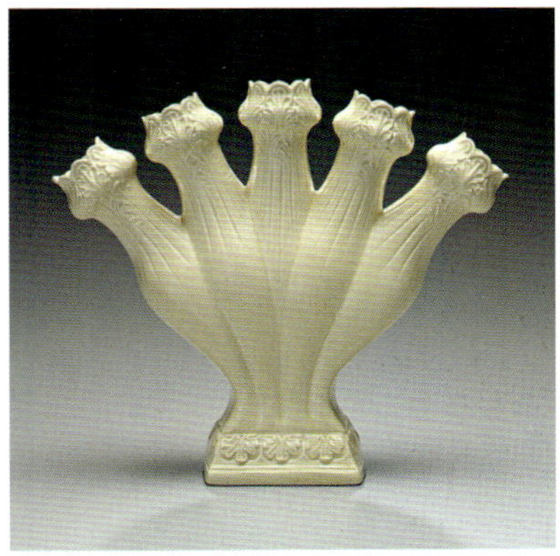

Figure 43 Posy holder, 1950–1970. Creamware. H. 8". Mark: In script, "Palin Thorley / Williamsburg." (Courtesy, Collins Collection.) A simplified version of the Colonial Williamsburg example, this is the third form of the five-finger vase produced by Thorley. It may have been inspired by or copied from an eighteenth-century example.

Leaf-shaped Sweetmeat Dishes

The leaf dish was the last Thorley form added to the Craft House catalog. It is shown in the undated edition (see fig. 12), which probably was first printed in 1953 or 1954, and appears in price lists through the spring of 1956.

The original, eighteenth-century sweetmeat dish from which this was reproduced was in Thorley's own collection. Objects sold at the Craft House either had to be in Colonial Williamsburg's collection of antiques or based on archaeological sherds found in the historic city, however, so in 1952 Thorley presented the leaf dish as a gift to Colonial Williamsburg. After he stopped providing ceramics to the Craft House, the dish was returned to him and remained in his home until shortly before his death.[4]

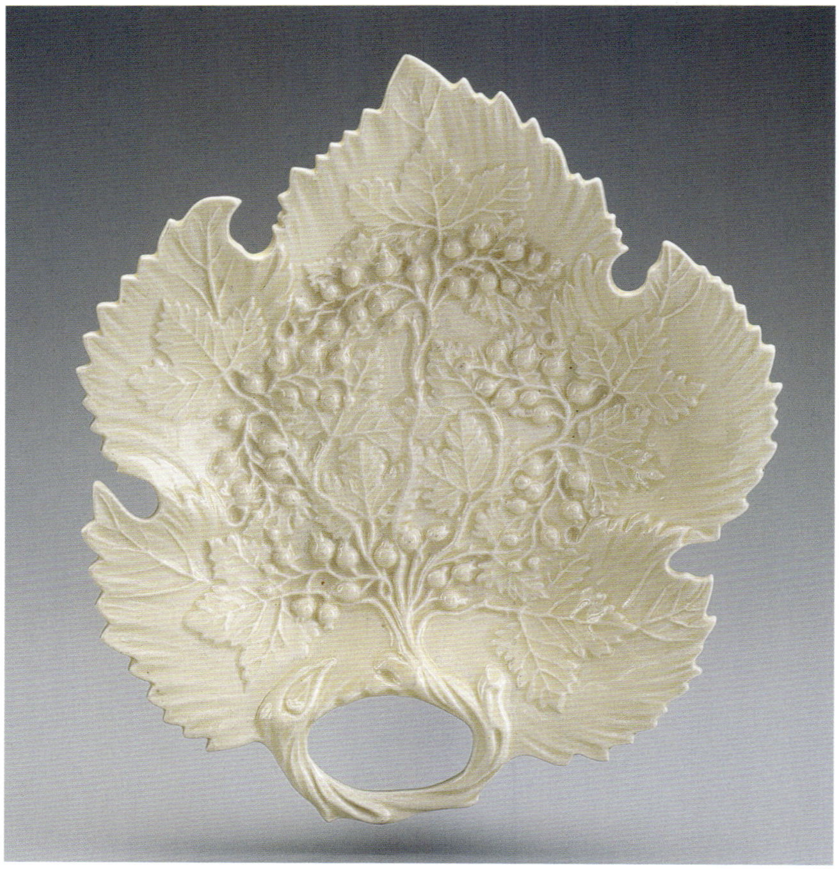

Figure 44 Leaf-shaped sweetmeat dish, 1956–1970. Creamware. L. 8". Mark: Thorley stamp (illustrated in fig. 10). (Courtesy, Peter and Jean Brown Collection.)

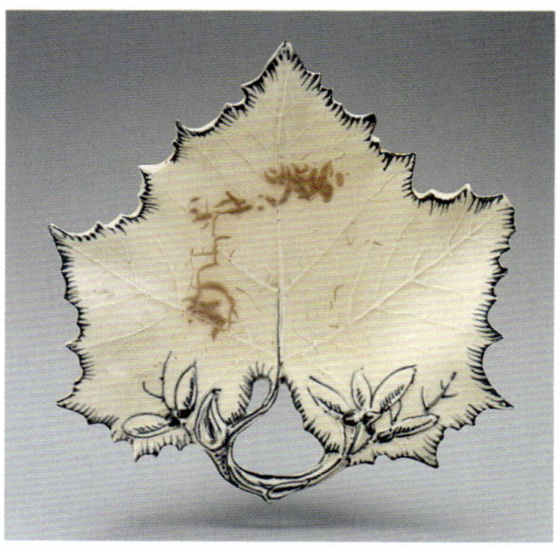

Figure 45 Sycamore leaf dish, 1955–1970. Whiteware. L. 7". Mark: Thorley stamp (illustrated in fig. 10).

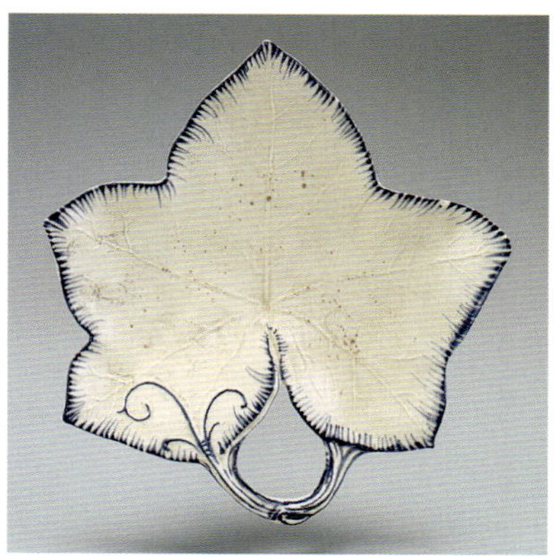

Figure 46 Ivy or vine leaf dish, 1955–1970. Whiteware. L. 6". Mark: Thorley stamp (illustrated in fig. 10).

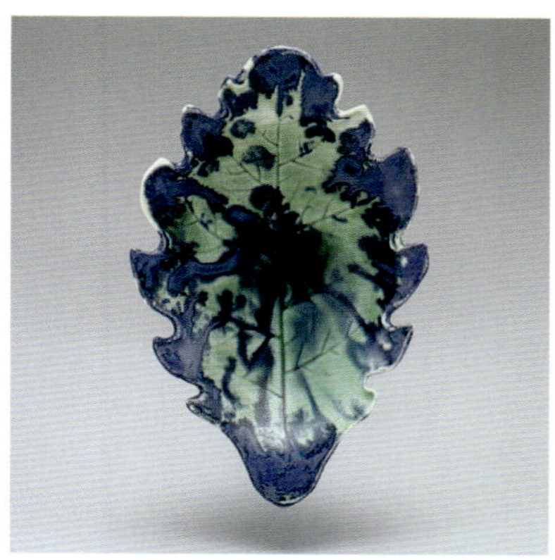

Figure 47 Oak leaf dish, 1955–1970. Whiteware. L. 6". Mark: Thorley stamp (illustrated in fig. 10).

The three other leaf-shaped sweetmeat dishes produced by Thorley are eighteenth-century in style but probably were his original designs. Unmarked examples or those marked with handwriting could have been made anytime after he came to Williamsburg (1949), but the four illustrated in figures 44–47 bear his printed mark and therefore have to postdate 1955.

Inkstands

Inkstands with inkwells and cover were reproduced from a French faience example in Colonial Williamsburg's antique collection (see, for example, fig. 48). Because of the similarities of French faience and English delft—both are technically of the same material, an opaque glaze made of tin and lead oxides over a rough earthenware body—it is sometimes difficult to differentiate the wares made in England from those made in other northern European countries.

In slip casting, a liquid clay with the consistency of heavy cream (called slip) is poured into a plaster of paris mold (see figs. 1, 50, 75). The plaster of paris draws out the water, leaving the clay along the sides. When enough clay is collected, the excess slip is poured out and the remaining clay is left to dry. Once the object is removed, the mold is allowed to dry thoroughly so can be used to form another object.

Thorley discontinued making the reproduction inkstands, which first appeared in 1951, after he and the Craft House parted ways. He experimented with another French heart-shaped inkwell design based on a blue-and-white example made at Marseilles and in Colonial Williamsburg's collection, but it is doubtful that it was produced.

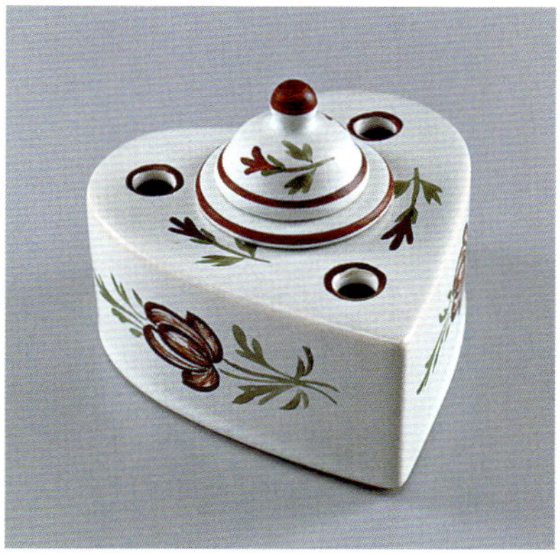

Figure 48 Inkstand, 1952. Whiteware. W. 4½". (Courtesy, Moyles Collection.) Mark: Thorley Craft House (illustrated in fig. 8). Inscribed, on base: "Tom Moyles of Williamsburg 1952." This example was made for Thomas Moyles, director of the Williamsburg Inn and a personal friend of Thorley. It is the earliest recorded Thorley presentation piece.

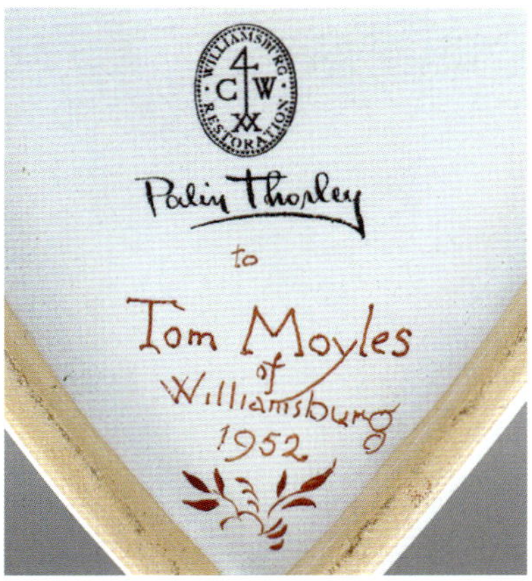

Figure 49 Mark and inscription on the underside of the inkstand illustrated in fig. 48.

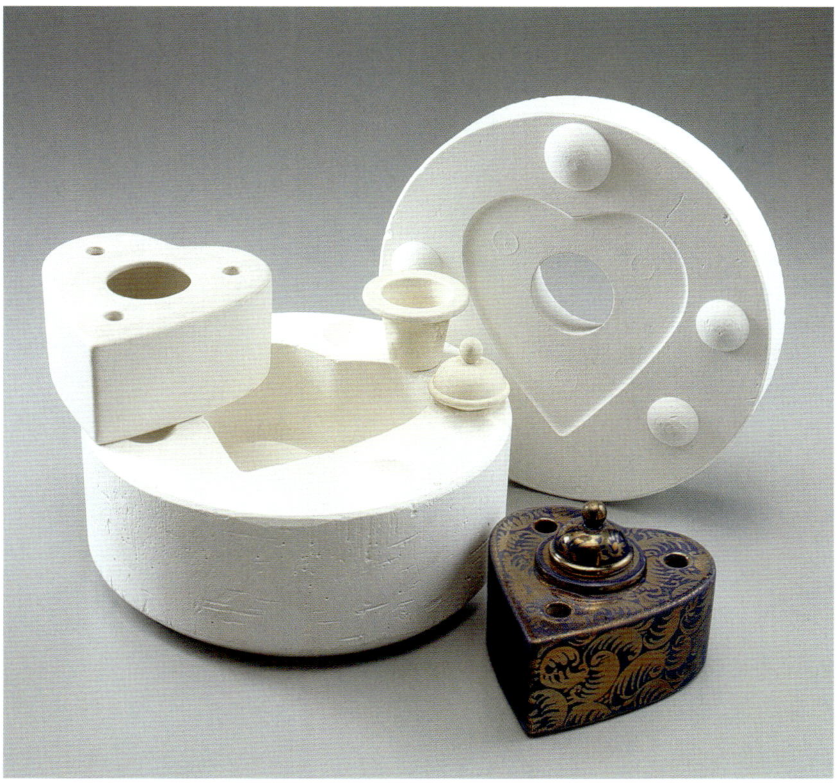

Figure 50 A plaster of paris mold for slip casting a heart-shaped inkstand, 1951–1970. A once-fired but unglazed example showing its three parts, and a finished example, glazed and decorated in a non–eighteenth-century design.

Thorley Delft

Delft is a term used to describe ware with a generally white opaque glaze applied to hide a crude gray or red earthenware body. In Spain and Italy it is generally referred to as majolica; in France, faience; in Holland, delft; and in England, delft, delftware, or tin-glazed earthenware. Some of the pieces produced by Thorley in Williamsburg—jardinieres, flower bricks, ointment pots, and tiles—are reproductions of English delft. The eighteenth-century inkstand was made in France and therefore, though technically the same as delft, is generally referred to as faience. In a letter to the author dated 1985 Thorley claimed that he was the first person in America to produce delft, although in actuality it was a whiteware with the appearance of eighteenth-century delft.

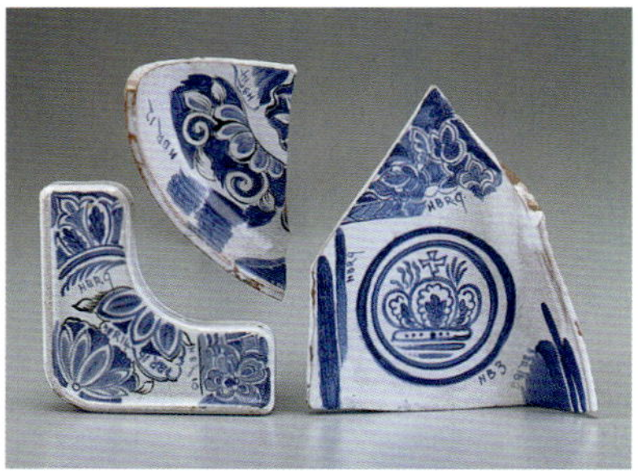

Figure 51 Three delft-like fragments, ca. 1942. These fragments show a crude dark red earthenware body and are covered with an opaque glaze. Thorley experimented with reproducing delft for Colonial Williamsburg, using his extensive knowledge of eighteenth-century English pottery.

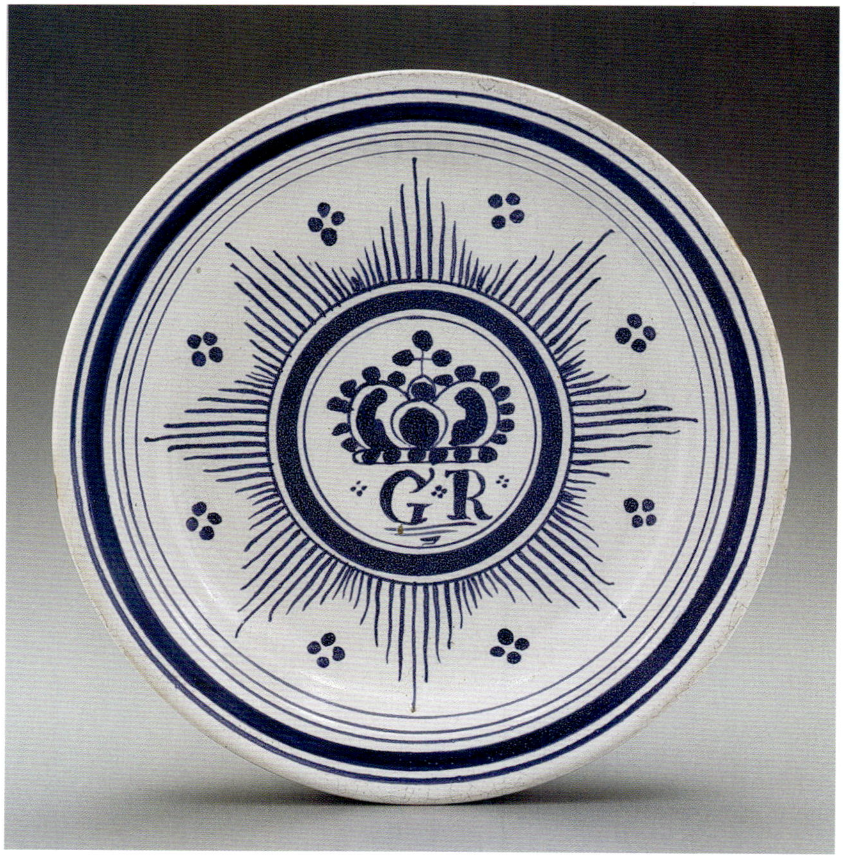

Figure 52 Plate, 1942–1945. Whiteware. D. 9". Mark: variation of the Colonial Williamsburg Craft House mark illustrated in fig. 8. (Courtesy, Colonial Williamsburg Collection.) This plate, reproduced from an eighteenth-century delft example, is listed in the Craft House price list only once (1942–1945) and does not appear in any of the catalogs. It is possible that it was produced for a very short time or that it was sold through other Colonial Williamsburg outlets; ointment pots such as the one illustrated in fig. 57, for example, were sold at the Galt Apothecary shop. The 1942–1945 price list for the Craft House catalog lists it as "Lambeth Delft Plate (blue & white)" with no mention of the "GR" cipher. The fragments illustrated in fig. 51 show related decoration, and the plates illustrated in figs. 54 and 55 are from the same mold. The original "GR" plate was made in England between 1714 and 1727 to honor King George I, who ruled Britain and the American colonies.

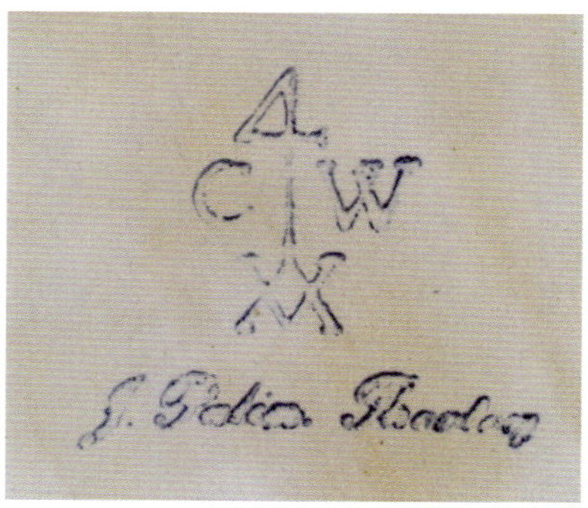

Figure 53 Detail of the mark on the underside of the plate illustrated in fig. 52. The significance of this rare version of the Craft House mark is unknown. It does not appear on the usual production.

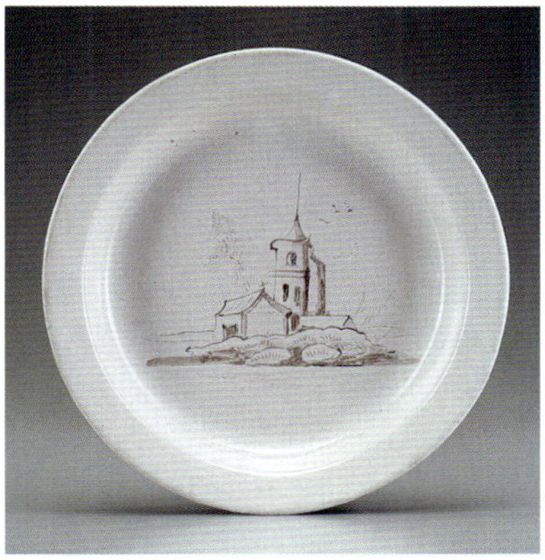

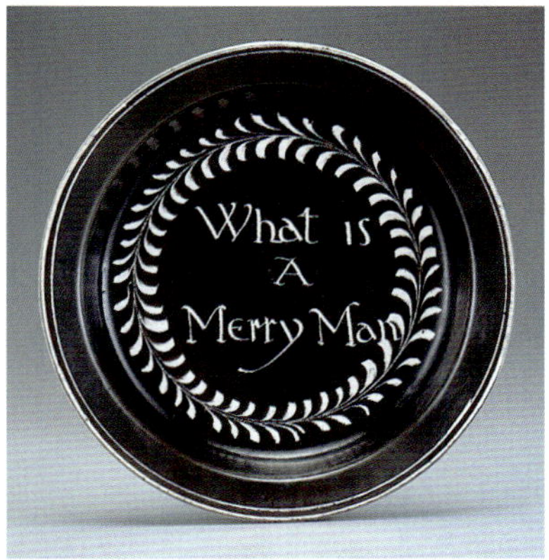

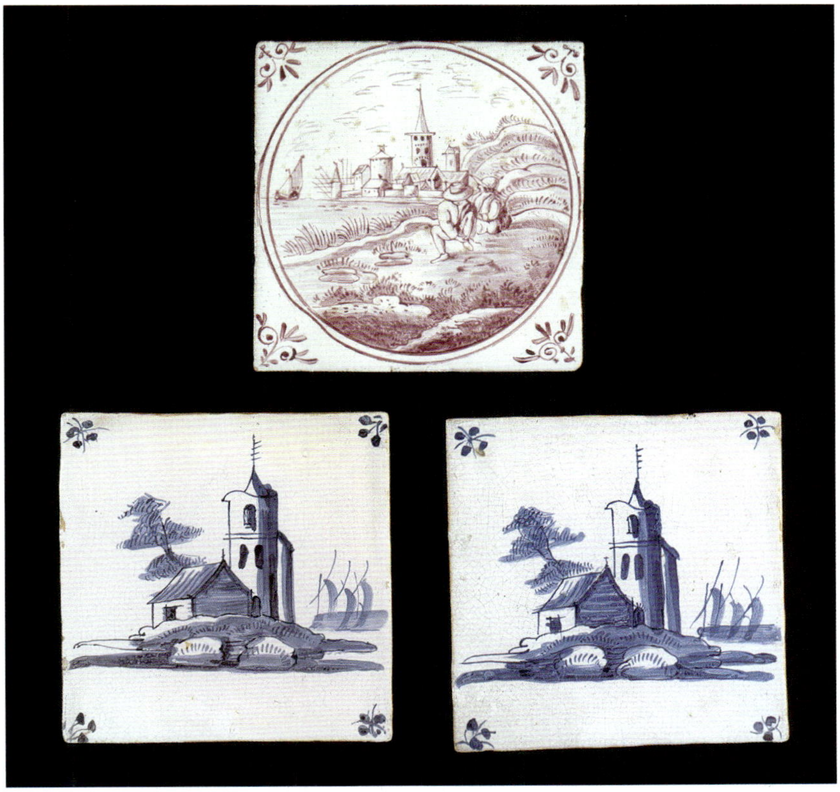

Figure 54 Plate, 1942–1950. Whiteware. D. 9". Mark: experimental mark only. Thorley appears to have experimented on this plate using motifs usually found on eighteenth-century delft tiles.

Figure 55 "Merry Man" plate, 1943–1950. Whiteware. D. 9". Unmarked. The source for the inspiration for this plate is unknown, as it dates from the 1940s and Colonial Williamsburg did not acquire its set of Merry Man plates until 1962. The verse, one line of which is displayed on each of six plates, reads: "What is a Merry Man, / Let him do what he can, / To entertain his guests, / With wine and merry gests, / But if his wife do frown, / All Merryment goes down." The rhyme is from the seventeenth century, although dated antique examples of this wreath version range from 1717 to 1745.

Figure 56 Tiles, 1942–1955. Whiteware. 5" x 5". Marks: experimental marks only. Hand-painted scenes in the eighteenth-century delft manner.

In the early 1940s Thorley discussed the possibility of producing tiles for the Craft House and development commenced. Thorley wanted to produce reproductions only, but the Craft House managers thought that would be too costly and asked him to do less-expensive, mass-produced "souvenirs." Thorley believed this would lower the quality of their other products. A compromise was reached and Thorley began developing line drawings of some of Williamsburg's buildings that were to be copperplate printed in the style of eighteenth-century English tiles. Correspondence shows that his drawings of Bruton Parish Church and the Capitol were returned to him for corrections to the architectural details. Other proposed drawings include depictions of the Wren building and the St. George Tucker House. A drawing of Williamsburg's Colonial Court House of 1770, sized correctly for a tile, has been discovered in the Thorley Papers.

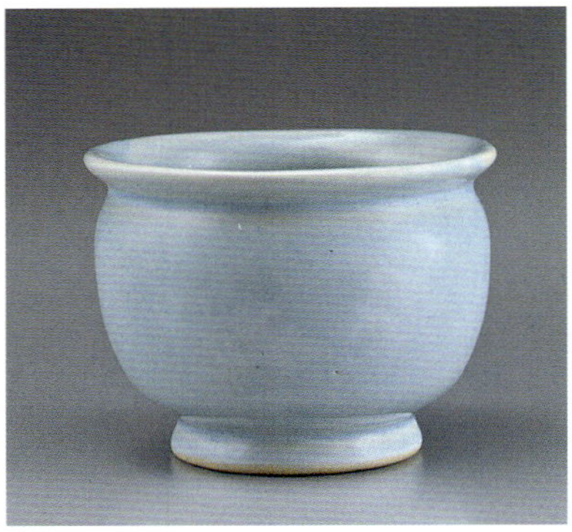

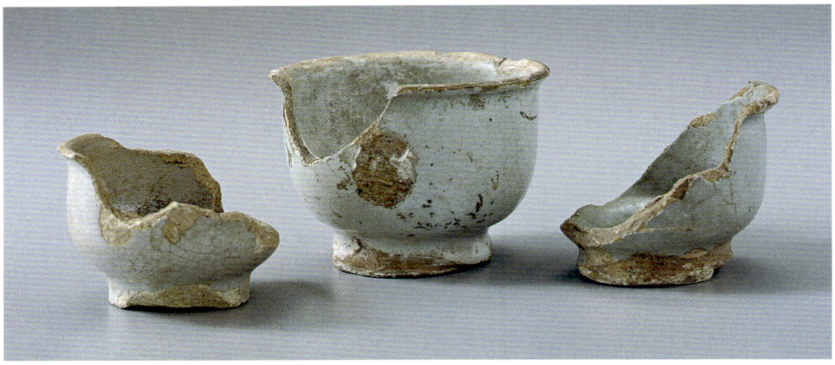

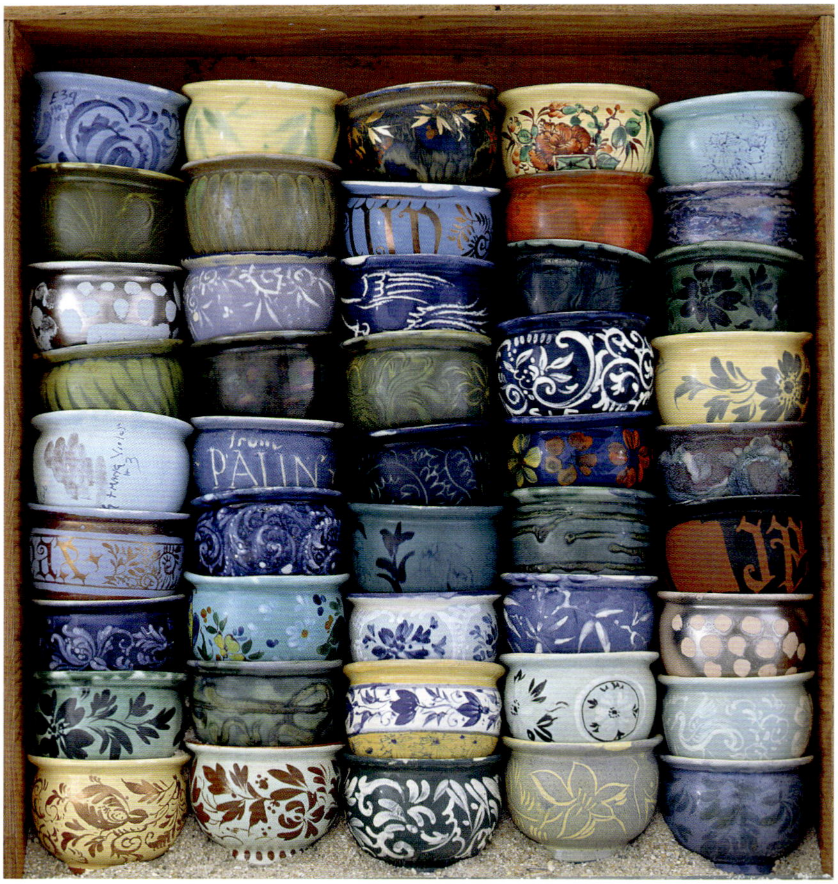

Figure 57 Ointment pot, 1955–1965. Earthenware, with Thorley's original glaze. H. 2¼". Mark: Thorley stamp (illustrated in fig. 10). (Courtesy, Moyles Collection.)

Figure 58 Ointment pots, England, eighteenth century. Tin-glazed earthenware. H. of center pot 2". These archaeological sherds, excavated in Williamsburg and loaned to Thorley in 1940 as examples of the opaque delft glaze, eventually became prototypes for his ointment pot. They were found in his studio in 1986 and subsequently returned to Colonial Williamsburg.

Figure 59 Ointment pots, 1954–1970. Whiteware. H. of pot 2½". These pots were decorated by Thorley and used for experiments and gifts.

Ointment Pots

Ointment pots (figs. 57–59) varying in size and generally undecorated were used by apothecaries in the eighteenth century, who typically would fill the pot with an ointment or medication, then cover it with parchment, which was held with a string, and give it to the patient. Often, when the pot was empty, it would be used for some other purpose in the household.

Beginning in 1954, Thorley produced pots for the Crafts Shop department of Colonial Williamsburg. The restored Galt Apothecary Shop on Duke of Gloucester Street sold reproductions of delft ointment pots—vessels that would have been used there in the eighteenth century. Many examples were sold containing a soap ball.

Although the Craft House had severed relations with Thorley in 1956, Williamsburg's Crafts Shop department was considered its own entity and therefore might have continued to receive ointment pots from Thorley. According to Thorley, he had a huge shipment ready to be delivered, but the order was canceled, leaving him with hundreds of ointment pots. These he used for experimenting with reglazing and decorating to give as small gifts. When he moved to a retirement home in 1986, at least thirty boxes of these colorful jars were found.

Plates with Unfired Decoration

Forty-seven plates with unfired decoration were discovered in Thorley's shop by David Tetrault. They were wrapped, as a few of them still are, in what appear to have been local newspapers dated March and April 1952. All but three of the plates are 10½ inches in diameter and painted on unmarked blanks; the remaining three are 8 inches in diameter and bear the mark of the Edward M. Knowles China Company. The mark includes "51-2," indicating the blank plates were made in 1951 or 1952. The fact that some of the plates were made at Knowles is not unusual, as Thorley used china blanks from wherever he could get them most easily and least expensively. With this evidence, it appears that Thorley decorated the plates and wrapped them in 1951 and/or 1952, shortly after he had moved to Williamsburg and when he was so busy that he was unable to fill the orders from the Craft House and other customers. Although many of the plates are decorated

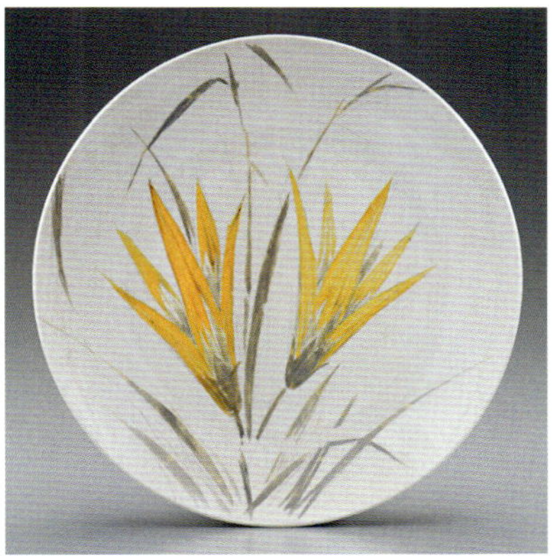

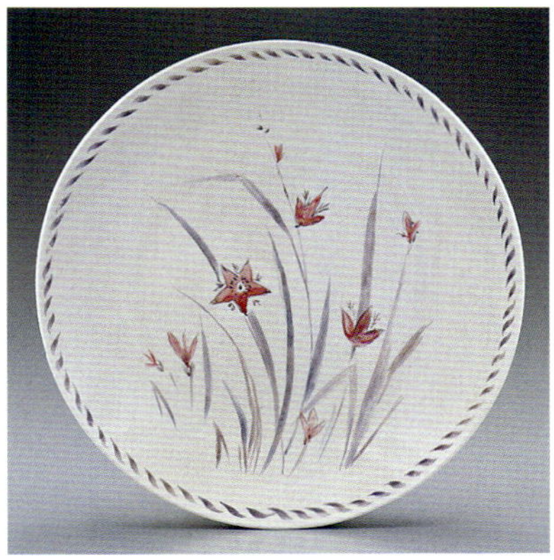

Figure 60 Plates with unfired decoration, 1951–1952. Whiteware. D. 10½".

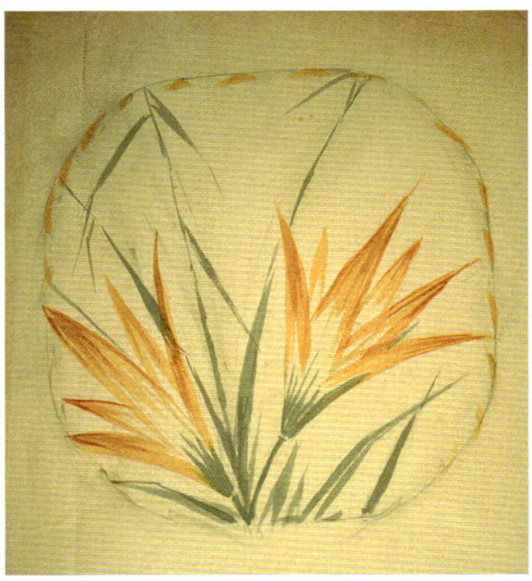

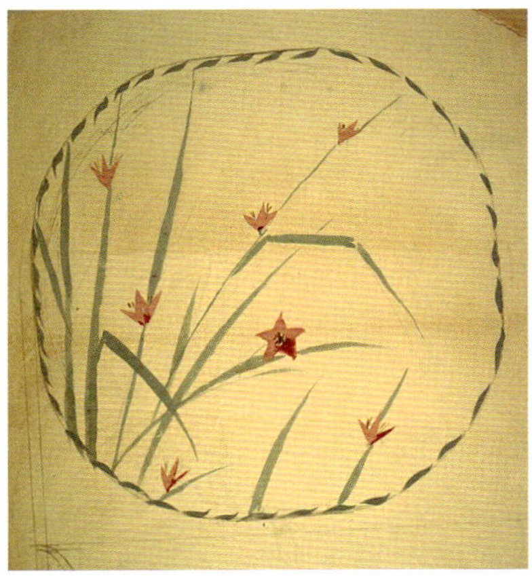

Figure 61 Drawings for the plates illustrated in fig. 60. Paint on paper.

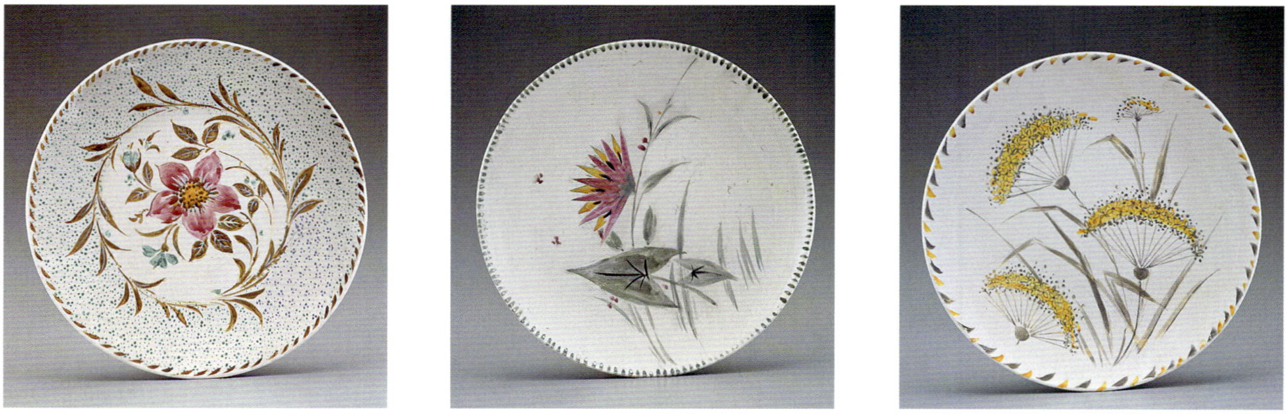

Figures 62 Plates with unfired decoration, 1951–1952. Whiteware. D. 10½".

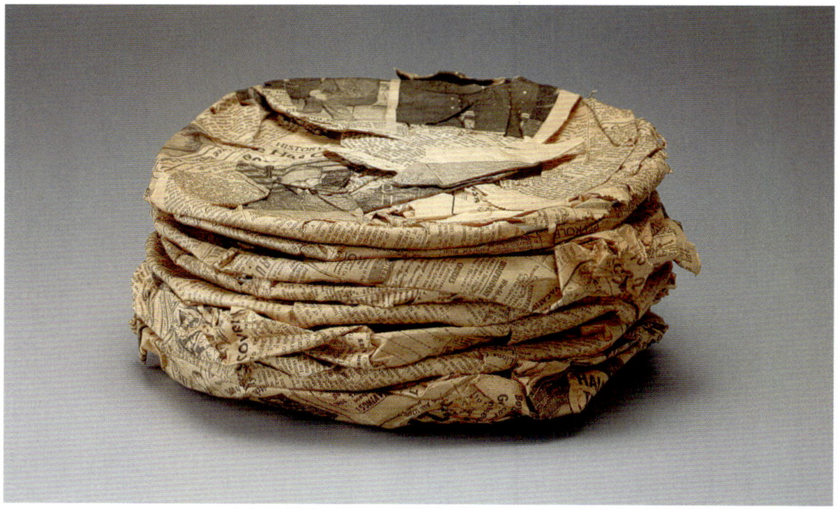

Figure 63 Stack of plates with unfired decoration still wrapped in local newspapers dated March and April 1952.

with exotic flowers, others illustrate scenes, figures, and geometric designs, among other motifs. The original drawings for some of the designs were found among Thorley's papers; some are attached to the unfinished plates. It is impossible, at this time, to determine whether this series of plates was intended for a special order from a large factory or whether Thorley planned to produce them himself.

Figures

Thorley was primarily a designer and potter of useful wares—tea and dinner services and vases. The few figures he created were probably done for his own pleasure. How many he made is unknown, but in addition to the ones illustrated here (figs. 64–66) and the chatelaine discussed previously (Austin, "Thorley, Part 1," figs. 28, 30), there are illustrations in the Thorley Papers of an elongated Chinese sage that Thorley might have created, as well as a seated cherub, which was sold at Christie's, New York (June 12, 1987, lot 42), and is said to bear Thorley's incised signature.

Thorley clearly was fond of the female form, as his papers contain many sketches—some elaborate, some just a few lines—of women, some classically drawn, while others (fig. 66) were created with the free-flowing lines of art nouveau.

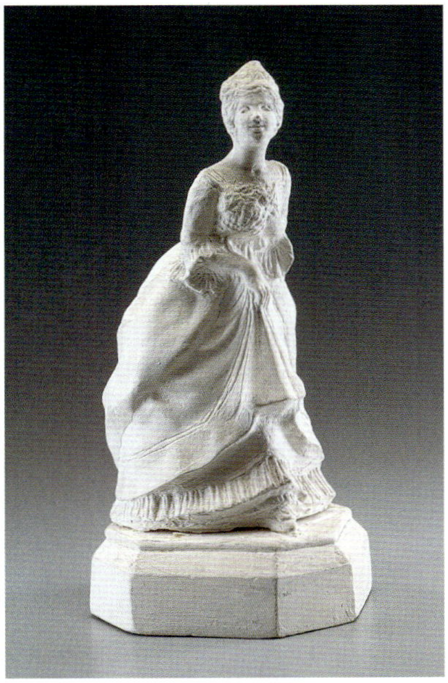

Figure 64 Figure, 1955–1970. Whiteware. H. 11". Glazed but undecorated figure of a lady in eighteenth-century dress, Mark: Thorley stamp (illustrated in fig. 10). (Courtesy, Tetrault Collection.) The figure has separated from its base.

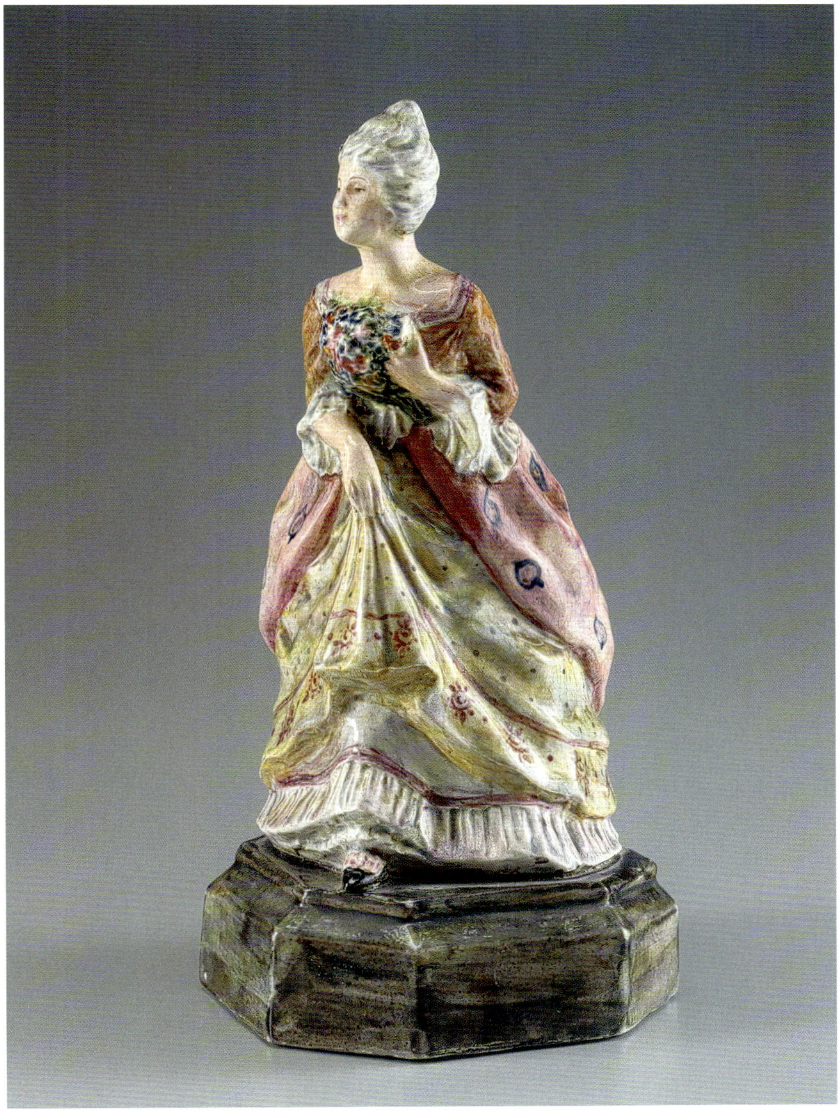

Figure 65 Figure, 1955–1970. Whiteware. H. 11". Unmarked. This is the same figure as that illustrated in fig. 64, but it is decorated in polychrome colors.

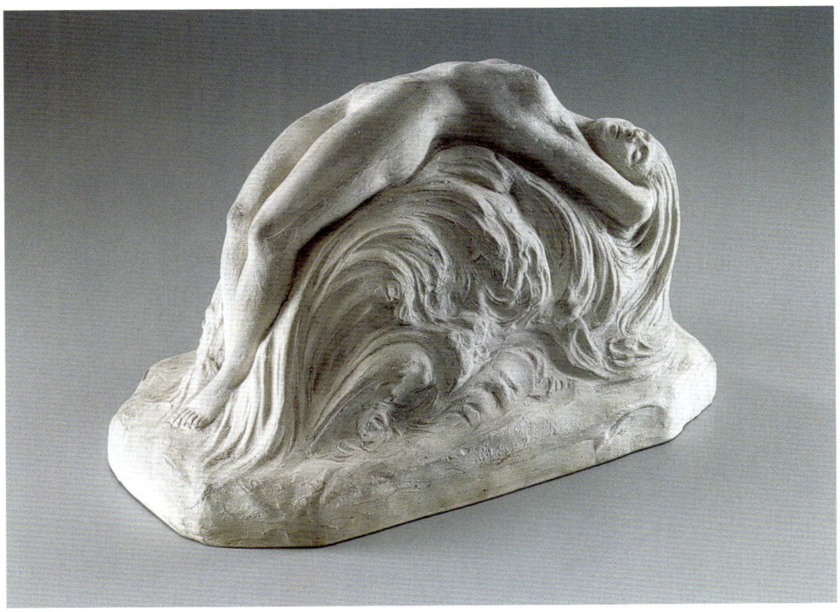

Figure 66 Unglazed figure of a reclining nude, 1930–1960. Earthenware. L. 10". Unmarked. (Courtesy, Tetrault Collection.)

Figure 67 *Left to right*: *Bottle,* thrown with sides pushed in, 1955–1970. Stone- or earthenware. H. 4". Unmarked. This piece probably was not made by Thorley and might not have been decorated by him. *Vase,* thrown, 1955–1970. Earthenware. H. 4½". Marks: experimental. Probably not made by Thorley but perhaps decorated by him. *Bottle,* molded, 1955–1970. Earthenware. H. 4¾". Marks: experimental numbers and handwritten "Palin Thorley / Williamsburg." *Vase,* molded, 1955–1970. Earthenware. H. 7½". Marks: experimental numbers and handwritten "Palin Thorley / Williamsburg."

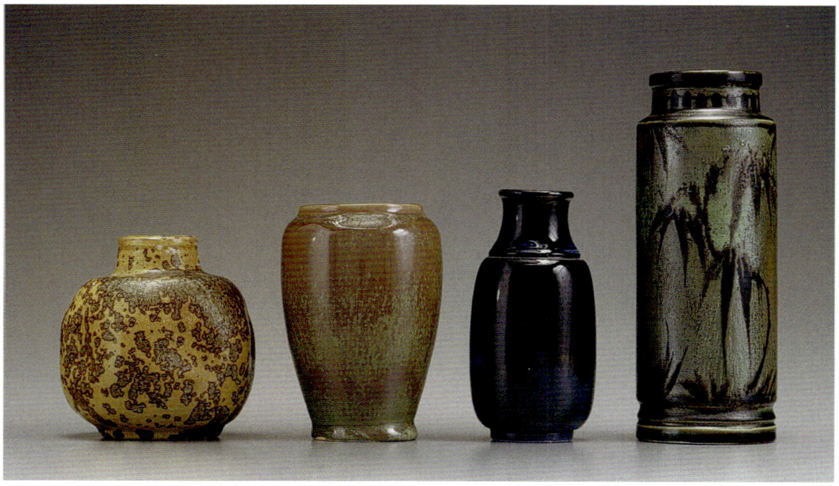

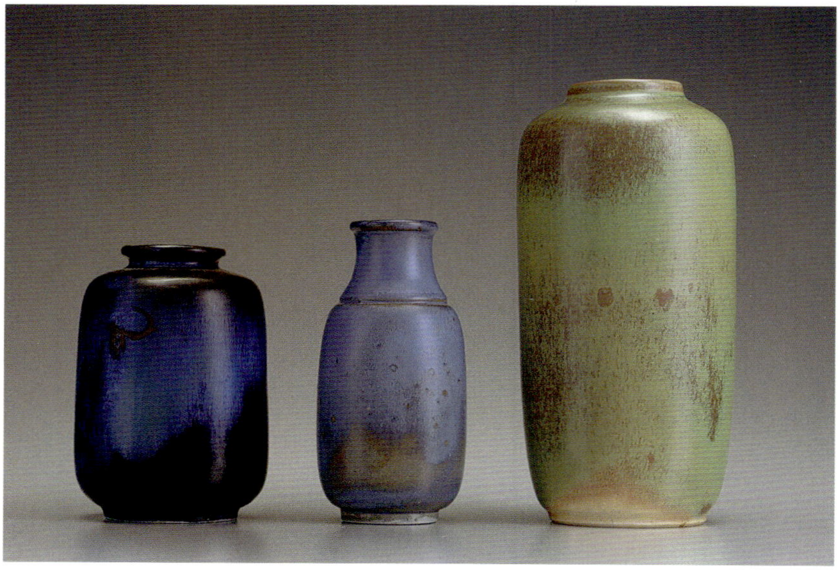

Figure 68 *Left to right*: *Vase,* molded, 1955–1970. Earthenware. H: 4½". Mark: Thorley stamp (illustrated in fig. 10). *Bottle,* molded, 1955–1970. Earthenware. H: 4¾". Marks: experiment numbers and handwritten "Palin Thorley." *Vase,* thrown, 1955–1970. Stone- or earthenware. H. 7¼". Unmarked. The vase was probably not made by Thorley but was perhaps decorated by him. The same hands appear to have made and decorated both it and the vase illustrated in fig. 67, second from the left.

Later Experimental Pieces

Among the several thousand pieces of ceramic found in Thorley's studio, by far the majority were molded by him. He decorated many of these, but others were unglazed and/or undecorated. (Thorley sometimes found it prudent, when experimenting, to buy blanks from other potteries and use those as a surface to work on, like the plates with unglazed decoration; see figs. 60, 62, 63). A small group of items was made entirely by other hands, perhaps gifts from friends or purchases for study. In addition, unfortunately, there are a few pieces in which Thorley's involvement cannot be determined.

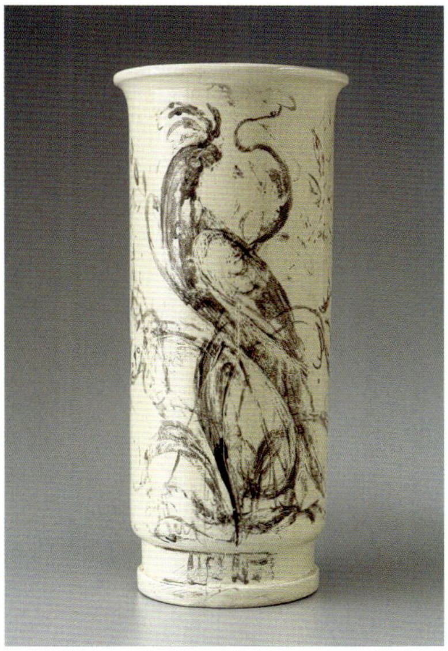

Figure 69 Vase, 1955–1970. H. 9". Mark: Thorley stamp (illustrated in fig. 10). One of two vases, from the same mold, found at Thorley's studio. They were fired twice and have a clear creamy ground. Thorley decorated both in black, with exotic birds, but they were not fired again. This example is in far better condition than the other, although it shows some loss due to handling.

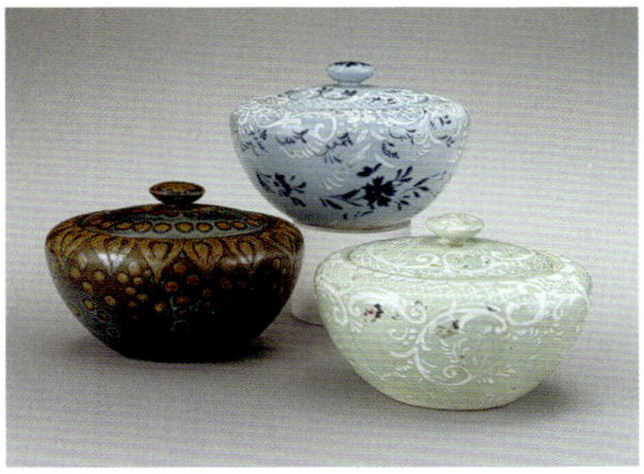

Figure 70 Covered containers, 1950–1970. Whiteware. L. 7". Marks: the brown example is unmarked; the other two have experimental numerals only. The poor quality of the body of these covered containers suggests that they were not made by Thorley. They were, however, decorated by him. The two on the right show the influence of eighteenth-century ceramics, where a thick white enamel is placed over a pale-colored ground, often giving the appearance of damask—a technique known as *bianco supra bianco.*

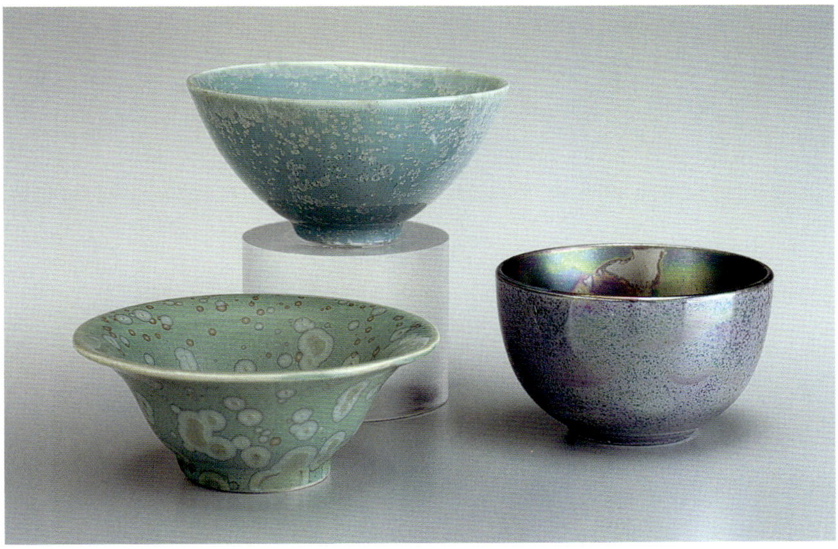

Figure 71 Bowls, 1950–1970. *Top:* Stoneware or possibly porcelain. H. 2". Mark: in script on bottom, "3C5 Thorley." *Lower left:* Whiteware. H. 1½". Marks: experimental marks only. *Lower right:* Whiteware. H. 2". Mark: in script on bottom: "Palin Thorley." Thorley must have acquired many small bowls, roughly the size of a modern teacup, on which to try his experiments. Although he may have molded some of the bowls, few are duplicates, and they range from what appears to be Chinese hard-paste porcelain to crude earthenware. No molds or unglazed examples have been found.

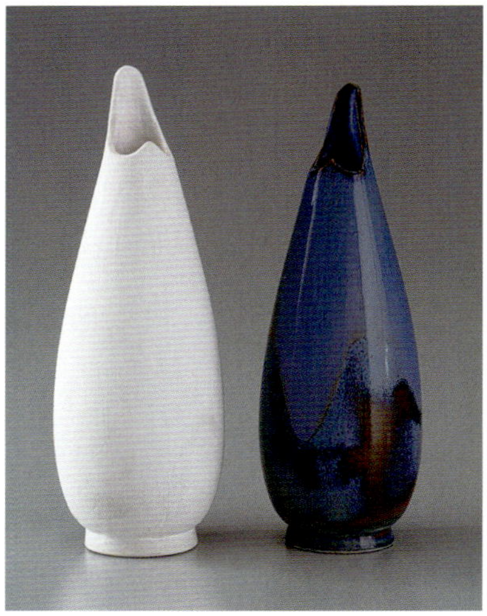

Figure 72 Vases. *Left:* 1955–1970. Earthenware. H. 7¾". *Right:* 1950–1970. Earthenware. H. 7½". One might think that these unusual modern-shaped vases were not molded by Thorley were it not for the fact that the white example is still in biscuit and bears the Thorley stamp (illustrated in fig. 10) while still in its green or unfired state.

The lustrous blue vase was a successful experiment in that its condition is good enough to have appeared in his shop. Only the series of numerals on the base explains that it was indeed experimental. His name, handwritten, also appears on the bottom.

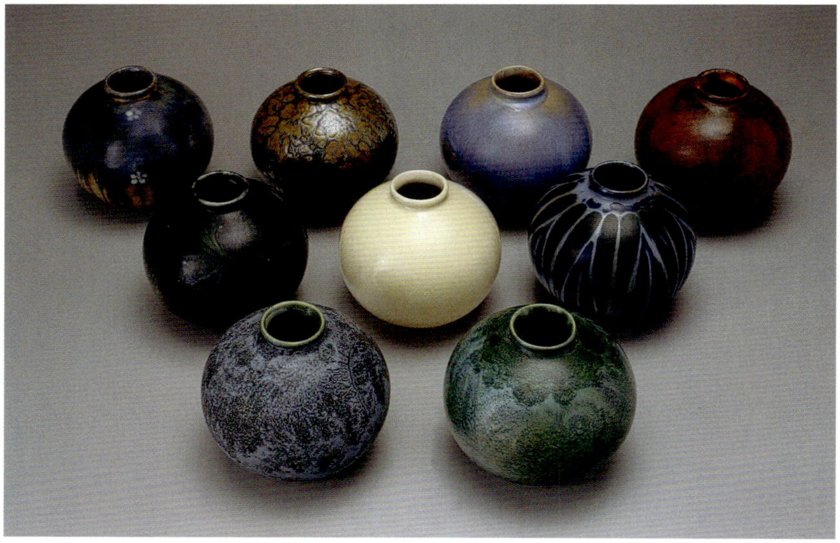

Figure 73 Vases, 1950–1970. Whiteware. H. 4¼". Many of these globular vases were found in Thorley's studio. Some—bearing printed or handwritten marks—were ready to be sold, whereas others were experimental pieces, showing only the numerals and letters of the formula he was trying. Also found was a master mold (a positive mold in plaster of paris that looks like a larger but solid version of the final piece) and an example that had been fired once, marked, and covered with a coat of glaze not yet fired.

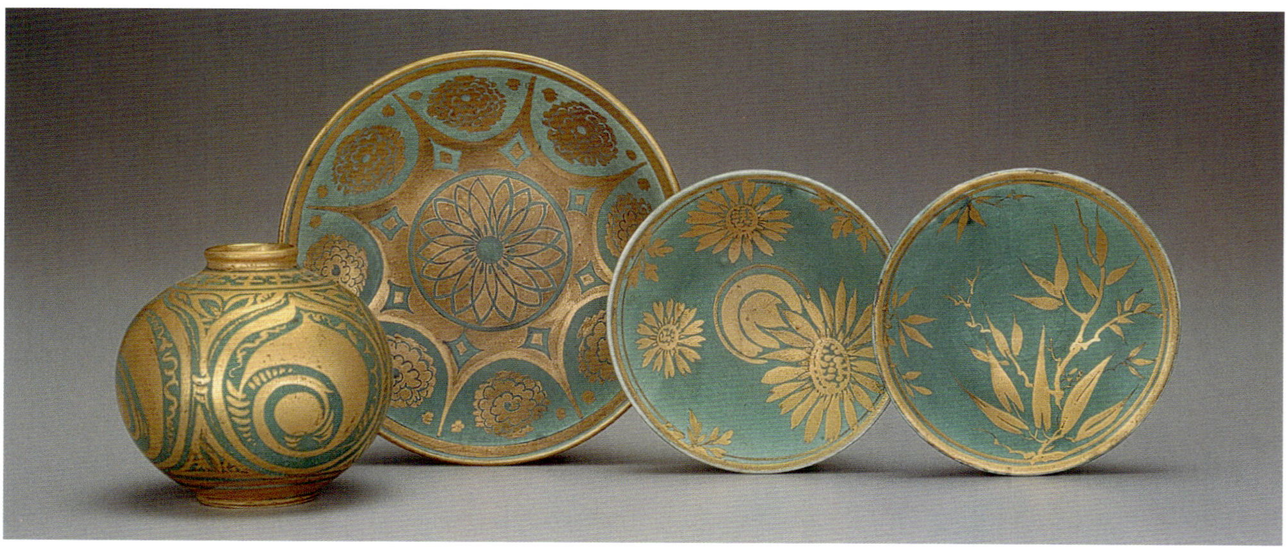

Figure 74 Green and gold objects, 1950–1970. Earthenware. Marks: handwritten, in gold, "Palin Thorley / Williamsburg." *Vase:* H. 4¼". *Large bowl:* H. 1½", D. 7½". *Small bowls:* H. 1¼", D. 5½".

The objects illustrated in figures 74 and 75—on which gold is applied to a green ground—are slightly matte in texture, which makes them quite different from the other Thorley pieces in metallic colors, and the luster is very shiny. The designs are Japanese as well as geometric.

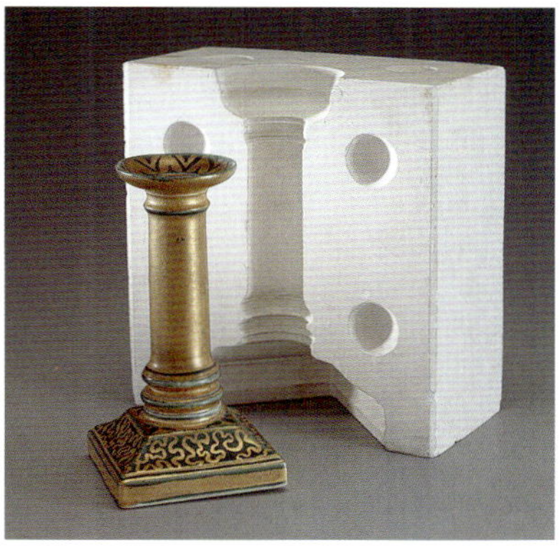

Figure 75 Candlestick, 1950–1970. Whiteware. H. 6". (Courtesy, White Hall Restaurant Collection.) Mark: handwritten, in gold, "Palin Thorley." The candlestick is shown with a plaster of paris mold.

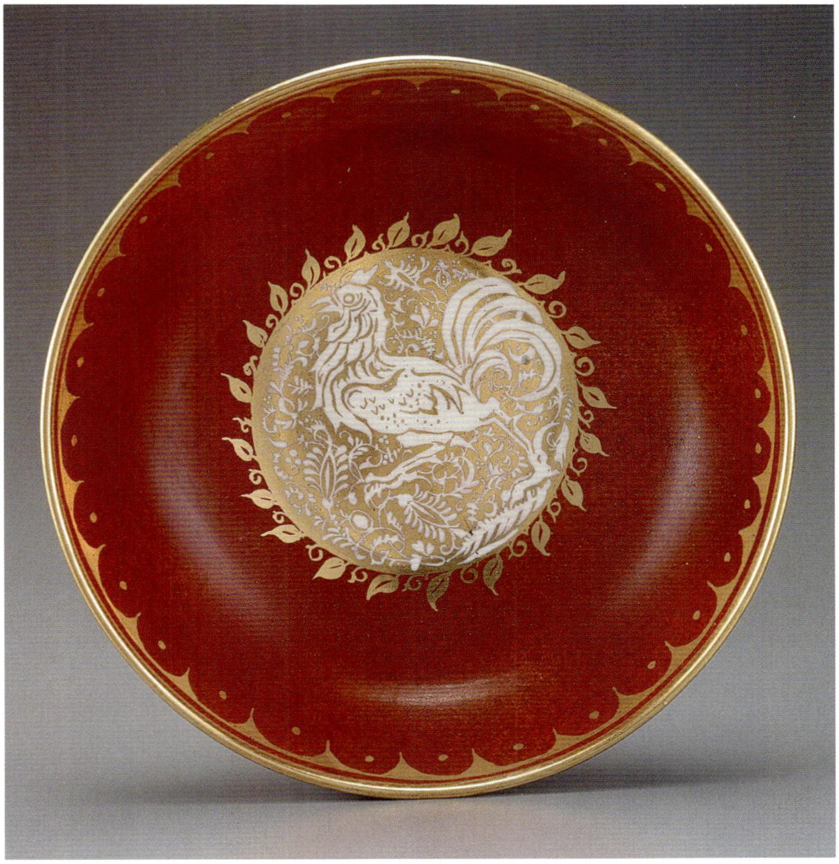

Figure 76 Bowl, 1950–1970. Whiteware. D. 9". Mark: handwritten "Palin Thorley/ Williamsburg / Virginia" and his monogram (see fig. 77). Thorley placed a very shiny ground color on both sides of this bowl, leaving a white reserve only on the interior bottom. The gold is quite lustrous and appears to have been burnished.

Figure 77 Mark and monogram on the underside of the bowl illustrated in fig. 76.

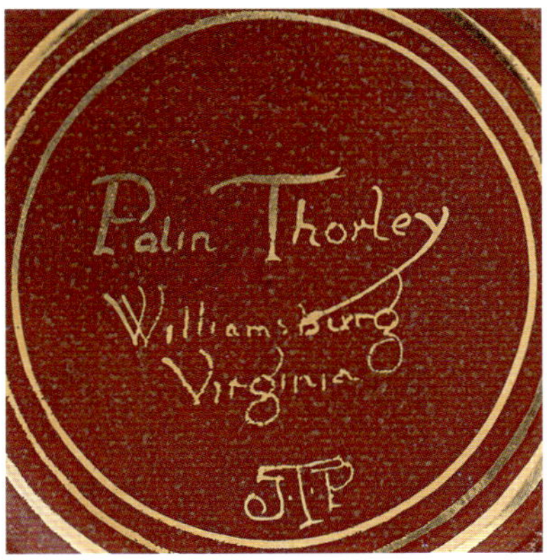

Crystalline Glazes

Thorley's papers indicate that in 1967 he was working on "opalescent glazes," and in February 1968 he shows tests for a crystalline glaze. These glazes were very popular in the first decades of the twentieth century, when Thorley was learning the art of ceramics. In his youth he had been fascinated with art nouveau, a style he had never gotten out of his system. Going through the sketches among his papers, it is, in many cases, impossible to ascertain whether his drawings were done in the 1920s or the 1960s. Thorley's interest in resurrecting crystalline glazes coincided with the art world's new appreciation of its beauty.

Figure 78 Vase, 1960–1970. Whiteware, with crystalline glaze. H. 9¾". Marks: experimental only. This tall vase might not have been made by Thorley, but it was decorated by him.

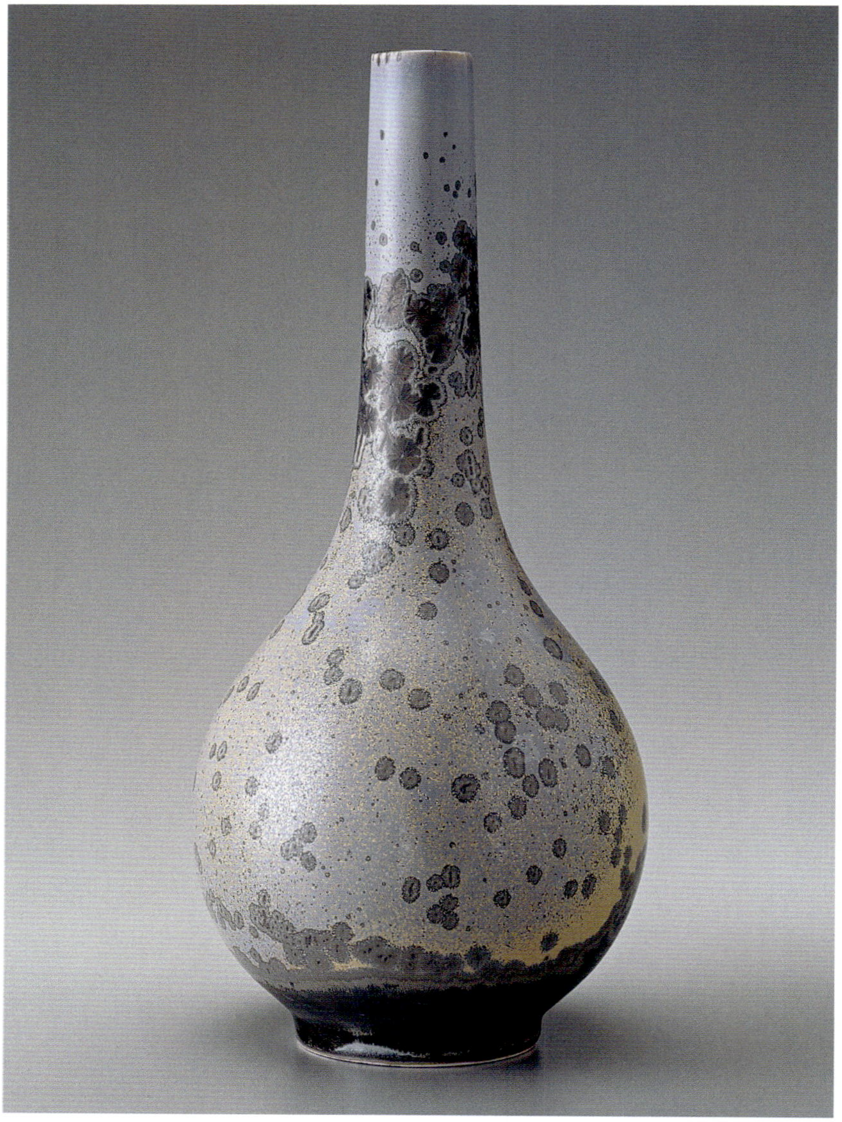

Figure 79 Vase, 1960–1970. Whiteware, with crystalline glaze. H. 5". Marks: experimental only. This vase might not have been made by Thorley but probably was decorated by him.

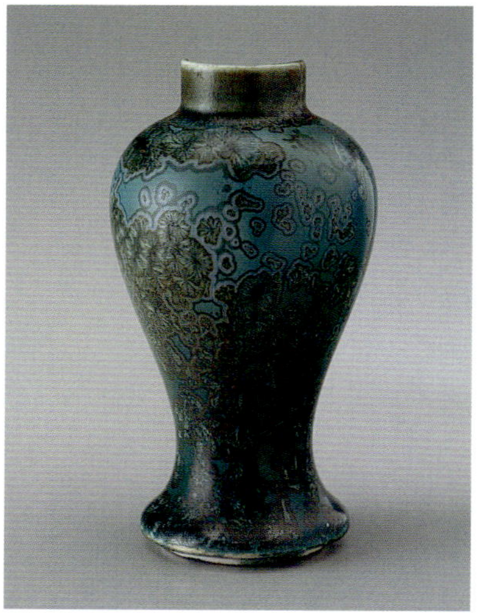

Figures 80 Plates, 1960–1970. Whiteware, with crystalline glaze. D. 6". Three examples from a large series of bread-and-butter plates from Thorley's studio. All are bisque (that is, unglazed) on the reverse and bear the series number "RC-10," with another number that differs according to the glaze.

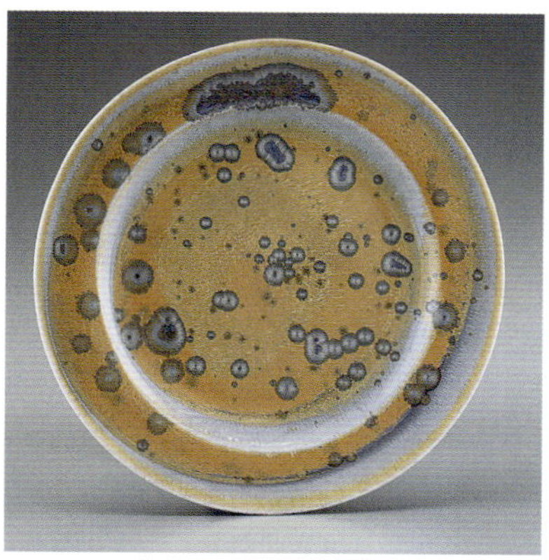

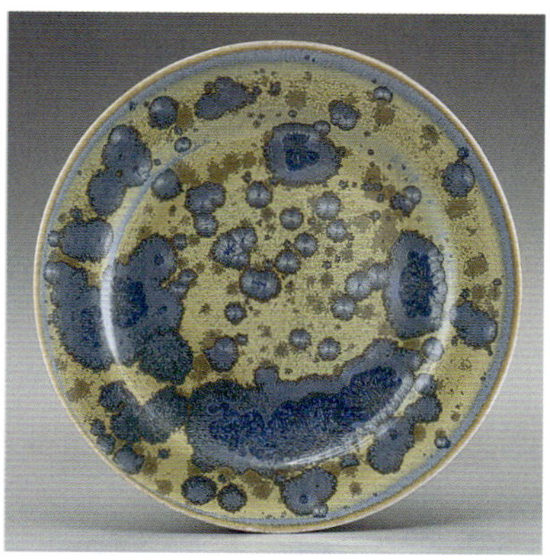

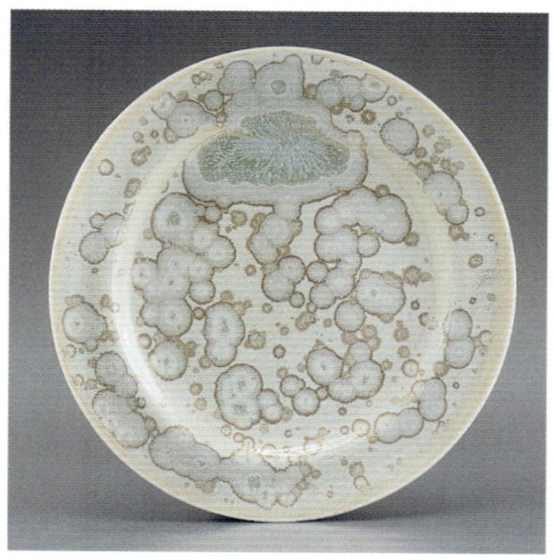

Luster Glazes

Palin Thorley was very proud of the luster glazes he produced in Williamsburg, and he used them in large quantities, both during and after his association with the Craft House. He produced copper luster, which had been very popular in the nineteenth century (fig. 81); pink luster, apparently his most favored (figs. 82–84), and silver luster on both white and yellow grounds (figs. 86–91). Yellow ground, often referred to as "canary," was commonly used at the end of the eighteenth century, and when lustrous glazes came into vogue in the nineteenth century, embellishing the yellow ground with silver luster was a natural progression. Thorley chose to apply

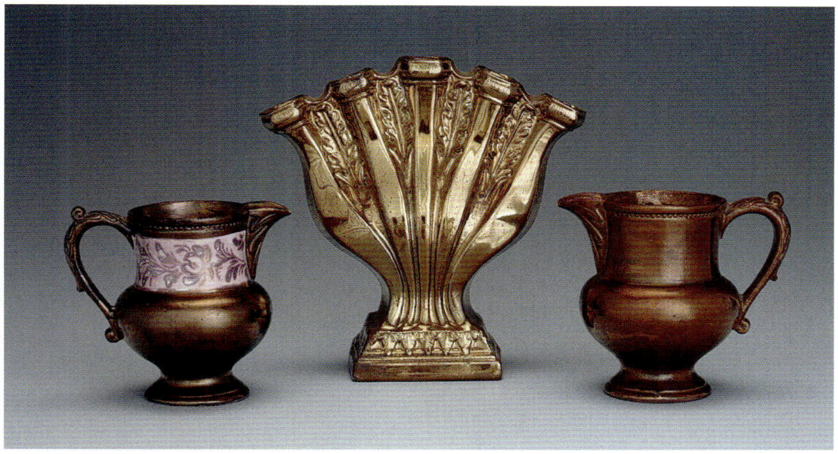

Figure 81 Copper luster jugs and posy holder, 1960–1970. Earthenware. *Jugs:* H. 4½". Unmarked. *Posy holder:* H. 7". Mark: handwritten, mostly hidden under the dark glaze, "Palin Thorley / Williamsburg."

the yellow over a solid silver ground (figs. 86, 89; see figs. 23, 24). The yellows of his pieces are often more of a muddy shade than the pure yellow of the earlier era.

Thorley was pleased that he was responsible for the increase in the popularity of luster in 1925 while at Allerton's in Staffordshire.[5] This prompted the continuation of its manufacture from perhaps the eighteenth century, as he claimed, but certainly from the early nineteenth century. In an interview he tells this story:

> I came to America in 1927 and in 30s I'm in, of all places, . . . Richmond. I'm walking down the street with a friend and I look in an antique shop window and there on the table is a very nice service of this luster ware was made by me at Allerton's. So I turned to my friend and said, "Harry, that's one of my pieces." "Oh," said he. . . . "It is, Harry. Let's go in and find out, have some fun." I went into the store and I walked all around and I looked at tables and chairs and so on, but finally I got round to this table where the luster set was. I asked the gentleman in attendance about it and he told me it was very rare to find such a complete set. One of the old families in Virginia had passed away and had a liquidation of their home and this had been found in an old cupboard . . . such a lovely story and told me it [was] discovered by these people who purchased the home and the rest of it . . . nice story. Well, I turned the plate up [and] pointed . . . there was no [factory] mark on the back, but there was impressed a pair of figures impressed in the clay. Now I rubbed my finger over it and said, "You see this? This particular plate was made by me at Allerton's in 1924. That's the mark," and I described it to this gentleman in detail. So he finally laughed and he said, "Well, strange stories do gather around

Figure 82 Pink luster lamp base, 1960–1970. Whiteware. H. 12". Mark: handwritten, in pink, "Palin Thorley / Williamsburg Virginia."

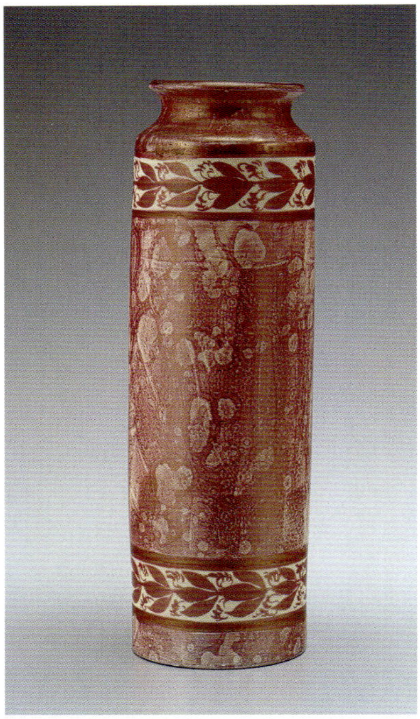

Figure 83 Five pink luster pieces made and decorated by Thorley, 1960–1970. *Vase:* H. 4½". Marks: Thorley stamp (illustrated in fig. 10). *Jugs in back:* H. 4½". Mark: handwritten, in pink, "Williamsburg," and Thorley's monogram. *Candlestick:* H. 6". Unmarked. *Jug at right:* H. 4½". Mark: Thorley stamp (illustrated in fig. 10).

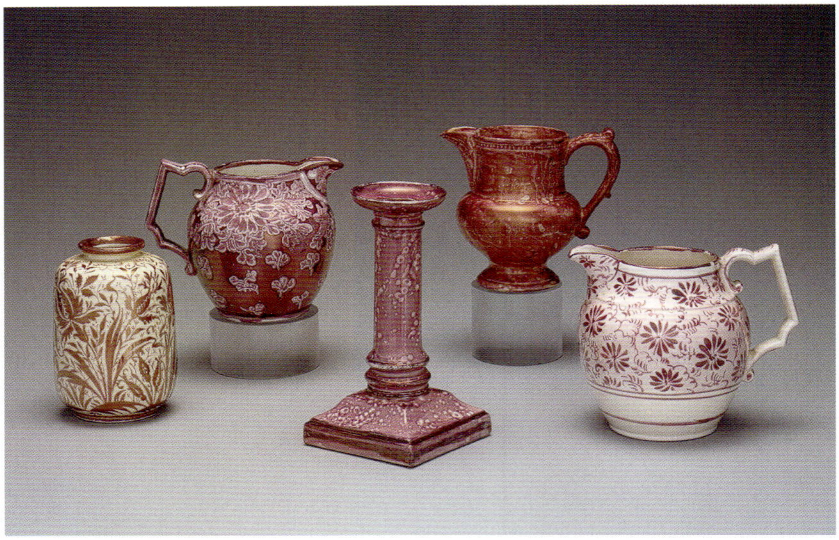

Figure 84 Pink luster plate, 1960–1970. Bone china. D. 10". Mark: handwritten, in pink, "Palin Thorley / Williamsburg." Although not made by Thorley, this plate was decorated and signed by him.

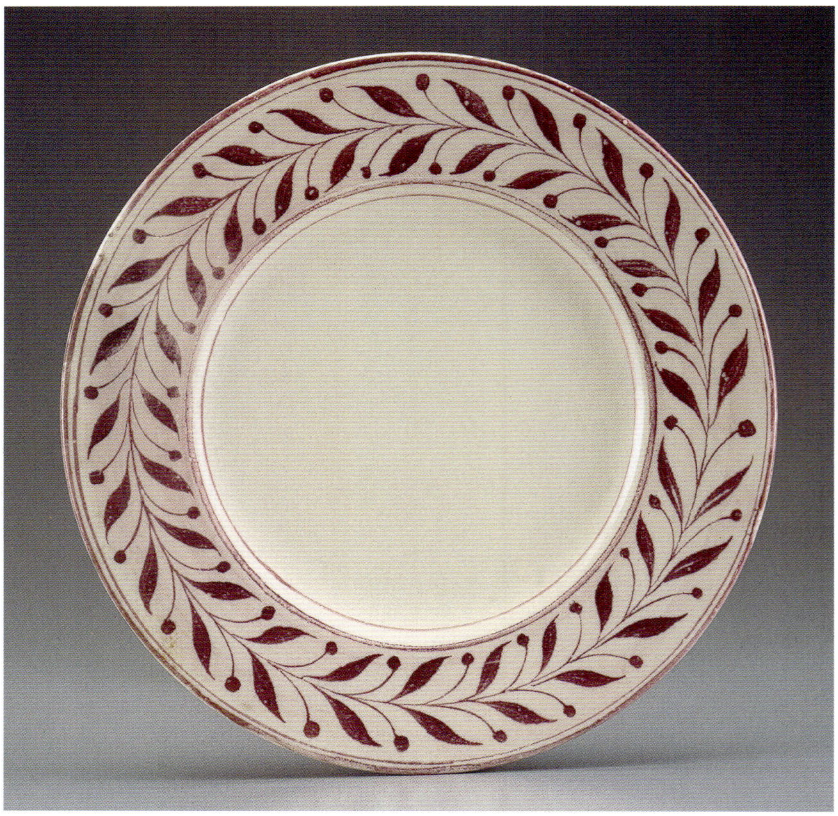

Figure 85 Mark on the underside of the plate illustrated in fig. 84.

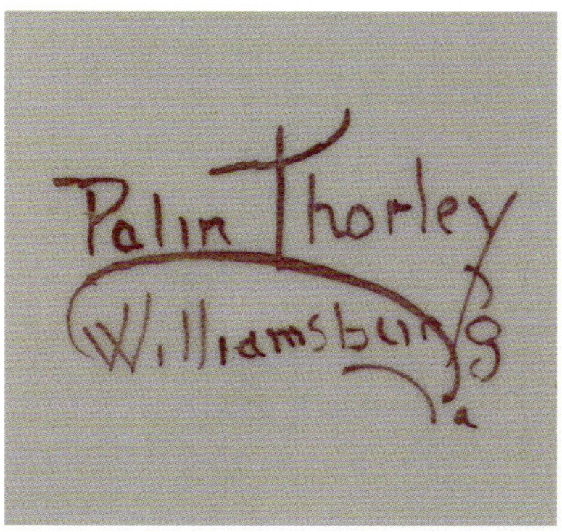

with most ceramic pieces, don't they?" I said, "They certainly do." We both had a good laugh. . . . So there it traced me through to Richmond. Now [several years] later, . . . in 1949, I established my pottery here in Williamsburg, Virginia, the great revival of eighteenth-century ceramics . . . wonderful, wonderful. So obviously here was a place where I should start to make this pink luster, there's lots [of examples] of it all around here in the homes. . . . So I then started to make it here and I've been making here ever since and still make it despite the fact that . . . for various reasons I haven't done so much lately, I still produce it here on Jamestown Road in Williamsburg and it was started by Mrs. Woolley, as I already mentioned, at Allerton's. So what I'm doing here is continuity, it's the unbroken line of eighteenth-century luster made in all respects identically . . . no modern changes in it whatsoever. That's why I get such a good [product] . . . which is just like the original. There's no difference except in the time it was done—one was done in 1830 and mine was done round about 1950 or later.

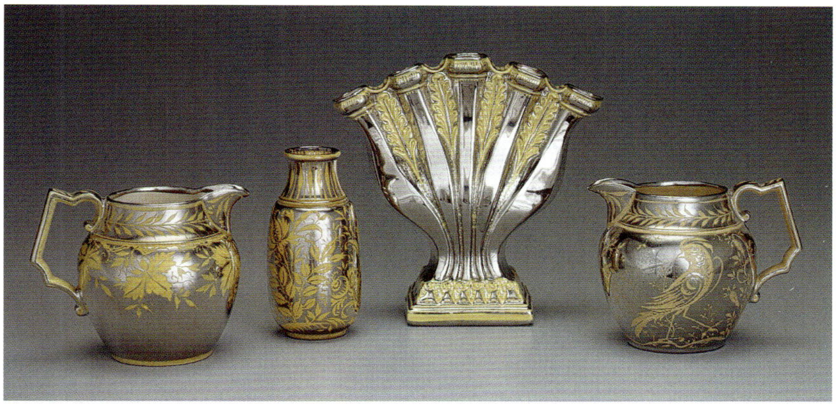

Figure 86 Silver luster objects, 1960–1970. Earthenware, with silver on a yellow ground. *Left to right: Thorley Jug:* H. 4½". Mark: Thorley stamp (illustrated in fig. 10). *Vase:* H. 5". Marks: handwritten, in red, "Palin Thorley / Williamsburg / Va." *Posy holder:* H. 7". Mark: handwritten, in red, "Palin Thorley / Williamsburg / Va." *Thorley Jug:* H. 4½". Mark: Thorley stamp (illustrated in fig. 10).

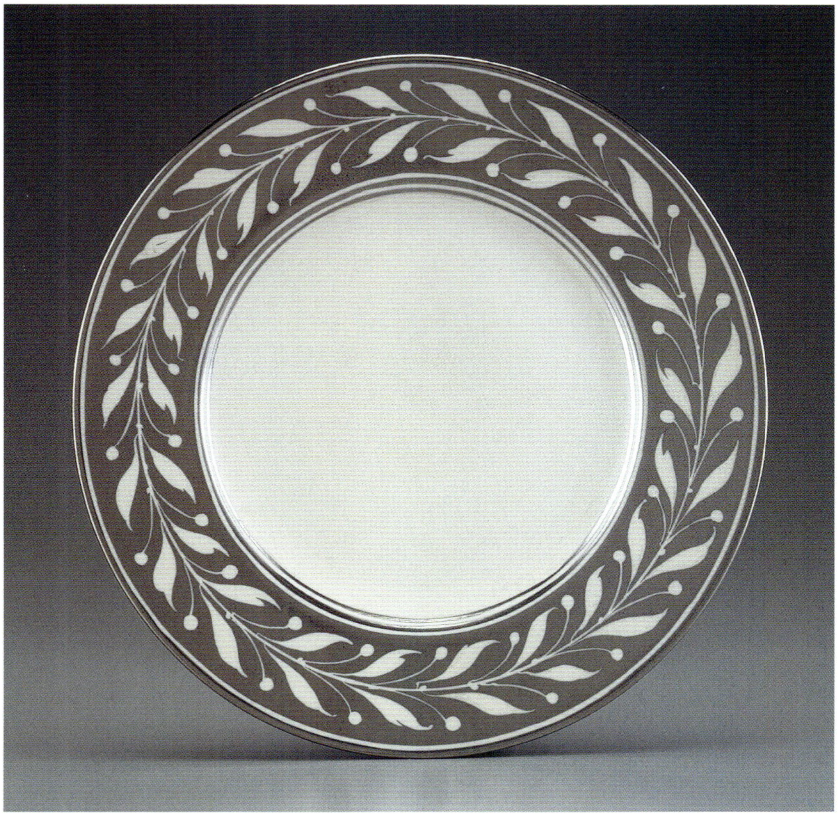

Figure 87 Silver luster plate, 1960–1970. Bone china, with silver on a white ground. D. 10". This example was not made by Thorley, but it was decorated by him.

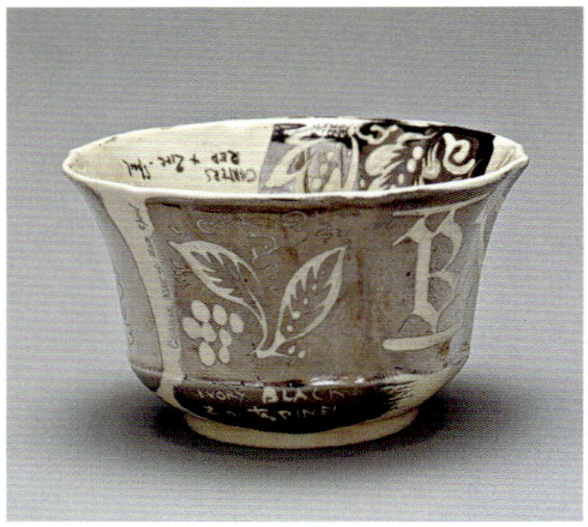

Figure 88 Silver luster sugar bowl, 1960–1970. Bone china, with experimental silver on white ground, covered with many unrelated designs. D. 3¾". This bowl has no maker's mark, but bears many formula names, such as "Kling Ivory black" and "Carters red + Zirc, Spinel."

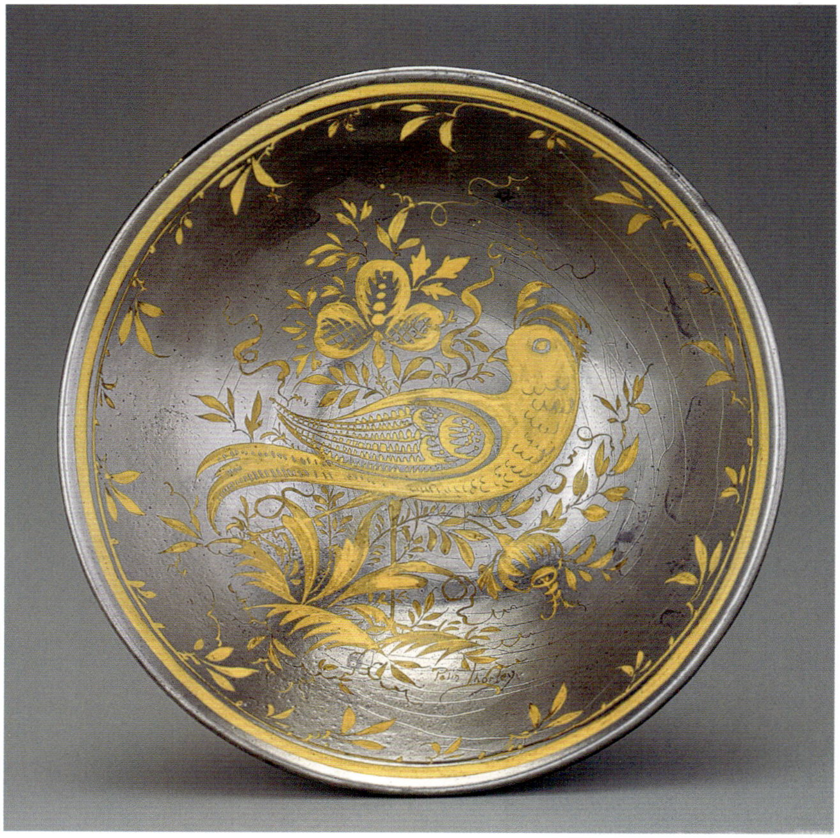

Figure 89 Silver luster bowl, 1960–1970. Earthenware, with silver on yellow ground. D. 8½". Mark: Thorley stamp (illustrated in fig. 10). Here Thorley employs the bird design found on the Thorley Jugs (see fig. 24).

Figure 90 Silver luster sugar and creamer, 1960–1970. Bone china, with silver on white ground. H. of creamer, to top of handle, 3½". Mark: (on both) handwritten, in pink, "Palin Thorley / Williamsburg." The image shows a finished example of the sugar bowl illustrated in fig. 88.

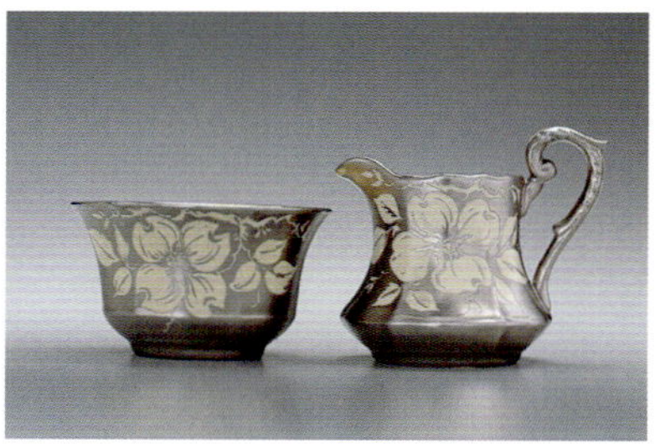

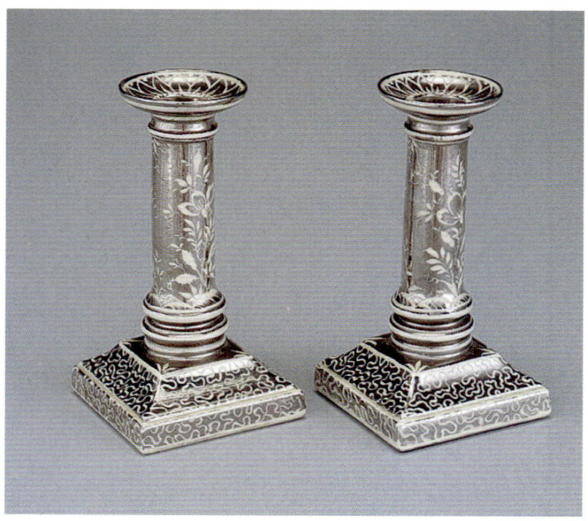

Figure 91 Silver luster candlesticks, 1960–1970. H. 6". Whiteware, with silver on white ground. Mark (on both): handwritten, in red, "Palin Thorley / Williamsburg."

Thorley never mentioned to me that he had siblings. In one of the taped interviews he did with Christine Sheridan, he does refer to a brother but only once, and as a child and not by name, so it was a surprise to learn after Thorley's death that two brothers and two sisters survived him. A letter received from the granddaughter of his brother John states that his brother Noel was also a pottery manufacturer and that his other brother, John, was a ceramic litho printer.

When it was revealed a few years ago that Noel Thorley had been involved with the manufacture of Wedgwood-style jasper and basalt wares being sold as eighteenth-century wares, several people asked me if Palin could have made fakes. My answer then, as it is now, was no. Palin Thorley was too proud of his abilities to want to sell what he made as the work of another.

This is not to say, however, that he did not enjoy fooling people. In this series about him I have mentioned three instances in which his work was believed to be of earlier manufacture than it was and the amusement he derived from the error.[6] He also often told me a story involving a friend in Pittsburgh who had an important collection of Chinese porcelain. Thorley copied one of his friend's blanc de chine pieces and secretly replaced the original with his copy, solely to trick his friend. His delight was not in fraud, but in revealing his expertise—especially to so-called experts.

In one of his taped interviews Thorley mentions that when he was at New Chelsea he produced ware decorated with the "Swansea rose" design, which had been popular with several British potteries in the nineteenth century. He sold these pieces as new, but was aware that it was likely, as he had not marked them, for them to be mistakenly dated a hundred years earlier.

He was also aware that the Greek-style vases—red figures on basalt—that he made for Wedgwood in the 1910s were being sold as eighteenth- and early-nineteenth-century pieces. Thorley had one in the corner cupboard in his living room. He was very proud of it and had me handle it several times. (On one of my last visits I noticed that the vase was gone, as were several other objects, among two in the form of Wedgwood basalt Grecian oil vessels that had been on his mantelpiece. Thorley had told me they were fakes, but would say nothing more about them.)

On the subject of fakes, I have to mention the mid-eighteenth-century Staffordshire figures usually referred to as "Astbury figures" that were copied in the second decade of the twentieth century by a potter named Fell. Thorley saw them being made when he was an apprentice at Wedgwood, and he was aware that they were being sold as eighteenth-century pieces even though they occasionally appeared with Fell's name on them. Thorley was not particularly bothered, however. He was always of the opinion that too much emphasis was placed on the date and provenance of an object, and that its true and sole value lay in its beauty, quality, and the enjoyment it gave.

Thorley's Legacy

Exactly when Thorley stopped producing, I cannot say. From his records it does not appear that much was being produced in the mid-1960s, when we first met. By the early 1970s he had written to former contacts to say he was producing, which brought in new orders. There is nothing to indicate that those orders were ever filled, however, and letters—much like the ones he received during his entire tenure with Colonial Williamsburg—claimed that the objects had not been delivered. In 1976 he said that he was "costing" the production of the posy holder, and even as late as 1980 he wrote, "I am engaged in organizing production Kilns, etc. to counteract present day inflationary costs in production."

During the many evenings I spent with Thorley, we typically would sit in his parlor, sometimes with a glass of port, and "talk pots." In spite of pointed hints from me, he rarely took me into his studio, so what he was producing, if anything, is uncertain. Among his papers are unsigned contracts with firms he was to do business with and potteries that wanted to buy him out and work with him. With his determined mind and artist's heart, Thorley never said quits. How many times did I hear him say, "John, I am going to fire up the old kiln"—pronouncing *kiln* with the final letter silent, of course, the way potters did and the preferred way in the dictionaries of his era—but it was not to be. He eventually had to turn over his affairs and care to a loyal friend, Charles Crone, and in 1987 Joseph Palin Thorley, a superior artist, great intellect, fascinating character, and good friend, passed away.

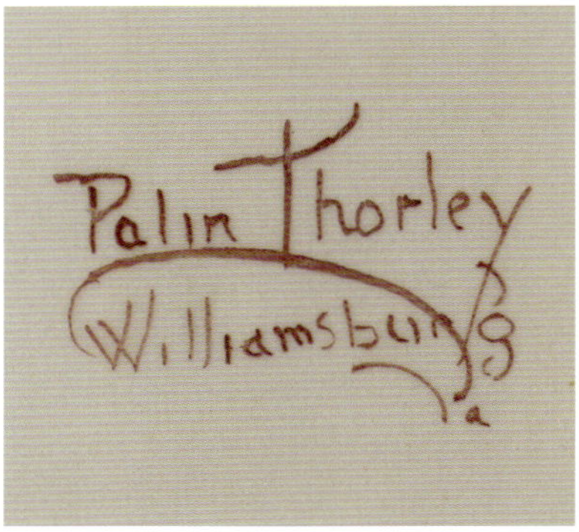

1. See John C. Austin, "J. Palin Thorley (1892–1987), Potter and Designer: Part 1," *Ceramics in America,* edited by Robert Hunter (Hanover, N.H.: University Press of New England for the Chipstone Foundation, 2005), pp. 160–201.

2. The plate was the "GR" plate, which must have been produced for a very short time as it appears only in that one price list. For further discussion on this design, see p. 196, fig. 52.

3. Louise Bang Fisher, *An Eighteenth-Century Garland: The Flower and Fruit Arrangements of Colonial Williamsburg* (Williamsburg, Va.: Colonial Williamsburg, 1951).

4. I noticed the dish was missing during one of my last visits to Palin but do not know whether he gave it away or sold it; perhaps it was stolen.

5. For more on this period in his career, see Austin, "Thorley, Part 1," pp. 179–81.

6. The first refers to a slipware piece he made in his youth (ibid., p. 166); the second to a trick he played on Frederick Rhead (ibid., p. 184); and the third to the luster tea service found in Richmond (see pp. 207, 209 herein).

Appendix 1

Making the Red for Pottery Decoration

In her taped interviews with Palin Thorley, Christine Sheridan recorded the following story about Thorley's father teaching him and his brother how to prepare the red color used for decorating pottery.

Yes, . . . I remember it full well. My brother and I were then about, say, about six or seven years of age, certainly not more than seven or eight years of age. When my father [wanted to] keep us quiet at home [he] taught us how to make colors, to make pottery colors, like his daddy used to make at Ridgway, he said. So to educate us we went to the drugstore—we call it here the chemist shop—and we bought two or three pennies' worth of iron sulfate crystals. . . . So [when] my young kid brother and I get home, father takes the shovel that we used to shovel the coal into the fireplace with—in England, you know, we have a kitchen fire in which they cook, [and] there's a little shovel there [for shoveling] coal onto the fire as required. Well, the first thing father did was to get this shovel and wipe it off with a clean rag, then we took the blow bellows, which also hung by the side of the fire in those days, and we blew the fire until all the coals were bright red and clear and hot. Then way back right under the chimney so the fumes went up . . . the flue of the chimney, we put [the] iron sulfate [crystals] on the shovel. They start to smell and smoke up the chimney and we watch it and turned it and father showed how to do it. The color changed remarkably from a green substance to white, then it began to change color and we let it go until it was almost a dark purple. Then father took that off and showed us that was wrong—you see, it's gone purple. That's because we overfired it. It had too much heat. Now we will do it again and we'll watch it and stop it earlier when it gets to a good red color. Well, with brother and I watching, you know, standing there holding it, quite excited seeing these colors change and we watched it. At last it came off. Then father scraped it, showed us how to scrape that onto a tile . . . a glass slab the tile was . . . and he gave us a muddler, a flat-headed muddler. We ground it. Crushed and ground and we ground all day, kept us quiet. . . . Each one in turn until our fingers got tired. But it finally passed father's supervision and then we washed it. Father told us how important that was. We washed it in hot water, stirring it, then we would let it settle in the bowl, then we poured off the water and decanted it, the water off. There at the bottom, finally, we had this bright red powder, which of course was red-eye and oxide . . . father described to us the difference between the red oxide and the crystals of sulfate, what iron really was, because brother and I didn't know of it. Having little lessons in chemistry we didn't know that, but we did see changes took place under heat, which was another lesson. Then father said, "Now we have to make a flux." Here is a good flux we read here in your grandfather's receipt book, there written in lead pencil was the flux he used for iron reds. So we weighed that again and put it in the muddler, we ground it with a pestle and mortar and we put that in the kiln again in the hot heat with the blow bellows going and melted that and that was our glass frit. It came out like glass because brother and

I were interested in that. Now we examined this, now we don't have those other materials, the lead, the borax, the flint, and so on. We've got glass options. We have to crunch that up with a hammer and wash that again and then grind it. Grinding was a tedious job. Pounding it and pounding it and grind[ing] it. When it [is ground] fine enough, we weighed them out in different proportions—so much flux, so much red, so much flux—and then ground them separately. Then father brought some plates from of one his friend's factories. My brother and I sat down with a plate each with a little decorative start at one end, one side of the plate, which we would have to repeat all the way around. Father did the little decoration on the side and we repeated all around the plate edge, we made a pattern. Then we took them down to my father's friend, William Wardle, who owned the Worlock Pottery Company in Shelton. To the best of my knowledge, Worlock Pottery may be still there, but of course, Billy Wardle isn't. Brother and I would walk this mile or two to the factory and we took down these plates we painted and Mr. Wardle would put them through his kiln and fire. Then we came back with some plates with a different tones of red all painted out. And they [were] all pretty good. I kept those plates for years. But that's how we learned to make colors. That was just the beginning, eye-red. Oh, the trouble came after my brother and I got to trying other things . . . it was alright to make the reds, of course, but my mother found out that the shovels were not lasting very long. . . . When we kept heating all kinds of oxides on the shovels, very soon [there was] a very, very big hole in the middle of the shovel. And it was some time before mother discovered the wastes in shovels was tied to the amusement of kids in making colors. So color making was off from then on. We were not permitted to run into the danger of getting burned, but when father was teaching us to make these colors, it did keep us quiet for days on end. The whole place was full of pieces of pottery we painted with these colors. But in the first attempts, of course, the colors weren't put on very well. As my brother and I did more and more there came a time when we put it on with some decency and some of the red-painted plates looked pretty good after a few weeks' struggling. But that was one way which to learn how to do something in the potters' craft. He starts at six or seven years of age. When I got to Wedgwood when I was fourteen [and] the people showed me how to do this, I [had] already done it and they couldn't understand that, and they thought it was instinct. It wasn't. It was because we had been around the kitchen fireplace on Sunday afternoons making color. That's all the story on coloring making, that's how I started to make it.

Appendix 2

Watching Mocha Ware Being Made

In her taped interviews with Palin Thorley, Christine Sheridan asked him to share his memories of mocha ware.

Were they . . . little trees that grew up the side of the slipped ware and spread out like ferns beautifully? Oh yes, I remember . . . there's a place in a very old part of the town, I forget the name, it was off of Hope Street, a side street. There are some very ancient old houses there . . . houses built for workmen in the late eighteenth century. They are in very bad state of repair, but as you walked along the street from school [there was] this little window, very small, little window [that you could look] through . . . into a basement under this house. You know, I would peek through the window because the window was on the level with the street pavement. It wasn't above the wall it was level with the street as you walked. And boylike I crawled on my hands and knees and peered through the window and in that room was a group of potters and women working making mocha ware—making ware, I didn't see much what it was, I just saw them doing it. There were a couple of men throwing and then two or three working on lathes, and women were treadling the lathes. All that makes me smile when I come to think of treadling the lathes. I [haven't] seen that for years. The old-fashioned eighteenth-century lathe wasn't worked with a motor, you know. They had no steam engines or motors in those days. They worked with a treadle. The treadle was on the floor just in front of the lathe and it . . . was a flat piece of wood and the woman put her foot on it, on the spring, [and as] she pressed the treadle up and down and up and down, the cable would work the wheels. If the turner wanted his lathe to go very fast he would give the signal and she would pedal with her right leg on this pedal rapidly or slowly as he required her. It would have to [be in] perfect harmony between them because when he was turning on his lathe he would increase the speed or slow it and she could do it perfectly. But I often wondered, if a young woman did that for many years, I bet her right leg was twice as big as her left leg, I used to say. Of course, in those days you couldn't see. They didn't wear short skirts. She would have quite the job if she's living these days, because she would have one leg with big muscles and the other leg very thin and scare-crowy, which would have looked odd. Oh, when I look back, I would smile at those . . . old ladies. Wedgwood still had one when I was a boy there. The old treadle lathe, it was eighteenth century, so I peer through the window, then I saw the man at the lathe drop a little can . . . drop some spots on it out of the can and I watched what they were doing. Well, I couldn't figure it out and I was curious. . . . In the coldest dark, thick with clay, very grimy, very dirty the shop was. I don't think they cleaned the clay fragments off the floor in years. All the people, poor people coming there making these pots. I was watching. Well, a woman came in and walked up and there was a box, kind of a square box in the corner of this room, and to my great amazement she lifted up her dress and sat on the box—well, kind of squatted over it. She didn't sit on it 'cause the box had narrow sides, just an ordinary box. I said, "Oh good gracious! My, my" . . . at this thing. Then . . . watching again another day a man comes up, he did something in the box and they weren't paying attention to him and I kept looking, [wondering what was] going on. Then the man would sit there and they all were chewing tobacco, their mouths—you'd see the jaw pulled out here while they worked—and they walked to the box and spit in it. So the tobacco piece went in. And this went on every time I peeked. Every time I came home from school, I peered through the window on my hands and knees. Grimy window—never been cleaned in years . . . you had to put your nose close to the panes to see through it. I'd get closer, closer up to it; what are they doing in the box is a mystery. One day I went home and I saw something that was rather shocking and I told my father, "Oh that place down there off Hope Street, with the shop underneath, they're doing this and the other," and he started to laugh. "Oh, they're making mocha ware." What they do, he said, they save urine until it ferments and gets very smelly and very unpleasant. Then they put tobacco, thick chewing tobacco when it's stirred up, then put some manganese, some iron in it, and stir that up. And that's the stuff they splatter on the ware. When the ware is wet on the lathe and dampened with a sponge, well a drop of it, it runs like a little street that makes the tree trunk, then it spreads out, just like a million branches of trees and looks very nice on jugs and beer mugs, but on nothing else. They don't use it much on plates. They just make jugs, mugs, bottles, and beer mugs. That's called mocha ware.

[When asked if he ever made mocha, Thorley claimed he never had. "Somehow it left me with a feeling of . . . I didn't feel for mocha ware."]

Ivor Noël Hume

Neither "Landskip" nor "Scripture": Collecting Dutch Maritime Tiles

▼ NO ONE COLLECTS ANYTHING without reason, without a catalyst. My introduction to Dutch tin-glazed maritime tiles came by way of a powder horn decorated by a British soldier stationed in Jamaica. An amateur naturalist, he had engraved the horn with the various creatures he saw there, among them a manatee, which he took to be a mermaid (fig. 1). That led me to Columbus's log and his mermaid sightings, and thence into the deep waters of Poseidon-related mythology.

The pleasures of collecting are driven by perception, in my case by a quirky way of looking at sea-related delftware tiles. Some may see them as attractive (or maybe peculiar) wall decoration, but to me each one is a picture from the past, painted by someone who transferred his thoughts and talents to a square of whitened clay. I reach out to him, trying to imagine whether he was hot or cold as he worked, whether he was well or ill fed, and how he got along with his wife. Indeed, was he even a he? Such musings, though perhaps not typical of most academically trained curators, add a dimension to each tile.

Using tiles for flooring material is as old as using clay slabs in roofing. But it was not until the Middle Ages in Europe that tiles became elements of paving decoration, at which time they were small, thick, and decorated by means of incuse ornamental stamps whose impressions were filled with white clay before being lead glazed. Being subjected to heavy traffic, these tiles would have quickly lost their decoration had it been painted on the surface of the clay. Besides, the white clay (later used in tobacco-pipe making) was expensive. By the fourteenth century the white clay was spread over the surface of the stamp and impressed onto the tile in the manner of block printing. One such printed tile was impressed with the figure of a mermaid and survives in a library floor in England's Litchfield Cathedral (fig. 2). It is, however, a rare exception, as the majority of medieval tiles are heraldic in style.

The art of decorating ceramic tiles with oxide of tin over a white engobe spread across Europe in the early sixteenth century as a result of trade with the Ottoman Empire. Ambassadors and traders saw newly built mosques and palaces decorated with gleaming tiles painted under the glaze in blue (and later with red and green) that were vastly superior to the Byzantine-inspired floor mosaics of European churches and abbeys. However, the Ottoman (Iznik) tiles covered walls, ceilings, domes—indeed, anything above the floor. Contemporary European tile makers were eager to learn the mystery of what eventually would be known as delftware, but were

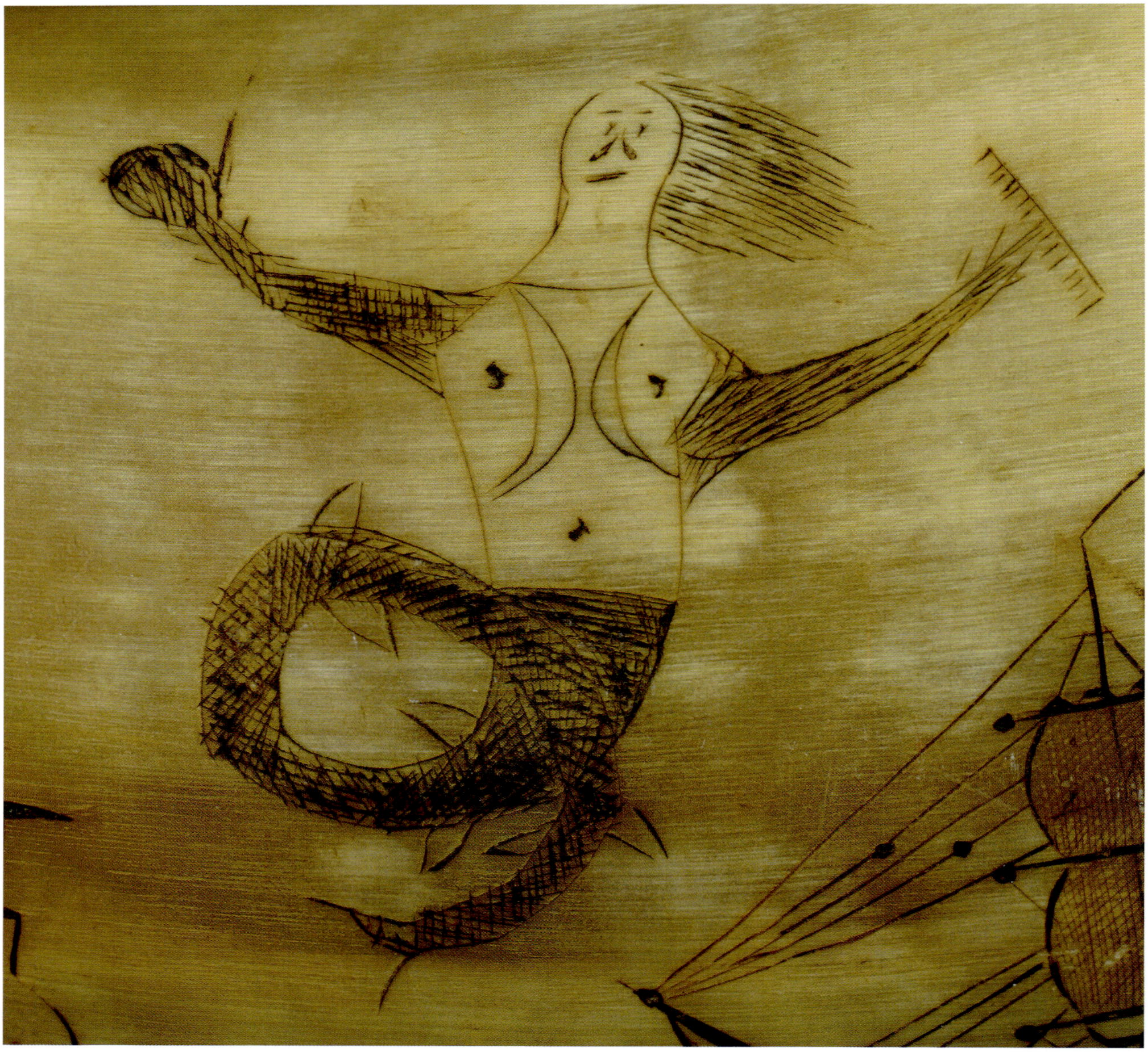

Figure 1 Detail from a powder horn, Jamaica, ca. 1820. (Noël Hume Collection; unless otherwise noted, all photos are by Gavin Ashworth.) This carving depicting a mermaid with the head of a manatee is probably the earliest "mermaid" rendering drawn from life.

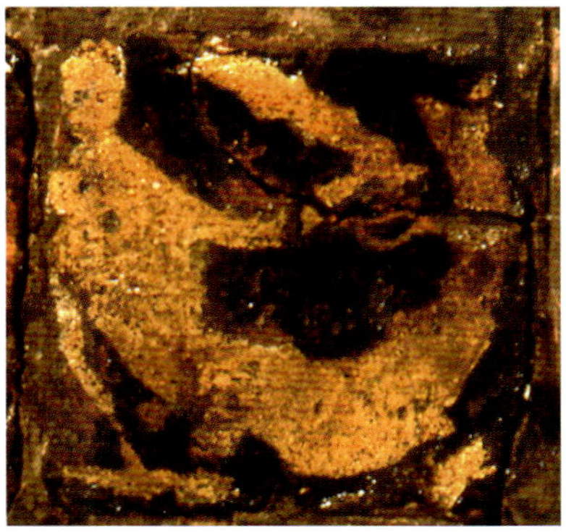

Figure 2 Paving tile, England, fourteenth century. Red brick with pipe-clay inlay. Decorated with a mermaid holding a fish, this tile is part of the floor of the library at Litchfield Cathedral. (Photo, courtesy W. W. Rodwell.)

content to make their tin-glazed tiles for use on floors.[1] Thus, the first potters in England to make the new tin-glazed ware were reported in 1567 to have "followed their Trade, making Gally Paving Tiles," which were thick but not very durable.[2]

Potters Jasper Andries and Jacob Janson were Protestant Flemings who fled to Norwich from Antwerp to escape Spanish-Catholic persecution. It was inevitable, therefore, that their products would emulate those they had been making in the Netherlands. Being intended as floor mosaics the tiles came in units of four to create repeating designs, commonly featuring pomegranates and grapes (both fruits being biblically associated with the seeds of fecundity) and Dutch tulips (fig. 3). Whether the potters who made

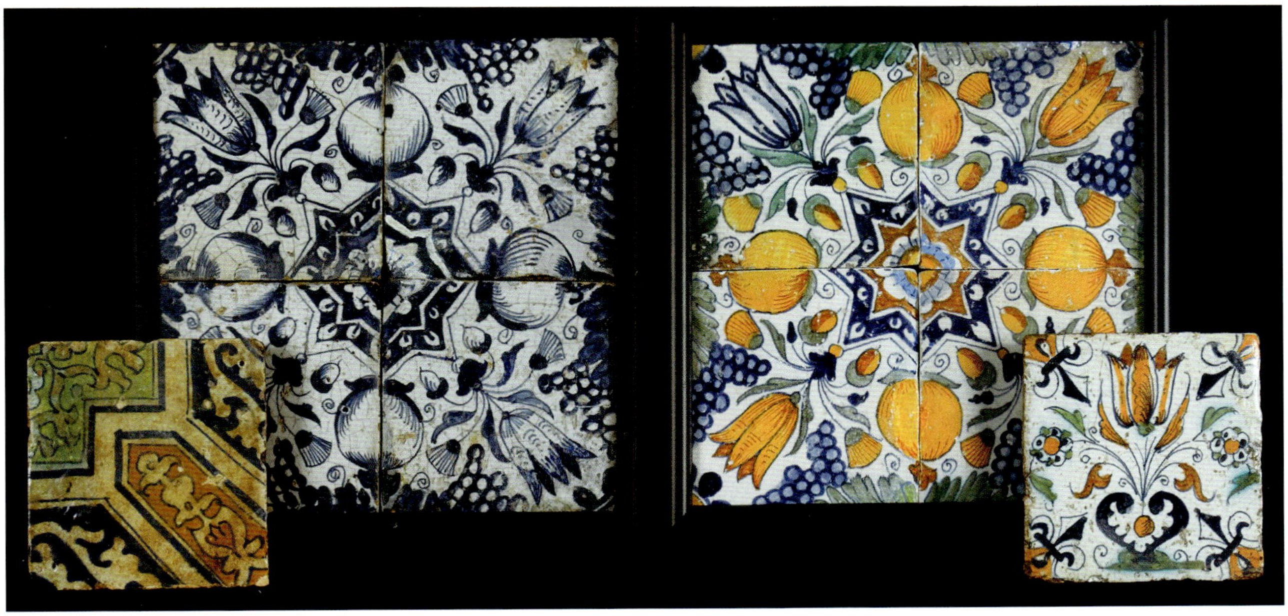

Figure 3 Paving tiles. *Left:* Netherlands, ca. 1570–1600. White engobe laid on coarse red brick clay; Flemish polychrome. 5" x 5" x 0.71". Found in 1919 in Tower Street, London. The two corner pinholes penetrate .55" into the tile, and the edges are slightly beveled. *Center:* Two four-tile panels, probably Rotterdam, ca. 1600–1630. Delftware; yellow clay, with tulip and pomegranate designs in monochrome and polychrome. 5.25" x 5.25" x .47". Two pinholes are visible in one of the tiles. *Right:* Netherlands, ca. 1600–1650. Delftware. 5.19" x 5.19" x .47". Slightly beveled, with polychrome tulip and fleur-de-lis corners and two pinholes.

the tiles or the people who walked on them were aware of the iconography is anybody's guess. It is, however, a fact that the decoration was used on both sides of the North Sea from about 1570 to 1620. By the latter date the Dutch and Flemish potters were emulating their Ottoman confreres and looking upward from the floor to walls, stoves, and fireplaces.

These tiles, having become purely decorative, could be made with less clay and their painted surfaces lent themselves to more imaginative designs, thereby opening up a huge industry that lasted through the eighteenth century. But not so in England, where production of decorative paving tiles came to an end about 1625 and did not reemerge as wall tiling until late in the 1670s. Documentary records show that even then the English manufacturers were having trouble producing tiles capable of competing with the burgeoning Dutch industry.[3] When eventually they were successful, contemporary taste called for a uniformity of biblical or landscape designs contained within two concentric circles. The common corners of early Dutch and, later, English tiles are classified by collectors as "oxhead" and "spider," though neither name really describes those often slapdash devices which when four tiles were placed together continued to create repeating patterns.[4] The large panel illustrated in figure 4 demonstrates how, even when tiles painted with differing subjects were mounted together, their corners provided a uniformity of design.

It is not my purpose, however, to pursue the evolving design elements and processes that are already well known to collectors of English tiles.

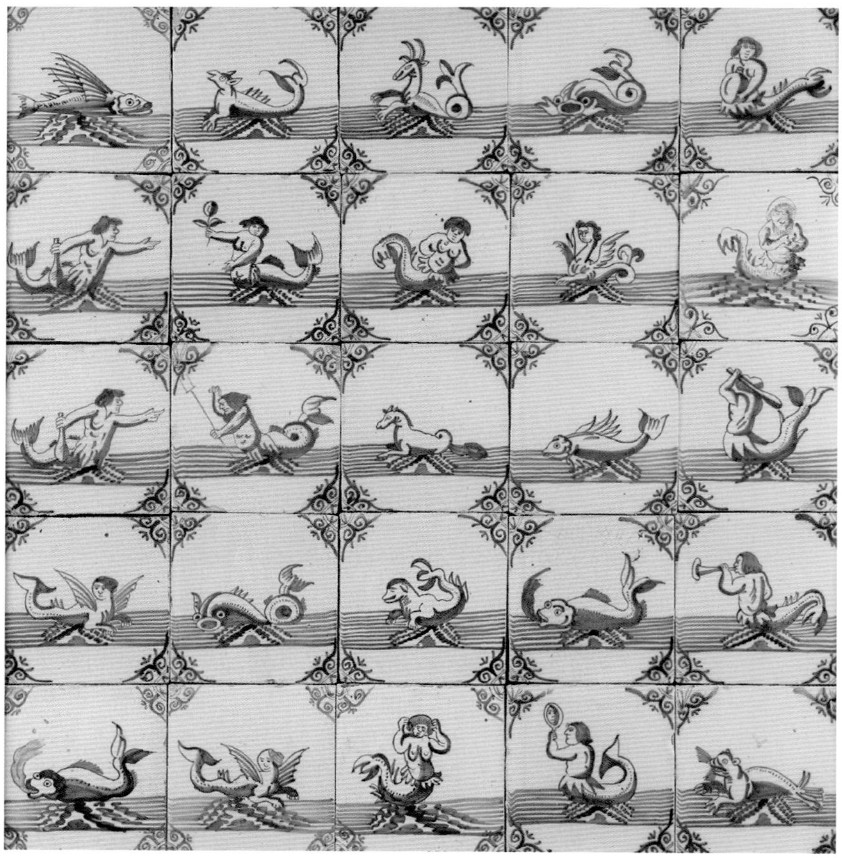

Figure 4 Tile panel, Holland, ca. 1650. Delftware. Dutch tiles were used in panels as well as surrounds.

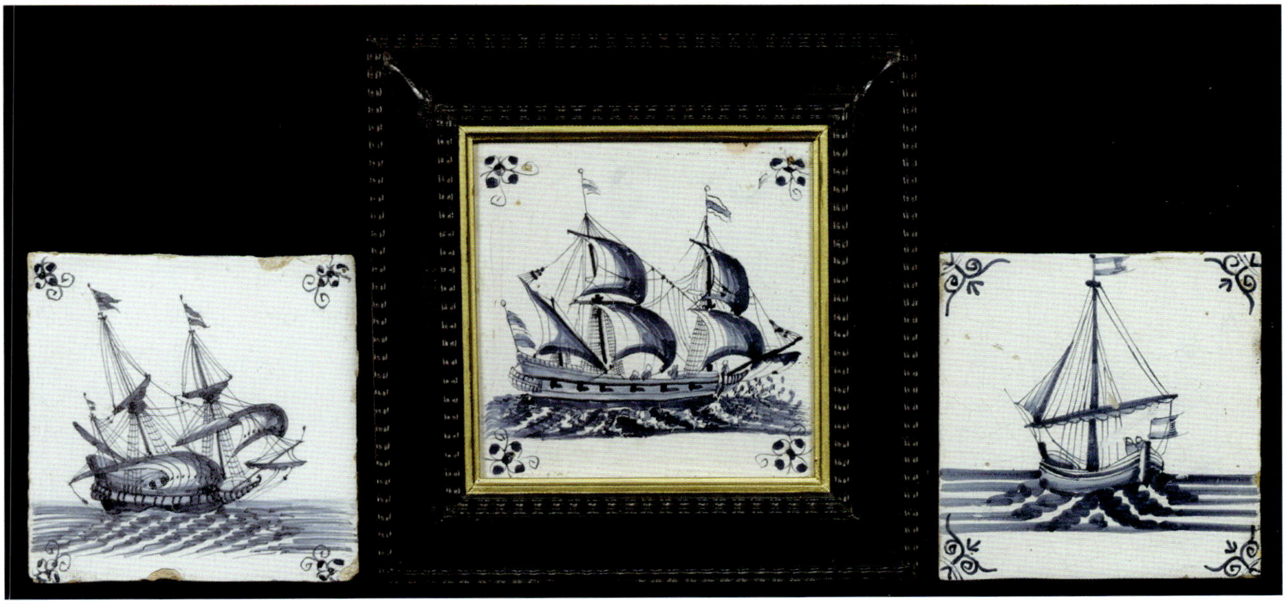

Figure 5 Ship tiles, Holland. Delftware. *Left:* Ca. 1680–1710. Yellow clay. 4.88" x 5" x .39". Showing a three-masted merchantman under half sail on a rippling sea, this tile has spider corners and two pinholes close to them. *Center:* Ca. 1650. Pale pink clay. Approx. 5" x 5" (thickness obscured by frame). This depiction of a three-masted warship in full sail with guns run out and four crewmen visible is striking for its realistic treatment of waves and spray and the care taken in drawing the rigging. The tile has spider corners with inset pinholes. *Right:* Ca. 1650–1700. Orange clay flecked with iron. 4.88" x 4.88" x .31". This tile illustrates a single-masted pinnace with two crewmen and a chevron sea. There is a pinhole .55" in from two of the oxhead corners.

Suffice it to say that merpeople are absent from English tile designs in the seventeenth and eighteenth centuries. In America the market preference is reflected in a New York dealer's 1764 advertisement for "A Few very neat Scripture and Landskip Chimney Tiles."[5] Even ships are few and far between, probably because the vast majority of the population had no contact with the sea. The Dutch, on the other hand, had difficulty keeping away from it. As their livelihoods depended on deep-sea fishing and an international carrying trade, it made sense that ships were a popular subject for tile painters.

Indeed, they continued to be so well into the nineteenth century, both for individual tiles and as elements in eight- and ten-tile panels.[6] The depictions were sometimes very carefully done, enabling the buyer to identify a wide range of sailing vessels, from galleons and coast-hugging ketches to rowboats (figs. 5, 6).[7] The ships, which can be identified by their national flags or pennants, are invariably Dutch. For seafaring collectors, ship tiles can be a category unto themselves. For me, however, what might lurk beneath the waves has been my primary fascination.

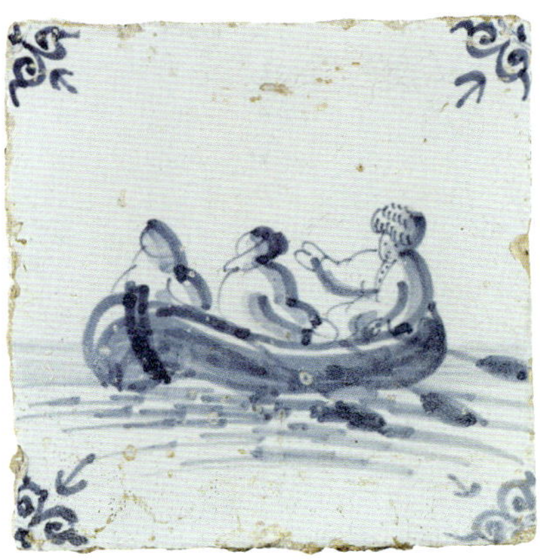

Figure 6 Tile, Holland, ca. 1650–1700. Delftware; yellow clay. 5" x 5" x .43". The tile, which depicts three men seated in a rowboat, one wearing a fur hat, has oxhead corners and no visible pinholes; the edges are slightly beveled. The painting is crude and the colors diffuse.

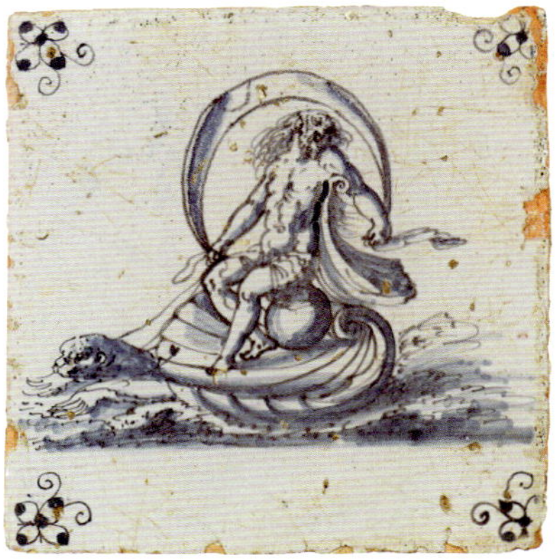

Figure 7 Tile, Holland, ca. 1625–1650. Delftware; orange clay with iron inclusions. 5.27" x 5.27" x .43". Represented here is an aged Poseidon looking somewhat distraught. He is seated on an orb (possibly a pearl) and holding a sail while riding on a seashell pulled by a reined otter. The subject is outlined in manganese and filled in with two tones of pale blue. The sea and its spray are similarly treated. At each of the spider corners is an inset pinhole.

Ever since we humans grouped ourselves into tribal units aspiring to a degree of civilization we have embraced the notion that there were gods to watch over our houses, our crops, and our children, and to take the blame when things did not work out the way we expected. We believed that these gods had to look like us—or at least partially like us. Thus, Poseidon, the god of the ocean, had a human head and torso (fig. 7), whereas the rest of him could be, but not always was, fish. It goes without saying that his wife, Amphitrite, was similarly constructed, her piscatory parts sometimes kept demurely below the surface of the sea. They had numerous progeny, the best known being Triton, whose nether parts were shaped like the tail of a dolphin and whose principal activity was blowing into a twisted seashell—violently, to summon storms, or gently, to calm the waves (fig. 8). Thus wrote the Roman poet Ovid:

> Already Triton at his call appears
> Above the waves; a Tyrian robe he wears,
> And in his hand a crooked trumpet bears;
> The sovereign bids him peaceful sounds inspire,
> And give the waves the signal to retire;
> His writhen shell he takes, whose narrow vent
> Grows by degrees into a large extent.[8]

Figure 8 Tile, Holland, ca. 1680–1710. Delftware. 5" x 5" x .43". The standard Triton, shown here blowing into a seashell, is painted in two-tone blue. The tile has oxhead corners, two of which have inset pinholes, on a pink to yellow body with beveled edges.

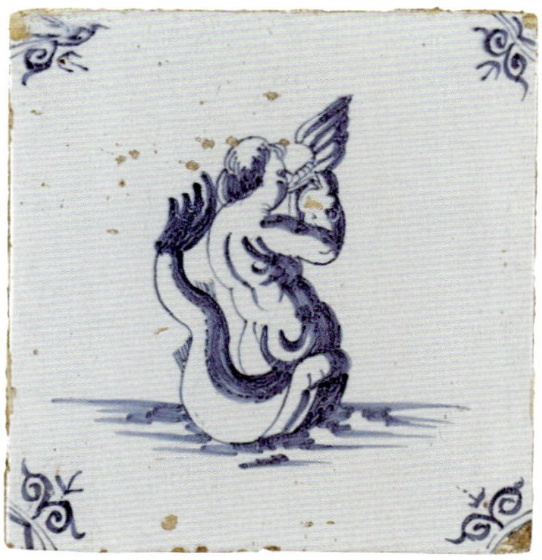

Over time there came to be many second-string gods named Triton, some of them possessing the forelegs of a horse. (Among his many other attributes, Poseidon was the god who smote a rock with his trident and thereby gave birth to creatures half horse and half human.)[9]

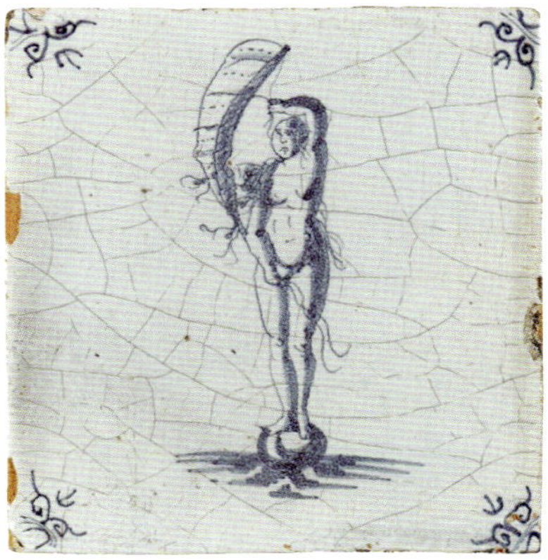

Figure 9 Tile, Holland, ca. 1650. Delftware. 6.1" x 6.1" x .39". Aphrodite (Venus) catching the wind in her sail. She stands on an orb, perhaps a pearl, on a modified chevron sea. The tile has oxhead corners, and one inset pinhole is visible through the thick, heavily crazed glaze. The clay body is sandy pink.

The deital ideas of the Greeks were repackaged by the Romans, who passed them off as their own. Thus, Poseidon became Neptune and Amphitrite became Salacia, goddess of salt water. Her name became synonymous with salaciousness and explains why supporters of Queen Elizabeth I drew cartoons of Mary Queen of Scots as a whoring mermaid. Also of Greek ancestry was Aphrodite (fig. 9)—Venus, to the Romans—who was one of an abundance of sea-dwelling daughters of god and goddess Nereus and Doris. These sisters came to be known as Nereids, nubile maidens given to cavorting amid the waves and helping sailors in perilous waters, but that is deeper than we need dive into remote mythology. My point is that between them the Greeks and Romans created an undersea world recognizable to Dutch tile painters and, presumably, to the good citizens of Holland who mounted them on their walls.

Being a practical people, the Dutch tile makers assumed that human relationships in the undersea (or on-the-sea) world behaved as they did in burghers' bedrooms (fig. 10). The Victoria and Albert Museum has examples of these tiles, but keeps them discreetly hidden in a drawer![10] The Dutch evidently did not consider them prurient, but simply the natural precursor to my next examples (fig. 11), which show what might be Salacia suckling her infant and Father Neptune holding up his Triton pup in a failing attempt at house-training. It can be difficult to distinguish between

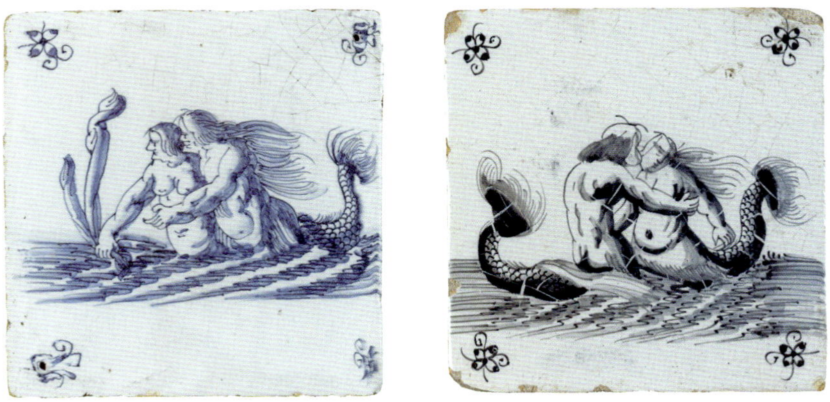

Figure 10 Tiles, Holland. Delftware. *Left:* Ca. 1650–1700. 4.92" x 4.92" x .39". A Triton is depicted with his windblown, elderly wife, who reaches for an unidentified object in a diagonal sea. The clay body is yellow with beveled edges, the glaze blue-tinted and somewhat crazed. There are spider corners with two inset pinholes. *Right:* Ca. 1640–1670. 5" x 5" x .47". The subject—a Triton kissing an unresponsive mermaid in a diagonal sea—is painted in an unusually dark blue color, augmented with a lighter wash. Two pinholes are inset .67" from spider corners. The clay body is pale pink with a yellowish gray glaze, which, like the tile illustrated at left, is crazed in the top right corner.

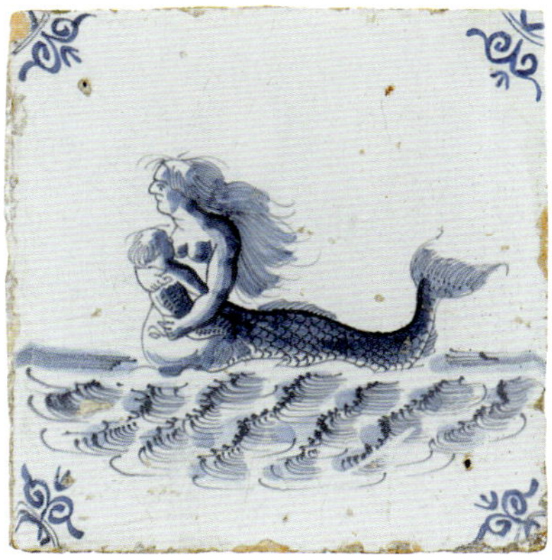
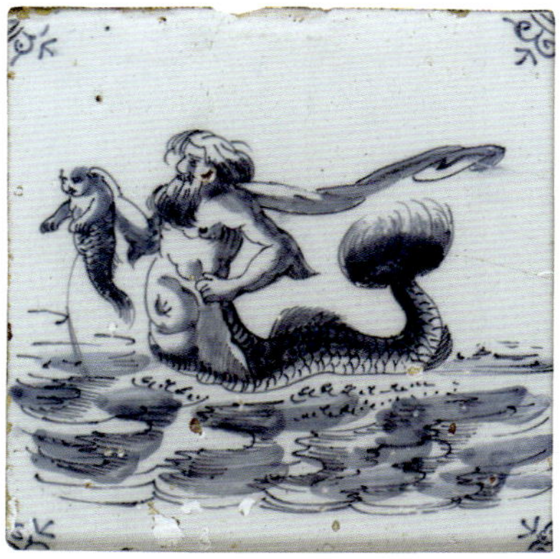
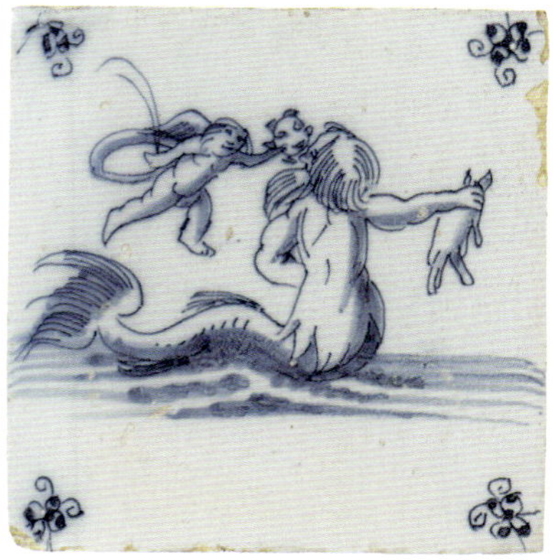

Figure 11 Tiles, Holland. Delftware. *Left:* Ca. 1640–1660. 5.11" x 5.11" x .39". The mermaid (Salacia?) shown on this tile suckles her merchild in a billowing sea. The clay is pale pink and beveled at the edges; two pinholes are set 1.02" in from the oxhead corners. *Center:* Ca. 1640–1660. 4.92" x 4.92" (thickness obscured by a frame). In this example a merman (Neptune?) endeavors to house-train his pup in a sea drawn in short blue dashes overlaid with darker hatching. This is a rare instance wherein the figure was drawn separately and the sea kept separate by a row of small dots and hooks. The tile has four pinholes set .78" in from the poorly defined oxhead corners. *Right:* Ca. 1690–1720. 4.88" x 4.88" x .39". A merperson of uncertain gender holds a small animal—perhaps a piglet—while reacting to a winged Nereid offering a diadem, all on a modified chevron sea. Compare the linear and cartoonlike treatment of the piscine body with that of the tile at far left, in which the scales were carefully drawn before applying the secondary wash. The bluish gray glazing is over a yellow body; spider corners, with two pinholes near beveled edges.

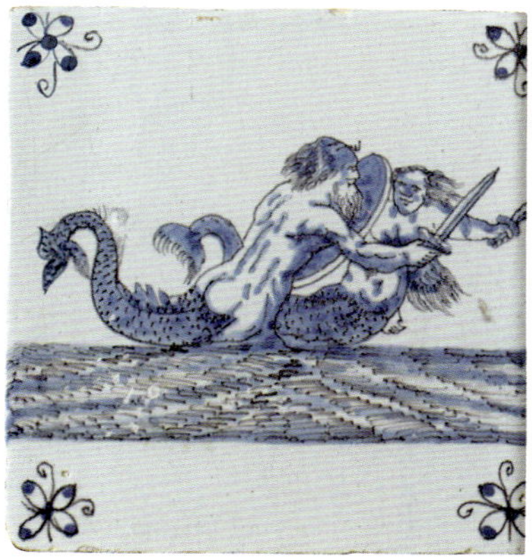

Figure 12 Tile, Holland, ca. 1630–1660. Delftware. Yellow clay. 5.11" x 5.11" (thickness obscured by a frame). A Triton and a mermaid are shown fighting with swords. One of them has the advantage of a shield. The figures are outlined in manganese, as are the ripples of the chevron-style sea. These figures (as well as mermaids on several other tiles) are distinguished by a stippled treatment of the fish scales under a solid blue wash. The spider corners are well executed and have two pinholes inset .66" from them. For another example from the same factory, see fig. 14, center.

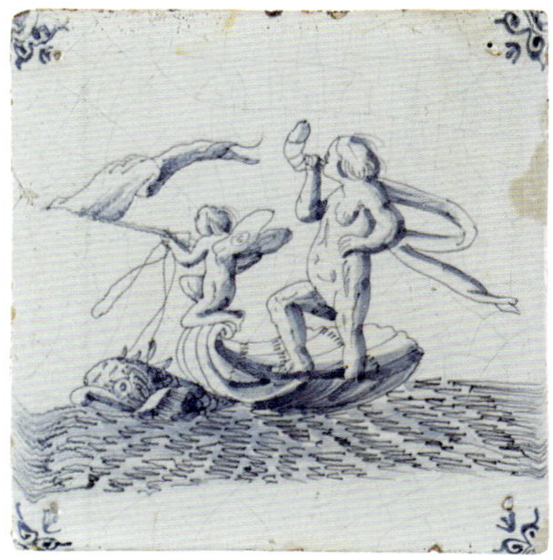

Figure 13 Tile, Holland, ca. 1640–1670. Delftware. 5" x 5" x .39". Here a young Triton is blowing a conch shell and a Nereid waves a large flag while riding on a seashell drawn by a none-too-happy dolphin. They appear to be racing along and may have been the seventeenth-century Dutch version of today's jet skiers. The figures are well drawn, as is the zigzag version of the chevron sea. In contrast, the oxhead corners are sloppily treated. The gray-blue glaze is crazed overall, and the clay is brick red; the edges are only slightly beveled. This tile is unusual in having four rather than two pinholes, set .59" in from the corners.

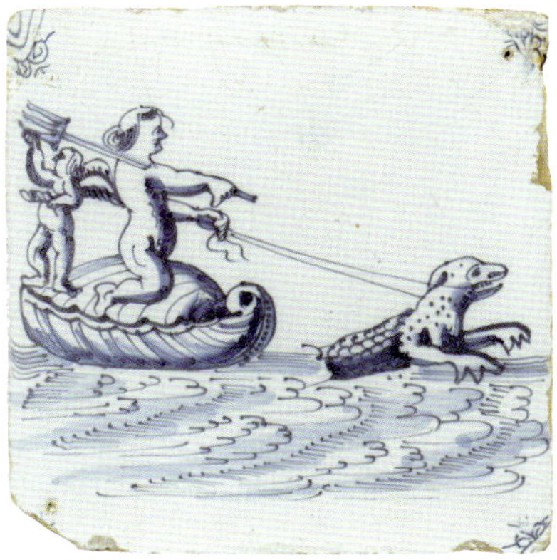

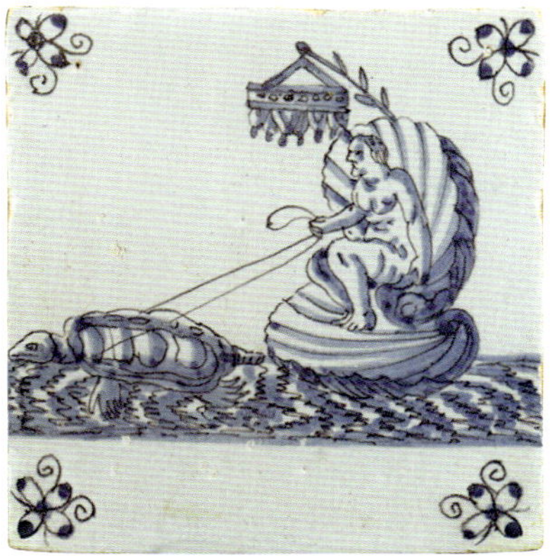

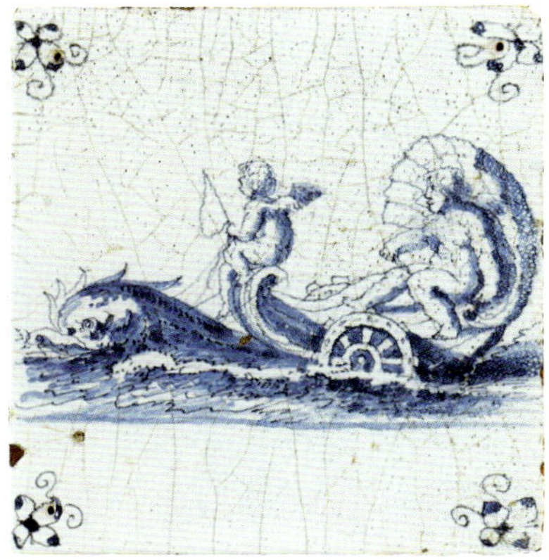

Figure 14 Tiles, Holland. Delftware. *Left:* Ca. 1660–1700. 4.92" x 4.91" x .35". Neptune bearing a trident is shown here riding on a closed seashell drawn by a scaly leopard whose head and shoulders are stippled. Behind Neptune stands a winged Nereid holding a small flag. The sea ripples to the right. The clay is yellow, the edges beveled; there are two pinholes .39" in from the oxhead corners. On the back, drawn in cobalt, is a large "64," probably a dealer's stock number. *Center:* Ca. 1630–1660. 5.11" x 5.11" (thickness concealed by a frame). Neptune rides an open seashell with an elaborate sunshade suspended over him. He holds the reins of a well-drawn turtle, the details of which are obscured by a lighter blue wash. The clay is yellow, and the figures are outlined in manganese—as is the sea, which is created from rows of zigzag lines. The spider corners are boldly executed, and there are two pinholes .66" in from them. See fig. 12 for a tile from the same factory. *Right:* Ca. 1650. 5.11" x 5.11" x .43". Represented here is a Triton riding on a wheeled sea chariot with a wind-baffling shell behind him, powered by a dolphin. In the chariot's bow sits a winged Nereid holding reins and a whip. The sea is realistically treated with the crests of the waves in a reserve. The surface of the glaze is heavily crazed, pitted, and black-speckled, and there are spider corners with three pinholes well in from them. This may be an example of the surface condition that preceded the addition of *kwaart*. The clay, which is orange and visibly wedged, is beveled at the edges.

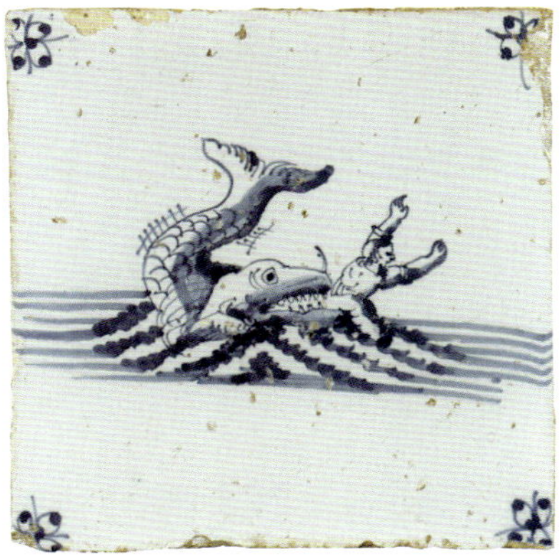

Figure 15 Tile, Holland, ca. 1680–1710. Delftware. 5.03" x 5.03" x .31". Jonah is shown in the jaws of a saber-toothed whale atop a chevron sea. The clay is yellow with beveled edges. Two pinholes are close to the spider corners.

Neptune and his sons, who were prone to fighting amongst themselves (fig. 12) or behaving as classical prototypes for jet ski hooligans (fig. 13).

Depictions of godly vehicles are relatively common. They usually take the form of seashells drawn by sea stags, sea horses, and turtles, or by the dolphins mythologically associated with Poseidon—he being a powerful, sometimes warlike figure, and his dolphins the embodiment of serenity at sea. Now and then the beasts had the benefit of wheels (fig. 14, right).

Unlike the more common terrestrial tile designs, recognizable biblical sources are few. The only such motif known to me is that of Jonah and the Whale (fig. 15), and only once have I seen a merperson carrying a Christian cross (fig. 16). Later, however, the story of Noah and the biblical flood

Figure 16 Tile, Holland, ca. 1650–1690. Delftware. 4.92" x 4.92" x .39". A merman with a forbidding face holds a cross (crucifix?) on a hooked sea; on the horizon are two yachts. The seventeenth-century Dutch, having thrown off Catholic idolatry, produced very few religiously oriented tiles. It is tempting, therefore, to see this grim merman as the personification of Catholic oppression. The clay is yellow, with beveled edges; two pinholes are set .59" in from the oxhead corners. There is some sub-surface crazing of the glaze.

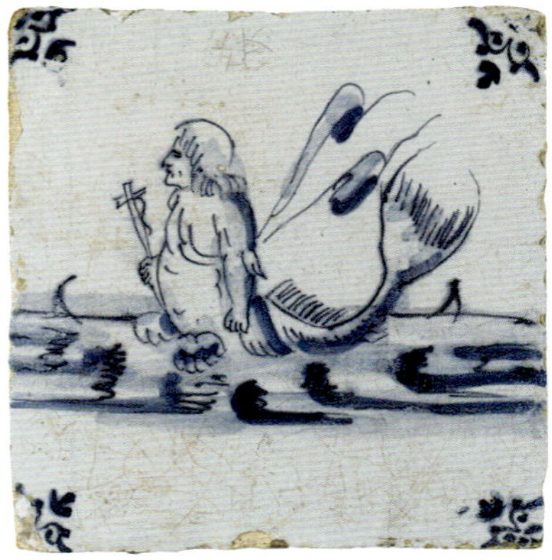

would be used as one among the many circle-enclosed scriptural subjects. On the left in figure 17 is the English version and on the right the Dutch—a demonstration of how difficult it can be to tell one from the other.[11] Both date between about 1720 and 1800 and do not really fit within my

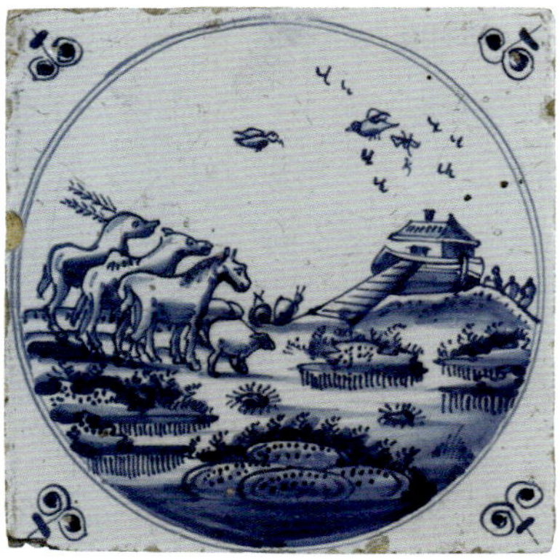

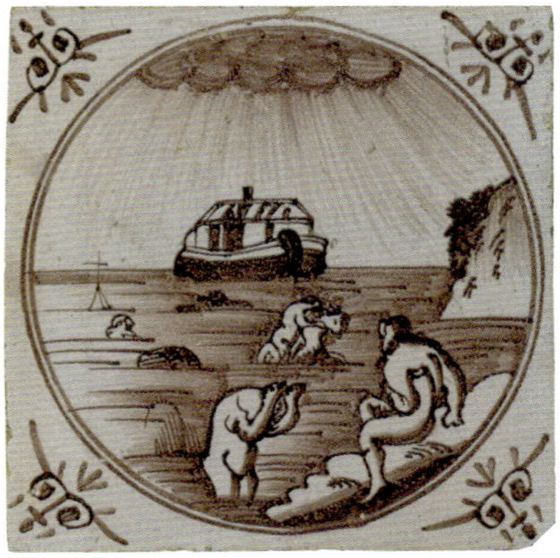

Figure 17 Tiles. Delftware. *Left:* Probably London, ca. 1720–1740. 4.96" x 5.04" x .31". An English approach to the biblical flood, executed in blue. Animals, birds, and people are on their way to the ark as water rises in the foreground. The design is framed within two concentric circles and the corners are a rare stylized and hornless version of the Dutch oxheads. The clay is a sandy yellow, the edges are not beveled. There are four pinholes inset from the corners.

Right: Holland, ca. 1800. 5.07" x 5.07" x .27". A less optimistic, Dutch version of the biblical flood executed in manganese, featuring the ark, the spire of a submerged church, miscellaneous floating debris, a man perhaps riding a sea horse, and, in the foreground, two despairing people—all within double concentric circles and oxhead corners. The clay is pinkish yellow, with beveled edges, and the glaze is tinted a pale purple by the manganese.

classification of maritime tiles, very few of which date any later than the end of the seventeenth century.

Dating within the seventeenth century is difficult once we get past the thick paving tiles, which, as I have noted, were over by about 1620. As none of the wall tiles bears a date, one has to rely on previous pundits, who might have been wrong. There are, however, distinct stylistic differences, particularly in the treatment of water, varying from narrow bars to flowing chevrons and individual waves. The painters' differing treatment of the waves can be clues to dating associations, as can their methods of depicting merpeople's piscine scales, although it is doubtful that they represent an evolutionary sequence. On the contrary, they are more likely the product of several designers and painters in different factories. Although corner motifs got smaller the more they got away from four-unit paving tile compositions, the ubiquitous oxheads continued on through the eighteenth century, as did the so-called spider elements. Thickness, however, is a relatively reliable guide—the later they dated the thinner they were made. Chamfered edges are often encountered from about 1630 through the nineteenth century, the beveling designed to enable the faces of the tiles to fit snugly together while allowing maximum space for mortaring the back edges.[12] However, these details are also more likely to represent products of different factories than of date.

Very rarely were tiles marked or painted on their backs, although exceptions occur in all periods when numbers were applied to identify each tile's placement within large panels, an important assistance when each tile represented, say, two letters in a trader's advertisement or no more than a tree branch in a landscape constructed of a hundred or more tiles.[13] In the eighteenth and nineteenth centuries, pattern and batch numbers were sometimes applied, and now and then bored painters doodled on the backs. Figure 18 is a rare early exception, on which the apparently religiously motivated artist drew a cross and the Christian IHS in retrograde—along with

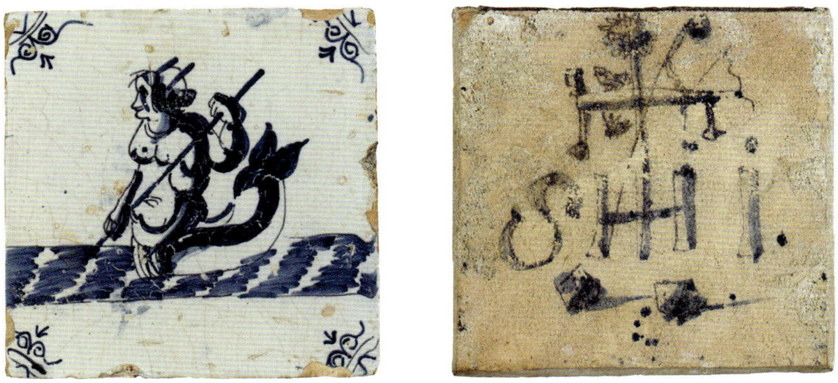

Figure 18 Tile, Holland, ca. 1660–1680. Delftware. 5.11" x 5.11" x .39". *Left:* The front shows a figure of uncertain sex who carries a spear and rides on a stylized strip of ocean. *Right:* The back is ornamented with enigmatic doodles, the largest being the letters *SHI*, which, if viewed in retrograde, would read *IHS* but for the *S* being reversed. The possible Christian symbolism is enhanced by reading the upper doodle as a cross drawn up out of the *H*. The fact that it is capped with a sunflower(?) and has one bird sitting on the lateral bar and another falling off it does not lend itself to such an interpretation, however. The two objects under the *SHI* might or might not be dice.

The front and back of this tile are poorly painted, the deep cobalt blue speckled and cracked beneath the glaze. The clay is pink with iron inclusions. The edges are sharply beveled, with two pinholes inset from the oxhead corners.

much else that now makes no sense. Another, perhaps unique, Dutch tile of about 1635 is inscribed "WIT SONDER QAERT," meaning without *kwaart*.[14]

The words provide a convenient segue to saying more about tile manufacturing. *Kwaart* in Dutch means to cover, in this case a covering of silver sand and lead oxide applied after painting to give the surface a bright, transparent lead glaze. By the mid-seventeenth century, however, the makers had discovered that by adding lead silicate to the underlying tin-enriched engobe, the second step could be avoided—hence the back notation "WIT SONDER QAERT." Experiments with different clays continued, but by the end of the century the early red-firing brick clays had been abandoned in favor of denser, yellow-firing clay. This thinner, first-fired (1,020°C) biscuit stage was then coated with the white engobe on which the decorator painted. That done, the tiles were refired standing on edge and back-to-back in a glost oven at a temperature of about 900°C. In theory the finished tile would emerge with both glaze and decoration bonded into the biscuit. But failures were frequent. Glost firing at too high a temperature causes the glaze to bubble and sag, distorting the painted design in the process. The image illustrated on page 253 of this volume (Beverly A. Straube, "A 'Wretched Tile' from Jamestown," fig. 1) demonstrates that misfires that should have been discarded as wasters reached Jamestown, Virginia, as scarcely acceptable seconds.

The first step in tile manufacturing called for flat sheets of the prepared clay to be air-dried to the leather-hard stage, at which time a square board edged with iron was laid on the clay as a sizing template. To prevent slipping on the still damp clay, small nails were driven through the corners of the template, creating tiny holes still visible in the corners of the finished tiles. Dutch authorities have stated that four securing nails set well in from the tiles' corners were the norm from about 1550 to 1650. The number of nails was later reduced to three or two and set closer to the edges. In the never-ending quest for dating criteria, one is prone to take the nail (or pin) hole feature as holy writ. Wrong again. When the holes were shallow, they could be obscured by the engobe. Three-hole examples occur in the second half of the seventeenth century, and two-hole ones on and off through the eighteenth. Some seventeenth-century tiles exhibit no holes at all. Nevertheless, it is fair to say that, in general, four holes in from the corners indicate an earlier date than fewer closer to the edges.

Seventeenth-century wall tiles usually measure about 5 by 5 inches, but some of the earliest, particularly those on red-firing clay, are slightly larger (5⅛ by 5⅛ inches). To achieve those dimensions there had to have been two stages of shrinkage, the first driving out 20 percent of the moisture from the leather-hard unfired clay, and the second during glost firing, together amounting to an 8 percent reduction. In the eighteenth and nineteenth centuries tiles became appreciably larger, averaging 6 by 6 inches (fig. 19).[15] The thickness was reduced from an average of .39 inches in the seventeenth century to .27 inches in the eighteenth.[16] The difference is a fairly reliable means of distinguishing between early and late—with emphasis on fairly.

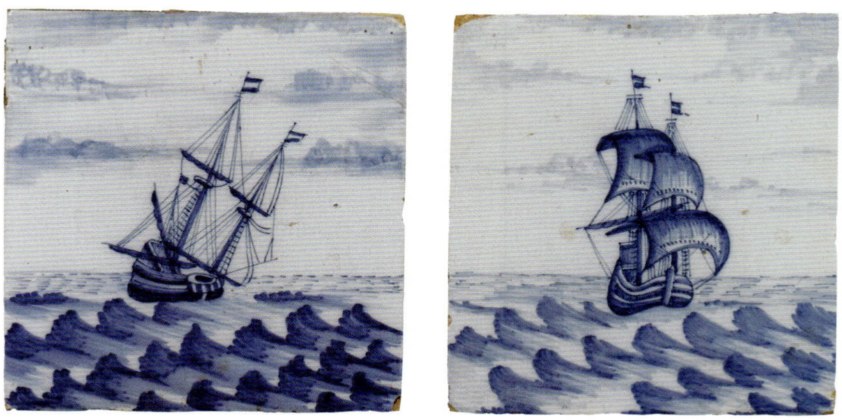

Figure 19 Tiles, Holland, 1850–1900. Approx. 6" x 6" x .27". Depicted is a pair of two-masted ketches, one with sails furled and the other under full sail; the foreground water is windswept, under realistically treated clouds. The clay is yellow with beveled edges; there are no pinholes or decorated corners.

From first to last the tiles' decorative designs were applied with the aid of stencils that enabled specks of charcoal to provide outlines. The painters then connected the dots and subsequently filled in the details by hand. However, I believe that an experienced tile painter could, and did, create the simpler mermaids and fish solely by eye. To my knowledge no pattern books from the seventeenth century survive, and it is uncertain who drew the designs that then became stencils. Some inspirations almost certainly came from ornamented sea charts, some from contemporary engravings, others from fish sold in the markets or salvaged from the memories of bibulous sailors. But more challenging are the multitude of figures holding incongruous objects, such as the old man brandishing two flaming torches (fig. 20). Is he, perhaps, Hephaestus, the Greek god of fire whose mother,

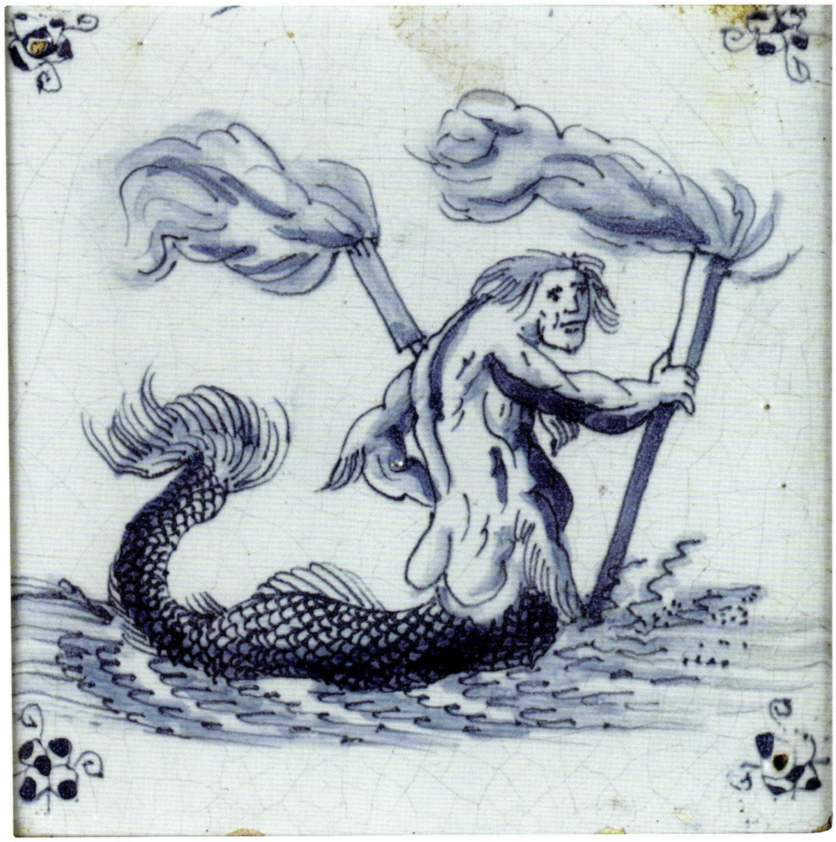

Figure 20 Tile, Holland, ca. 1650–1680. Delftware. 4.72" x 4.72" (thickness and body color obscured by a frame). A distraught merman is shown searching for something with a torch in each hand. He is anatomically odd in that he has a fin growing out of his right thigh and sagging buttocks resembling female breasts. The remainder of his piscine body, however, is very carefully drawn, as are the ripples around him. The tile has spider corners, two of which are pierced by pinholes.

Hera, threw him into the ocean? Or is he a brother who was said to have scared off assailants by waving firebrands? Who, too, is the person with the curious hat and hairstyle who appears to be squirting milk from his/her breasts (fig. 21)?[17] Then again, the dark-visaged and horned character wielding a club may be a sea devil or simply Neptune in a black and stormy mood (fig. 22).

Figure 21 Tile, Holland, ca. 1650–1690. Delftware. 4.96" x 4.96" x .19". A determined-looking merperson with an attenuated neck, straight hair, and a dish-shaped hat, is squirting water (or milk?) from his (her?) breasts. The sea is drawn with straight lines, and possibly there is a weed-covered rock in the foreground. This most unusual figure is painted in dark blue outline and filled in with two tones of blue. Oxhead corners and two pinholes. The glaze is thick, *kwaart*-enriched, and heavily crazed. The clay body is a pale pink and very sandy, with slightly beveled edges. This tile is probably a product of an inexperienced factory.

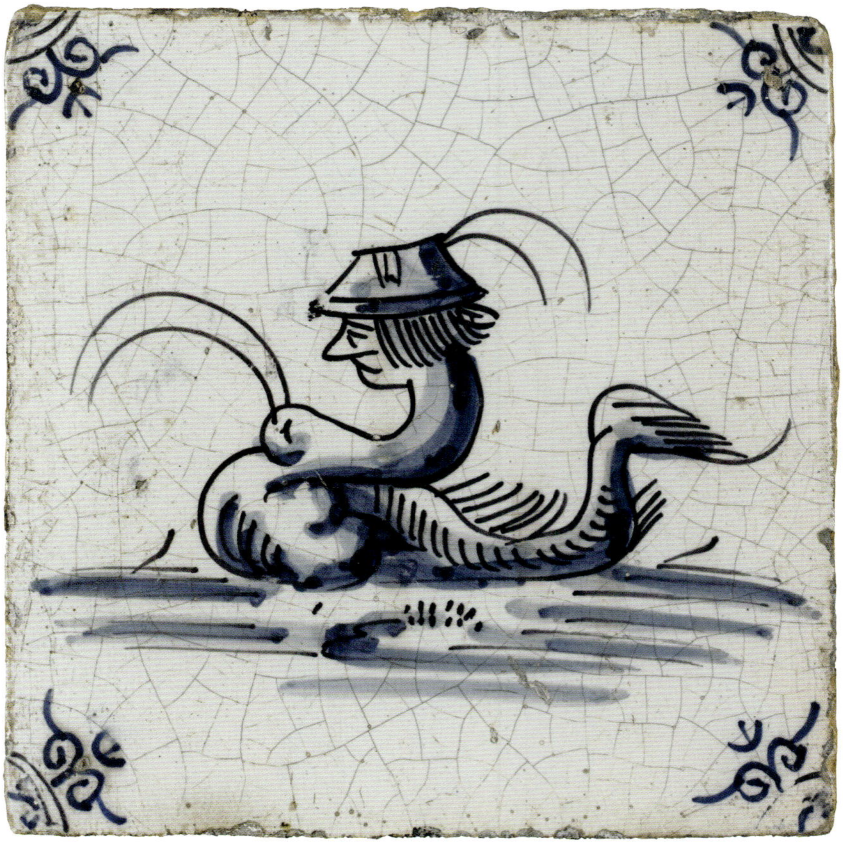

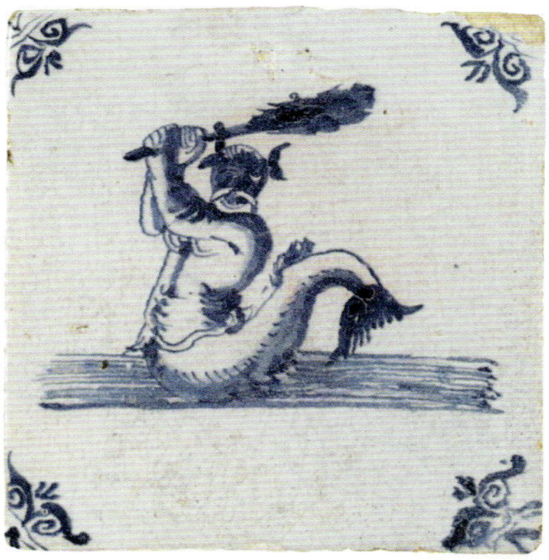

Figure 22 Tile, Holland, ca. 1680–1710. Delftware. 5.03" x 5.03" x .39". A demoniac merman with horns and blue-filled face is wielding a club or broom. This forbidding monster rides on a straight-line sea. The clay is yellow; oxhead corners and two inset pinholes. There is crazing in the engobe under the glaze.

There is, however, one classical figure about whom there can be no doubt. Although Aphrodite is remembered as the goddess of love—some of it sublime and some venal—she was born in the sea and stepped out of it onto the island of Cyprus. As goddess of the wind—and thereby of sailors—she is depicted standing on a pearl or a seashell with a sail billowing behind her (see fig. 9). Aphrodite is one of the few standard tile figures to occur on delftware plates (fig. 23). Mermaids, on the other hand, are more common, albeit divorced from any classical, mythological context (fig. 24). The illustrated tile dates from about 1720, by which time the

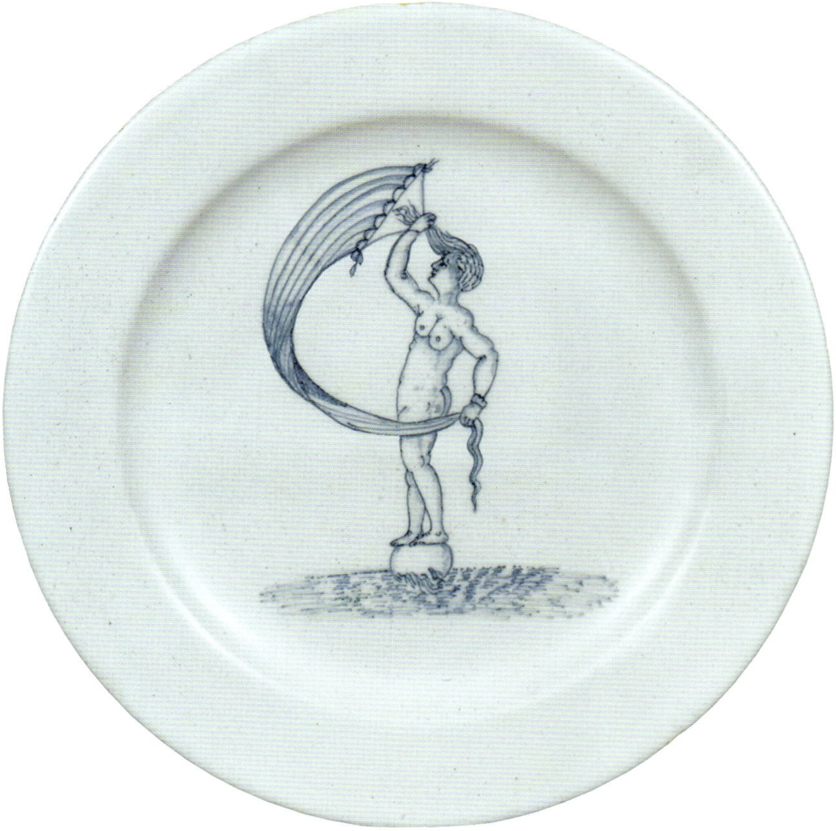

Figure 23 Plate, Holland, ca. 1650–1700. Delftware. D. 8.42". This shallow plate has a grayish glaze and is decorated with a standing figure of Aphrodite holding a billowing sail in the same hand that clutches a tress of her hair. She stands on a pearl in the midst of a calm sea. The back of the plate exhibits three sagger pin scars. See fig. 9 for a tile that shows a comparable image but of inferior quality.

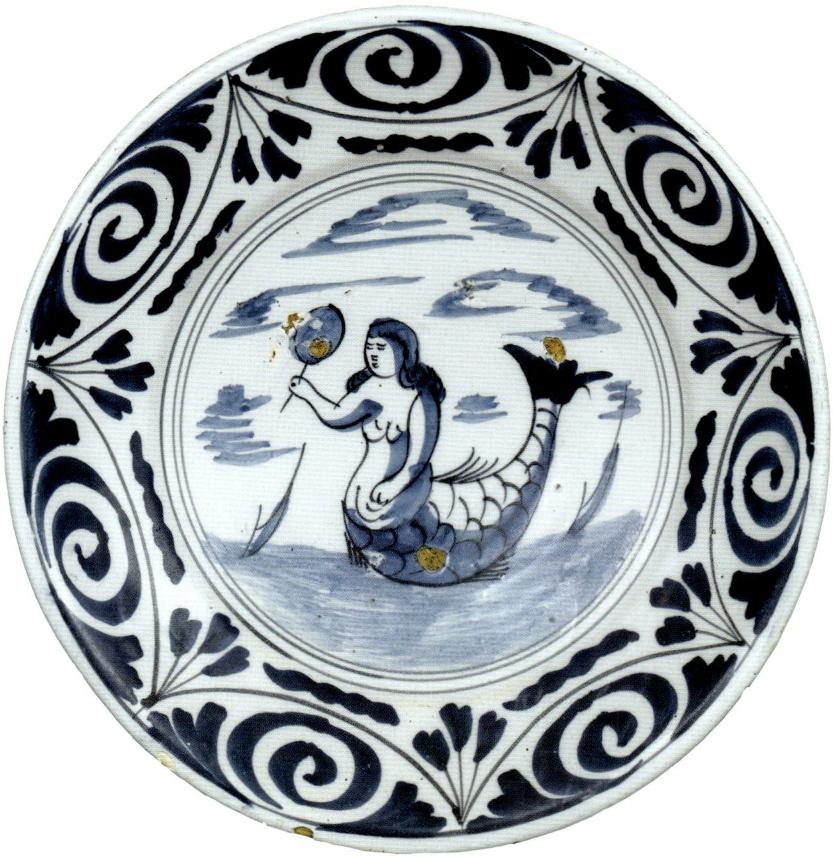

Figure 24 *Left:* Dish, probably North Netherlands, ca. 1690–1710. Delftware. D. 10.62". This dish is decorated with a mermaid in a standard pose flanked by stylized sailing vessels, painted in a rich, dark blue. In the center are triangular stilt scars, and the stepped back and foot ring are lead-glazed. *Below:* Tile, Holland, ca. 1720–1750. 5.11" x 5.11" x .23". Here the basic tile mermaid holds a mirror but no comb; chevron water treatment. The tile has spider corners and two pinholes close to the slightly beveled edges. Its tin glaze is a pale bluish gray and the design a weak blue, with some discoloration due to having been mounted inside a fireplace.

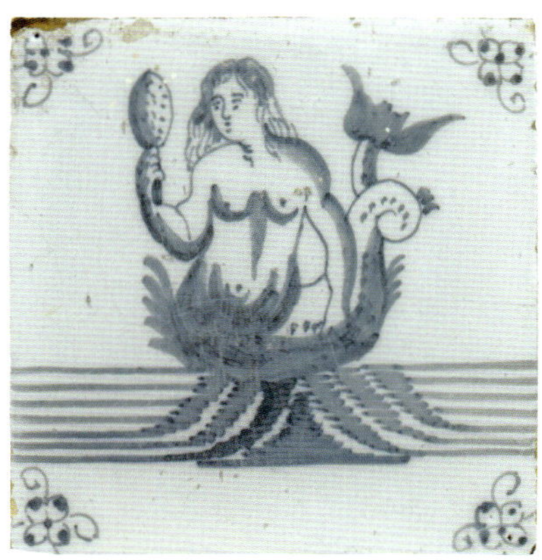

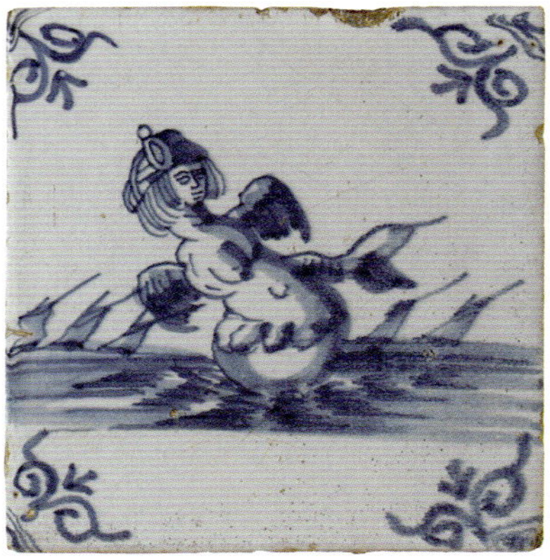

Figure 25 Tile, Holland, ca. 1700–1740. Delftware. 5.04" x 5.04" x .31". A winged but armless siren is wearing an elaborate cap and, below her waist, something that could be a very short skirt or four bovine udders. The figure is flanked by five sailboats, and the sea is a modified chevron treatment, presumably portraying a parting of the waves. The palette is a pale, somewhat diffuse blue. The edges are slightly beveled, and there are oxhead corners with two near-to-edge pinholes.

mirror-holding mermaid has settled into a conventional pose to be flanked by the kinds of distant sailing vessels frequently seen on eighteenth- and nineteenth-century sea- and landscape tiles (fig. 25).[18] Another demonstrates how it was possible for a mermaid to spin flax, with and without the help of feet to grip the distaff (figs. 26, 27).

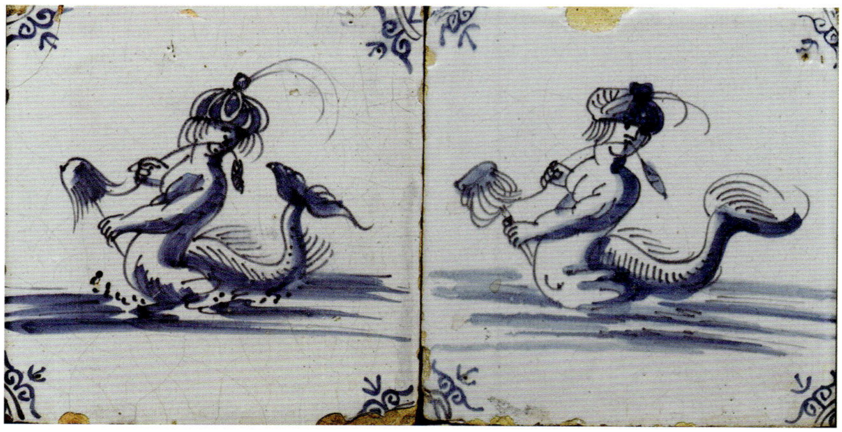

Figure 26 Tiles, Holland, ca. 1650–1700. Delftware; yellow clay. *Left:* 5.11" x 5.11" (thickness concealed by framing). *Right:* 5" x 5" (thickness concealed by a frame). Two treatments of the same subject, namely merpeople wearing elaborate bonnets and demonstrating how to spin flax without legs to grip the distaff (see fig. 27). The sea is treated as lateral stripes, one with spray and the other without it. There are dual pinholes set .59" in from the oxhead corners.

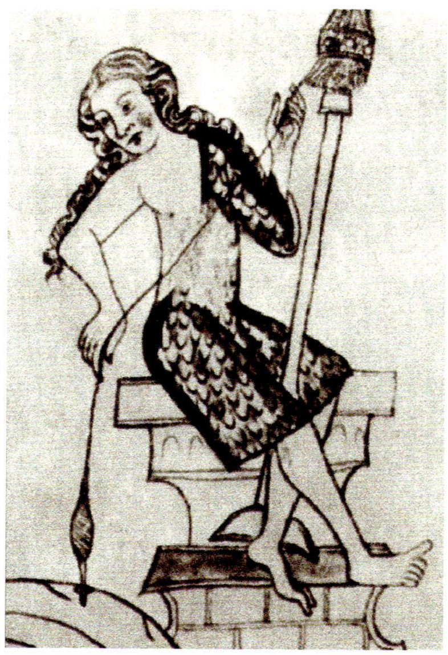

Figure 27 Woodcut, central Europe, ca. 1340. A woman spins with spindle and whorl, gripping the distaff between her thighs; at its top is a bundle of flax. Detail from the *Velislae biblica picta,* National Library of the Czech Republic, Prague.

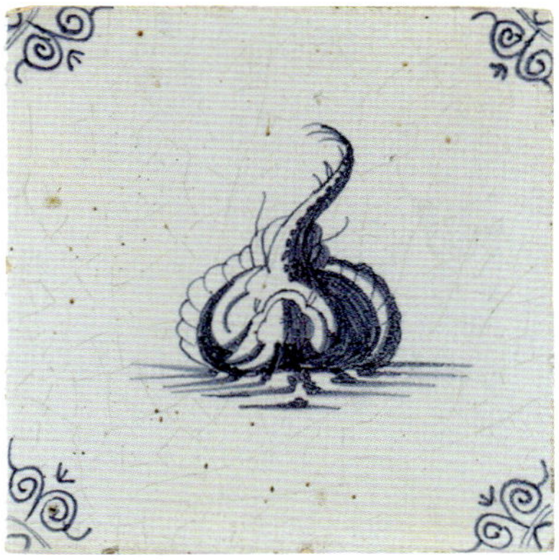

Figure 28 Tile, Holland, ca. 1700. Delftware. 4.92" x 4.92" x .33". Yellow clay with beveled edges; possibly *kwaart* sealed. The tile is decorated with a stingray in a bellicose pose, the glazed ground a pale gray and crazed beneath the surface. There are oxhead corners and only one visible corner pinhole.

As common as depictions of merpeople are those of aquatic mammals and fish—some of them are readily recognizable, such as eels, cod, swordfish, and militant stingrays (fig. 28), but more speak to the mythological idea that Poseidon/Neptune's underwater kingdom included every form of terrestrial life, among them giraffes, horses, unicorns, and an elephantine dragon (fig. 29).

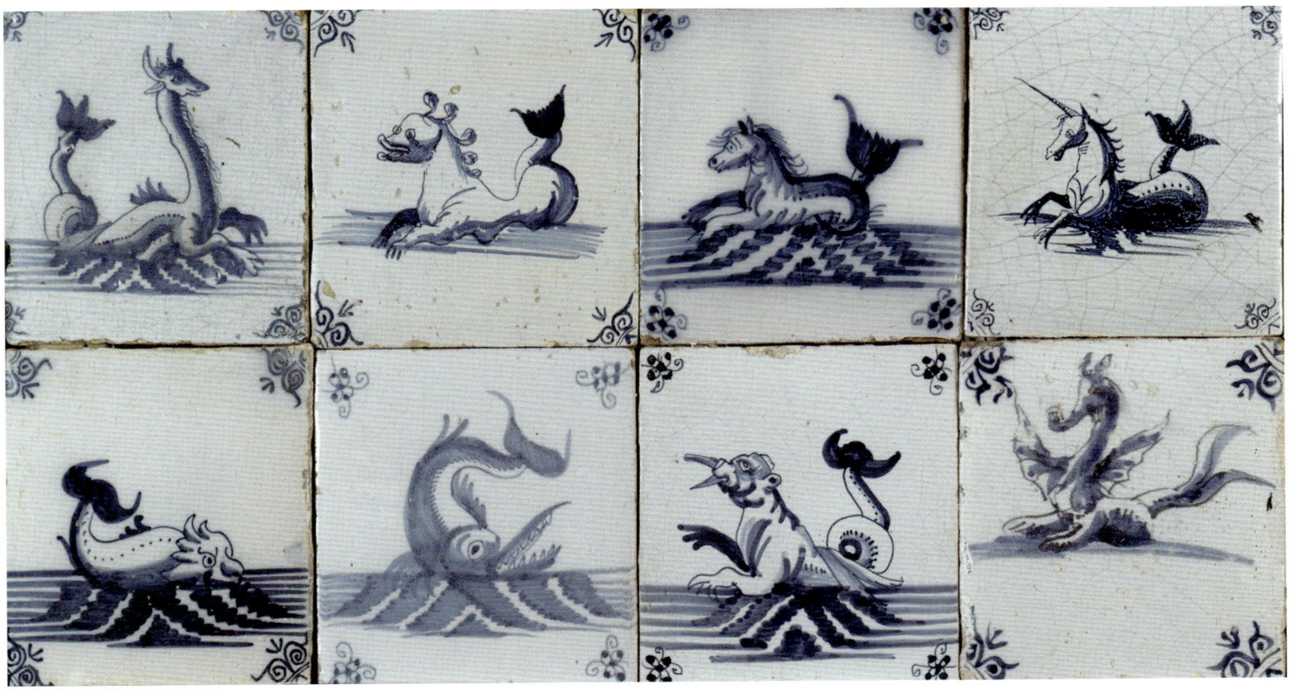

Figure 29 Tiles, Holland, 1640–1700. Delftware. 5"–5.11" x 5"–5.11" (thickness concealed by a frame). This eight-tile aquarium comprises a sea giraffe, two horses (one with braided mane), a unicorn, a dolphin, a shark, a walrus, and a winged dragon. Three of these yellow-body tiles have spider corners, the rest have oxheads; most have dual pinholes. The dragon has its pinholes far in from the corners (.98") and is arguably the earliest of the group.

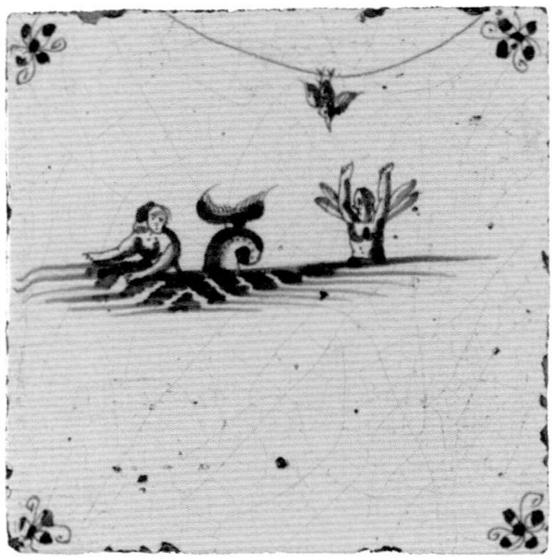

Figure 30 Tile, Holland, ca. 1680–1710. Delftware. 5.04" x 5.04" x .35". A dramatic scene at sea occupies only the upper half of this tile. A departing mermaid leaves behind a drowning female wearing a black bra and water wings. A bird, probably dead, dangles from an overhead loop of string. Two of the spider corners have pinholes set .39" in from them. The clay is yellow with slightly beveled edges; the glaze is crazed and tinted blue.

And then there are those that steadfastly defy identification. This collector has no idea what to make of a bird dangling from a celestial string above a surprised mermaid responding to a drowning person wearing water wings and a black bra (fig. 30). But, as I suggested at the outset, trying to figure out what each painter had in mind is the warp if not the weft of maritime tile collecting. Figure 31 suggests that its painter had tired of drawing standard monsters and so built his own, multi-propellant form—with fish body, semihuman head, bird wings, animal legs and feet, and a description-defying tail.

Figure 31 Tile, Holland, ca. 1680–1700. Delftware. 4.84" x 4.96" x .31". The head of this sea creature resembles several in the engravings of Peter Bruegel the Elder, with its pointed ears, spouting nose, and straight hair. The wings are like those of a griffin, the chest is scaled, the legs are somewhat doglike, and the tail is reminiscent of peacock feathers. A lower fin has a similarly scrolled terminal. The sea is calm and small craft sail on it, a detail not seen until the late seventeenth century. The corners show rudimentary oxheads; there are no pinholes. The pink body has slightly beveled edges; the glaze is tinted blue.

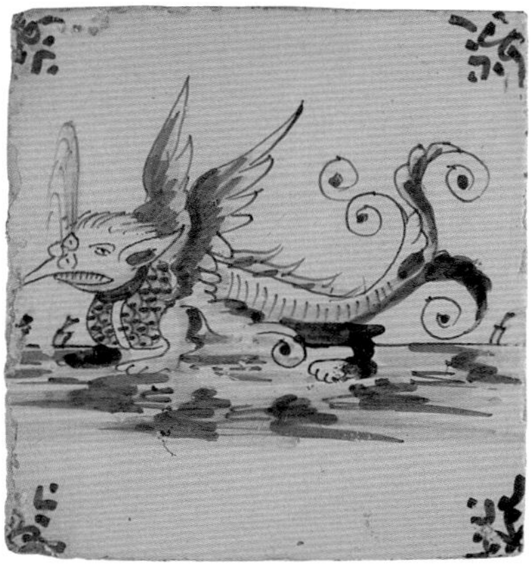

It is unfortunate that few tile-lined or tile-framed fireplaces survive in American homes, most having been stripped out when smaller stoves and grates were installed. There is, however, a splendid example of the use of maritime tiles surrounding a fireplace in the Warner House in Portsmouth, New Hampshire (fig. 32). The house was built in 1716–1718; if the tiles were new when installed they can provide a unique dating guide for the wide range of merpeople and creatures used there. One has to be reminded, however, that antique Dutch tiles are still traded and used by collectors; a

Figure 32 Tile-lined fireplace at the Warner House, Portsmouth, New Hampshire, built 1716–1718. These fireplace tiles, attributable to ca. 1680–1725, were probably new when they were installed. Those in the top three rows are not originally from this fireplace, although they are contemporary with it. (Courtesy, Warner House Association; photo, Douglas Armsden.)

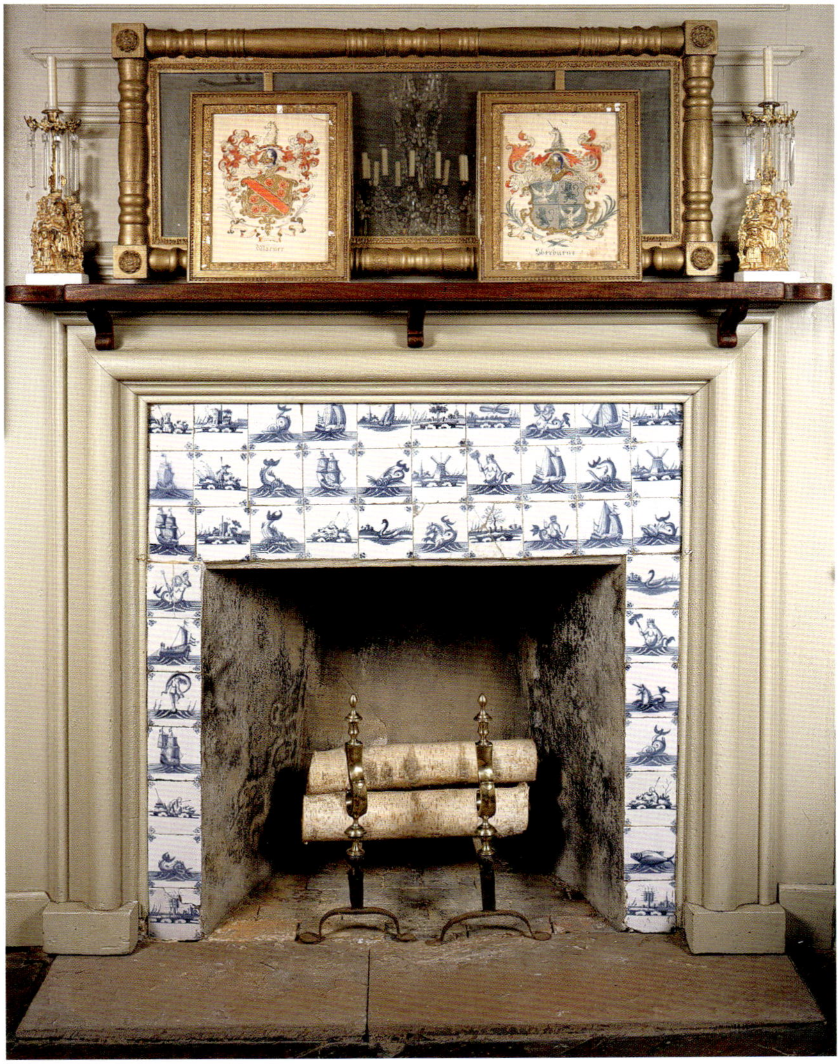

couple in Pennsylvania, for example, has used them to good effect in their "Dutch" sitting room (fig. 33). Others, such as myself, have had them reproduced to provide a satisfyingly aquatic theme for a bathroom (fig. 34) and a summer fire screen. Although the reuse of antique tiles would have pleased their seventeenth-century makers, modern antiquarian purists may

Figure 33 Original Dutch maritime and piscine tiles used in the re-created Dutch sitting room of a modern Pennsylvania home. (Photo, Ivor Noël Hume.)

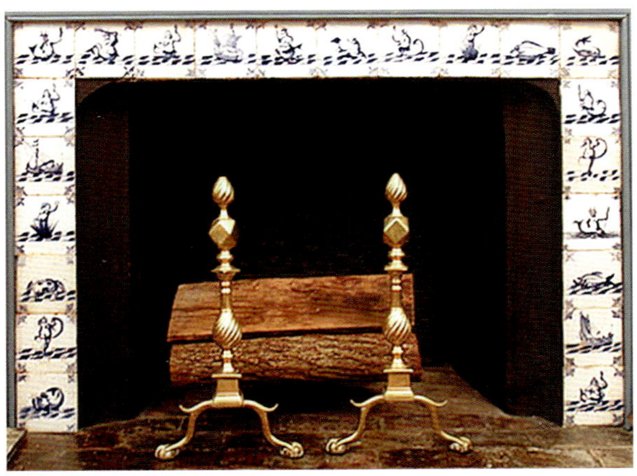

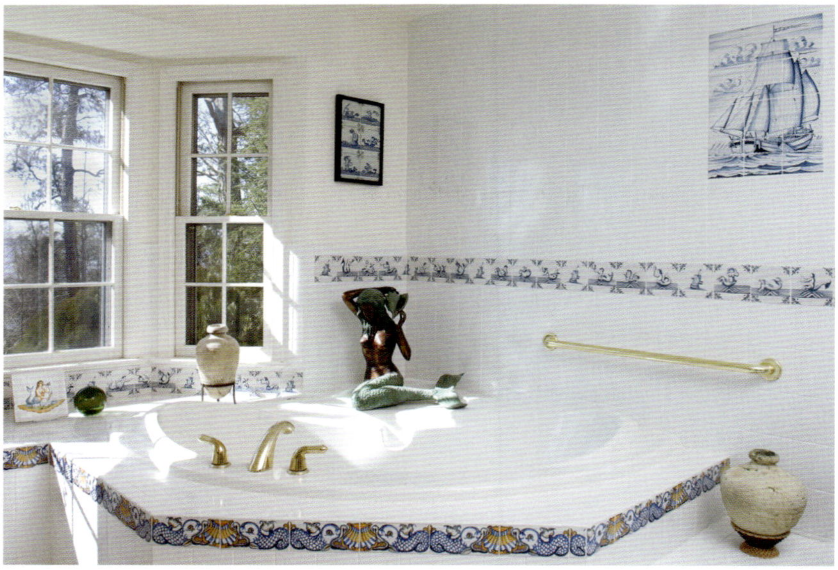

Figure 34 Bathroom at the home of the author, decorated with Dutch delftware tiles that were reproduced from originals in the collection. All are attributable to ca. 1650–1700, with the exception of the ship panel, which was copied from a ca. 1800 pair removed from a Dutch farmhouse.

be less enthusiastic. Indeed, there is much to be said for seeing the tiles as legitimate works of folk art, to be framed and enjoyed individually. Although few of us have the means to acquire a genre painting on canvas by David Teniers or a harbor scene by Willem van de Velde, a ship in full sail or a portrait of an amorous mermaid painted on clay can be ours for a relatively short song.

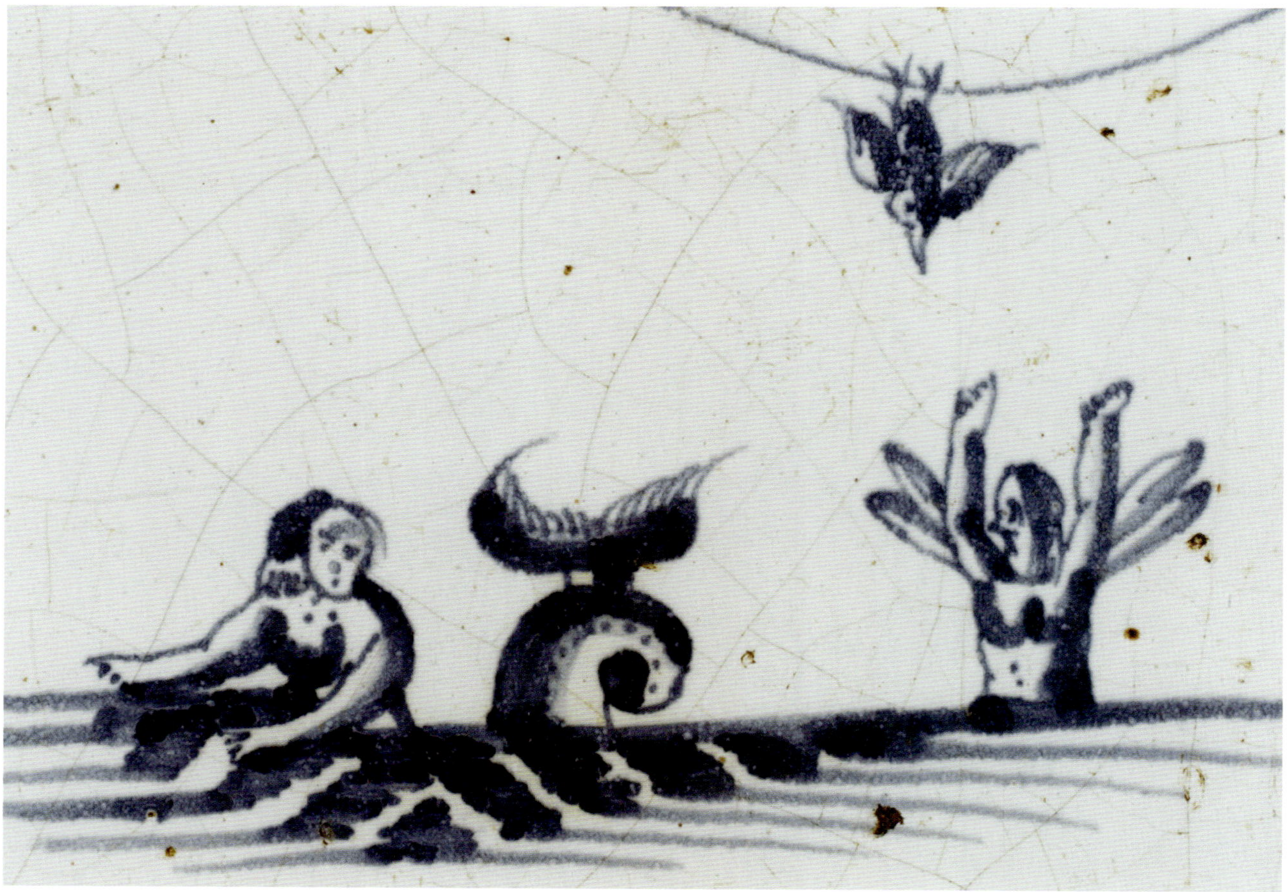

Figure 35 Detail of the tile illustrated in fig. 30.

ACKNOWLEDGMENTS

I want to express my thanks to tile authority Jan Holtkamp in the Netherlands, who kindly reviewed my captions. In some instances his dating is more precise than mine, and what follows are those wherein we differ: fig. 3 (right), 1600–1630; fig. 5 (left), ca. 1650; fig. 8, 1650–1700; fig. 11 (left), ca. 1650; fig. 11 (center), 1650–1660; fig. 11 (right), ca. 1650–1700; fig. 12, ca. 1650; fig. 13, ca. 1650; fig. 14 (left), 1650–1660; fig. 14 (center), ca. 1650; fig. 15, 1680–1700; fig. 16, 1670–1690; fig. 17 (left), ca. 1800; fig. 17 (right), 1680–1700; fig. 21, 1660–1680; fig. 22, 1660–1680; fig. 26 (left), 1660–1680; fig. 30, 1680–1700.

1. I am here using the term *delftware* with a lowercase *d* unless the tiles are known to have been made at Delft.

2. John Stow, *A Survey of the Cities of London and Westminster and the Borough of Southwark . . .*, 2 vols. (London: Printed for W. Innys, 1754–1755), 2: 327.

3. Rhoda Edwards, "London Potters circa 1570–1710," *Journal of Ceramic History* 6 (1974): 22. In 1676 a Dutchman named John Ariens van Hamme arrived in London and obtained a patent for making "Tiles and Porcelaine and othere earthen wares." He died three years later, and nothing is known about his products. Ibid., p. 115.

4. Other early corner designs are a spearhead-like device described in Holland as a "lily" but in English heraldry as a fleur-de-lis (fig. 3, right), and another called "fretwork." However, I have not seen either of those used on maritime tiles. For spiders, see fig. 5, left and center; for oxheads, see figs. 5, right, and 6.

5. Rita Susswein Gottesman, *The Arts and Crafts in New York, 1726–1776: Advertisements and News Items from New York City Newspapers* (New York: Printed for the New-York Historical Society, 1938), p. 90; Andrew Marschale, *New-York Mercury*, December 17, 1764. See my fig. 17.

6. Some eighteenth-century Dutch tile pictures required as many as four hundred, whereas the largest known English panel employed a mere ninety-six. Jonathan Horne, *English Tin-glazed Tiles* (London: J. Horne, 1989), p. 116.

7. By comparison with the painters of Renaissance majolica or the artists who would work on eighteenth-century delftwares, the tile painters were a pretty unassuming lot, their artistry at best described as naive. But therein lies its appeal. See fig. 34 for a reproduced, twelve-tile ship of about 1800.

8. John Potter, *Archaeologia Graeca, or The Antiquities of Greece*, 2 vols. (Edinburgh: Stirling and Kenny, 1832), 2: 82.

9. These are not to be confused with another sea creature, known as a hippocampus, which was half horse and half fish.

10. Ivor Noël Hume, "The Mermaid Mystique," in *In Search of This and That* (Williamsburg, Va.: Colonial Williamsburg Foundation, 1996), p. 89.

11. Eighteenth-century London shopkeepers were happy to sell both together. On August 17, 1745, Hannah Ashburner at the Sign of the Rose at the corner of Fleet-Bridge, advertised that her stock included "English & Dutch Tyles." Aubrey J. Toppin, "The China Trade and Some London Chinamen," *English Ceramic Circle Transactions* 3 (1935), pl. XIX, opp. p. 45.

12. Even this is not an infallible rule. In my collection is a pair of printed Liverpool tiles (ca. 1775–1780) depicting the difference between the plump British butcher and his emaciated French competitor; one has right-angled edges and the other chamfered. To further muddy the waters, beveled edges occur on some very early Dutch tiles.

13. Horne, *English Tin-glazed Tiles*, p. 119.

14. Jan Pluis, *De Nederlandse Tegel: Decors en benamingen 1570–1930* (Leiden: Nederlands Tegelmuseum, 1997), p. 94, fig. 88b. See also Dingeman Korf Sr., "Wat doen we met onze tegels," *Antiek* (November 1966): 12–14.

15. In a letter of 1677 to Viscount Conway, Dr. William Johnson wrote about a supply of "Holland" tiles: "The way of sale of them is so much a foot, counting always four tiles to the foot, 10 inches making the Dutch foot." Edwards, "London Potters circa 1570–1710," p. 115, citing the *Calendar of State Papers, Domestic Series, of the Reign of Charles II*, 28 vols. (London: Longman, Green, 1860–1939), vol. 19 (March 1, 1677–February 28, 1678), p. 289.

16. The thickness of paving tiles from the late sixteenth and early seventeenth centuries could be as much as .59 inches.

17. Dutch hats and caps of the seventeenth century came in a great variety of shapes, but this does not appear to be one of them.

18. Several merperson plates were made in London (Lambeth) delftware about 1720–1730 (see, e.g., Noël Hume, "The Mermaid Mystique," p. 90), but the figure drawing is reminiscent of late Adam and Eve chargers. Another found in excavations at London Towne in Maryland is akin to the representations of Georgian kings. Fragments of three such plates were found in a tavern site, prompting one to speculate that the tavern or inn was named the Mermaid. The figure is more Tritonesque than mermaid, lacks both comb and mirror, and wears an air of surprised puzzlement. Al Luckenbach, "Ceramics from the Edward Rumney/Stephen West Tavern, London Town, Maryland, Circa 1725," in *Ceramics in America,* edited by Robert Hunter (Hanover, N.H.: University Press of New England for the Chipstone Foundation, 2002), p. 145, fig. 25.

Merry Abbitt Outlaw

New Discoveries Introduction

▼ IT WAS *not important that they survive.*
What mattered was that they should bear
Some lineament or character,

Some affluence, if only half-perceived,
In the poverty of their words,
Of the planet of which they were part.

Wallace Stevens, "The Planet on the Table"

The great twentieth-century American poet Wallace Stevens wrote those lines about poetry, but they could easily speak to the ceramics discussed in this year's New Discoveries. In this issue we learn of several survivors from our past and why they endured. Their importance lies in what they tell us of their worlds.

Al Luckenbach, Susan Tunick and Jay Shockley, and Bly Straube write about architectural ceramics dating from the mid-seventeenth to the mid-nineteenth centuries. From them we learn about decorative two-dimensional tiles made for interior spaces, as well as elaborate, relief-molded terracotta elements manufactured for building exteriors. All are rare witnesses to America's architectural past and testify to the role of clay in providing both embellishment and structural support.

Kathryn Boxhorn, Rob Hunter and Hank Lutton, John Kille, and Ron Napier report on utilitarian ceramic survivors that represent the widely varied tastes and values of nineteenth-century Americans. Why did these containers—among them an archaeological example, a stoneware bottle already of considerable age when it was left behind in a Civil War hut—remain intact for so long? Was it to do with the type of ware, or the shape, or perhaps even the inscription? All of these vessels are marked in some way, and their markings reveal much about the worlds from which they came.

Two contemporary American clay artists are also represented this year. Warren Bakley and Gerald Thiebolt have been working with clay for many years as associates in a San Diego clay studio. Strongly influenced by their backgrounds in design and architecture, their unique works are created to enhance interior and exterior spaces. Unlike the architectural examples from our past, their ceramics speak to us of the now. But in times to come, it is hoped that their sensibilities will continue to speak "of the planet of which they were part."

Robert Hunter and Hank D. Lutton

A Yankee Jug in Dixie

In almost all instances, when utilitarian ceramic vessels survive the passage of time we lose the history of their everyday use. We may ask many questions about a simple stoneware jug—Who were the original owners? How was it used? Did its use change through time? Why did it survive ?—but expect few answers. However, an opportunity to address some of these questions arose from the discovery of an early-nineteenth-century New Jersey example (fig. 1).

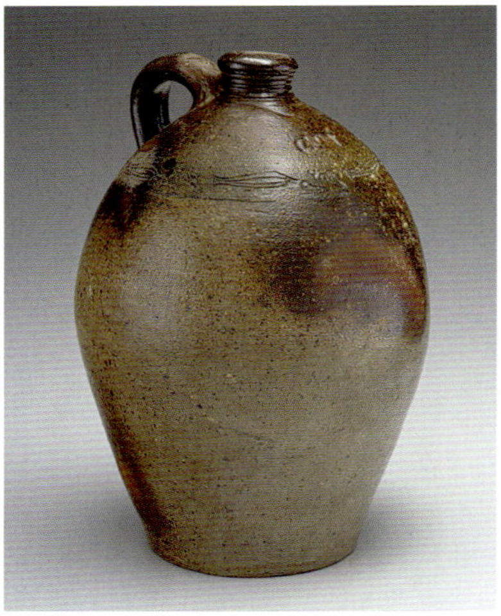

Figure 1 Jug, Old Bridge, New Jersey, 1805–1827. Salt-glazed stoneware. H. 12". (Courtesy, York County Historical Museum; photo, Gavin Ashworth.)

The jug was found in 1989 during archaeological excavations of a Civil War encampment in Yorktown, Virginia, prior to the development of a housing complex. Archaeologists from the James River Institute for Archaeology uncovered a roughly rectangular feature filled with dark brown and gray sandy loam that stood out against the surrounding orange clay subsoil (fig. 2). Further excavation revealed a small, semi-subterranean "dugout" dwelling no more than 7.5 feet long and 6 feet wide with a partially intact brick chimney projecting from one end (fig. 3). The feature was interpreted as the remains of a Civil War–era winter hut.[1]

Among the artifacts recovered in the fill of this feature is an intact one-gallon salt-glazed jug with rouletted decoration and impressed with the word "GIN" (fig. 4). The vessel is attributed to one of the potteries in Old Bridge, New Jersey. The potting operations at Old Bridge date as early as 1805, when James Morgan Jr., Jacob van Wickle, Nicholas van Wickle, and Branch Green established a "Manufactory of Stone Ware," which continued under various owners to at least 1823 and possibly as late as 1827. The

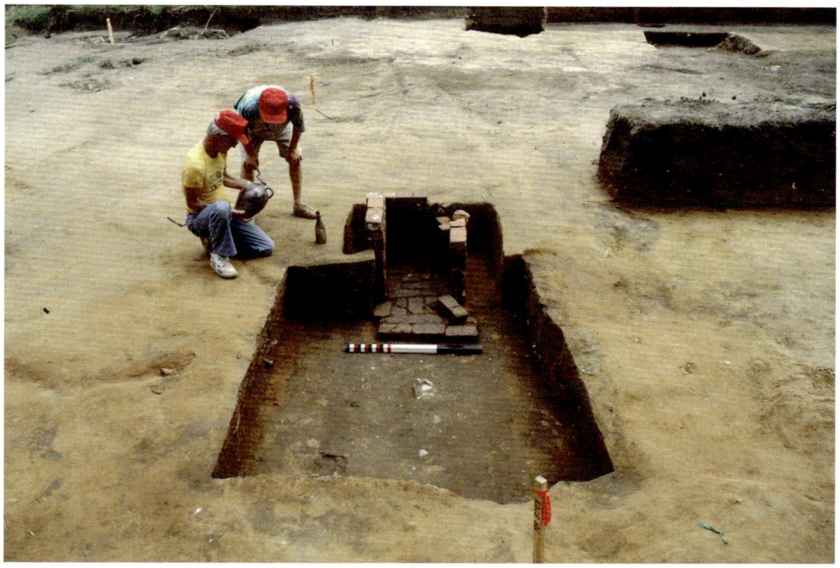

Figure 2 Overview of the excavated Civil War hut with excavators examining the recovered New Jersey stoneware jug. Note the firebox in the upper part of the feature. (Photo, James River Institute for Archaeology.)

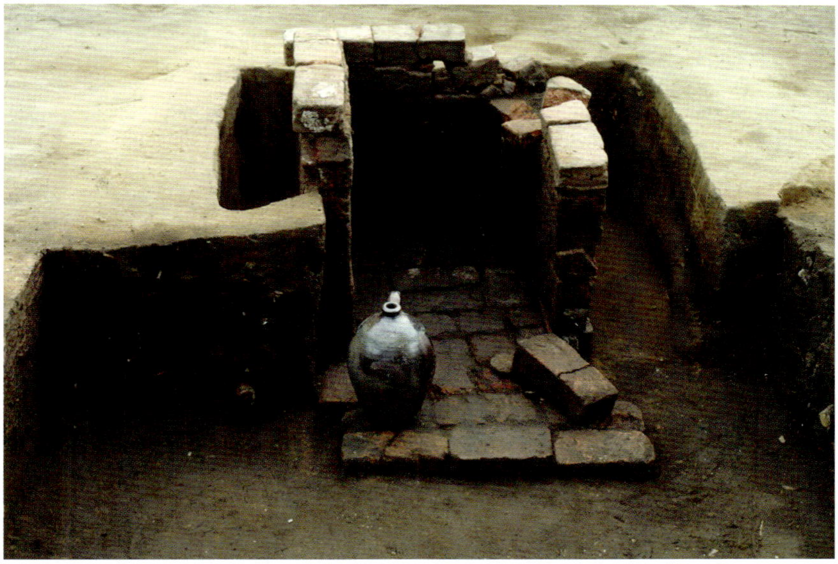

Figure 3 Detail of the hearth base and firebox of the hut illustrated in fig. 2. (Photo, James River Institute for Archaeology.)

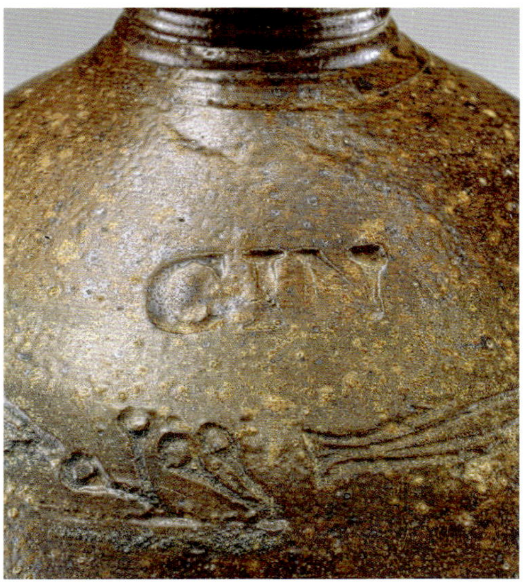

Figure 4 Detail of the impressed "GIN" mark and the coggle wheel decoration of fish and berries on the jug illustrated in fig. 1.

stoneware is distinguished by fish and berry coggle wheel ornamentation, the decoration seen on the excavated example.[2] For comparison, a similar antique jug, without any history and from the same New Jersey pottery, is shown in figures 5 and 6.

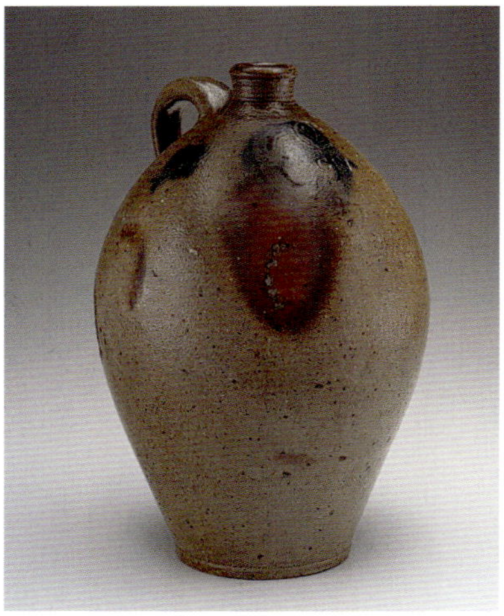

Figure 5 Jug, Old Bridge, New Jersey, 1805–1827. Salt-glazed stoneware. H. 14". (Private collection; photo, Gavin Ashworth.)

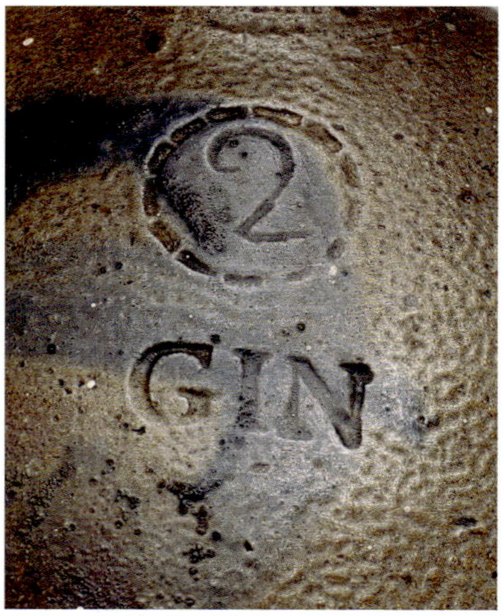

Figure 6 Detail of the impressed marks on the jug illustrated in fig. 5.

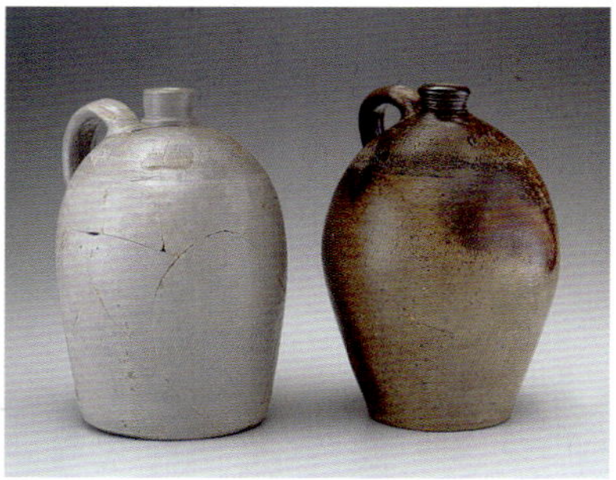

Figure 7 Stoneware jugs. *Left:* Baltimore, ca. 1860. H. 11½". *Right:* Old Bridge, New Jersey, 1805–1827. H. 12". (Courtesy, York County Historical Museum; photo, Gavin Ashworth.) Both of these complete jugs were found in the excavation.

The historical record of Civil War activity in Yorktown suggests that the hut was built and occupied by Confederate forces during the winter of 1861–1862 and subsequently occupied and/or backfilled by newly arrived Union troops in May 1862. The Union troops occupied Yorktown through July 1864. Which troop might have owned the jug is unclear in the archaeological record, but the best inference suggests a Union origin.

Most interesting about the jug, which was made in the 1820s or possibly even earlier, is that it had been in active use for at least forty years. In contrast to the newfangled cylindrical shape of another one-gallon jug (fig. 7, left), found in the same feature and probably made in Baltimore after 1860, the graceful ovoid form of the New Jersey gin jug (fig. 7, right) was certainly out of style.[3] In addition, the base of the jug exhibits quite a bit of heavy wear, indicative of constant use.

It is tempting to speculate as to the contents of either jug in the military context. Did the New Jersey jug actually contain a winter's ration of gin, or did it merely hold cooking oil? Perhaps future residue analysis can provide additional fodder for both military and ceramic historians to ponder. For now, the simple artifact provides confirmation of the relative long-term use of an everyday object—in contrast to the disposable orientation of twenty-first-century America.

Robert Hunter, Editor, *Ceramics in America;* somerwell@aol.com
Hank D. Lutton, Department of Archaeology, Boston University; hdlutton@bu.edu

1. Garrett R. Fesler, Matthew R. Laird, and Hank D. Lutton, "'Beautiful Confusion': The Archaeology of Civil War Camp Life in an Urban Context," in *Huts and History: The Historical Archaeology of Military Encampment during the American Civil War,* edited by Clarence R. Geier, David Gerald Orr, Matthew B. Reeves (Gainesville: University Press of Florida, 2006).

2. M. Lelyn Branin, *The Early Makers of Handcrafted Earthenware and Stoneware in Central and Southern New Jersey* (Rutherford, N.J.: Fairleigh Dickinson University Press, 1988), pp. 97–98.

3. See John E. Kille, "Distinguishing Marks and Flowering Designs: Baltimore's Utilitarian Stoneware Industry," in *Ceramics in America*, edited by Robert Hunter (Hanover, N.H.: University Press of New England for the Chipstone Foundation, 2005), p. 115.

John E. Kille

Mixing Religion and Drink: Personified Expressions in Rockingham Glaze

The recent discovery of two unusual, inscribed yellow ware vessels gives credence to the old adage that books should not be judged by their cover—and suggests that their titles should be carefully read, too.

An intriguing Rockingham-glazed "man-on-a-barrel" spirits flask and a novelty bank in the shape of a jug appear to depict merriment and the consumption of alcohol (fig. 1). However, the inscriptions on the flask and the bank indicate that these objects were more than conventional representations of familiar forms. Given the period in which they originated, it seems likely that both were made by potters responding to the proselytizing of religious missionaries active in the United States and England during the second half of the nineteenth century.

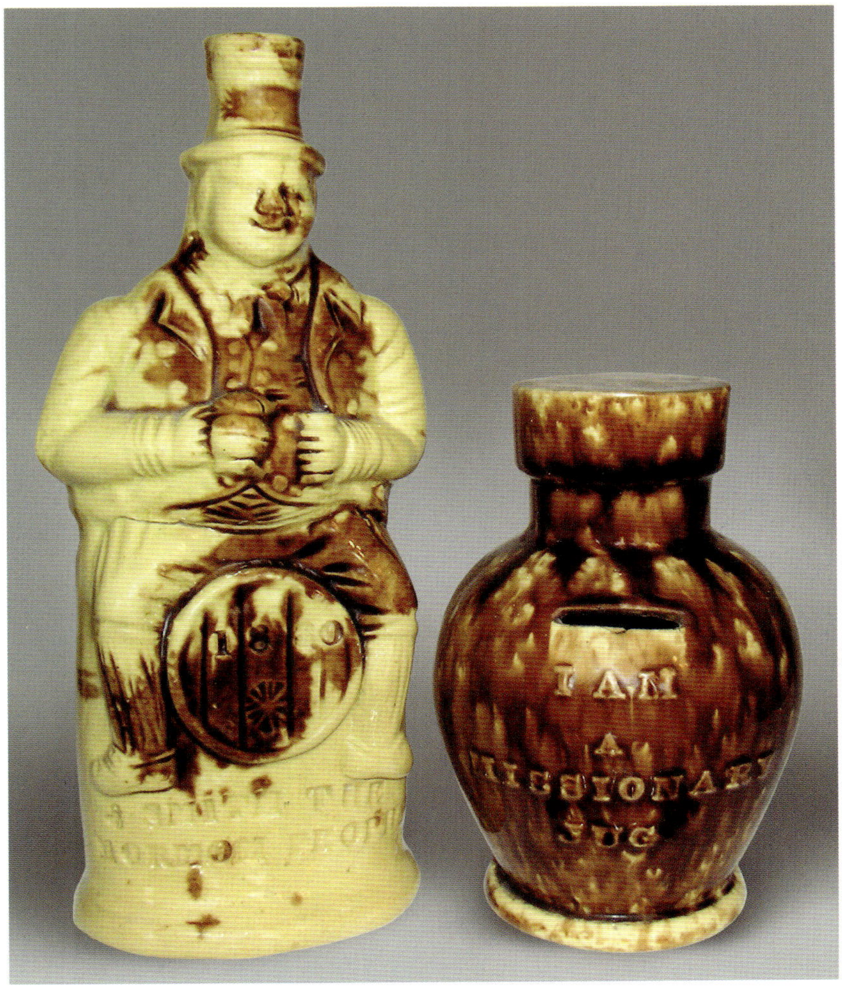

Figure 1 Flask and bank, mid-nineteenth century. Rockingham-glazed earthenware. *Left:* Probably English. H. 9". *Right:* English or American. H. 5½". (Author's collection; photo by author.)

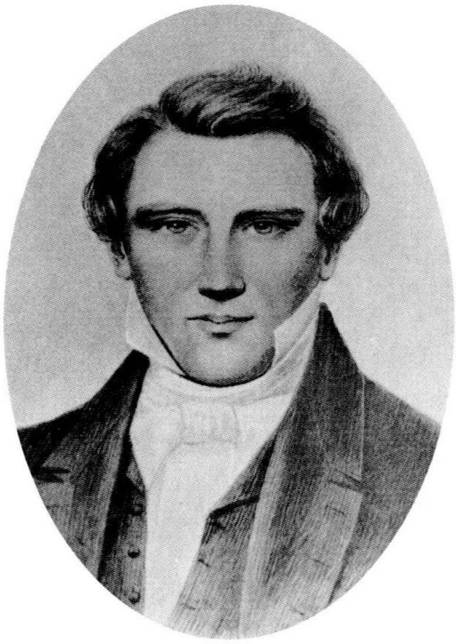

Figure 2 Charles William Carter, *Joseph Smith Jr.* Etching, 1886. 5" x 7". (Courtesy, Church Archives, The Church of Jesus Christ of Latter-day Saints.)

Each vessel bears a written message applied by the maker in order to advocate a particular cause or point of view, which is highly unusual for Rockingham-glazed ware. Even though the messages deal with serious matters, they are presented with veiled humor or irony.

The impressed inscription "J. Smith/The Mormon Prophet/1830" seems a straightforward commemoration of Joseph Smith Jr., a religious visionary from Palmyra, New York, who in 1830 organized the Church of Jesus Christ of Latter-day Saints and published the revolutionary *Book of Mormon* (fig. 2). That the potter inscribed the message on a toper flask, however, makes clear the intent to lampoon rather than commemorate. The piece is a direct assault on "The Word of Wisdom," a revelation Smith purportedly received at Kirtland, Ohio, in 1833, which advised against the consumption of wine, tobacco, hot drinks, and strong drinks.

Revered by millions today, this influential religious leader and his followers were driven from several Midwest towns prior to settling in Salt Lake City, Utah. The open hostility and violence they experienced along the way culminated in 1844, when Smith and his brother Hyrum, who were imprisoned in a county jail in Carthage, Illinois, were brutally shot and killed by an angry mob (fig. 3). This flask is a lasting reminder of resistance to the controversial teachings of Smith and his Mormon missionaries, many of whom sought converts in England who would relocate to Utah.

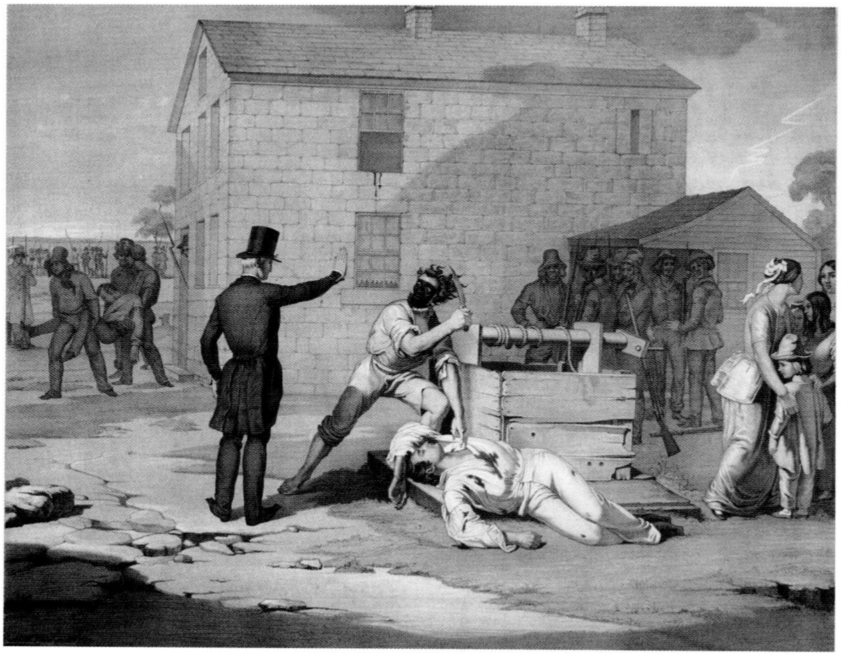

Figure 3 Lithograph by Nagel and Weingaertner after C. G. Crehen, *Martyrdom of Joseph and Hiram Smith in Carthage Jail, June 27, 1844* (New York, 1851). (Courtesy, Library of Congress, Prints and Photographs Division.)

Another potter created the unusual Rockingham-glazed bank using an ovoid jug form and applying to it the words "I AM A MISSIONARY JUG" with molded, raised letters. In contrast to the flask, this self-promoting

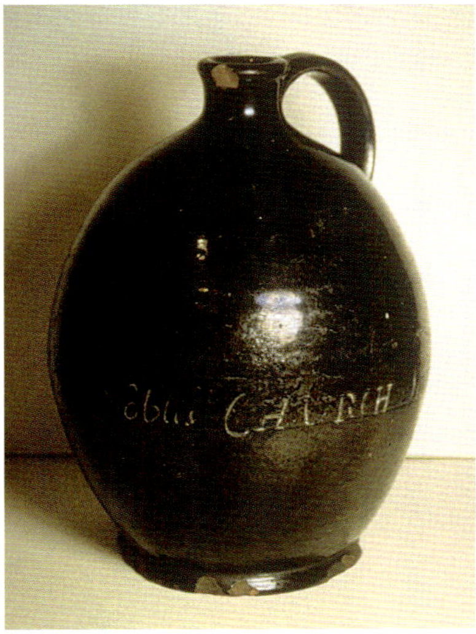

Figure 4 Communion jug, attributed to Daniel Bayley, Newburyport, Massachusetts, 1763. Manganese-glazed earthenware. H. 10½". Inscribed: "For the use of Mr. Noble's CHURCH July 10, 1763." (Courtesy, Old York Historical Society, York, Me.) This jug—one of the earliest dated pieces of American redware—was used to store wine for communion services at the Fifth Parish Congregational Church of Newbury, headed by Reverend Oliver Noble. It is the only documented example of eighteenth-century ecclesiastical redware produced in New England.

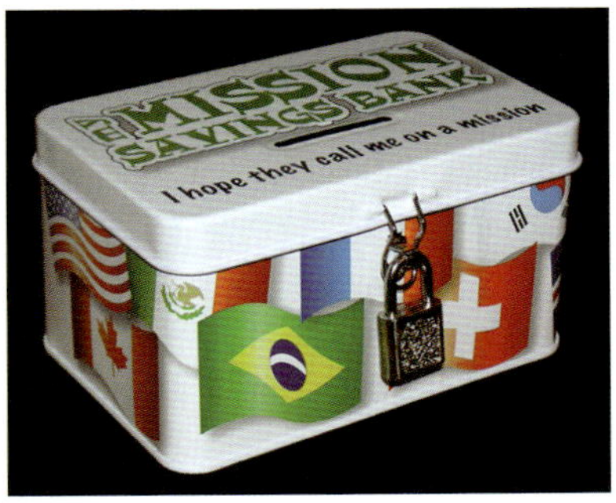

Figure 5 Bank, 2006. Tin. H. 4". Mark: on lid, "MISSION SAVINGS BANK / I hope they call me on a mission"; on underside, "Made in China." This modern version is currently being sold at the Brigham Young University Book Store in Salt Lake City, Utah. (Author's collection; photo by author.)

money holder was personified in order to collect funds in support of the outreach of religious missionaries, who often traveled to distant lands. Appropriating a jug form for this benign purpose was not necessarily controversial, as these vessels were used for nonalcoholic liquids as well as spirits; they were also used by churches to hold communion wine (fig. 4).[1]

People of all ages, backgrounds, and religious denominations during this period actively supported the charitable work of missionaries; children in particular were encouraged to set aside a portion of their earnings to help these workers. Even Abraham Lincoln's son, Willie, who was known for his thoughtful and studious nature, reportedly donated his allowance to a missionary society.[2] Traditional missionary savings banks continue to be made today, although mass-produced metal has replaced handcrafted pottery (fig. 5).

It should be noted that nineteenth-century Rockingham-glazed vessels associated with religious matters are relatively rare (neither of the objects under discussion is in any of four exhaustive collector guides covering a wide range of yellow ware forms).[3] One notable exception is the well-known Rebecca-at-the-Well teapot. Based on a biblical story, this popular motif was created by potter Edwin Bennett of Baltimore and later copied by several East Coast competitors.[4]

Different types of ceramic vessels with human forms and characteristics have existed for centuries, and many were intended to personify, commemorate, or satirize a wide range of historical figures.[5] Potters in general are infamous for interjecting crass humor into their craft. The more celebrated examples include dribbling puzzle jugs, mugs with frogs, chamber pots containing excrement, and, of course, vessels depicting the English drinking character Toby Fillpot.

The two Rockingham-glazed novelty items shown here offer commentary on the work of religious missionaries during the nineteenth century and reflect how potters artfully participated in social discourse during this period. In turn, while these vessels no longer retain their intended function, they do continue a dialogue that should now be considered historically significant.

John E. Kille, Assistant Director, The Lost Towns Archaeology Project; jkille@aacounty.org

1. John Kille, "Fire the Kiln, Raise the Flag, and Praise God: The Bayley Jug and the Convergence of Craftmaking, Politics, and Religion in Eighteenth-Century New England," in *Piecing It Together: Ceramics and the Social History of Early New England* (York, Me.: Old York Historical Society, 1997), pp. 25–42. The subject of this study is the rare earthenware communion jug illustrated in fig. 4 of this article.

2. Philip B. Kunhardt Jr., Philip B. Kunhardt III, and Peter W. Kunhardt, *Lincoln: An Illustrated Biography* (New York: Random House, 1992), p. 290. Willie died of typhoid fever in 1862 at age eleven.

3. See Joan Leibowitz, *Yellow Ware: The Transitional Ceramic* (Atglen, Pa.: Schiffer Publishing, 1985); Lisa S. McAllister and John L. Michel, *Collecting Yellow Ware* (Paducah, Ky.: Collector Books, 1995); Lisa S. McAllister, *Collector's Guide to Yellow Ware Book II* (Paducah, Ky.: Collector Books, 1997); and Lisa S. McAllister, *Collector's Guide to Yellow Ware Book III* (Paducah, Ky.: Collector Books, 2003). A similar flask is pictured in McAllister and Michel, *Collecting Yellow Ware,* p. 101. However, it is incised with the name "JIM CROW," a blackface

character performed by American entertainer Thomas Dartmouth Rice, who brought his act to English theaters in 1836.

4. Jane Perkins Claney, *Rockingham Ware in American Culture, 1830–1930: Reading Historical Artifacts* (Hanover, N.H.: University Press of New England, 2004), pp. 81–84, in which the popularity and successful marketing of Rebecca-at-the-Well teapots are explored.

5. Ivor Noël Hume, *If These Pots Could Talk: Collecting 2,000 Years of British Household Pottery* (Milwaukee, Wis.: Chipstone Foundation, 2001), pp. 301–27. This chapter provides an excellent discussion of nineteenth-century English portrait and Toby flasks, as well as illustrations of several of these vessels.

Al Luckenbach

Painted Tiles from Tulip Hill, Anne Arundel County, Maryland

The use of decorated tin-glazed earthenware tiles to decorate the interior of European dwellings was quite fashionable from the late sixteenth through the mid-eighteenth centuries. As is often depicted in period genre paintings, these tiles were used both as baseboard or washboard treatments and, most frequently, as decorative fireplace surrounds. Similar "gally paving tyles" were used as decorative flooring stretching back to medieval times. By the eighteenth century, tin-glazed earthenware tiles "for chimnies" were being produced, principally in England, Holland, Italy, and Portugal, in countless stylistic varieties.

The fashion of decorating with tin-glazed earthenware tiles was transported to the Chesapeake region with the first colonists. Archaeological examples have been recovered from a number of colonial town sites and outlying plantation sites in the Chesapeake, including several locations in Virginia—Jamestown, nearby James City County, and Williamsburg—and in St. Mary's City, Maryland. In Anne Arundel County, Maryland, tiles have been excavated at both Providence and London Town (fig. 1).[1]

Despite apparent widespread use, extant buildings in this country with in situ fireplace tile treatments are surprisingly rare. Four examples are known from Charleston, South Carolina, and several from areas farther north—Marblehead and Portsmouth, Massachusetts; Newport, Rhode Island (fig. 2); West Coxsackie, New York; and Philadelphia.

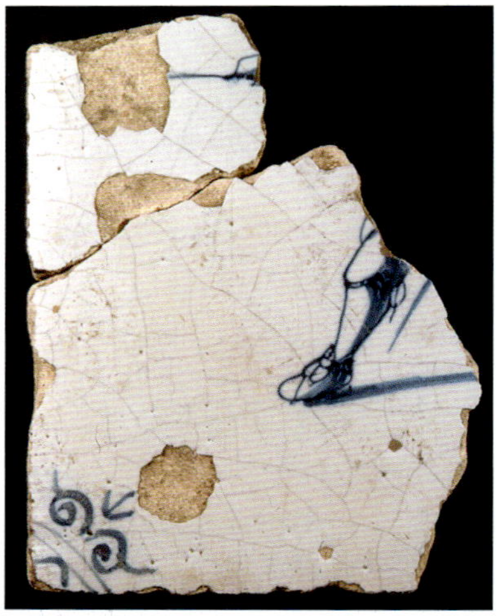

Figure 1 Tile fragment, Holland, ca. 1650. Tin-glazed earthenware. H. 3½". (Courtesy, Anne Arundel County; photo, Al Luckenbach.) Pikeman tile recovered from Burle's Town Land (18AN818).

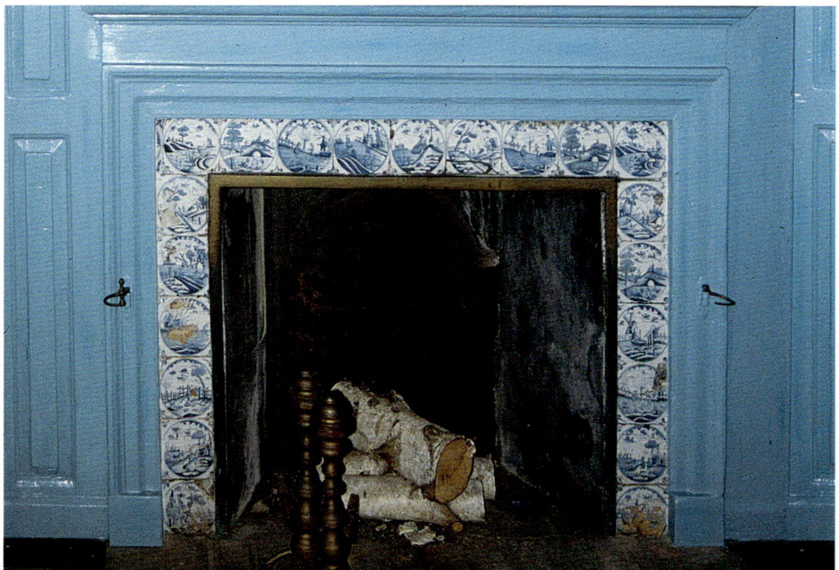

Figure 2 Fireplace with hearth surround of delftware tiles at the Wanton-Lyman-Hazard House, Newport, Rhode Island, ca. 1700. (Photo, Carl Lounsbury.)

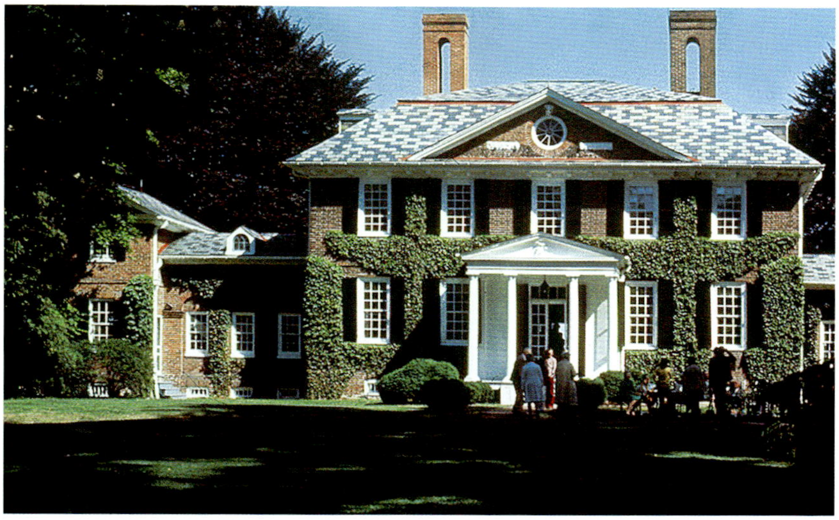

Figure 3 Tulip Hill, southern Anne Arundel County, Maryland, built between 1756 and 1762. (Photo, Donna Ware.)

Carl Lounsbury, an architectural historian with the Colonial Williamsburg Foundation, has been assembling a database of documentary references to specific architectural features for nearly twenty years.[2] His work contains numerous references to the use of fireplace tiles, beginning with a 1702 travel narrative about New York City that states: "the hearths were laid with the finest tiles I ever saw. . . ,"[3] followed by thirteen separate entries between 1720 and 1751 that mention what are variously described as "Dutch," "white," or "chimney tiles," often using a combination of these terms. A single reference in 1752 to "galley tiles" is followed in 1756 by a lengthy description in *A Complete Body of Architecture,* which states: "These Dutch tiles were once in great reputation in ordinary houses but at present they are grown into neglect, and that not without reason . . . being tenderer and more easily damaged."[4]

After a noticeable gap in the documentary record, fireplace tiles underwent a period of resurgence. Between 1765 and 1784 seven references to tiles often include the descriptors "neat copper-plate chimney tiles," "copper plate printed tiles," and "Liverpool tile." The new fashion was undoubtedly related to the production during this period of transfer-printed tin-glazed earthenware and creamware tiles.

Among standing structures in the Chesapeake Bay region, only one remains that has original colonial fireplace tiles in situ: Tulip Hill, in Anne Arundel County, Maryland (fig. 3). Built between 1756 and 1762 and currently owned by the Chaney family, who have been prominent county residents since the 1650s, Tulip Hill is one of the finest eighteenth-century, five-part Georgian houses in the United States. An account book kept during the construction by its owner, Samuel Galloway, offers valuable information about the building,[5] although it does not specifically mention the tiles. Suggesting their origin, however, interesting new documentary evidence reveals that Samuel Galloway's father, John Galloway, who married Jane Roberts Fishbourne in 1744, purchased ten dozen "Holland tyles" from the estate of her former husband, William Fishbourne, of the same year.[6]

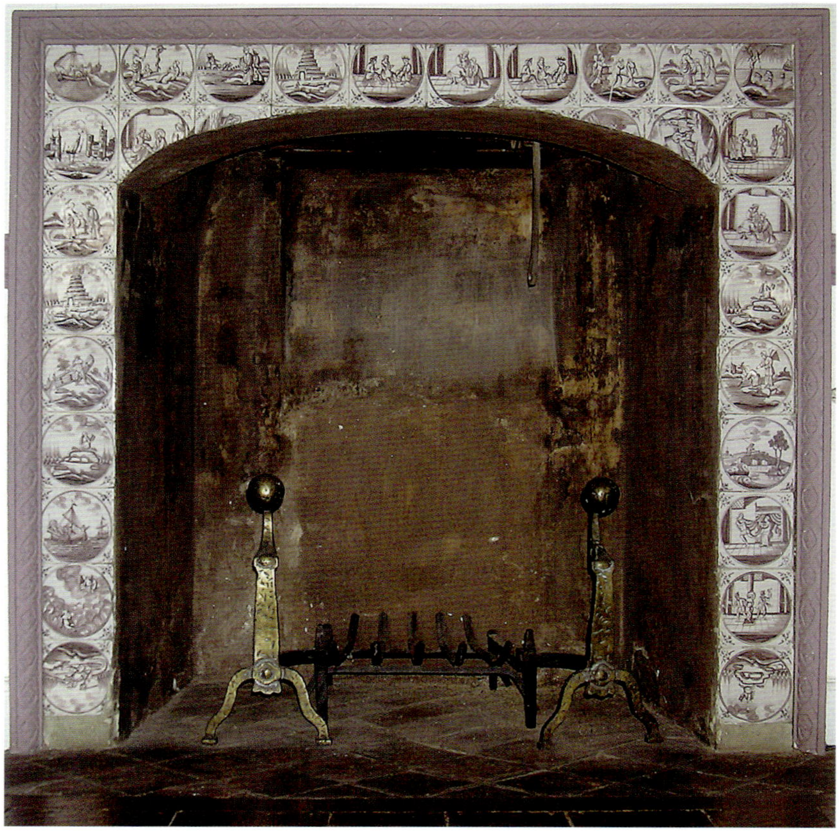

Figure 4 Fireplace at Tulip Hill, with hearth surround of tiles and mid-eighteenth-century wooden overmantel and moldings. (Photo, Al Luckenbach.)

This sole remaining tiled fireplace surround is located in a second-floor bedchamber in the southeast corner of the mansion's main block. If other examples ever existed in the home, they were updated in later renovations, particularly with the circa 1790 addition of wings and hyphens to the structure. The construction of the surviving fireplace is typical of the earlier eighteenth century, with a shelf mantel, an overmantel, and earthenware floor tiles used as the hearth base (fig. 4).

The surround is ornamented with thirty-two Anglo-Netherlandish, manganese-decorated tiles whose decoration is principally biblical in nature (figs. 5–8). Depicted are sixteen Old Testament scenes, eleven New Testament scenes, two landscapes, and three unidentifiable scenes. Six tiles were altered to reflect the arched top of the hearth opening; some were cut

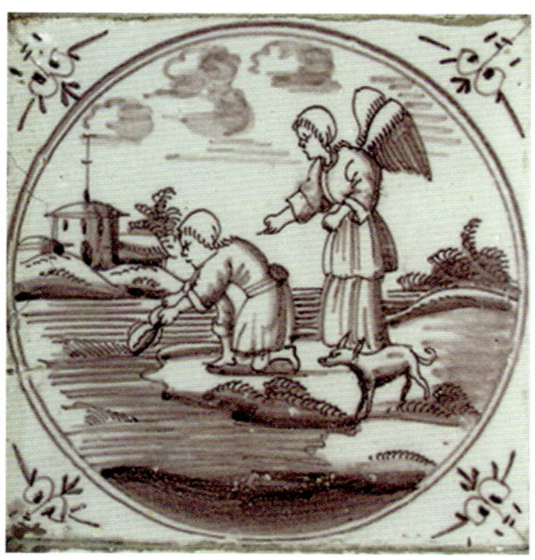

Figure 5 One of the ca. 1740 delftware tiles bordering the hearth illustrated in fig. 4. (Photo, John Kille.) This tile depicts Tobias Catches the Fish, a biblical scene.

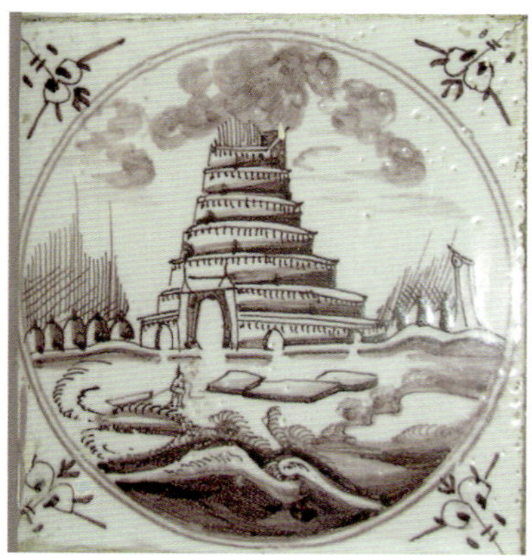

Figure 6 One of the ca. 1740 delftware tiles bordering the hearth illustrated in fig. 4. (Photo, John Kille.) The tile pictures the Tower of Babel.

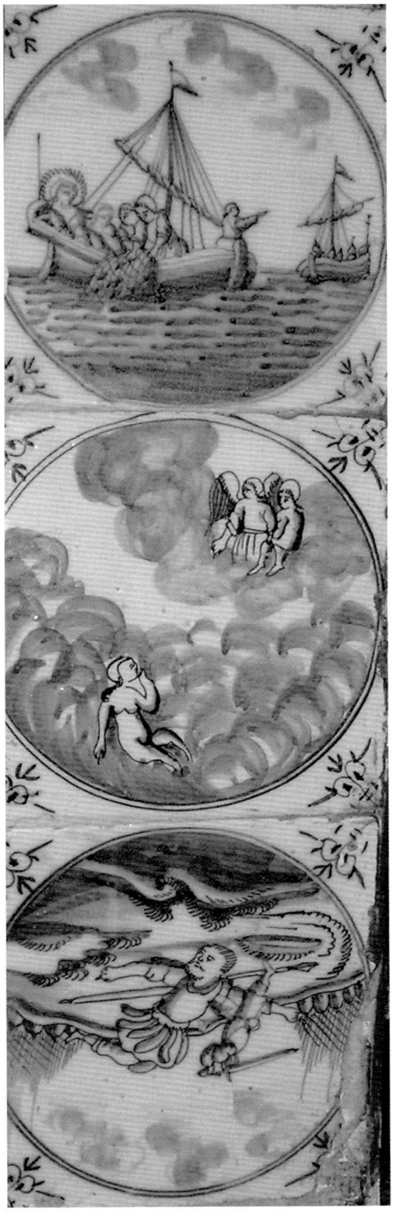

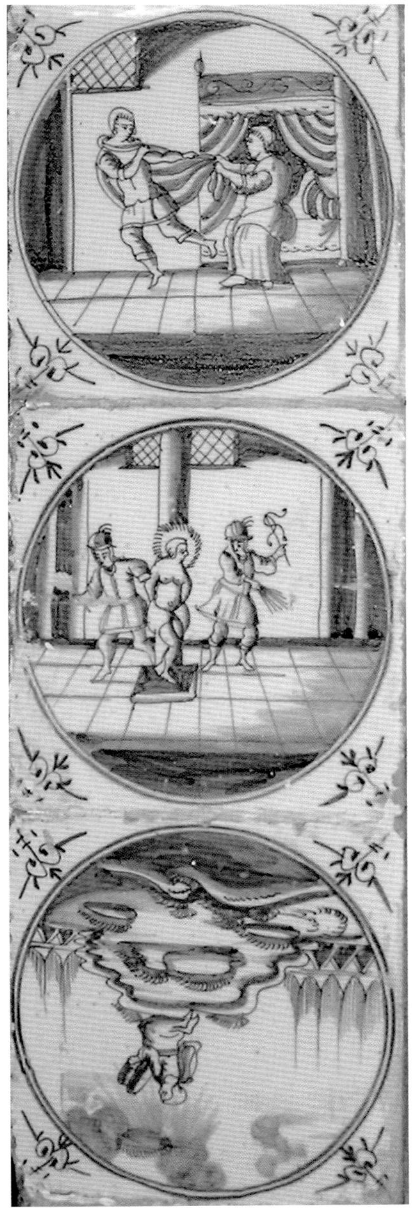

Figure 7 Detail of ca. 1740 delftware tiles bordering the lower left portion of the hearth illustrated in fig. 4. (Photo, John Kille.) The tiles depict (top to bottom) Christ and Fishermen; Lazarus in Heaven; and David Beheads Goliath (upside down).

Figure 8 Detail of ca. 1740 delftware tiles bordering the lower right portion of the hearth illustrated in fig. 4. (Photo, John Kille.) The tiles depict (top to bottom) Joseph and Potiphar's Wife; Christ at the Column; and Moses Receiving the Laws (upside down). (Photo, John Kille.)

to near slivers, thus accounting for the unidentified scenes, and two were placed on their sides. Curiously, the bottom tile on each side was inserted upside down (see figs. 7, 8). This peculiarity is unexplained, as no evidence suggests that there were later repairs.

A list of the motif depicted on each tile and its biblical reference (if one exists) is presented below.

Top Row	*Motif*	*Biblical Passage*
A1	Jonah and the Whale	Jonah 2:10
A2	Moses and Bronze Serpent	Numbers 21:9
A3	Moses and Pharaoh's Army?	Exodus
A4	Tower of Babel	Genesis 11
A5	Joseph Interprets for the Cupbearer and the Baker	Genesis 40:8
A6	The Circumcision	Luke 2:21

A7	Joseph Interprets for the Cupbearer and the Baker	Genesis 40:8
A8	Expulsion from Garden of Eden	Genesis 3:24
A9	Christ Heals the Blind Man	John 9:6
A10	Noah and the Great Flood	Genesis 7:19
Second Row		
B1	Coastal Landscape Scene	n/a
B2	Christ Washes Disciples' Feet	John 13:5
B3	Christ and the Fishermen	Luke 5:5
B4	Unidentified—fragmentary	n/a
B7	Unidentified—fragmentary	n/a
B8	Unidentified (stone wall)	n/a
B9	Christ Carries the Cross	John 19:17
B10	Pilate Washes His Hands	Matthew 27:24
Left Column		
C1	Christ Heals the Blind Man	John 9:6
D1	Tower of Babel	Genesis 11
E1	Samson Takes the Doors of Gaza	Judges 16:3
F1	Moses Receives the Laws	Deuteronomy 5:22
G1	Christ and the Fishermen	Luke 5:5
H1	Lazarus in Heaven	Luke 16:22
I1	David Beheads Goliath	1 Samuel 17:51
Right Column		
C10	The Circumcision	Luke 2:21
D10	Moses Receives the Laws	Deuteronomy 5:22
E10	Tobit Catches the Fish	Tobit 6:3
F10	Pastoral Landscape Scene	n/a
G10	Joseph and Potiphar's Wife	Genesis 39:12
H10	Christ at the Column	John 19:1
I10	Moses Receives the Laws	Deuteronomy 5:22

The disappearance of tile fireplace surrounds in colonial-period homes in the Chesapeake might be due to renovation, as updating them and their mantels was common—and it might also be related to their fragility. Given that wooden—and later slate, and perhaps even marble—panels became the fashionable medium of fireplace surrounds in later years, it is intriguing to contemplate that renovators might have entombed rather than removed the original tiles. Until such hidden treasures come to light, however, Tulip Hill remains the sole illustration of this once widespread treatment in the Chesapeake Bay region.

Al Luckenbach, Anne Arundel County Archaeologist, Director, Lost Towns Project; alluck@aol.com

1. Gary Wheeler Stone, *Seventeenth-Century Wall Tile from the St. Mary's City Excavations, 1971–1985,* St. Mary's City Research Series, no. 3 (St. Mary's City, Md.: Historic St. Mary's City, 1987); John L. Cotter, *Archaeological Excavations at Jamestown Colonial National Historical Park and Jamestown National Historic Site,* Archeological Research Series, no. 4 (Washington, D.C.: National Park Service, U.S. Department of the Interior, 1958); Al Luckenbach, *Providence 1649: The History and Archaeology of Anne Arundel County, Maryland's First European Settlement* (Annapolis: Maryland State Archives; Crownsville: Maryland Historical Trust, 1995); John C. Austin, *British Delft at Williamsburg* (Williamsburg, Va.: Colonial Williamsburg Foundation, 1994).

2. Carl R. Lounsbury, *An Illustrated Glossary of Early Southern Architecture and Landscape* (New York: Oxford University Press, 1994).

3. Howard Henry Peckham, "Madam Knight's Journal: A Woman Travels to New York, 1704," in *Narratives of Colonial America, 1704–1765,* edited by Howard H. Peckham (Chicago: R. R. Donnelley and Sons, 1971), p. 37.

4. Isaac Ware, *A Complete Body of Architecture: Adorned with Plans and Elevations from Original Designs, in Which Are Interspersed Some Designs of Inigo Jones Never Before Published* (London: T. Osborne and J. Shipton, 1756), p. 63.

5. Donna M. Ware, *Anne Arundel's Legacy: The Historic Properties of Anne Arundel County* (Annapolis, Md.: Anne Arundel County, 1990), p. 59.

6. J. Reaney Kelly, "'Tulip Hill': Its History and Its People," *Maryland Historical Magazine* 60, no. 4 (December 1965): 349–403.

Ronald L. Napier

Peter Bell Jar

The earthenware jar (fig. 1) sat among nondescript, unimportant pots in a basement in Front Royal, Virginia, twenty miles from where it had been turned, glazed, and fired. Two collectors who came to view and enjoy the antiques in the house were invited to the basement almost as an afterthought. One of them picked up and examined each of the pieces in the basement but initially failed to notice that the earthenware jar was stamped with a potter's mark. A second inspection, however, revealed a mark of true significance. The piece was stamped "PBELL" twice (figs. 2, 3), the mark of Peter Bell, patriarch of the Bell family of potters.

Peter Bell was born on June 1, 1775, in Hagerstown, Maryland. The first record of his employment was in 1798, in a partnership known as Leisinger and Bell. He remained in Hagerstown until 1824, when he and his family moved to Winchester, Virginia. He potted there until his retirement in the spring of 1845. That September he returned to Hagerstown,[1] where he died in his sleep on June 18, 1847.[2] Three of his sons became master potters. The eldest, John Bell, potted in Hagerstown and Winchester at his father's shops, then at his own shop in Chambersburg, Pennsylvania, and, ultimately—and for the bulk of his career—at his shop in Waynesboro, Pennsylvania. Peter Bell's son Samuel moved to Strasburg, Virginia, and established a pottery there in 1843. Samuel's brother Solomon joined him in 1845. Several of the sons of John and Samuel also became potters.[3]

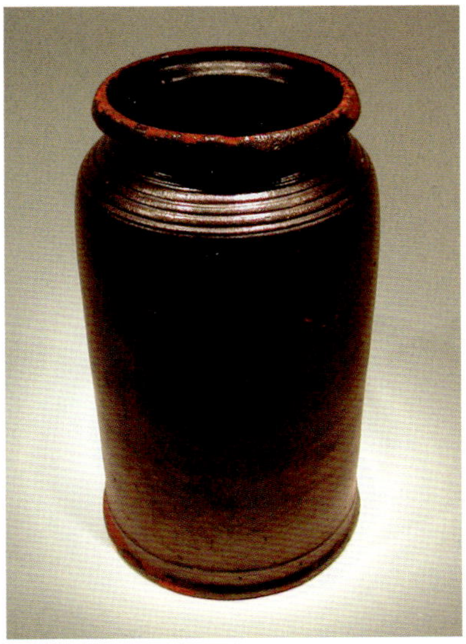

Figure 1 Jar, Peter Bell, Winchester, Virginia, 1825–1845. Lead and manganese-glazed earthenware. H. 10⅝". (All photos, Bret Headley.)

Figure 2 Detail of the stamped mark on the shoulder of the jar illustrated in fig. 1.

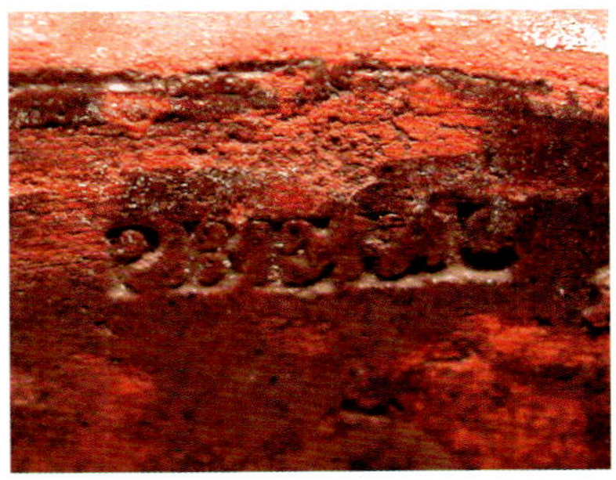

Figure 3 Detail of the stamped mark on the base of the jar illustrated in fig. 1.

The earthenware jar has a lead and manganese glaze, which feels satiny to the touch. The interior and exterior are glazed but the base is not. The step-down, rib-molded beading above the shoulder reflects the skill of a master potter. The beveled foot serves to increase the stability of the piece. It is stamped "PBELL" on its shoulder and base, an extremely rare mark. In his *Pottery of the Shenandoah Valley Region* H. E. Comstock notes that only two examples bearing this mark are known.[4] The other is a large bowl in the collection of the Museum of Early Southern Decorative Arts in Winston-Salem, North Carolina.[5] The earthenware jar illustrated here is currently on loan to the Museum of the Shenandoah Valley in Winchester, Virginia.

Ronald L. Napier, Chair, Collections Committee, Museum of the Shenandoah Valley; ronnapier@hotmail.com

1. H. E. Comstock, *The Pottery of the Shenandoah Valley Region* (Winston-Salem, N.C.: Museum of Early Southern Decorative Arts, 1994), pp. 100–102.
2. Alvin H. Rice and John Baer Stoudt, *The Shenandoah Pottery* (Strasburg, Va.: Shenandoah Publishing House, 1929), p. 28.
3. Comstock, *Pottery of the Shenandoah Valley Region,* pp. 100–102.
4. Ibid., p. 105. This piece was discovered too late to be included in that publication.
5. William E. Wiltshire, *Folk Pottery of the Shenandoah Valley* (New York: E. P. Dutton, 1975), pp. 24–25.

Beverly A. Straube

A "Wretched Tile" from Jamestown

The word went out among the staff to gather at the fort site. This is a standard ritual of the Jamestown Rediscovery project when something special is found in situ, and it is an event that happens often. The archaeologists are uncovering the artifacts and architecture of James Fort, the site of our nation's beginnings four hundred years ago on the banks of the James River in Virginia. Now owned by the Association for the Preservation of Virginia Antiquities, the sponsor of the excavations, the site has yielded nearly one million artifacts that have led to the rewriting of Jamestown's history.

The circle around the squatting archaeologist grew as the staff arrived one by one. The young man squinted up at us with a grin. "So what is it?" he asked, as he carefully brushed the object with his whisk broom. We could tell the object was a complete Dutch wall tile, but the tile was upside down so we could not see the motif on its face.

Lots of fragments of these tin-glazed earthenware tiles had been showing up during the recent excavations. We were working within the perimeter of James Fort's palisade walls but we were uncovering evidence of a substantial building that had been built somewhere nearby, after the palisades had come down in the second quarter of the seventeenth century. This evidence included clay roofing and floor tiles as well as the decorative tin-glazed wall tiles from the Netherlands.

Speculation concerning the possible decoration on the tile face flew quickly among those gathered. All of the other recovered tile fragments had been of Dutch ladies and gentlemen in fancy dress, tradesmen and artisans wielding their crafts, and soldiers in various poses taken from the prints of Jacob de Gheyn.[1]

The archaeologist carefully used the tip of his trowel to separate the tile from the earth that had sheltered it for almost four hundred years and cautiously turned it over for all to see. Faces fell and there was a collective sigh of disappointment. No one had anticipated that our only complete tile in over three hundred tile fragments would be what the Dutch called a *wrakke tegel,* or "wretched tile" (fig. 1).[2] On the face of it was a very blurry image

Figure 1 Tile, Holland, ca. mid-seventeenth century. Tin-glazed earthenware. H. 4⅞". (Courtesy, Association for the Preservation of Virginia Antiquities; photo, Michael Lavin.)

of a man in a wide-brimmed hat holding a walking stick. The blurriness had occurred from overfiring in the glost kiln,[3] which had caused the pigments of the design to flow. Tiles are fired standing upright on bats about ⅕ inches apart and supported by soft rolls of clay.[4] The direction of the glaze flow on our tile reveals how it stood in the kiln.

"Wretched tiles" are seconds or thirds and they made up the majority of a Dutch tile workshop's inventory—surprising, considering their infrequent appearance at archaeological sites. It is estimated that only 15 percent of fired tiles in the seventeenth century were graded first quality.[5] According to Dutch delftware historian Dingeman Korf, "even if tiles had 'run' they were sometimes added to the batch that was to be delivered. This happened often with tiles of second class quality and so they did find a place on the wall."[6]

The consumers at Jamestown were obviously somewhat discriminating. The reason the wretched tile is our only complete tile is because it was thrown out before it ever found "a place on the wall"!

Beverly A. Straube, Senior Curator, Jamestown Rediscovery, Association for the Preservation of Virginia Antiquities; bly@apva.org

1. Jacob de Gheyn, *The Exercise of Armes for Calivres, Muskettes, and Pikes* (The Hague, 1607).
2. *Dutch Tiles in the Philadelphia Museum of Art,* with essays by Jan Daniël van Dam and Pieter Jan Tichelaar; catalogue by Ella Schaap et al. (Philadelphia: Philadelphia Museum of Art, 1984), p. 22.
3. This is where the second firing, which fuses glaze with the pigments, takes place.
4. Jan Pluis, *De Nederlandse Tegel: Decors en Benamingen, 1570–1930 (The Dutch Tile: Designs and Names, 1570–1930),* exh. cat. (Leiden: Nederlands Tegelmuseum, 1997), p. 77.
5. Ibid., p. 41.
6. Dingeman Korf, *Dutch Tiles* (New York: Universe Books, 1964), pp. 16–17.

Gerald L. Thiebolt

Creating Architectural Elements in Clay

During my years studying architecture I always enjoyed working in three dimensions to communicate a design, as with models and sculpture, more than in two-dimensional drawings showing various views. The look and feel of various materials, their textures, patterns, and colors, appealed to my visual and tactile senses. As I look back, these interests played an important role in formulating my work in clay.

I wanted to use my architectural skills in a positive way, so in the 1960s, before I graduated from Virginia Polytechnic Institute and State University, I signed up for the Peace Corps and was sent to northern Iran as an architect for a small city. It was a remarkable experience, being dropped into an environment in which a few simple materials—brick, glass, steel, and sometimes concrete—are used to create an amazing variety of building forms.

A view from a minaret in the desert city of Yazd presents itself as a vast sculpture of khaki-colored domes, rectangles, squares, undulating roofs, dark courtyards—all variations of the same color, positive and negative, dark and light. Beautifully tiled mosques and public buildings, covered in glazed tiles of turquoise and arabesque designs of yellow, white, and blue, glisten in the distance. Part of this sculptural phenomenon was shaped by continuity of materials. It created a "language," whereby domes, half-domed arches, and cube shapes—all simple geometric forms—became the nouns; the rich surface decoration in bricks and tiles were the verbs; and the rhythms and counterrhythms of pattern, texture, and color were the adjectives. Even the rhythm of calligraphy was used to emphasize archways and intense patterns in squinches pointed to the transition of a square frame to a circle, which became the base of the dome. Arabesque designs in the form of plants inside some of the domes create an ethereal world of abundance with the heavens as a backdrop.

This approach to design was quite different from the "less is more" concept of design I studied in school. The worldview of American architecture was a technical one, with no ornamentation and virtually no concept of a cultural component other than its own. What I experienced in the Middle East was more humanistic and enriching in scope and scale. The use of simple materials was reinforced during my six-month trip to India and, in a very different way, in Japan, where I lived for one and a half years.

Returning to the United States during a recession was challenging and I had two short-lived jobs in architecture before I was put out of commission by a bicycle accident. That event instigated my career in clay. I learned to hand build from a friend who was taking classes and began working night and day in an explosion of creativity. I loved the plastic quality of clay, the ability to shape and texture the material at will, and I decided to get a degree in ceramics.

After three months in school, my professor referred me to Rhoda Lopez, a ceramic artist working in an architectural scale who also had a teaching studio affiliated with the University of California at San Diego. After a few years Rhoda and I became partners when she initiated work on her largest project—the Memorial Wall, nine feet high and thirty-three feet long, for the First Unitarian Universalist Church of San Diego (fig. 1). She designed

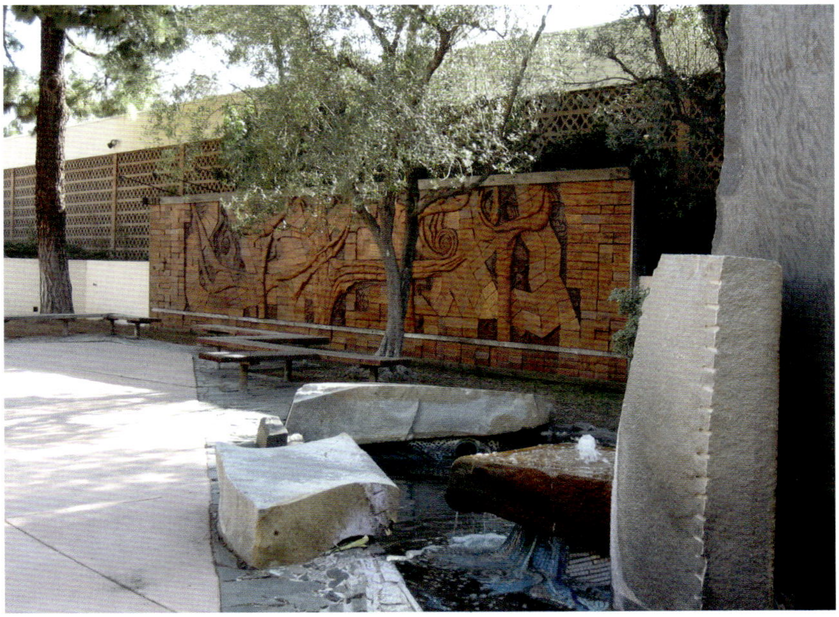

Figure 1 Rhoda Lopez with Gerald Thiebolt, Memorial Wall, First Unitarian Universalist Church of San Diego, begun in 1978. Stoneware. L. 40' (originally 33'). (Unless otherwise noted, photos by Gerald Thiebolt.)

a joyful, prominently placed work that would relate to the church community on a daily basis. Her design contained many abstract forms and featured a series of plaques impressed with the names of deceased church members.

I started the project by designing and constructing the simple A-frame structure to support the clay. With the help of assistants Rhoda began installing the clay pieces, much like a puzzle, while I was finishing graduate school. The project stalled temporarily when Rhoda fell ill, but after she recovered the two of us resumed working together and created the Memorial Wall with three levels of relief: soft, undulating tree forms in the highest relief; lower-relief plaques contrasting crisp angular forms; and a darker, lowest-relief textured background (fig. 2). To produce the textures, we impressed five tools and/or found objects in the clay to create new patterns in never-ending designs in an infinite number of motifs.

Rhoda and I continued working together during a fifteen-year partnership that ended with her untimely death. We collaborated on many projects, including murals for other public buildings, sculptural fireplaces, and addi-

Figure 2 Closer view showing the three levels of relief modeling: undulating trees, stamped designs, and memorial plaques.

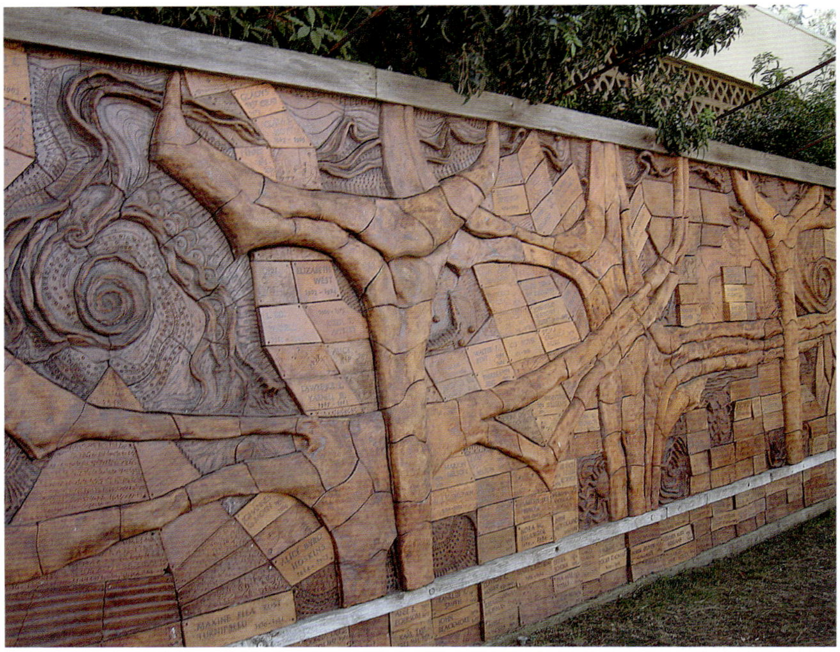

Figure 3 As a joyful memorial for a living church, the wall is continuously updated, as shown in this close-up view.

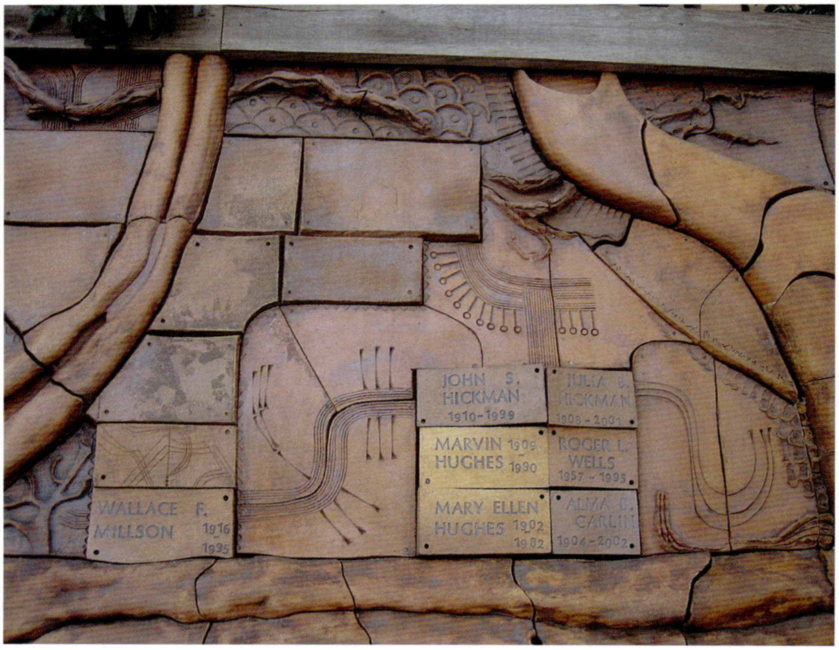

Figure 4 Gerald Thiebolt, Chalice, 2003. Stoneware. H. 33", D. 21".

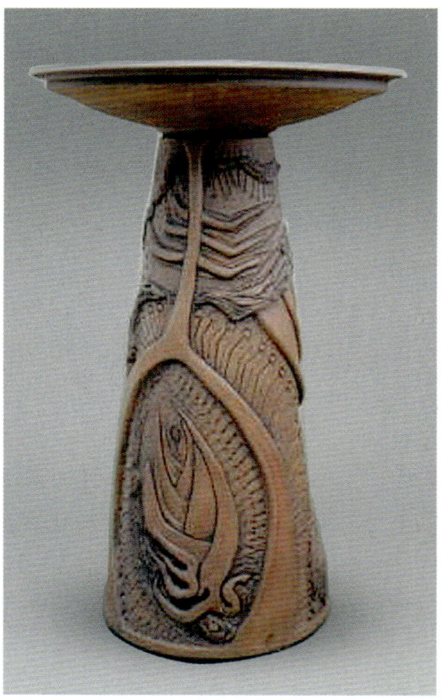

tional projects for the First Unitarian Church. In 1992 two additions to the Memorial Wall—one on each side—were completed, increasing its length to forty feet. It continues to be a vital part of the First Unitarian Church community, as names are continually added (fig. 3), and it is viewed as a joyful place that unites the congregation; indeed, many couples arrange to be married before it.

Since making the wall I have continued producing pottery for this church, including a "chalice" used to hold a flame during services (fig. 4). Measuring twenty-one inches in diameter by thirty-three inches high, the piece contains sculptural elements that directly relate to the Memorial Wall.

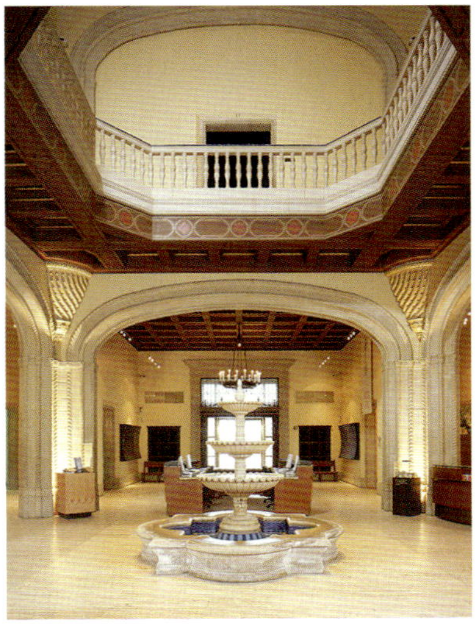

Figure 5 Rotunda, San Diego Museum of Art. H. of fountain 7'. (Courtesy, San Diego Museum of Art, The John M. and Sally B. Thornton Rotunda.)

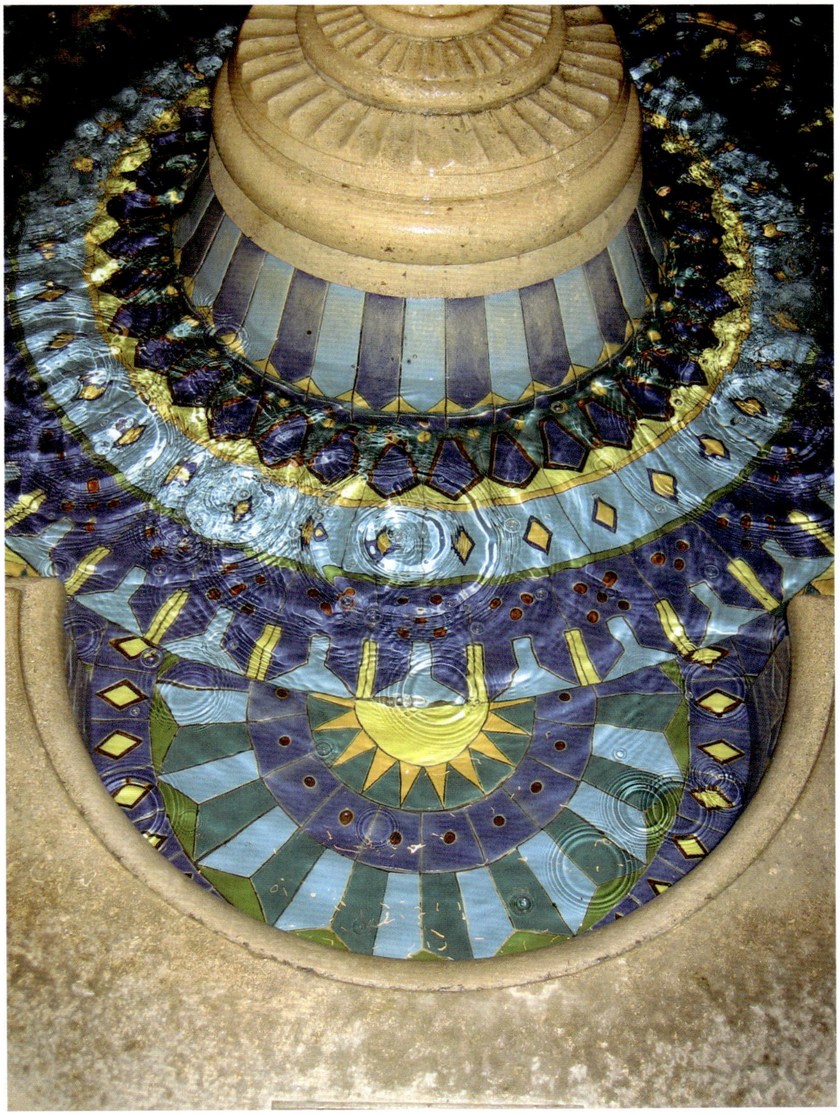

Figure 6 Detail, central fountain basin, San Diego Museum of Art. The tiles were designed, made, and installed by the author. In this view the basin resembles an elegant turquoise necklace.

It was designed with ease of use in mind; it has wheels hidden inside the base so it can be moved when not in service.

Another project of mine is on view in the rotunda of the San Diego Museum of Art. As part of the centrally located fountain designed by architects Robert Thiele and Spencer Lake, I designed, made, and installed the tiles for its basin (figs. 5, 6).

Although I do wheel-throw pots, my creation of architectural clay designs continues to this day, currently including two that are yet to be built: a memorial wall that is seventy feet long and varies in height from seven to eleven feet; and five shaped walls around a gathering space with a brick surface and a stream of handmade textured brick. My interest in designing architectural spaces has led me to integrate it in unusual ways into environments using manufactured materials, such as terracotta bricks, and then combining it with my own work.

Recently, in an architectural landscape, I created a flow of lines and patterns that extends through the courtyard to the outside entry and to the

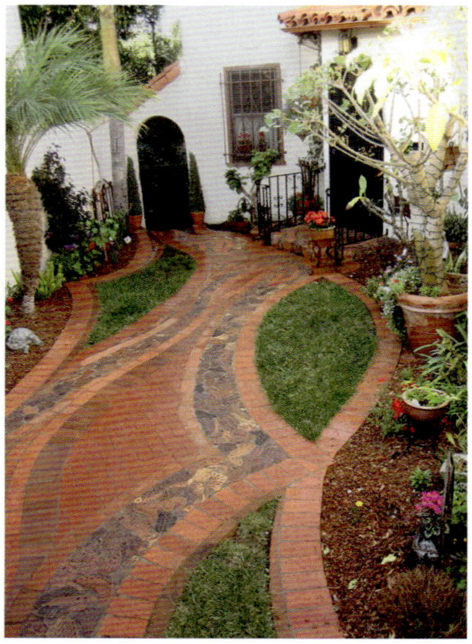

Figure 7 Gerald Thiebolt, landscape design, 1996. Handmade decorated stoneware bricks are combined with manufactured terracotta bricks in this San Diego yard.

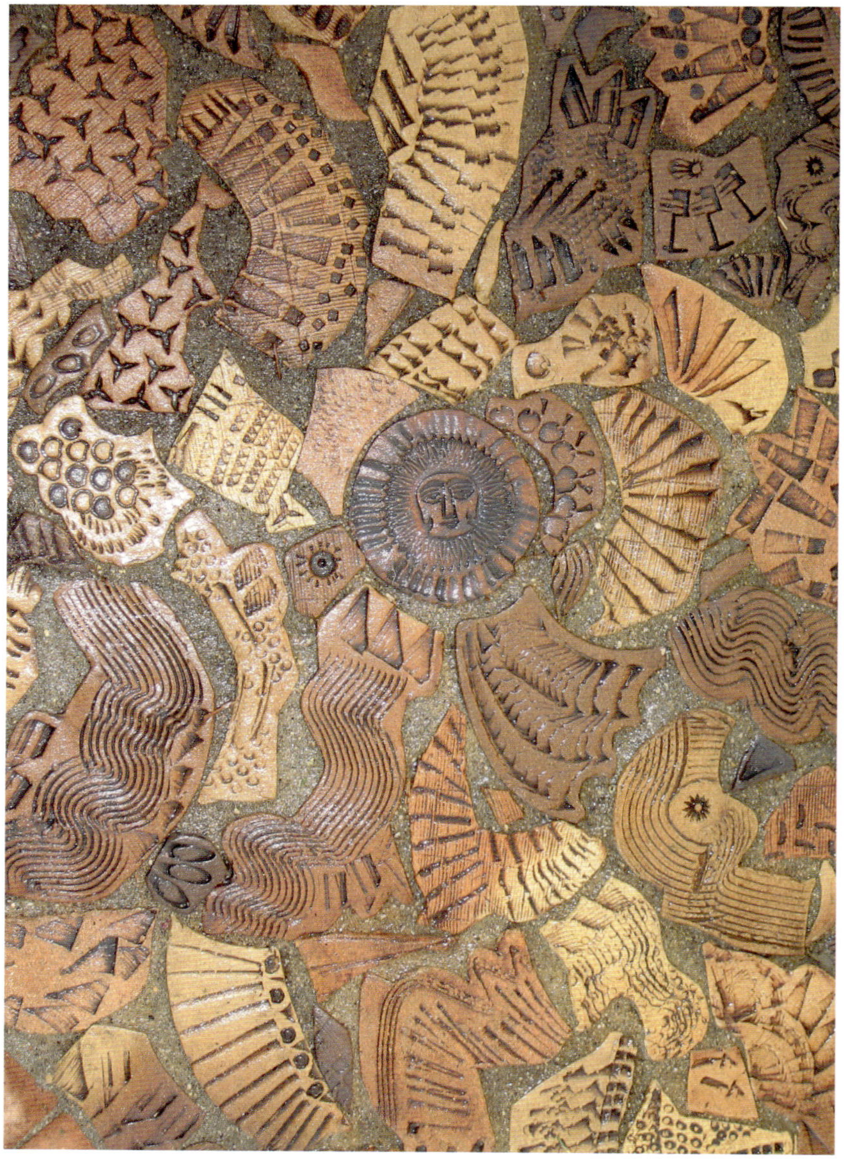

Figure 8 Close-up view of motifs stamped into stoneware clay and fired for use in the walkway of the yard illustrated in fig. 7.

Figure 9 The manufactured brick, chosen for its color and texture, was reshaped for use in the landscape illustrated in fig. 7.

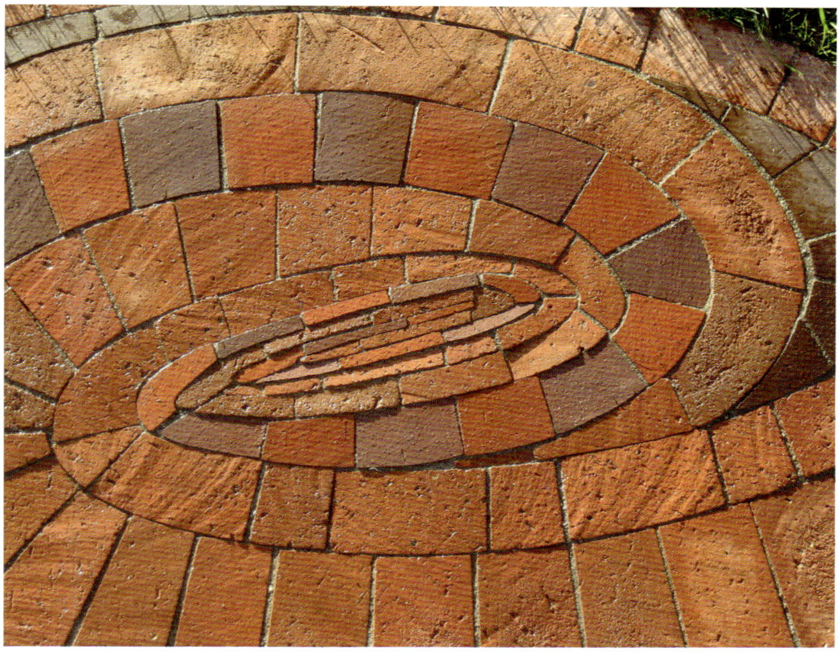

drive (figs. 7, 8). I included five different colors of brick and blended them with my handmade stoneware elements (fig. 9). From an agave-like desert plant form at the entryway flow primordial roots of handmade textured stoneware bricks that wind their way throughout the project, representing the root from which life springs and is sustained. My experiences with deserts in southern California and abroad have taught me to appreciate the tenacity of life, which has been a theme in much of my work. This particular commission took me full circle, because I embellished flat surfaces with color, pattern, and texture, similar to the surface designs in Persian architecture that were revealed to me at the beginning of my career.

Gerald L. Thiebolt, Studio Associate, Clay Associates;
info@clayassociates.org

Warren Bakley

A Designer Touches Clay

I have come to see my creativity with clays as part of a ceramic tradition that in this century is showing movement away from utilitarian objects toward more decorative ones. I would describe myself as a designer doing sculpture within the ceramic process. It is from my role as a designer of spaces, interiors, and furnishings that my ceramics take their architectural structure and form. My intent is to express with available indigenous materials something that speaks not only of the time but also, I hope, speaks something of myself (figs. 1, 2).

Traditionally, the role of the ceramist as an independent potter was to supply utilitarian objects. With plastics, mass-produced ceramics, and other man-made materials replacing functional handmade clay items, that role has evolved to the production of objects with very little purpose other than to delight the eye and/or embellish a specific singular lifestyle. Current

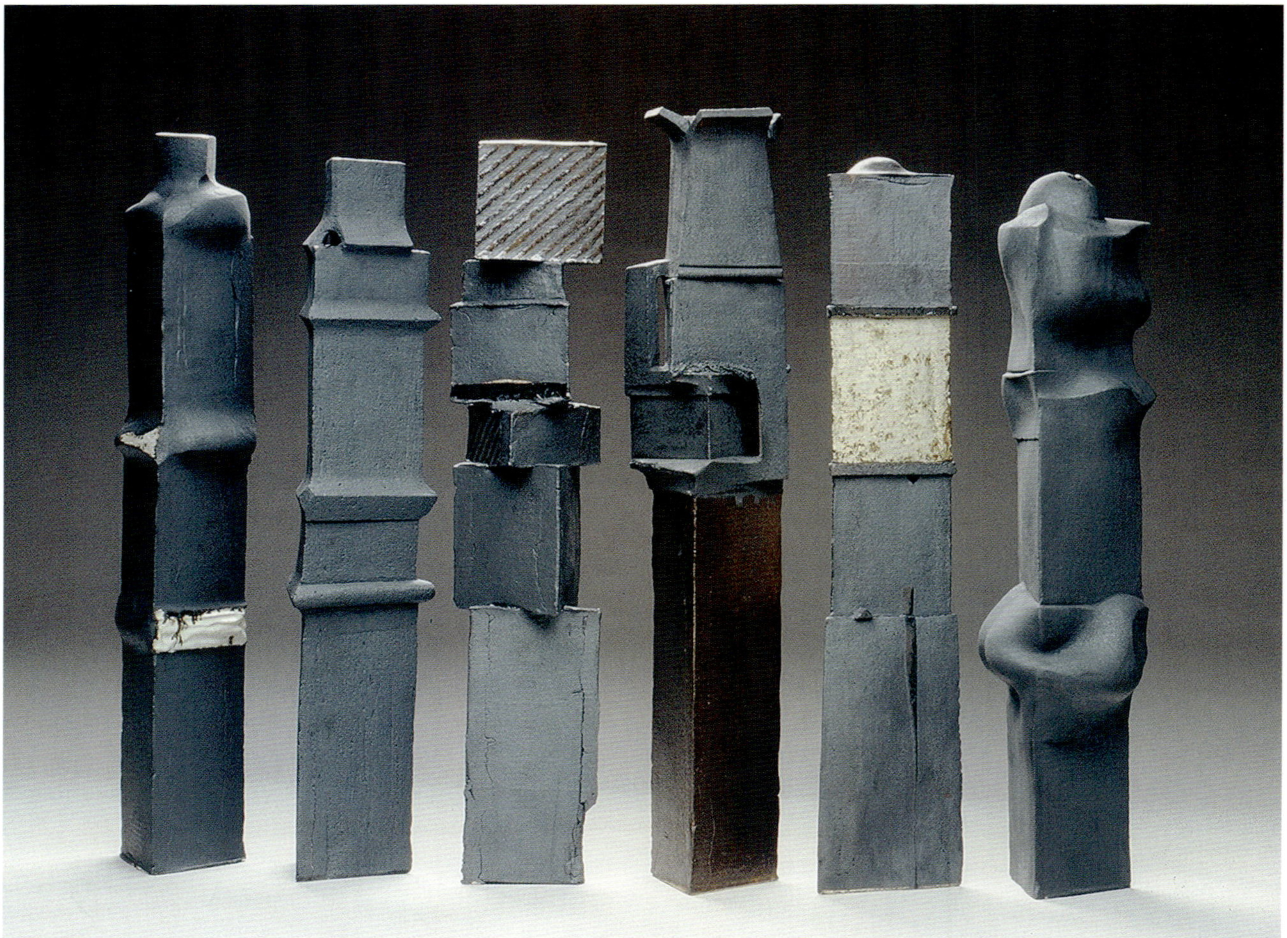

Figure 1 Warren Bakley, Construction, 1994. Stoneware. H. 30". (Private collection; photo, Martin Trailer.) Mixed clays were used to create the six architectural columns in this sculpture. The exteriors of the columns are decorated with a black iron oxide slip.

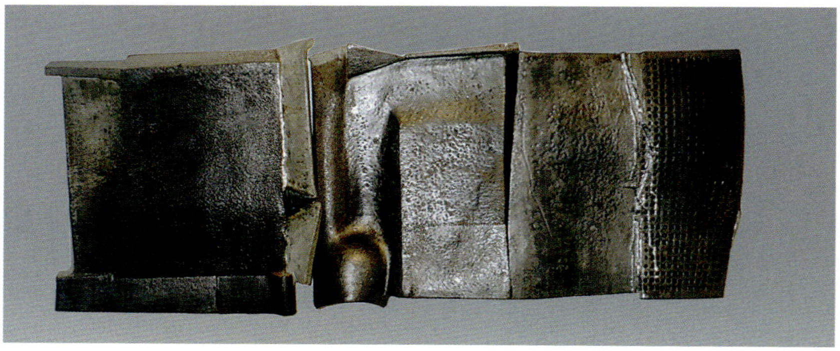

Figure 2 Warren Bakley, Construction, 2003. Stoneware. L. 42". (Private collection; photo, Warren Bakley.) This elongated wall hanging was slab formed and partially sandblasted to create texture and variation. Kiln-fired metallic oxides give the sculpture its high sheen.

popular publications attest to an affluent population that seeks ceramics for their artistic appeal, their eccentricity, or merely to fill a void in a larger space. This suggests the notion that a clay object can be valued for its ability to satisfy something of the inner self, like a found shell or pottery sherd that has no specific usefulness but engenders in the beholder a sense of security in that it cannot become obsolete.

My introduction to clay was in New Jersey at age seven. The clay lay luminous white beneath the clear water of a natural spring surrounded by grasses growing through a sandy reddish soil. From this clay I made a small pinch pot that was fired in the oven of a woodstove. The pot did not survive, but my interest in clay and its possibilities did. The education of designers and architects very often includes classes in ceramics as an introduction to three-dimensional form. It was in a class such as this that clay again captured my attention and sparked an intense interest in the ceramic arts that has lasted a lifetime.

Throughout my college years and until I moved to California, I worked with clays dug from fields, river bottoms, and erosion ditches in areas of southern New Jersey. These clays were unique in that they needed no processing other than wedging. Depending on the sources, they offered a satisfying variety of mature firing ranges. They were excellent for potting and structurally good for hand building. In their raw state they possessed exceptional color inspiration. This, for me, is an important part of the creative process.

Naturally formed clays are not readily available in California. The commercial clays appear to be imposters, man-made parodies manufactured

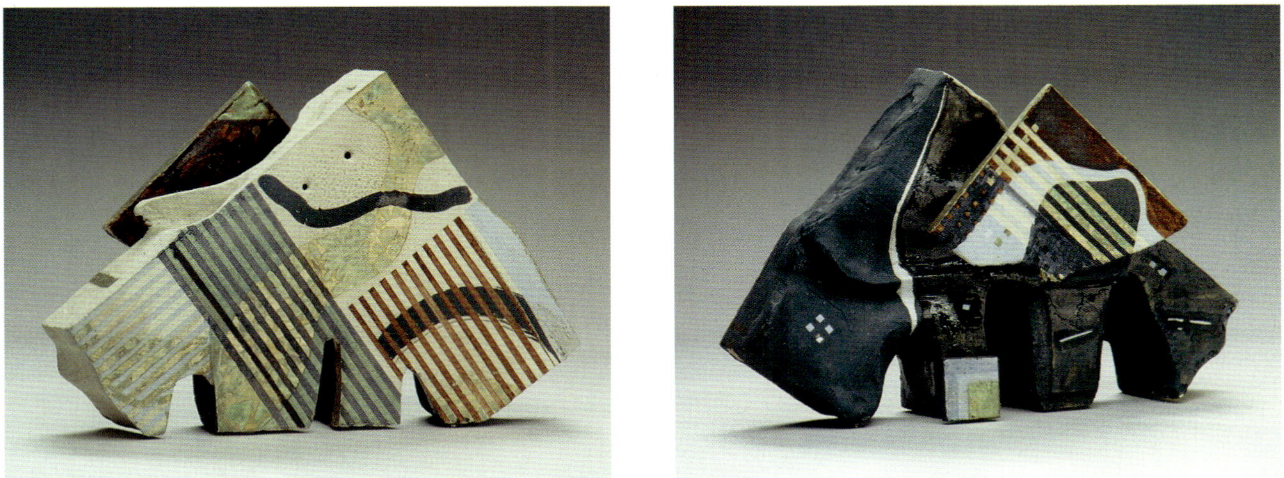

Figure 3 Warren Bakley, Construction, front and back, 1986. Stoneware. H. 10". (Private collection; photo, Martin Trailer.) This sculptural construction, slab formed from mixed clays, was decorated with iron oxide and partial glaze.

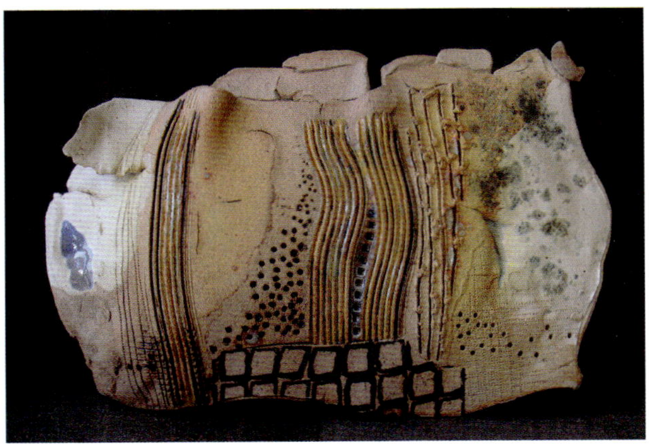

Figure 4 Warren Bakley, Construction, 2005. Earthenware. L. 23". (Private collection; photo, Warren Bakley.) A partially glazed slab-formed wall tile, this incised and stamped wall hanging was created using mixed earthenware.

from a boringly consistent recipe for color, texture, and sculptural properties. They have little indigenous presence and, sadly, seem robbed of their rightful alluvial lineage. Despite this reality, I am still affected by the natural coloring and tactile qualities of raw clays, and these qualities play a major role in how I am feeling during the creation of each piece (figs. 3, 4). In

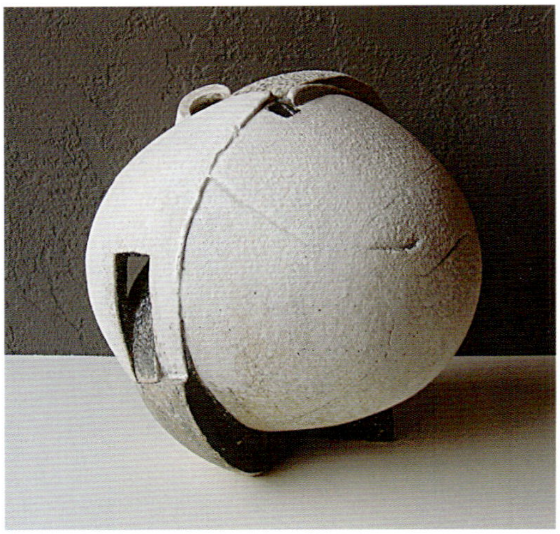

Figure 5 Warren Bakley, Construction, 2004. Stoneware. H. 12". (Bakley collection; photo, Warren Bakley.) Slab formed, this construction displays partial glaze and vitrified kiln washes.

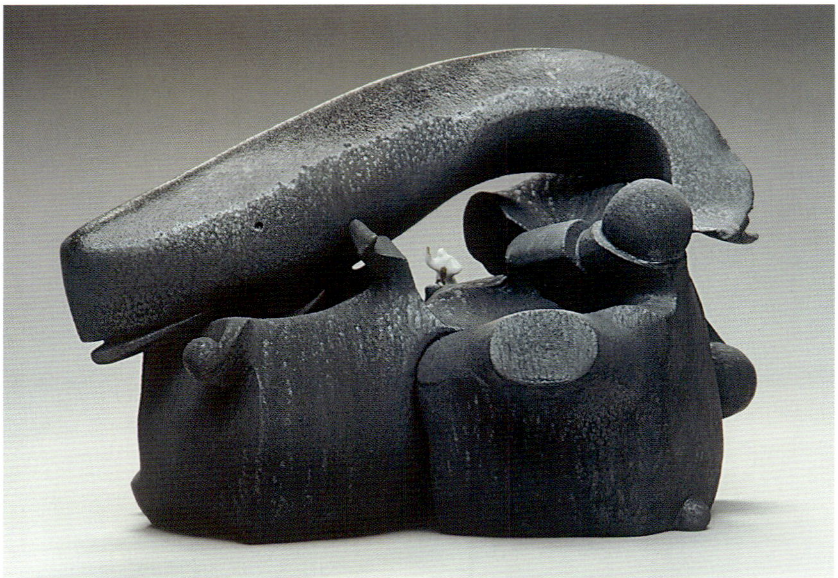

Figure 6 Warren Bakley, Sculpture, 1993. Stoneware. H. 10". (Private collection; photo, Martin Trailer.) This sculpture was inspired by Dr. Jonas Salk's declaration in 1993 that he would infect himself with the AIDS virus to prove the effectiveness of his experimental antiviral procedure. Jonas is the little white figure on a surfboard attempting to ride a turbulent ocean while taming a large whale.

completing each piece the application of the glaze is minimal. This is deliberate, to create a natural continuation of the adjacent unglazed surfaces. This allows the sculptural form to dominate, with the surface remaining open to the play of light and shadow (fig. 5).

I see my use of clay not only as the means to express my creative thoughts but also as being part of objects that transcend the bareness of plastics and promote the use of natural materials. I believe that when used with integrity, traditional materials—just like shells and sherds—do speak for their time, as well as to the soul of the viewer (fig. 6).

Warren Bakley, Designer; sbakley@san.rr.com

Susan Tunick and Jay Shockley

Early Examples of American Architectural Terracotta in Charleston, South Carolina

There are more extant buildings ornamented with American-made terracotta from the 1850s in Charleston, South Carolina, than in any other location in the United States. Eight structures have been identified thus far, and pieces of ornament from another have been salvaged. The terracotta on these buildings was manufactured in Worcester, Massachusetts, by the earliest known American firm. What led Charleston architects to contact this particular firm and take an early interest in a relatively fragile material that needed to be carefully crated and shipped a long distance from New England—sometimes with disastrous results—is unclear.

In 1848–1849 Henry Tolman Jr. established a business in Worcester manufacturing ready-made terracotta ornament. He was influenced to do so by Elbridge Boyden, an innovative local architect. Tolman joined with Jonathan Luther in 1851 to form Tolman, Luther and Company for the exclusive manufacture of architectural terracotta. Boyden, through his firm, Boyden and Ball, served as an agent for the sale of this material. By 1853 Tolman was joined by Samuel Hathaway and William Stone, and the company became Tolman, Hathaway and Stone. Exterior architectural terracotta ornament was first supplied for a number of Worcester buildings designed by Elbridge Boyden. It was later used for various buildings throughout New England and as far away as Philadelphia and Charleston, South Carolina.

The earliest known order for Worcester terracotta for a building in Charleston was placed in 1850 by Edward Brickell White, the city's leading architect. Records indicate that he contacted Boyden, ordering material for porticoes he had designed for two educational institutions: Corinthian capitals for the High School of Charleston (1840–1842, White), 55 Society Street, and Ionic capitals for the College of Charleston (1828, William Strickland), 66 George Street (fig. 1).[1] The high school's terracotta was damaged in the earthquake of 1886 and subsequently removed (later replaced with replicas of the capitals). The earthquake also damaged the college, but there the terracotta survives, and a recent examination (using

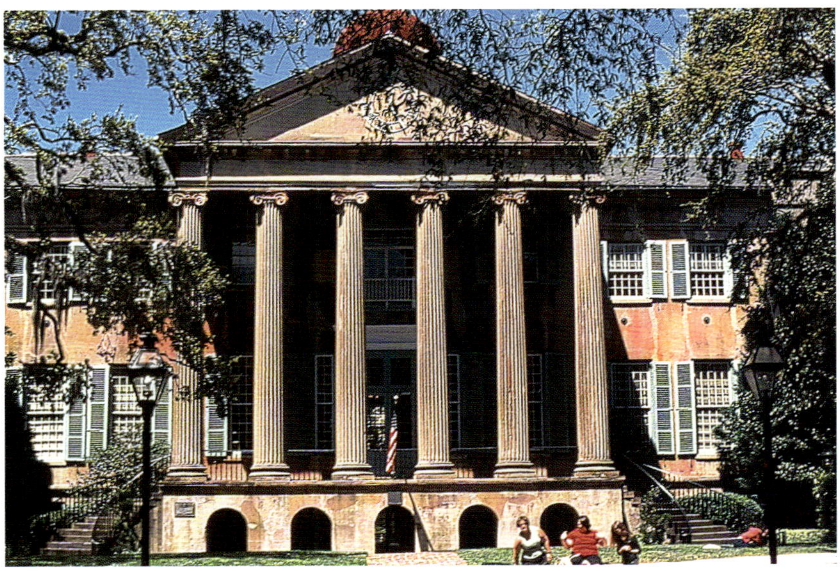

Figure 1 College of Charleston (now Randolph Hall). (Courtesy, Friends of Terra Cotta Collection; photo, Paul Tunick.) The portico was added in 1850 by architect Edward Brickell White.

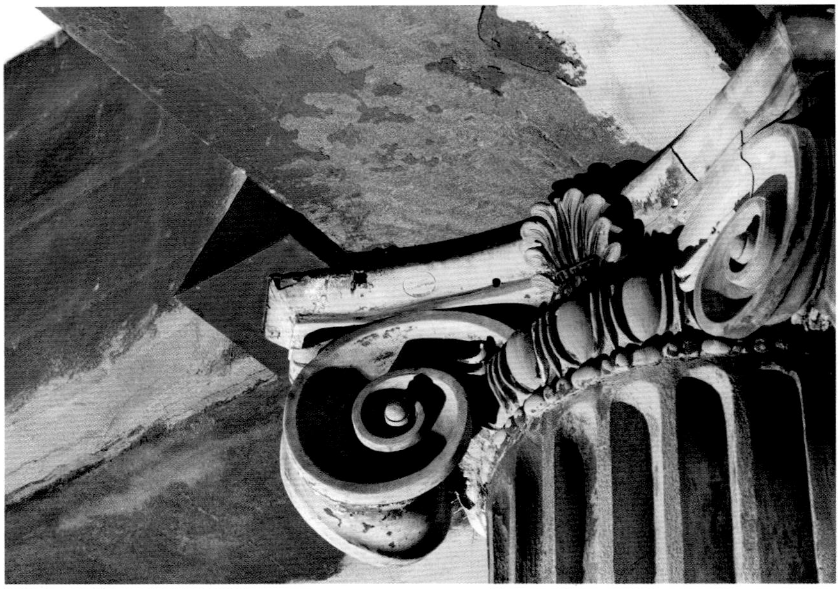

Figure 2 Terracotta capital on the portico of Randolph Hall. (Courtesy, Friends of Terra Cotta Collection; photo, Rick Rhodes.)

Figure 3 Detail of the stamp on the Randolph Hall capital, which reads: "MANUFACTURERS, TOLMAN, LUTHER & CO., CANAL STREET, WORCESTER, MASS." (Courtesy, Friends of Terra Cotta Collection; photo, Rick Rhodes.) This is the only known stamp from the company.

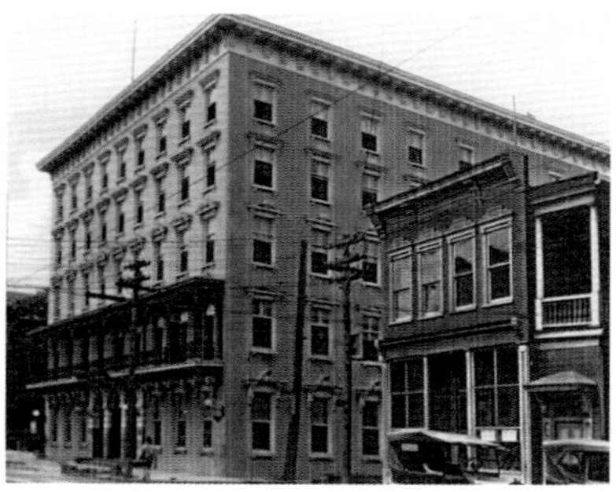

Figure 4 Mills House Hotel, undated photograph, illustrated in C. Patton and W. Eric Emerson, eds., *Charleston Alone among the Cities* (N.p.: Arcadia Publishing, 2000), p. 57. (Courtesy, South Carolina Historical Society, Charleston, SC.) This building used extensive terracotta ornament on the window pediments and lintels.

binoculars) of the Ionic capitals (fig. 2) yielded a surprising discovery: upside down on one of the capitals is an oval stamp that reads: "MANUFACTURERS, TOLMAN, LUTHER & CO., CANAL STREET, WORCESTER, MASS." (fig. 3). Because it was standard practice in this era for terracotta to be painted, it is doubtful that the manufacturer had any expectation of its stamp being visible (it is unknown at what point the paint was removed). This is the only known surviving example from this company.

Contemporary accounts in the *Charleston Courier*—cited in Beatrice St. Julien Ravenel's 1945 book *Architects of Charleston*—identified Worcester as the source for the terracotta on the Mills House Hotel (1853, John E. Earle), 115 Meeting Street (fig. 4). The first structure in Charleston with such amenities as running water and steam heat, the hotel's architectural features and elaborate cast-iron balconies added to its reputation as one of the city's premier hotels and most impressive structures. Otis Mills, the hotel's owner, was a Massachusetts native, and John Earle may have been connected with Elbridge Boyden in Worcester.[2] This 5-story, 125-room hotel, with its numerous terracotta window pediments and lintels, probably featured the most extensive execution of terracotta ornament on an American building to that time.[3] Ravenel noted that "'Mr. Hathaway of Worcester,

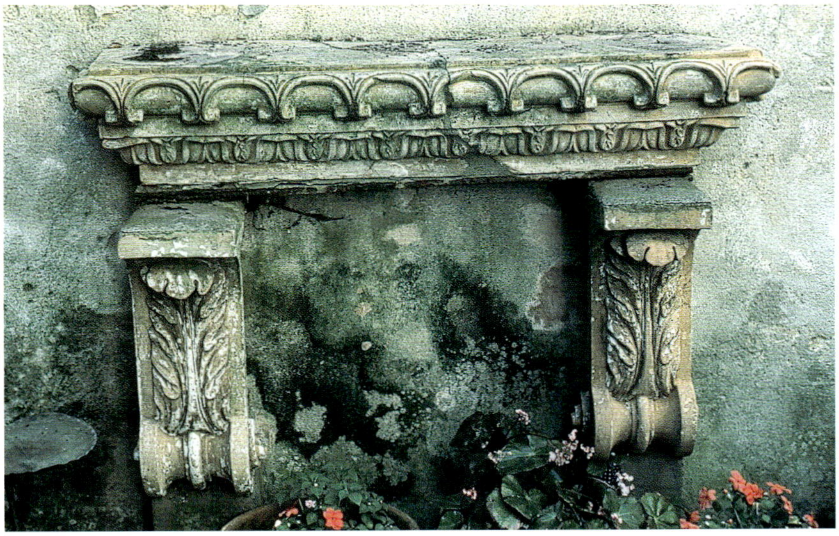

Figure 5 Salvaged terracotta from the Mills House Hotel. (Courtesy, Friends of Terra Cotta Collection; photo, Paul Tunick.) These pieces of terracotta were saved when the hotel was demolished and reconstructed in 1968 with fiberglass ornament instead of terracotta. This reassemblage, in a garden in Charleston, might not be the original configuration.

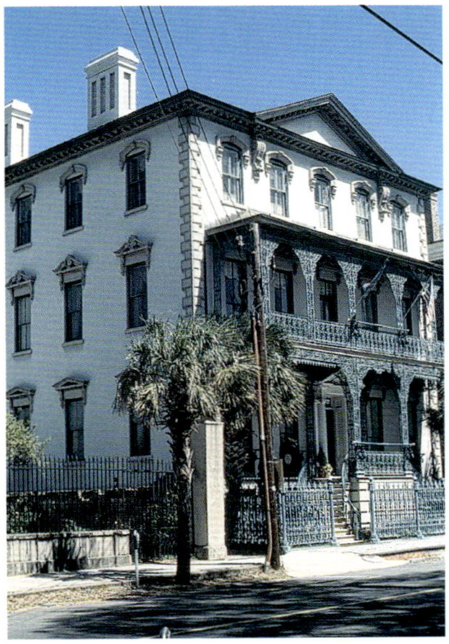

Figure 6 The Governor John Rutledge House. (Courtesy, Friends of Terra Cotta Collection; photo, Paul Tunick.) All the window pediments on this house are made of terracotta from Tolman, Luther and Company. The ornament varies between floors and facades.

Mass.' . . . furnished the terra-cotta window cornices and pediments. These were considered fireproof and created a furor, so that similar ones were placed on houses all over the city."[4] In 1968 the greatly deteriorated Mills House Hotel was demolished and reconstructed, but at least one original window lintel and several brackets were salvaged (fig. 5).

There are seven extant buildings in Charleston with Worcester terracotta, showing the direct influence of the Mills House Hotel. In 1853 one of them, the Governor John Rutledge House (ca. 1763), 116 Broad Street, was either replaced or remodeled and enlarged by architect P. H. Hammarskold (fig. 6). The elaborate second-story window pediments found on the front facade (fig. 7) are identical to those originally found on the Harrington Corner Block (ca. 1849, Boyden) in Worcester (fig. 8). In the early 1970s the ornamentation on the Harrington building was removed, but Susan McDaniel Ceccacci, a local architectural historian who did pioneering work on Boyden and the Tolman firm, salvaged several pieces that are

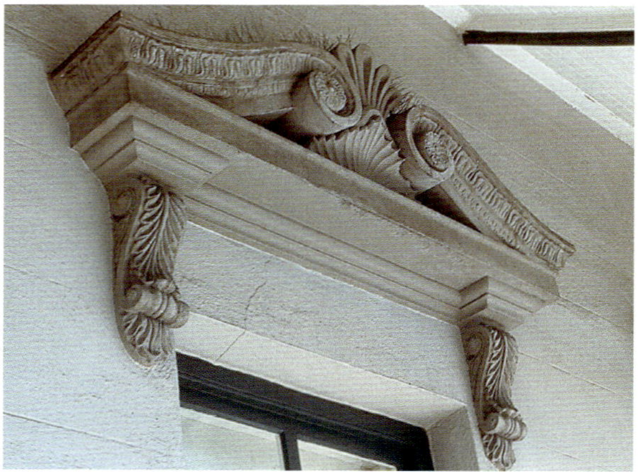

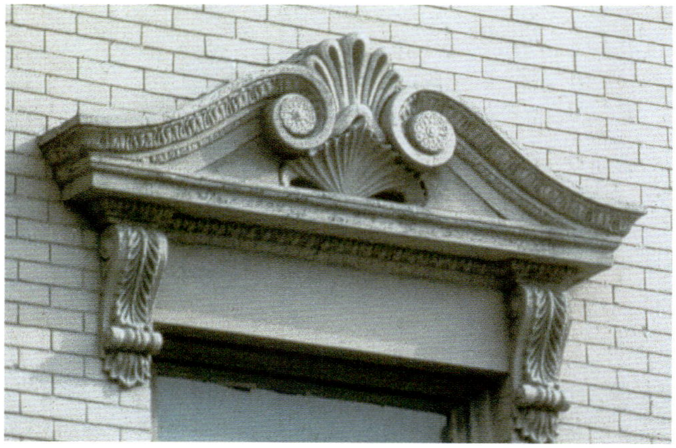

Figure 7 Window pediment detail on the second story of the front facade of the Governor John Rutledge House. (Courtesy, Friends of Terra Cotta Collection; photo, Paul Tunick.) This terracotta pediment is identical to ones removed from the circa 1849 Harrington Corner Block, Worcester, Mass.

Figure 8 Terracotta window pediment on the Harrington Corner Block, Worcester, Mass. (Photo, Susan McDaniel Ceccacci.) This unit, salvaged when the building was stripped in the early 1970s, was donated to the Worcester Historical Museum.

now in the collection of the Worcester Historical Museum. The museum also owns a copy of an 1852 Tolman, Luther and Company catalog, which shows a window bracket identical to those found on the Rutledge House (fig. 9).[5] Charleston's "Bee's Row," the five rowhouses at 101–109 Bull Street (ca. 1853–1854, architect unknown), and the Smith Building (ca. 1801–1815, Hugh Smith; alteration mid-1850s), 15 Broad Street, also display this bracket. In addition, the Smith Building's top-story window lintels match the salvaged lintel from the Mills House Hotel.

Three other important Charleston buildings are known to have incorporated architectural terracotta in the 1850s that did not survive the 1886 earthquake. The Unitarian Church (1772–1787), 6 Archdale Street, was remodeled in 1852–1854 by Jones and Lee, who incorporated Gothic terracotta pinnacles from Worcester on the tower and buttresses (fig. 10).[6] A similar pinnacle by the Tolman firm was illustrated in a publication on the 1853 Crystal Palace Exhibition of All Nations in New York (fig. 11).[7] The Unitarian Church survives, but the terracotta was removed after the earthquake. Ravenel quoted an article in the *Charleston Courier* that mentioned

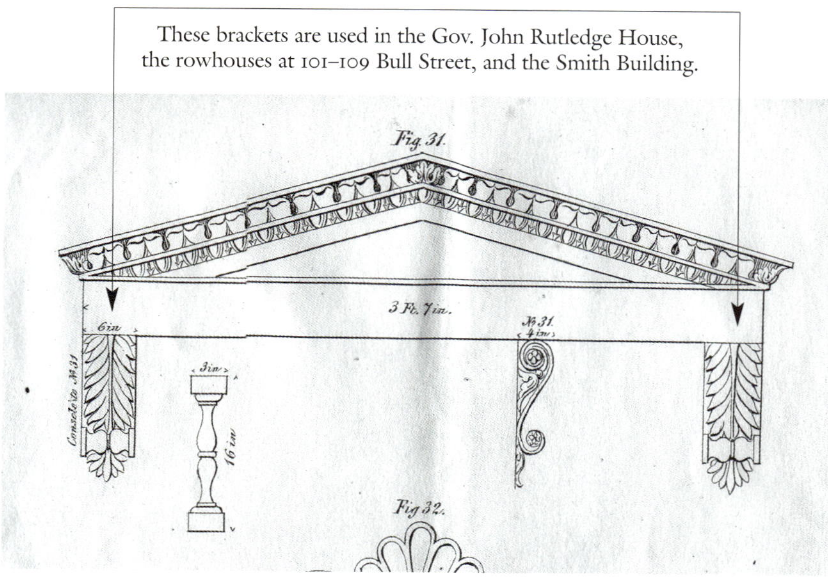

Figure 9 Tolman, Luther and Company, *New Designs of Terra Cotta Ornaments, Manufactured by Tolman, Luther and Co.* (Worcester, Mass.: By the company, 1852), n.p. 12" x 9". (From the collections of the Worcester Historical Museum.)

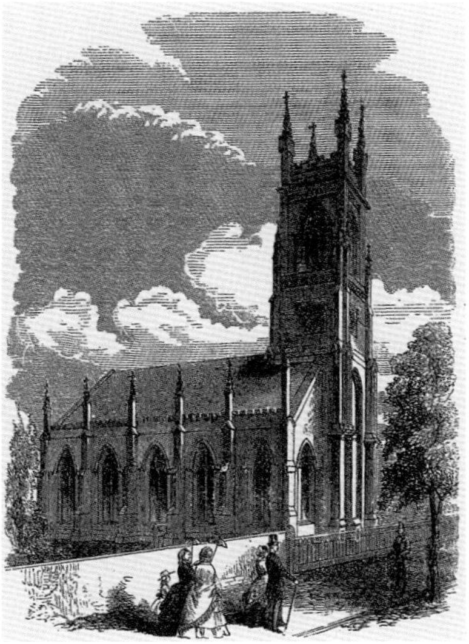

Figure 10 First Unitarian Church, illustrated in *Harper's New Monthly Magazine,* June 1857, p. 14. (Courtesy, Cornell University Library, Making of America Digital Collection.)

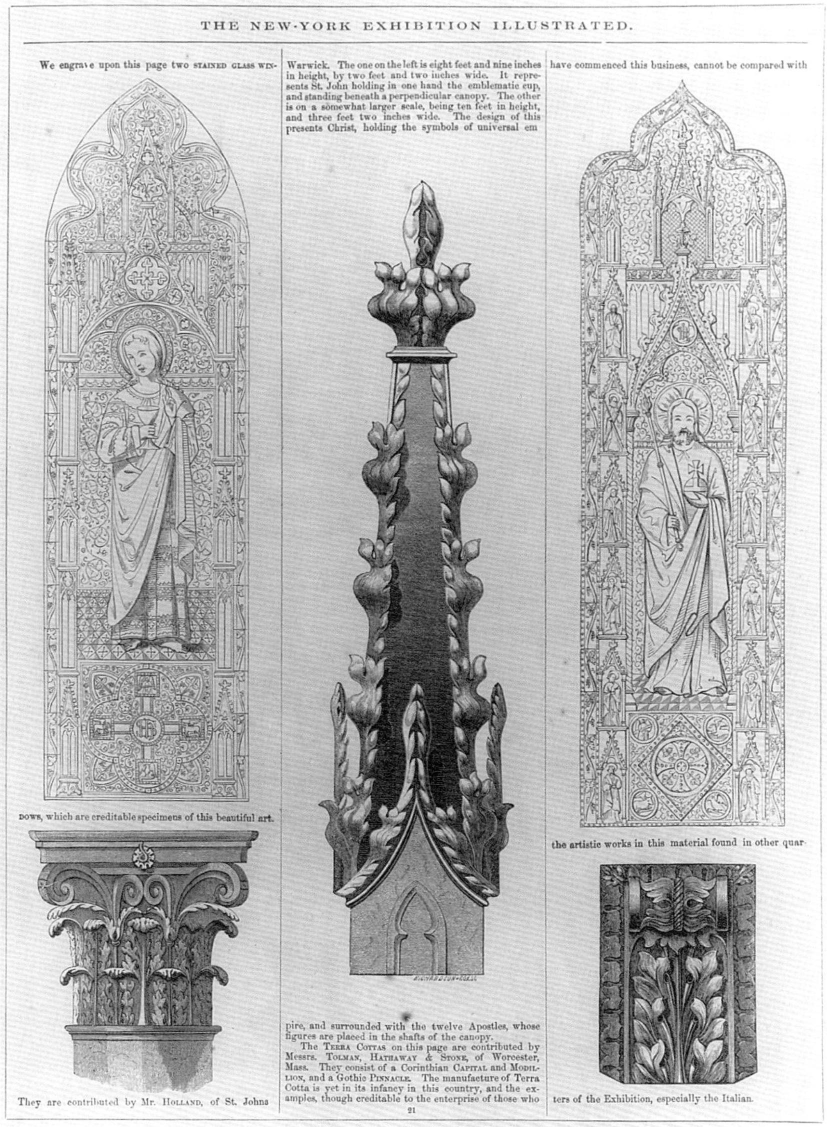

THE NEW-YORK EXHIBITION ILLUSTRATED.

We engrave upon this page two STAINED GLASS WINDOWS, which are creditable specimens of this beautiful art. They are contributed by Mr. HOLLAND, of St. Johns Warwick. The one on the left is eight feet and nine inches in height, by two feet and two inches wide. It represents St. John holding in one hand the emblematic cup, and standing beneath a perpendicular canopy. The other is on a somewhat larger scale, being ten feet in height, and three feet two inches wide. The design of this presents Christ, holding the symbols of universal empire, and surrounded with the twelve Apostles, whose figures are placed in the shafts of the canopy.

The TERRA COTTAS on this page are contributed by Messrs. TOLMAN, HATHAWAY & STONE, of Worcester, Mass. They consist of a Corinthian CAPITAL and MODILLION, and a Gothic PINNACLE. The manufacture of Terra Cotta is yet in its infancy in this country, and the examples, though creditable to the enterprise of those who have commenced this business, cannot be compared with the artistic works in this material found in other quarters of the Exhibition, especially the Italian.

21

Figure 11 Terracotta capital, Gothic pinnacle, and modillion by Tolman, Hathaway and Stone. From Benjamin Silliman, ed., *The World of Science, Art and Industry: Illustrated from Examples in the New-York Exhibition, 1853–54* (New York: G. P. Putnam, 1854), p. 21. The Gothic pinnacle is similar to those used in the 1852–1854 remodeling of the First Unitarian Church.

the terracotta on the Charleston Club (1853–1854, Barbot and Seyle; demolished), Meeting Street, as being "somewhat in the style of the Mills House."[8] And the Guard House (1838–1839, Charles F. Reichardt; demolished), the city's police station, also on Meeting Street, was fashionably updated with terracotta window ornament when its front colonnade was removed in 1856.[9]

It is difficult to ascertain just how widespread the use of terracotta was in Charleston in the 1850s since so few relevant records have come to light. Also, since most of the buildings of the period have been repainted, very close examination and/or invasive procedures are necessary to determine the actual historic material. Given Charleston's history of natural disasters, the existence of eight buildings ornamented with very early American terracotta is impressive and adds considerably to the city's stature as the municipal pioneer in American historic preservation. Future research may yield more examples of terracotta in Charleston and in the South.

Susan Tunick, Friends of Terra Cotta; pstunick@worldnet.att.net
Jay Shockley, NYC Landmarks Preservation Commission; jshockley@lpc.nyc.gov

1. The *Charleston Courier,* April 25, 1851, noted that White had received some of the terracotta in broken condition.

2. Susan McDaniel Ceccacci, "Architectural Terracotta in the United States before 1870" (master's thesis, Boston University, 1991), chap. 2, note 22.

3. Beatrice St. Julien Ravenel, *Architects of Charleston* (Charleston: Carolina Art Association, 1945), p. 249.

4. Ibid.

5. The window pediment illustrated in figure 32 in the catalog has two elements that were used above the second-story windows on the sides of the Rutledge House. In addition, the brackets shown in figure 32 are identical to some of the brackets salvaged from the Mills House Hotel, shown in figure 31.

6. George M. Fiske, "Clay Building Material," *Brickbuilder* (May 1892), p. 36. Fiske refers to "the pinnacle of the First Unitarian Church, Charlestown [*sic*], North Carolina" as made in Worcester and taken down after the "1885 [*sic*]" earthquake. Fiske reported that he had an "excellent" piece of that terracotta in his office.

7. Benjamin Silliman, ed., *The World of Science, Art and Industry: Illustrated from Examples in the New-York Exhibition, 1853–54* (New York: G. P. Putnam, 1854), p. 21.

8. Ravenel, *Architects of Charleston,* p. 231.

9. Kenneth Severens, *Charleston: Antebellum Architecture and Civic Destiny* (Knoxville: University of Tennessee Press, 1988), pp. 161–62.

Kathryn Boxhorn

A Pot for Abby Cross off Westerly, Rhode Island, and Stonington, Connecticut

Salt-glazed stoneware was made in America beginning in the early years of the eighteenth century and reached its peak of development in the first quarter of the nineteenth century. By the early years of the twentieth century its importance as useful utilitarian ware, folk art form and decoration, and production had declined to near extinction.

The vast majority of American stoneware made during this two-hundred-year period was produced in quantity and for utilitarian purposes. Periodically, however, potters would create a single piece in honor of a specific individual. The name of the individual was inscribed on the piece in some manner and, presumably, given to him or her as a gift. These special objects, termed "presentation" pieces, number among the most beautifully decorated and historically intriguing stoneware items ever produced.

One such presentation piece, made for Abby Hinckley Cross by the Adam States II pottery in Stonington, Connecticut, is believed to date from between 1820, the approximate year of Abby's marriage, and 1826, the year of Adam's death. Impressed on the front is "ABBY, CROSS," beneath which are two deeply incised flowers (fig. 1). One flower is daisylike in appearance, having six petals and an impressed circle in the center. The other is more like a carnation, with many deeply incised petals. The stems have large incised leaves and are joined at their bases. On the reverse is impressed "A.STATES" (figs. 2, 3). Cobalt decoration was applied at the base of the handles, over Abby Cross's name, within the floral decoration, and inside the potter's stamp.

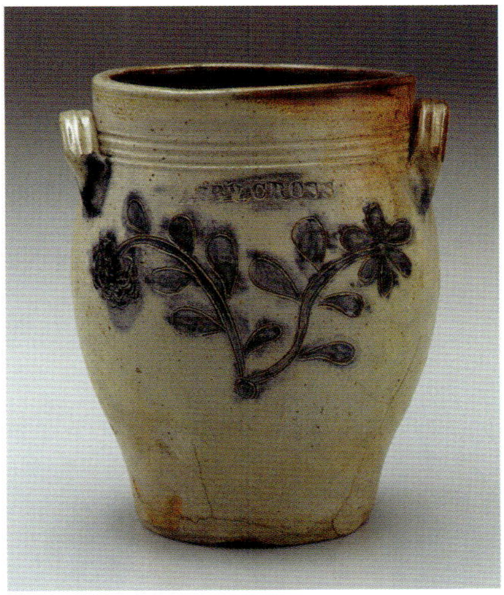

Figure 1 Storage jar, Adam States II, Stonington, Connecticut, 1820–1826. Salt-glazed stoneware. H. 9". This ovoid jar has applied, flattened, and ribbed handles, incised lines approximately one inch below the rim, and an early footed base. The interior has a dark Albany slip glaze. Small clay imperfections can be seen on the surface of the pot.

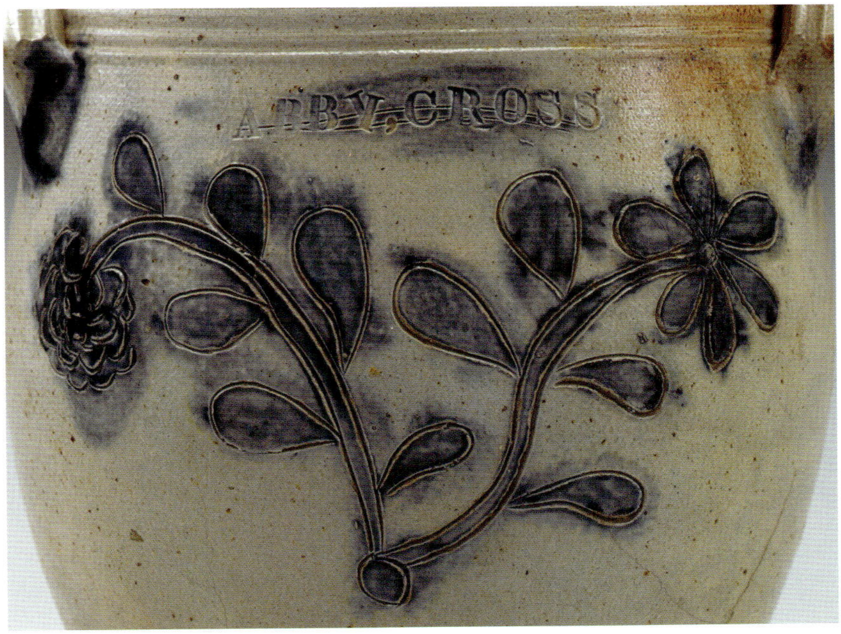

Figure 2 Detail of the stamp on the pot illustrated in fig. 1.

Figure 3 Detail of the stamp on the back of the pot illustrated in fig. 1.

Abby Cross

Abby Cross's family tree includes members of families that were direct descendants of James Babcock, Allen Breed, George Lawton, Thomas Hazard, and Reverend John Lothrup, some of the earliest settlers and founders of some of the most prominent seventeenth- and eighteenth-century families of New England.

Abby was born in Stonington, Connecticut, in 1802, the second child of Samuel Hobart Hinckley (b. 1772), a farmer, and Abigail Helme (b. 1772), both of whom were also born in Stonington.[1] Abby married George Dilwyn Cross (b. 1799) in Westerly, Rhode Island, and gave birth to seven sons and four daughters between 1821 and 1843.

George Cross (1799–1872) was a lawyer and entrepreneur. He served six terms in the state's general assembly, served as a state senator, chief justice of the Court of Common Pleas for Washington County, and, in 1840, was appointed by the State of Rhode Island to be one of the commissioners to survey and adjust the long-disputed boundary line between Connecticut and Rhode Island.[2]

Samuel Hinckley Sr. (1589–1662), the progenitor of the Hinckley family, came to America from Trenterden, County Kent, England, arriving at Boston in 1634. In 1635 he moved to Scituate, Massachusetts, where in 1637 he was made a freeman.[3] In 1640 he moved to Barnstable, Massachusetts, where he remained until his death. His eldest son, Thomas, became the governor of Plymouth Colony; his youngest son, John (b. 1644), was the second progenitor of the branch of the family to which Abby belongs.

John's second child, Samuel Hinckley (b. 1670), was the first of the Hinckley family to settle in Stonington. His son Samuel Jr. (b. 1705/6) had a son, Abel. Abel's second son and sixth child was Samuel Hobart Hinckley, Abby's father.[4]

Abby's maternal grandfather, Oliver Helme, was a descendant of Christopher Helme (b. ca. 1615) of Long Sutton, Lincolnshire, England. Christopher immigrated to Boston, where it is believed he arrived in July 1637 as a member of a party led by Reverend John Wheelwright, a clergyman from Lincolnshire. Wheelwright, banished from Massachusetts for his religious beliefs, established himself and his followers on land near the Piscatoquake River now known as Exeter.

Christopher Helme was one of the thirty-four signers of the Exeter combination on May, 4, 1639. By 1644 he had settled in Warwick, Rhode Island, where he was received as a free inhabitant.[5]

Christopher's second son, Rouse Helm (b. 1641/44), was one of the early settlers of Kingstown, Rhode Island. Rouse's son Rouse II (b. 1683), Oliver's father, was clerk of the state assembly in 1720, an assistant justice of the peace, and a superior court judge. Oliver was born in South Kingston where, in 1753, he was made a freeman. He moved to Connecticut in 1760.

Abby's grandmother Katharine's great-great-great-grandfather John Greenman (b. 1602, Somersetshire, England) immigrated to America and was admitted as a freeman in Dorchester (now Boston) in 1631. In November 1636 he was one of the forty-six original purchasers of land from the Indians in Taunton, Massachusetts. On May, 20, 1638, he was admitted as a freeman in Newport, Rhode Island.

John's son Edward (b. ca. 1630), who was born either in England or in Dorchester, Massachusetts, was admitted as a freeman in Newport in 1655 and in Westerly in 1661. In 1661 a tract of land called Misquamicut (now Westerly) was purchased from Sosooa, a Native American. Edward purchased a quarter of a share of this tract. Edward served as a commissioner and deputy to the General Assembly. Edward's son Edward Jr. (b. 1663) was admitted as a freeman in Kingstown, Rhode Island, in 1696. He served as a deputy to the General Assembly for Kingston and speaker of the House of Deputies. Edward Jr.'s son, Silas (b. ca. 1690), Katharine's father, was a merchant and a judge. He also served as the town clerk of Westerly and as deputy to the General Assembly.[6]

Adam States II

Adam States II was the second of three Adam States who produced stoneware in Connecticut in the eighteenth and early nineteenth centuries.

Adam States I emigrated from Holland and initially settled in South Amboy, New Jersey. By 1743 he was in New York, in 1746 he was in Pennsylvania, then he moved to Rye, New York, and by 1750 was in Greenwich, Connecticut, where he established a pottery and remained until his death, in 1769.

Adam II, born June 8, 1756, in Greenwich, was the sixth child of Adam States I. He was indentured to and learned the pottery trade from his uncle Peter. Soon after his marriage in 1778 to Esther Noyes of Stonington, Adam developed a pottery business in Stonington, living and working there until his death in 1826. He was married three times and fathered fifteen children.

Adam States III, the first child of Adam II, was born in Stonington in 1781. He owned one-half interest in a pottery at Stonington Borough/Long Point, which he operated during the years of 1812–1816 and 1819–1821 with his cousin William (son of Peter, to whom his father had been indentured).[7]

Connecting the States and the Hinckley Families

Westerly, Rhode Island, and Stonington, Connecticut, were closely knit neighboring communities. Although information has not been found linking the States family of Westerly and Stonington and the Samuel Hinckley families through marriage, it can be shown that they were certain to have known each other as neighbors and fellow church members.

Adam States II, Nathan and Elijah Hinckley (brothers of Abby's grandfather Abel), and Oliver Helme were members of Westerly's Hill Church, for which Helme served as clerk beginning in June 1784. During this time, in order to avoid state taxes, residents of Connecticut were issued certificates verifying their membership in the church.[8]

Peter States (also known as Staats) (b. ca. 1720), Adam II's uncle, owned land in Stonington that bordered property owned by Oliver Helme and John and Thomas Hinckley.[9]

Benjamin States

Another pot exists (not pictured here) that was made by the same hand and during the same time period as Abby's jar. This second piece is a two-gallon, straight-sided pot, decorated on one side with deeply incised flowers with long stems that meet and are joined at their bases with several small leaves. There are a number of incised lines approximately one inch below the rim. The handles are flattened and ribbed. The inside has an Albany slip glaze, and clay imperfections can be seen on the vessel's surface. The flowers, stems, and leaves are filled with cobalt, and cobalt brushstrokes appear at the ends of the handles. The front of the pot is marked "B STATES"—upside down—in two places, one below the incised lines near the rim, the other above the incised lines. Both marks are brushed with cobalt.

It is reasonable to believe that this piece was made either for or with the help of Benjamin States (b. 1809), Adam State II's fifteenth child and youngest son. If he had trained as a potter in his younger years, Benjamin ultimately chose to become a merchant and civic leader. He held many public positions, including deputy sheriff, constable, tax collector, captain of the local Light Artillery, notary public, commissioner of the New York and Stonington Railroad, and collector of customs.[10]

Two stoneware vessels made by members of the States family's long line of potters come to life as information about those who made them and those who received them becomes known. These two pieces, both made in Stonington, each tell a different story, separate yet intertwined.

Adam and Benjamin's story is of three generations of potters who produced beautiful objects under relatively primitive conditions. Their pots serve as a fine reminder of these talented men. Abby's story is of a woman, wife, and mother. We see her jar and think of her, her family, the immigrants she descended from, and how important they were to the founding and development of America. Two pieces of stoneware, each a product and symbol of the rich American heritage that is ours.

Kathryn Boxhorn; jfentonboston@aol.com

1. Her grandparents on her mother's side were Oliver Helme (b. June 17, 1731) and Katharine Greenman (b. April 22, 1738), of the well-known shipbuilding family of Westerly and Mystic. On her father's side her grandparents were Abel Hinckley Sr. (b. April 10, 1743) and Sarah Hubbard (Hobart) (b. ca. 1745).

2. Thomas William Herringshaw, *Herringshaw's Encyclopedia of American Biography of the Nineteenth Century* (Chicago: American Publisher's Association, 1902), p. 206.

3. "One enjoying civil or political liberty; one having the full rights of a citizen." *Merriam-Webster's Collegiate Dictionary,* 10th ed., s.v. "freeman."

4. *Biographical Review Containing Life Sketches of Leading Citizens of New London County, Connecticut* (Boston: Biographical Review Publishing Company, 1898), vol. 26, p. 115.

5. Howard M. Chapin, *The Early Records of the Town of Warwick* (Providence, R.I.: E. A. Johnson Company, 1926), p. 217.

6. Elisha R. Potter Jr., *The Early History of Narragansett* (Providence, R.I.: Marshall, Brown and Company, 1835), pp. 250, 255.

7. Lura Woodside Watkins, *Early New England Potters and Their Wares* (Hamden, Conn.: Archon Books, 1968), pp. 178–82.

8. In his book *Genealogy of the States Family* (New Haven, Conn., 1913), James Noyes States includes the following information:

> Adam, the sixth child, was born at Greenwich, Conn. in 1756, and died at Stonington, Conn., near Westerly, R.I. in 1826, aged 70 years. His grave is in a family burial place not far from Westerly on the west side of the Pawcatuch River, on the west side of the road going north towards North Stonington. In 1909, the place was owned by a man by the name of Hall, whose grandmother, Phoebe C. (States) Bliven, was the daughter of Adam States (3rd) of this paper, and Fanny (Chesebrough) States, his wife. The house was struck by lightning and burned down. It was on the west side of the road about opposite the house of Thomas Hinckley.

9. Harold Staats, *Genealogy of the Staats Family* (Ripley, W. Va.: National Staats Reunion Association, 1921), p. 204.

10. Ibid., pp. 182, 183.

Book Reviews

Alasdair Brooks. *An Archaeological Guide to British Ceramics in Australia, 1788–1901.* Sydney and Melbourne: Australasian Society for Historical Archaeology and La Trobe University Archaeology Program, 2005. vi + 87 pp.; 62 bw illus., 60 color illus. on CD-ROM, 9 tables, bibliography, 4 appendixes. $35 Australian (softcover).

The careful use of data from archaeological ceramic specimens can provide insight into a vast number of topics of interest to collectors, archaeologists, and social historians. Avenues of inquiry could include archaeological site dating and chronology, ceramic history and technology, trade, social status, gender, foodways, or any of a number of other topics a researcher might choose to pursue. However, ceramics must be accurately identified at a basic level in order to serve as the foundation on which subsequent work can be based. Such is the belief of Alasdair Brooks, and to that end he offers guidelines for making both accurate ceramic identifications and thoughtful analytical studies.

Quite a task for a slim volume, but Brooks's scope is limited in such a way as to make it feasible. As the title implies, the book has both physical and temporal parameters. It focuses on British ceramics in Australia between 1788 and 1901, with a solid emphasis on nineteenth-century ceramics. Instead of serving as a negative constraint to the work, these boundaries allow Brooks to set out well-organized data and a tightly reasoned model for analysis. Brooks also provides a thorough overview and comparison of ceramics studies in North America, the United Kingdom, and Australia.

The book is organized into five chapters and four appendixes. Chapter 4, on ware, decoration, form, and date, is the heart of the book. Appendix A provides a list of British pottery manufacturers known in Australia. Appendix B is a "pottery timeline" that includes important innovations in ceramics alongside important dates in Australian history. It also supplies a guide to reading registration marks on ceramics. Appendix C discusses further reading, and appendix D reproduces the Society for Historical Archaeology's Standards and Guidelines for the Curation of Archaeological Collections.

Brooks makes a number of important contributions to ceramics studies. In chapter 1 he reviews the history of work in both Britain and North America (with major emphasis on the United States and little specific discussion of Canadian research) and provides a context in which to understand past and current work in Australia. Admittedly oversimplifying his material,

Brooks observes, "British work is descriptive and atheoretical, American work is analytical and theory-informed." Despite this statement, he takes care to point out examples of British analytical work based on their descriptive tradition, as well as descriptive ceramics works that have laid the foundations for American analytical studies. Situated between these two approaches lies Australian research. The author laments the shortcomings of past ceramics studies in Australia but at the same time points out the huge improvements that have taken place there in recent years. Australia suffers primarily from little expert knowledge spread thinly across a vast country. In addition, Australia cannot simply co-opt either British or American research because of historical differences. Many British ceramics studies focus on the period before Australian colonization and are therefore not relevant. Much of the American work is based on the American ceramics market, which was on a much different scale than that of Australia and therefore not necessarily comparable. Thus Australia must build on previous scholarship when possible, but much original, substantive research is necessary.

In chapter 2 Brooks offers his model for ceramic analysis, which combines the best of British and American approaches. The new model is organized into two levels. Level 1, identification, encompasses ware, form, decoration, and date. At this level, a researcher would attempt to determine a ceramic sherd's (or possibly cross-mended vessel's) ware type, vessel shape, form(s) of decoration present, and, based on those data, a manufacturing date range particular to that specimen. Level 2, analysis, consists of economy, status, function, and meaning. Having identified and dated the sherd or vessel as accurately as possible, the researcher can use combined data from assemblages to address more complex issues. It is only possible to touch briefly on Brooks's four analytical topics. The economy of ceramics tends to center on trade, particularly between the British imperial source of ceramics and the colonial or other consumers of those goods (such as Australia, the United States, Canada, South Africa). Questions of status and ceramics focus on combinations of real economic buying power and efforts to display goods of a perceived status within ethnic, gender, or other defined groups. Function addresses a number of complex questions facing researchers. The author's excellent example points out the difference between a plate manufactured for the intended function of food consumption versus the intention of the purchaser to display that same plate on a dresser. In addition, the function of that same plate can change over time. The final category, meaning, is the most complex analytical subject of the group. It challenges the researcher to tease out ideological issues (such as gender or national identity) as expressed in ceramic acquisition and use.

In chapter 3 Brooks moves on to the basics for collections management (not necessarily limited to ceramics): processing, cataloging, and curation. As he notes in his acknowledgments, he hopes his book will be a volume that "an Australian historical archaeologist could just open and use" (p. v). As such, chapters 3 and 4 in particular stand as solid reference materials that should be used by Australians and anyone else interested in British ceramics. Rather than reiterating in this review the details of lab work discussed in chapter 3, it can simply be said that Brooks outlines what most researchers would consider "basic good practice." As support, appendix D provides the standards and guidelines for the curation of archaeological collections adopted by the Society for Historical Archaeology.

In an effort to fill Australia's void in expert knowledge, the longest portion of Brooks's book, chapter 4, provides detailed descriptions of British ceramics. He deals separately with ceramic ware types (the body or fabric of the ceramic), decoration (which can be applied to a number of different wares), and the most common ceramic forms. Ware and decoration can be particularly diagnostic. Brooks's dictionary-style definitions are thorough, useful, and well documented with citations from the best sources the ceramics literature has to offer. Many of the entries are illustrated with excellent black-and-white photographs in the text. In addition, the volume includes a CD-ROM with color versions (in Adobe Acrobat PDF format) of the same photos that are featured in the text. The color photos are exquisite in their high resolution, quality, and trueness of color, and will serve well as visual aids to researchers learning to distinguish differences among the various wares and decorative techniques. However, although the CD-ROM has much to recommend it, technology does have its shortcomings. Researchers with limited computer skills or access, or even just a limited computer, might not be able to take full advantage of the color images in this format. This detracts not at all, however, from the overall superior quality of this volume and the information presented in it. Chapter 4 closes with a discussion of dating techniques as they pertain to ceramics: maker's marks, registry marks, the Mean Ceramic Date formula, and stylistic change.

The body of the book ends with chapter 5, "Analysis and Interpretation." Brooks uses this chapter to provide discussions of and an example of level 2 ceramics analysis, without attempting to dictate "one sole straight and narrow 'correct' analytical path" (p. 56). The main example that Brooks presents is a brilliant miniature study in which he examines the impact of the American Civil War on worldwide ceramics exports from Liverpool and reports on its direct effects in Australia. This examination on the topic of economy is sure to inspire other researchers to look into historical cause and effect. With reference to status, he discusses some of the major previous works on that topic, such as Miller's CC Index and its applicability to the Australian context.[1] On the very complex topic of function, Brooks offers a number of cautions for attempting to ascertain the potential multiple uses of any one artifact. For meaning, he discusses a number of excellent examples of work where researchers have explored "myriad social, ideological, and symbolic implications of an artefact" (p. 65). The author closes this final chapter with a hopeful note for the great potential of ceramics studies in Australia. Simply put, if Brooks's volume is used as a reference and guide, and other researchers produce work of the same caliber as his, then Australian research does indeed have great promise.

Marie Blake Morton, M.A. R.P.A., Bluestone Research

1. George L. Miller, "A Revised Set of CC Index Values for Classification and Economic Scaling of English Ceramics from 1787 to 1800," *Historical Archaeology* 25, no. 1 (1991): 1–25.

Diana Edwards and Rodney Hampson. *White Salt-Glazed Stoneware of the British Isles*. Woodbridge, Suffolk, Eng.: Antique Collectors' Club, 2005. 336 pp.; 241 bw and 200 color illus., 5 appendixes, bibliography, index. $89.95 (hardcover).

I approach this book from the viewpoint of a collector with an interest in white salt-glazed stoneware and its contemporary British pottery.[1] Consequently, I am looking for a readily accessible and comprehensive guide that brings order to current, but scattered, information. The standard references by Charles Luxmoore (primarily on his collection) and by Arnold Mountford (on the holdings at what is now the Potteries Museum and Art Gallery, Stoke-on-Trent)[2] have been overshadowed by a range of in-depth archival inquiries and recent scrupulously recorded archaeological finds at factory sites, waster tips, and domestic-use areas. These new revelations usually have made their way into specialty ceramics publications rather than books intended for a general audience. Quite often such works have brought enough light into dark cupboards and attics to reveal hitherto unrecorded examples of this salt-glazed pottery. Rarely displayed objects in this book include a snuffbox (color plate 65) and a teapot master mold (color plate 56).

Authors Edwards and Hampson previously collaborated on *English Dry-Bodied Stoneware* and each has published several scholarly studies and papers in ceramics history that have added to our knowledge of pottery identification, production methods and sources, marketing, and distribution. Diana Edwards has written six books and is best known for those on black basalt, feldspathic stoneware, and Neale pottery and porcelain. Rodney Hampson has written four books, whose subjects included Longton potters and Churchill china; he has provided historical research for many projects in addition to editing premier journals.[3]

This volume integrates data from disparate sources—such as academic journals, old newspaper accounts, technical reports, historical records, and ceramics publications—as well as providing extensive footnotes. Still, the sheer quantity of information presented can obscure for future analysts the wealth of data provided. These new contributions include summary histories of selected potworks, a list of documented producers of white stoneware, correlation of modest numbers of sherds with extant objects, and clarification and identification of individuals in pottery-owning families sharing the same or similar surnames.

The volume is bound well in boards following the tradition of the Antique Collectors' Club, and the price is suitable for a book with many photographs. The Winterthur Museum and Ceramica Stiftung helped underwrite publishing costs.

Now, to open the covers of this new compendium to disclose what treasure might be at hand. A good beginning! The frontispiece gives a monumental centerpiece the public attention that it deserves.

The photographs (especially color ones) lend valuable insight into new discoveries (or rediscoveries). Pictures of inaccessible private collections and infrequently recorded objects are particularly welcome. Book design integrating color illustrations with the text aids in checking figures, but it would have been even easier if figure and plate numbers were highlighted in bold type in both text and captions. Black-and-white photographs that duplicate the same views in color should have been omitted. Altogether, the selections give a sense of the range of white salt-glazed stoneware produced, but not the relative frequency of each category.

New photography of far-flung pieces may not have been feasible. As a result, the resolution and orientation of various objects may thwart showing details mentioned in the text. Notwithstanding efforts toward consistency, these shortfalls in the illustration program are irritating. For instance, the text (p. 20) mentions knife-cut terminals and directs the reader to color plate 10, showing three mugs. Once there, though, the reader sees that the handle of the truly knife-cut mug is obscured while the other two terminals appear hand-manipulated, that is, thumbed or rolled. Further, the caption is incorrect, both as a result of reversing the sequence of the mugs as they are pictured and of using outdated interpretations.[4] Captions consisting only of source credits, not new information, may leave the reader with questions. Why, for example, is a *blanc de chine* porcelain mug paired with a salt-glazed one in figure 2? It also leaves open a potential chicken-or-egg question. Although globular jugs with short, often ribbed, cylindrical necks are thought by some to be based on Chinese wares produced for the western European market, the source for the profile could well be Rhineland stoneware imports to England.[5] In addition, the shoe model in color plate 9 has been recently proposed as Fulham, circa 1700–1710.[6]

The writers' history of British white salt-glazed stoneware manufacture begins with John Dwight, the late-seventeenth-century entrepreneur who built on German stoneware techniques to establish the British stoneware industry. His startlingly clever, detailed white figures are shown in side-by-side, comparable-quality photographs. Although this book is dedicated to white ware, it would have been helpful for a novice to be directed to other spectacular pottery busts and figures in grayish color or finished "bronze."[7]

The researchers describe economic and social contexts for the development of markets for white salt-glazed stoneware in the eighteenth century. To prepare the reader for considering the actual formative methods used in pottery designing, shaping, burning, and decorating, Edwards and Hampson give an overview of the sources, manipulations, and transportation of the fundamental ingredients—clay, flint, and salt.[8] The networking necessary to obtain raw materials in conjunction with attendant transportation modes is emphasized.

The spectrum of means and equipment used to mass-produce white stoneware is discussed extensively and portrayed from clay cleansing to surface decorating. The progression and endurance of design elements are traced. Nonetheless, a more clear-cut example of blue-colored applied relief would improve the coverage of forms and patterns. In color plate 49, for example, it is hard to tell whether the relief consists of blue-color clay or of blue color only on the surface. Applying relief was a troublesome technique, and some potters had difficulty removing air bubbles. Another move toward completeness would have been to include more specimens with reticulated walls and borders.

Quaint nomenclature to describe clay fabrics and the subsequent wares was too often used loosely and inaccurately in the ceramics literature of the early twentieth century. This work clarifies some of those misused terms, such as Crouch ware (pp. 65, 73) and ball clay (pp. 64, 66, 71).

The authors are unable to confirm whether plaster of paris mold-making was first introduced from France in the mid-eighteenth century, a traditional view. Semantics in the records may prevent our ever being able to distinguish this material from the closely related alabaster and calcined gypsum, cited in earlier documents, that might have been used in a similar way, but named differently. A defining sentence would have helped the reader separate the terms and realize the complexity of the situation.[9]

Despite the research at hand, unfounded tales persist. For example, the contribution and activity of mid-eighteenth-century Dutch enamelers, a Mr. Daniel, and the Warburton family—all in Staffordshire (p. 125)—remain uncertain. Another small quibble: The text establishes that particular salt-glazed stoneware with a possible lead glaze is inscribed "Swansea." Sparse documentation, but not archaeology, hints at salt-glaze potting in that place. It seems hasty to imply (pp. 120, 232) on such sketchy evidence that a referenced teapot could have been made there.

New World archaeological reports, newspaper accounts, and probate inventories compensate somewhat for the loss of archival material through time. As a special bonus, controlled excavations have brought forth remarkably complete American-used thin-body pieces; these were buried soon after production circa 1720. Such early ware as shown here (color plates 12 and 139) is seldom encountered elsewhere in dated contexts.

Juxtapositions of extant objects with archaeological finds and a description of the colonial trade are helpful for attribution and interpretation. Also appreciated are thumbnail sketches of ownership and potting activities at several known work areas correlated with sherd discoveries. The limited number of sherds included still manages to indicate the typical extent and quality of evidence—match, near match, or no match—on which attributions could be suggested. As is now clear, the presence of a sherd does not always a factory make. The authors note a blue-dipped fragment from South Carolina (p. 165, color plate 137) with blisters and blowouts in the glaze. There is at least one extant complete teapot (private collection) with very similar blemishes. In the absence of more definitive pot-making indicators from Carolina, it seems that the remainder could well be a British product, rather than an American waster. After all, there was the world of "seconds" and "worse seconds," as is amply indicated in appendix 3.

The book integrates the activities surrounding pot making with those of distributing wares for domestic sale and export. These operations extended to the management of work sites with the growing division of labor, capital formation, and vertical organization by ambitious owners who controlled supplies, transportation, production, and salesrooms. These areas of inquiry are fruitful for grasping the development of a national industry.

The Park Lane pothouse of John Eccles and Co. (1756–1767) seems out of place among Liverpool white stoneware occurrences in appendix 2 (p. 263). Only delftware was made at Park Lane, Eccles's sole known location during that period. Salt-glazed oven spacers, which were recovered on the opposite side of town at Dale Street, are not traced to Park Lane or any other specific pottery. The text shows Eccles's 1756 notice to make and sell all sorts of blue and white along with "black and white earthenware, being the first of the black and white *Colours ever brought to Perfection in* Liverpool." The former colors were prevalent for delft decoration, but the other set might refer to an unusual "black enamel" version, rather than to "Jackfield" blackware and white salt-glazed stoneware, as construed by the authors.[10]

Pottery inventories in the appendixes facilitate the comprehension of colloquial terms, and the companion customer list reveals social hierarchy and potter interactions. The bibliography offers a wealth of primary material organized by chapter for future investigations.

The eighteenth-century phenomenon of white salt-glazed stoneware was fully English, and this practical and scholarly reference promotes the understanding of the manufacture, economic importance, and present-day identification of the ware. This book condenses much of the current knowledge of the subject and consolidates extensive pictorial material. It is a sober and worthy monograph bolstered by copious notes that should stimulate new research. This study takes its place alongside the Luxmoore volume, with its valuable visual summary of molds, and the Mountford book, with its emphasis on the regional Staffordshire industry. Together they are essential for interpreting white salt-glazed stoneware.

Troy D. Chappell, Collector

1. The authors had intended to consider pieces from my collection for this book, but publication deadlines made that impossible.

2. Charles F. C. Luxmoore, *"Saltglaze" with the Notes of a Collector* (Exeter, Eng.: Pollard, 1924; reprint, London: Holland Press, 1971); Arnold R. Mountford, *The Illustrated Guide to Staffordshire Salt-Glazed Stoneware* (London: Barrie and Jenkins, 1971).

3. Diana Edwards and Rodney Hampson, *English Dry-Bodied Stoneware: Wedgwood and Contemporary Manufacturers, 1774 to 1830* (Woodbridge, Suffolk, Eng.: Antique Collectors' Club, 1998).

4. The mug on the left should read "Fulham (Dwight's factory), ca. 1705"; the middle one, "Staffordshire, ca. 1700"; and the one on the right, "probably Fulham, ca. 1750."

5. Soame Jenyns, *Later Chinese Porcelain: The Ch'ing Dynasty, 1644–1912* (London: Faber and Faber, 1951), p. 82 and pl. CXIX (1a), for information and example.

6. Peter Williams and Patricia Halfpenny, *A Passion for Pottery: Further Selections from the Henry H. Weldon Collection* (London: Sotheby's Publications, 2000), p. 122.

7. Robert L. Hobson, *Catalogue of the Collection of English Pottery in the Department of Mediaeval Antiquities and Ethnography of the British Museum* (London: Bemrose and Sons, 1903), pl. XIV for grayish and gilt bust of Prince Rupert (24 in. high), and pl. XV, for "bronze" figures of Meleager (12 in. high) and Mars (13¾ in. high).

8. For the unsatisfactory thermal shock property of delftware (p. 53), "fracture" meaning "to crack" seems a better descriptor than "friable," often used to describe a powdery surface condition before oven firing.

9. Gypsum—the basic calcareous substance—has a fine-grain solid form known as alabaster. Plaster of paris is ground gypsum from which water has been driven off.

10. For a synopsis of Park Lane products, see Frederic H. Garner and Michael Archer, *English Delftware,* 2nd ed. (London: Faber and Faber, 1972), p. 62. For locations and map, see E. Myra Brown and Terence A. Lockett, *Made in Liverpool, Liverpool Pottery and Porcelain, 1700–1850* (Liverpool: National Museums and Galleries on Merseyside, 1993), pp. 23–24, 28. For advertisement with original emphasis, see Lionel Burman, "Excerpts from Williamson's Liverpool Advertiser and Mercantile Register—1756," *English Ceramic Circle Transactions* 17, pt. 1 (1999): 34–46. For "black enamel" delft example, see Troy D. Chappell, "An Adventure with Early English Pottery," in *Ceramics in America,* edited by Robert Hunter (Hanover, N.H.: University Press of New England for the Chipstone Foundation, 2001), p. 203, fig. 29.

Gordon Elliott. *Potters: Oral History in the Staffordshire Ceramic Industry.* Leek, Staffordshire, Eng.: Churnet Valley Books, 2004. £9.95 (softcover).

These twentieth-century oral histories about work in Staffordshire potteries and service trades reveal insights into how pots were made and decorated and offer remarkable glimpses of nineteenth-century material culture survivals. Pottery production practices are a long-standing research interest of the author, Gordon Elliott, former Keeper of Ceramics at the Stoke-on-Trent City Museum (now Potteries Museum and Art Gallery) and ceramics history lecturer at Staffordshire University.[1] This book presents Dr. Elliott's interviews with residents who responded to his mid-1970s appeal in the *Staffordshire Sentinel* for oral histories originally intended to accompany displays in a planned museum extension.

At the time of the interviews the author was studying nineteenth-century child labor in the potteries and realized that the children described in those government reports were of the same generation as the grandparents of the people whom he was interviewing. Accounts such as Alice Morris's memories as a thirteen year old of wedging clay and providing foot power for the turner's lathe led the author to consider parallels between nineteenth- and twentieth-century conditions and to question whether worker welfare had improved in the intervening seventy years. Selections from the Samuel Scriven (1842) and Robert Baker (1865) government reports are included as appendix 1 so that readers can compare testimony between these different generations of pottery workers. Appendix 2 lists weekly wages in the ceramics industry from 1841 to 1938.

The nine interviews consist of five potters or assistants (all in clay-forming specialties), one laborer, one owner, and two artists and teachers. The tenth interview reveals glimpses of a flourishing pawnbroker trade often central to the gritty rhythm of the work week for potters whose jobs depended on foreign markets and who lived from one paycheck to the next.

Black-and-white photographs and drawings illustrate the workers, clay-forming processes, and individuals and ware types mentioned in the text. The potbank interior views are particularly well chosen and include a remarkable series of photographs from the early 1930s showing John Lockett's factory, its management, and earthenware thrower George Myatt using one of the last of the great wheels, powered not by steam or electricity but by his assistant.

The use of hand (or foot) power well into the twentieth century is one of the revelations of this book. Thrower Myatt preferred the two-legged stance that was possible with an assistant to power the great wheel, which he used to throw druggist and medical wares at Lockett's between 1919 and circa 1941. Lockett's, "a remarkable example in the 1950s of a 19th century time capsule with its workshops converted from a row of cottages and a pub" (p. 11), probably was able to stay in business because of its specialized hospital and box (covered jars) wares market. When box ware production was lost to glassmakers during World War II and the passage of air pollution regulations in the 1950s required kilns to be upgraded, this small family firm with its outdated equipment and layout and long-term employees went out of business.

Great wheels also survived in some bone china factories. Alice Morris turned a great wheel for a jiggerman and was lathe treader on a great wheel for a turner. She liked being on her feet and compared the treading to dancing. When slip-casting machines were introduced in the 1950s, turners and their assistants had to train for other jobs, so in her fifties she learned scolloping (scalloping cups and sugars). Still working at age seventy-five, she said that the young people's work "is rubbish. My scolloping is done twenty times slower but it's really good work, not like theirs. I don't know how the firm can sell most of it, I don't" (p. 77). This pride in their craft is a recurring element in the potters' accounts even as the difficulty in competing against more cheaply machine-produced wares is obvious.

Other material culture gems include information on potters' tools. The loss of jobs during the Depression and the shrinking number of throwing jobs as the twentieth century progressed led thrower Myatt to branch out from medical wares to general earthenware production. The general earthenware production was easier, not so heavy as the fifty-six-pound mortars and three-gallon barrel jugs that he had made at Lockett's, but he had to learn to throw a greater variety of shapes. Thrower's ribs usually had to be custom-made by each worker to fit his own hand as well as the particular shape. When, during the Depression, Myatt worked at as many as five firms a week to make a living, he had to have hundreds of ribs, one for each vessel shape and size at the different factories. Many of the ribs were marked so that the owner could keep track of them.

Platemaker Leonard Potts describes making bone china plates at Spode using a different hand-held profile for each plate shape. "A ten inch [unscalloped] plate profile would not make a ten inch scalloped plate" (p. 84). Although Potts attributed the twentieth-century use of hand-held profiles at Spode to owner Gresham Copeland's keenness "to keep everything that was old and good" and to a shortage of capital for buying jolly machines, the author points out that the bone china body was less plastic than the earthenware body and thus not as easily adapted to machine production.

Reginald Haggar, recently the subject of an affectionate Northern Ceramic Society centenary newsletter, was a key figure on the Staffordshire ceramics scene as founding tutor of the ceramics summer school at Keele University, founding president of the Northern Ceramics Society and artist, educator, designer, author, and collector.[2] Haggar vividly describes the art director's studio at Minton's, where he worked briefly at the beginning of the Depression:

> It was exactly like it was in the time of Leon Arnoux [Minton's art director from 1848 to 1892], whose portrait hung on the wall. . . . The water jars that were used were the same jars that were on the table in the portrait of Arnoux. . . . All the furnishings were exactly as they were in the 19th century. . . . The old traditions were upheld and maintained without alteration into the period when I came to the factory. (pp. 21–22)

Another picturesque scene from the Haggar interview is his description of Alfred Meigh of the Meigh potting family, "with his cape and his cap, his little carpet bag, his big boots and his knickerbockers. I remember that he had all the stories he liked to relate on his cuff so that he wouldn't forget them" (p. 25). The author notes that it was Meigh's collection of potters' marks and backstamps that Geoffrey Godden used as the basis for his 1964 encyclopedia of marks. Haggar compared Staffordshire characters such as Meigh to the individualistic appearance of the potbanks and the potteries' built environment.

Finally, as a white granite collector, I appreciated Haggar's comment that "elaborately gilt patterns" such as those by Minton were the sort that covered up "a multitude of sins. The sorting you know for an expensive plate is not the kind of sorting that you need for a plain white plate[,] which is more difficult to make" (p. 21). Such insider quotes make it easy to see why Elliott felt that these histories deserved publication. This book is the equivalent of a good Eastfield Village ("dish camp") lecture and demonstration or a *Ceramics in America* technology article, with working potters talking about how and sometimes why they did things the way they did and pictures illustrating the process.[3] The audience learns to recognize how the pot was made and the types of labor that went into it. As George Myatt explained:

> there are a lot of alleged connoisseurs of pottery that can't tell a cast or jollied article from something that has been thrown, whereas the thrower needs only to put his fingers inside an article and he can tell you the way it's been made. In my own little tinpot way, only on making, I always look inside being a thrower, to see if it's a good article because if it's made poorly, it's a poor article no matter what's done to it afterwards. If it's made wrong[,] it's wrong right through. (p. 64)

It is this collective memory of pottery making that is being lost as factories and equipment are sold, potteries jobs are outsourced to foreign countries, and workers die. By publishing these interviews, Gordon Elliott has preserved some of those memories and paid tribute to the workers who make the pots that we study and love. Anyone with an interest in Staffordshire pottery making will find treasures in this book.

Amy C. Earls, *Ceramics in America*

1. Gordon Elliott is author of *The Design Process in British Ceramic Manufacture, 1750–1850* ([Stoke-on-Trent]: Staffordshire University Press, [2002]), and *John and David Elers and Their Contemporaries* (London: Jonathan Horne, 1998), as well as compiler of *Some Descriptions of Pottery Making and Working Conditions, 1557–1844* (Stoke-on-Trent, Staffordshire: Ivy House Publications, 1970).

2. Haggar shared with his contemporary, Paul Johnson's headmaster father (see review in this issue, pp. 290–92), a love of painting and sketching crumbling old pottery buildings.

3. *Dish camp* is Joy Hanes's term for Don Carpentier's annual Eastfield Village summer ceramics workshop and seminar in the Albany, New York, area.

John Griffin, *The Leeds Pottery, 1770–1881*. Leeds, Eng.: Leeds Art Collection Fund, 2005. 2 vols., 639 pp.; 1,134 color plates and 186 other color illus. £70 (hardcover).

This landmark publication incorporates newly discovered sources, original factory design books, and the expertise of a dedicated collector. In two volumes, John Griffin revises and expands our knowledge of one of England's foremost earthenware factories. Volume 1 is mainly devoted to company history and a survey of production, predominantly creamware and pearlware. Volume 2 concentrates on analyzing and reproducing twelve surviving design, drawing, and pattern books from the late-eighteenth- and early-nineteenth-century Leeds Pottery. He concludes with details of a revival of factory products in the period 1880 to 1950.

Volume 1 begins with chapters devoted to a critical history of the factory. We read in great depth about the founding partners, their personal investment, and share ownership. The partnership articles of agreement are quoted in part, analyzed, and a complete transcript is appended. Details of retirement, death, and inheritance are given in full. One might wonder why all of this needs to be said. On reflection, I deduced that the author was cognizant of the need not only to explain why past discussions about the Leeds Pottery were inaccurate but also to prove his position beyond reasonable doubt. He does this convincingly by providing primary sources to ensure that there could be no further excuse for not knowing when the factory started, who founded it, and when and why it had financial successes and difficulties.

The inclusion of Court of Chancery proceedings, annual coal consumption, and insurance policies might be considered unnecessarily pedantic. Books on pottery are often written exclusively from the collectors' point of view, but Griffin's book offers so much more. It is essential reading for everyone interested in material culture. He discusses partners' salaries and workers' wages and also quotes the factory's changing value and prices realized when partners' shares were sold by their heirs. Often we wonder what the present-day equivalents of these amounts would be. When citing a newly discovered Hartley, Greens and Co. Leeds Pottery price list of 1796, the author indicates that £71.52 is the 2003 equivalent of £1 in 1796.[1] In his discussion of the price list he notes that the left side of the list features "A Scale of Twelve Inches English Measure," an important guide as every nation had its own measuring standard. An English foot at twelve inches was apparently different from the measurements of a foot in Denmark, France, Germany, and Russia. This is just one of the pitfalls identified in surviving letters and documents quoted by Griffin to illustrate the problems associated with foreign trade, particularly with Russia, in the earliest years of the nineteenth century.

Before surveying the factory's products, the author reviews factory marks, thirty-one of which illustrate an incredibly useful bookmark that accompanies the volumes. Discussion of the wares opens with a reproduction of a copy of the 1794 design book, annotated with the range of sizes and prices—usually per dozen—and generally known as the agents' book. In an invaluable exercise, using the published design books of five other factories for reference, the author's captions for the agents' book pages indicate which Leeds pieces are identical or similar to the products of their contemporaries. For those of us interested in identifying makers of creamware, Griffin generously shares with us this result of his extensive research.

The author illustrates a range of extant Leeds Pottery pieces with excellent photographs (most by John R. Griffin of London). He concentrates on marked examples, including creamware, pearlware, feldspathic stoneware, black basalt, and porcelain. Griffin spends little time discussing the specific types of ceramic produced at the Leeds Pottery, only giving a brief description before each set of illustrations. He seems to feel that his expertise lies in Yorkshire ceramic history, not in the history of the wares. This may be the only place where I take exception to the author's work, where I feel he does not hold to the same burden of proof he demands for an understanding of the Leeds Pottery. He provides no primary source for his emphatic assertion that "[p]earlware was introduced by Josiah Wedgwood in 1779" (p. 191; Wedgwood's name for the ware was "Pearl White"). Noting that "it has also been claimed that other Staffordshire potters had introduced what they called 'China Glaze' as early as 1775" (p. 191) hardly compensates for perpetuating the myth that Wedgwood introduced pearlware.

The first volume concludes with a chapter on the factory from 1827 to 1881. The author expresses his frustration at the lack of sources, reflecting that they do not allow him to discuss this period of the Leeds Pottery in the same detail as the 1770–1826 period. In fewer than ten pages he surveys this later history and the known products, concluding that the wares, no longer fashionable, were confined to the home market and were indistinguishable from the general run of Staffordshire products.

Volume 2 is primarily devoted to the surviving factory design, drawing, and pattern books. These are not the published books that have been made widely available through reprinting but are factory documents, sometimes a chronological illustration of painted patterns, sometimes a collation of shapes, sometimes a chronology interspersed with pasted "scraps." Many of the sketches have comments written on them by customers submitting orders, some of which include suggestions for improvements to the shape. These comments shed light on the way in which the English pottery trade was conducted in the late eighteenth and early nineteenth centuries. The fascinating terminology, the requests that Leeds Pottery make wares formerly obtained from Dutch manufacturers, and correspondence commissioning special designs all tell of a factory ready to respond to consumer needs. Understanding these books, synthesizing the story they are trying to tell us, as well as appreciating the artistry—all these complex issues are tackled by the author. It will take a truly dedicated reader to thoroughly understand his analysis, though. This reviewer needed to read and reread the associated text a number of times to assimilate Griffin's efforts to communicate both his insights and his reservations about what can be deduced from the design books. This is less a reflection on the author's writing than an observation on the complexity of these kinds of sources.

In closing, Griffin reviews the work of potters in the Leeds area who revived Leeds Pottery creamware. He illustrates the well-known 1913 catalog of the dealer W. W. Slee, in which we see pieces that were made by the Senior family of potters. These creamwares fueled the American market for colonial revival ceramics and can be found in many collections where they are said to be eighteenth-century examples.

Griffin is a meticulous researcher, determined to offer a convincing retelling of the Leeds Pottery story. He painstakingly cites every source, and his use of marked examples offers a real basis for understanding the wares of this prominent factory. The support of many Yorkshire-based charitable organizations made this publication possible. We owe a debt of gratitude to them and to John Griffin for producing what will surely be a standard work of reference for the foreseeable future.

Pat Halfpenny, Director of Museum Collections, Winterthur Museum, Garden, and Library

1. J. O'Donoghue, L. Goulding, and G. Allen, "Composite Price Index, 1750 to 2003," *Economic Trends* (London: Office for National Statistics, March 2004), pp. 41, 43.

David S. Howard. *The Unforgiving Minute.* Durham, Eng.: Memoir Club, 2004. 464 pp.; bw illus. throughout; index. £25 (hardcover).

It was at the suggestion of his wife Angela that David Howard wrote his memoirs. *The Unforgiving Minute* and his other two major publications of the past decade, *A Tale of Three Cities* and the monumental *Chinese Armorial Porcelain II,* were written while David was fighting cancer.[1] These represent three remarkable publications by one remarkable man. He began work on the present book in September 2003 and began transfusions in October; by the end of November he was told he had two or three months to live. It is extraordinary that he was able to persevere, despite leukemia, for sixteen months and produce a book of 464 pages, as well as work on other projects. Given these circumstances, a certain lack of editing in two or three of the sections can be forgiven. As Angela has stated, it was more important for the book to be published while David was alive than to tackle revisions that were beyond his physical limits.

The Unforgiving Minute begins with David's early years. Born in 1928, the first child of a successful businessman and his wife who were both from old-line English families, he is brief in his descriptions and mentions only the highlights of a happy childhood, shared with one other sibling. While he was at the preparatory school Belmont, World War II broke out, and concern for student safety led to the closing of the school on the coast and its removal to a pre-Georgian house farther north. At the close of the term, the suggestion was made to move the school to the Bahamas in case of invasion. At the time of the Battle of Britain, the school was evacuated to Clerihew House in Nassau, owned by an American entrepreneur named Harry Oakes, knighted in England, who generously offered it to the school.[2] The story of an English schoolboy's life in Nassau is perhaps the most delightful part of this book. The return to England in January 1944 is described modestly but must have been a hair-raising experience with war raging in the Atlantic Ocean. The route home was circuitous, flying from Nassau to Miami, traveling by train to New York for visas, back to New Orleans by train, boarding a neutral ship heading for Lisbon, again by plane to Ireland, and thence to England. What an adventure in the middle of the war for a sixteen year old!

After two years at Stowe, the war just over, David was commissioned into the Coldstream Guards, the oldest regiment in the prestigious Brigade of Guards, with the plan of going to university after he had served his time of duty. He was sent to Palestine, where the ever-interested, ever-observant young platoon commander enjoyed the extensive archaeological ruins of the Middle East and explored Petra. He returned to England two years later and wrote to Trinity College, Cambridge, that he would not take up the scholarship held for him. Rather, he had decided to go into business and travel. David joined Bridport Industries, a family business with his uncle and other relations being chairman and directors. Bridport was a manufacturer of cording and twine, especially the sports nets used for tennis and cricket. For outside interests, he became an active member of the Amateur Dramatic and Operatic Society and captained the Dorchester Rugby Club, again illustrating the vast breadth of his accomplishments.

In 1952 David married his first wife, Elizabeth, with whom he had four children. Throughout this period David's interest in heraldry burgeoned, first working on his own ancestry and then identifying coats of arms for others. It was in this manner that he was introduced to armorial Chinese export porcelain. He became fascinated by the collection of porcelains owned by Cecil Bullivant, who, after David had identified the arms on some six hundred pieces in his possession, introduced him to other collectors and owners. It was Bullivant who convinced David he should write a book on the subject. David drove throughout England photographing and cataloging armorial export porcelain in anticipation of publishing. While working on his future tome, he traveled extensively throughout Europe for Bridport. In 1962 he decided to leave the company and join Arnolds, the largest school supply business in England. This job would lead to extensive travel all over the world, including Africa, India, and the Middle East. It is difficult to understand how a man who traveled so much and was raising a family was able to write the most comprehensive volume on Chinese armorial porcelain ever printed.

David's first major volume, *Chinese Armorial Porcelain,*[3] had been greeted with great surprise and enthusiasm by collectors and scholars of Chinese export porcelain. No such comprehensive volume on the subject of armorial porcelains for the British market had ever been produced, and its sheer size reflected years of research and writing. The last book published by Oxford University Press in letterpress format, it broke new ground with David's insistence that each illustration be placed with the relevant text.[4] Containing 1,034 pages and hundreds upon hundreds of illustrations, it became an invaluable resource; the cover of my own copy has broken away and the edges of the pages are soiled from use.

David was well aware at the time that perhaps as many as three or four thousand additional armorial services might have been produced for which he had no visual record. In 1974, after the publication of the first volume, he began collecting photographs and documentation to extend his study. Thanks to his broad contacts with collectors, dealers, old aristocratic English families, auction houses, and museums, he was able to amass the material for a second volume. His enthusiasm, modesty, and energy infected all he met, and the ceramics world was more than willing to supply illustrations and information. Through his shop—first in London and later at Manor Farm near Chippenham, Wiltshire—the Internet, and his extensive correspondence, he was able to glean more than 1,300 new armorial examples for the British market. The second volume is almost entirely in color with some 2,000 illustrations, greatly adding to the ease of identification.

Both volumes reflect an innate sense of organization. The idea of arranging the services by border patterns, with introductory examples at the beginning of the book, was brilliant. Only someone with the most comprehensive knowledge of heraldry could write such entries for each example of porcelain. David's ability to compare services owned by different families and by members within the same families reveals an understanding of his subject that few other authors have attained.[5] And to think that all of this research and writing were done while running a business and lecturing extensively in the United States and Great Britain! The appendixes in both volumes are as enlightening and fascinating as the text and illustrations.[6] The appendix of family mottoes, for example, saves hours of time looking through the various services for the correct coat of arms.

When the first book was published in 1974, David decided to leave Arnolds and open his own heraldry shop at Hay Hill in London. This would enable him to make lecture tours, especially to the United States, which was to be the source of a large number of his clients. In the meantime, David had been approached by Mildred and Rafi Mottahedeh in New York. Their collection of Chinese export porcelain was legendary, and they wanted it published as a reference source for other collectors and curators. It was David who, with John Ayers, was to write the two-volume *China for the West,* published in 1978, which records this extraordinary collection that has now been dispersed throughout the world.[7]

David's *A Tale of Three Cities* was produced for the Sotheby's London exhibition commemorating the transfer of Hong Kong to the Chinese in 1997. The exhibition followed in the tradition of a number of significant China trade exhibitions that contributed greatly to the understanding of the imported exotic and practical products and the impact of the China trade on the West, beginning with the one at the Metropolitan Museum of

Art in New York in 1941.[8] David's first major China trade exhibition in 1984, including all of the objects made for export to the West, not just ceramics, was "New York and the China Trade" at the New-York Historical Society.[9] Later that same year David was guest curator for an impressive heraldry exhibition commemorating the five-hundredth anniversary of the College of Arms, London. The catalog accompanying Patrick Conner's exhibition "The China Trade, 1600–1860," held two years later at the Royal Pavilion in Brighton, England, included contributions from David.[10] Thus it was only natural that he should be the person to organize the show and write the *Three Cities* catalog for Sotheby's. David was the honorary curator of the project, while others in the field contributed to the catalog. The exhibition and catalog were equally comprehensive in their coverage of the subject, with sections on porcelain, lacquer, furniture, paintings, textiles, and metals, all tied together with documents, historical background, and survivals from dated shipwrecks. David did the research and writing for this immense production while undergoing treatment for cancer. The exhibition opened to rave reviews, and the beautifully illustrated and well-arranged catalog is a required addition to any library on the China trade.

David became a consultant for Stair and Company and began his extensive and well-known lecture tours throughout the United States. I was fortunate enough to lecture in a number of symposia with David. He always had a fresh approach to his subject and infected the audience with his enthusiasm. His recollections of the lecture tours in the book drag a bit and would have benefited from more careful editing to focus on the highlights alternating with the author's writing, publishing, and business pursuits.

Over the years David moved his business from Hay Hill to Grafton Street, then to Bath, and finally to Wiltshire. He and Angela generously and graciously entertained house guests from the museum and collecting world; the list of their guests reads as a who's who of the ceramics world. The last section of the autobiography relates the stories behind the exhibitions and publications, the extensive lecture tours, the recording of a full and busy life. In 1994 he published *The Choice of the Private Trader,* a beautifully illustrated catalog from the collection of Leo and Doris Hodroff, part of which is now at the Winterthur Museum.[11] This included in its introductory chapters previously unpublished information from the East India Company records on the role of the private trader and the way in which the company auctions worked.

One of the most enduring qualities of this recollection is the modesty with which it is told. There is no effort to impress or overwhelm the reader with the importance of the author's accomplishments. The story is told in the same self-effacing manner with which he had led his entire life. Although there are times one might put down the book, saturated with information and names, until another reading, it is a fascinating record of a fully led life—a life of diversity, far-flung interests, intellectual pursuits, research, exhibition talents, sports, the arts, acting, gardening, travel. I know of few people who can say they have used their time so well, for their own pleasure and the pleasure of others. I am only thankful that David was able to leave us this memoir as a reminder of what can be accomplished in a single person's life, and that he had the dedication and drive to produce such a work under the direst of circumstances. A fitting reward for his heroic efforts was for him to have seen it published.

Carl L. Crossman, Northeast Auctions, Portsmouth, New Hampshire

1. David S. Howard, *A Tale of Three Cities: Canton, Shanghai, and Hong Kong; Sino-British Trade in the Decorative Arts* (London: Sotheby's, 1997); idem, *Chinese Armorial Porcelain II* (Wiltshire, Eng.: Heirloom and Howard, 2003).

2. Within a year, the murder of Sir Harry Oakes was to become a cause célèbre, implicating even the governor, the duke of Windsor.

3. David S. Howard, *Chinese Armorial Porcelain* (London: Faber and Faber, 1974).

4. Traditionally, for the sake of economy, the photos in art books had been placed at the back.

5. The two volumes are necessary companions. In the first, the introductory chapters explain the history of the trade and Chinese export porcelain with such headings as "The Hon. East India Company," "Canton," "The Porcelain and Its Painting," "The Armigerous Families," "Heraldry and Chinese Painters," and "Collections and the Salesroom." Only a dealer very familiar with the trade could have written the last chapter and the similar but more comprehensive one on the analysis of prices paid in the second volume.

Chinese Armorial Porcelain II does not go over the old introductory territory of the history of the trade and porcelain but adds new and equally important chapters: "Reflections, Replacements, Reproductions and Revival," "Of Counters, Carriages and Bookplates"—the sources of inspiration for the Chinese decorators of porcelain—"Scotland and the Swedish East India Company," and "The Market during the Last Century and in the Future." Therefore, for a comprehensive overview of the trade and manufacture as well as later history of sales, both volumes are mandatory, yet each can stand alone. Although the first volume has long been out of print, Angela's intention is still to fully revise and publish, in the next few years, a second edition of volume 1, in color, to match volume 2. This will include a Supplement 3 for all the newly discovered services (more than two hundred to date) not in either volume.

6. Appendixes cover the governors of India, the directors of the Hon. East India Company, the captains, supercargoes, Bombay Country ships, lords mayor of London, City of London livery societies, analysis of occupations of owners of services, analysis by county of families with services, known bookplate originals of armorial services, and—last but not least—the invaluable list of mottoes on armorial porcelains.

7. David S. Howard and John Ayers, *China for the West: Chinese Porcelain and Other Decorative Arts for Export Illustrated from the Mottahedeh Collection*, 2 vols. (London and New York: Sotheby Parke-Bernet, 1978).

8. *The China Trade and Its Influences*, exh. cat. (New York: Metropolitan Museum of Art, 1941). This show was followed in 1954 by an exhibition and catalog at the Los Angeles County Museum, *American Ships in the China Trade* (Los Angeles: Los Angeles County Museum, 1955). My exhibition of China trade paintings and objects at the Peabody Museum and book, Carl L. Crossman, *The China Trade: Export Paintings, Furniture, Silver, and Other Objects* (Princeton, N.J.: Pyne Press) followed in 1972. My second book, *The Decorative Arts of the China Trade* (Woodbridge, Suffolk, Eng.: Antique Collectors' Club) was published in 1991. A comprehensive show at the DeCordova Museum in Lincoln, Massachusetts, in the late 1970s (Kee Il Choi Jr., *The China Trade: Romance and Reality* [Lincoln, Mass.: DeCordova Museum, 1979]), emphasized life in Canton and the China trade. The exhibit, which ran from June 22 to September 16, 1979, was organized with the Museum of the American China Trade, Milton, Massachusetts. From the early 1960s until its merger with the Peabody Essex Museum, the Museum of the American China Trade, founded by Crosby Forbes, made a major impact on the field with its rotating exhibitions and outstanding collections.

9. David S. Howard, *New York and the China Trade*, exh. cat. (New York: New-York Historical Society; Frenchtown, N.J.: Columbia Publishing, 1984).

10. Patrick Conner, with contributions by David S. Howard and Rosemary R. Wallis, *The China Trade, 1600–1860* (Brighton: Royal Pavilion, Art Gallery, and Museums, 1986).

11. David S. Howard, *The Choice of the Private Trader: The Private Market in Chinese Export Porcelain Illustrated from the Hodroff Collection* (London: Zwemmer, 1994).

Paul Johnson. *The Vanished Landscape: A 1930s Childhood in the Potteries.* London: Weidenfeld and Nicolson, 2004. 200 pp.; 18 bw pen-and-wash illus. by the author. £16.99 (hardcover).

How could I resist a book by noted historian Paul Johnson about growing up in Staffordshire? Johnson's account of life in Tunstall in the 1930s is not about potters, pottery, or potteries, although these provide local color to his affectionate sketches of the interwar Staffordshire scene. Although written by a historian, the book is not about how or why the industrial landscape vanished, either. Rather, this is a tribute to a traditional potteries way of life on the verge of modernization.[1]

As the son of the Burslem art school headmaster ("Willy") and his wife (unnamed), Johnson's middle-class perspective complements the better-known Staffordshire voices of Arnold Bennett (b. 1867), with his grim-smile ironic view of the "Five Towns," or Charles Shaw, a mold runner and handle maker in Tunstall from ages seven to ten in the 1830s.[2] Johnson's account begins at age three and ends seven years later with the beginning of World War II. His memories are in rough chronological order, with topics ranging from cinema to park to countryside and beyond, and emphasize the social institutions that maintained the status quo—home, school, chapel, and work.

Little Paul seems to have shared his artistic father's strong visual sense, memorizing and copying book illustrations and prints, such as Nelson dying at Trafalgar, before he could read. Gazing down on the clutter of potbanks and dwellings at night from his attic bedroom, Paul was fascinated by the variety of fires and their reflections in the dark waters of "the Sytch," a stygian stretch of furnaces and polluted waste along the Scotia River valley.

Members of Johnson's family are the principal characters. The youngest of four children in a family of educators and readers, Paul attempts to untangle the meanings of the conversations that he overhears (e.g., the Catholic church "going great guns" suggested to him Nelsonian cannons being dragged up the tower and bombarding the sullen Protestant mobs). This wordplay and the scattering of potteries' dialect and forthright speech are chief among the book's delights.

The elder Johnson's headmaster job entailed bringing art to the people and finding jobs for his students. During the Depression this meant hours spent outside class making friends with the managers of the works, learning about new product lines that might provide openings, and praising his students to potential employers. In the author's family, art was for everyday use. The headmaster believed that if people "eat off plates and drink from cups which have fine shapes and beautiful patterns and colours, their characters will be softened and their behaviour will improve. So it's up to us to provide these things and that's what the Potteries ought to do in its particular sphere" (p. 117). As a result, Paul's toys included Doulton figures—probably gifts from the works to his father.

In his spare time, Paul's father loved drawing buildings and landscapes. The potteries gave his father plenty of subject matter ready to hand. He told his son: "The Potteries is hideous, dirty, wasteful and, I suppose, inefficient these days. But it's beautiful. . . . The French have a word for it, as they do for most things involving art—*jolie laide.* The Potteries is an ugly woman who has a strange kind of beauty. I shall be sorry when they kill her off" (p. 196).

It may be that distance from the potbanks improved one's appreciation for this visual poetry. Paul's mother looked on the district with less benevolence and left after her husband died. The potbanks made more work for her. She complained that

> the pot-banks are so low that smoke just tumbles down into the houses. And those horrible pot-banks are everywhere, close up to where people live. They burn disgusting cheap slack instead of real Wigan coal, and you can hear it exploding in the ovens and see it jumping up the pot-banks in great showers of sparks like nasty Japanese fireworks, and coming down everywhere in smuts. . . . What is to be done in a place where a beautifully washed man's shirt comes back in the evening battle-ship grey, *or worse*? (p. 5).

Paul's sisters, Clare the tomboy and Elfride the bookworm, were his frequent companions. When the children were introduced to the Monopoly board game from America around Christmas 1936, the girls made their own versions, one with London and another with potteries street names replacing the Atlantic City, New Jersey, originals. Wanting to teach their brother to swim but thwarted by municipal bath rules against mixed swimming, they once smuggled him in on ladies' day disguised as a girl ("Polly"). With his sisters Paul rode the old Loop Line from Tunstall to Stoke for his early schooling, practicing drawing caricatures on the misted windows and learning to respond politely to strangers who commented on his efforts or asked about his lessons.

Encounters with potteries people are presented with the same affectionate touch and plenty of humor. These include portraits of Father Ryan, who built the Catholic church with its high Gothic towers and three and a half domes, and Alderman George Barber, who unabashedly speared the cold ham fat from the plate of visiting dignitary Queen Mary, who apparently did not appreciate this potteries delectable.

Although not the book's focus, pots and potteries are marks of the district's uniqueness. Inspired by his sisters' book that featured a fourteen-year-old girl hiding in a giant Italian garden vase, Paul inquired about local pots' possibilities as hiding places. Paul's father told him that "Wedgwood's made an even bigger vase for Queen Victoria but small boys are certainly not allowed anywhere near it" (p. 35). Father Ryan persuaded the Spode works to make "ritual pottery vessels of alarming size" (p. 57) for his great church. Cinema owner George Barber decorated the lobby with potted palms set in brass tubs encased within Spode jardinieres and the ladies' and gents' rooms with Staffordshire tiles. The duke of Sutherland was said to have got Copeland's to make a special set of stoneware bowls with his coat of arms on them for his numerous dogs.

The chapter "Smoke into Art" provides the obligatory visit to a potbank (Royal Pottery), as well as a visit to the Burslem art school. Johnson remembers peering over the brim of the clay vat, "smelling delicious" (p. 121), and picturing the clay as the raw material for fire and smoke, trade, profit, and art.

The final character is the landscape itself. Careful observation while sketching with his father showed that no two kilns were exactly alike, "even if built at the same time for the same purpose. The curve of the shoulder, the width of the bases, the height of the chimney top—all were variable and each combination was unique. The odd straight chimney provided a contrast" (p. 116). Seven of the author's eighteen illustrations, plus the cover and endpapers, include potbanks, supplying visual evidence of this discernment.

The built environment would have been a suitable subject for a series of romantic transfer-printed views, given sufficient distance from the gritty reality. When Paul was old enough to travel by bike from the northernmost of the towns, he could reach large parts of Cheshire and Derbyshire. From the highlands the potteries "appeared a vast smudge on the south-west horizon, a band of dirty grey and brown smoke, through which poked the tall chimneys and such landmarks as Father Ryan's great church tower and the clock tower of our park" (p. 164).

During Johnson's childhood green fields and Knypersley Lake lay beyond the stone dumps, abandoned clay pits, and coal mines. Paul and his siblings could pick their way along the mineral railways and duck down under the small brick bridges where vegetation was encroaching on the industrial detritus. Mow Cop, a picnic destination topped with a mid-eighteenth-century folly consisting of an artfully "ruined" tower and arch, would have been the perfect subject for a transfer-printed table service.

Like these sentimental views, Johnson's scenes are frozen in memory. Returning for a brief visit fifty years after leaving the district, he saw that both heavy industrial activity and slums were missing from the landscape: smoke and soot, coal mining, and secondary rail lines were gone. "Most of all, a thousand bottle-shaped pot-banks, the main and essential ingredient of that unique landscape, had been demolished" (p. 200). The potteries' *jolie laide* was dead, and although the forensic investigation is not presented here, Johnson's book provides an engaging portrait of the lady.

Amy C. Earls, *Ceramics in America*

1. The description resembles anthropology's "ethnographic present" approach, used to describe tribal cultures supposedly before contact with Europeans and their technology.

2. Arnold Bennett, *The Grim Smile of the Five Towns* (1907) (1946; reprint, Harmondsworth, Middlesex, Eng.: Penguin Books, 1985); Charles Shaw, *When I Was a Child* (1903; facsimile ed., London: Caliban Books, 1977).

James R. Koterski. *Early Potters and Potteries of Delaware: Historical and Commercial Perspectives, 1760–1890*. Wilmington, Del.: Cedar Tree Books, 2005. vii + 208 pp.; bw illus., bibliography, index. $30.00 (hardcover).

To give this book a fair and honest review was just one reason that I felt obligated to read it from cover to cover. Before reading, I was surprised to find how little has been written about Delaware ceramics history. The only information that I found in my library consists of brief mentions in two books by Bill Ketchum. The reference from 1971 states that "Delaware had at least one kiln. Abner Marshall of Hockessin advertised both Rockingham and yellow ware between 1859–1866."[1] The 1983 book refers to William Hare operating as a redware and stoneware manufacturer in Wilmington, Delaware, from 1857 to 1887.[2]

My second rationale for careful reading was my curiosity about how Delaware pottery production could be significant enough to warrant an entire book. In stark contrast to previous authors, Koterski identifies more than eighty individuals who were connected with Delaware's pottery industry, and he provides a checklist of these names in chapter 11. Near the end of his book he theorizes, "it is difficult to approximate the number of pieces of ceramics that were crafted in Delaware's potteries over this 130 year period, though it must have been several million" (p. 183).

This brings us to the third reason for my interest in scrutinizing this book, namely, how did Koterski come up with so much information—especially about short-lived attempts to produce porcelain products in Wilmington—that others had failed to discover? The answer is as complex as the vast variety of sources he used.

Koterski honed his research skills while obtaining a doctorate in organic chemistry and in manufacturing, marketing, and research and development positions at DuPont. He describes at great length the source of the clay that was used and the different chemical compositions chosen depending on the type of ceramic desired. Chapter by chapter he enumerates in detail the eleven pottery locations, the potters, their families, owners, and output. In little more than two years of research he uncovered population and manufacturing censuses, land deeds, tax records, records of indentures, newspapers, city directories, account books, wills and estate inventories, minutes from Friends meetings, invoices, family Bibles, maps, photographs, local histories, and numerous other obscure references. Despite his meticulous attention to every detail, Koterski admits that he was unable to find, throughout Delaware's long history of ceramics production, many surviving examples that were marked by their makers. Limited archaeological excavations have been conducted on only three sites. The small number of redware and stoneware fragments that were found has done little to help attribute unsigned examples in local collections. Because of the paucity of material evidence, the author had to rely on documents in historical archives and state records.

Mathew Crips is identified as the earliest "potter who turned clay into ceramic wares in Delaware" (p. 20), beginning in 1760. Using land indentures and a business ledger, Koterski attempts to provide a more complete history of this potter, including discovering the type of ware he produced and the types of glazes he used. Koterski compares the prices of pottery made at different points in Crips's business history and estimates that "nearly 500,000 saleable pieces of pottery were likely manufactured by Mathew during his lifetime" (p. 25). The author further identifies the distribution of Crips's earthenware to six states, discusses how the American

Revolution affected pottery prices, and describes who the probable employees were. Meticulous details include the types of wood burned in the kiln, Crips's business practices, and his relative wealth.

In outlining the property ownership of the pottery on Water Street in Wilmington by John Jones, who ran the pottery from 1783 to 1825, Koterski identifies other activities that some potters engaged in for their community. "Jones was the first in a line of potters to play a role in the city's fire protection. No doubt they felt a sense of duty since their kilns provided ignition sources in their neighborhoods" (p. 48). The minutiae about this early potter's life are unfortunately coupled with the statement that "no examples of pottery attributed to Jones's factory are known to exist today" (p. 45), a refrain repeated for most of Delaware's potteries.

Even by 1862, when the partnership of Ziegler and Neumayer was advertising the manufacture of earthenware and Rockingham ware, Koterski has to admit that "as with most other Delaware potters, no wares attributed to them are known today" (p. 57). Even Branch Green, who Koterski mentions as having marked stoneware from Philadelphia before leaving to go to Wilmington, is not known to have left any marked pottery from Delaware. This lack of identifiable ware would tend to discourage most researchers and collectors from acknowledging Delaware's place in ceramics history. While Koterski's book attempts to detail and document the existence of these potteries, his enthusiastic research skills may not generate enough interest in collecting Delaware ceramics. Unless more pottery sites are discovered with fragments that help to identify typical Delaware forms or a greater number of marked ware comes to light, this research will be limited in its relevance to collectors and material culture specialists.

The genealogical information Koterski has assembled would be valuable to local historians primarily in Delaware and parts of New Jersey and Pennsylvania. Some of the information for the average ceramics collector or researcher might seem unimportant and tedious, especially when various family members who had no direct contact with ceramics operations are fully documented as to occupation and movements later in life. Using primary and secondary sources Koterski attempts to establish the various potters' birth places and occupations before and after they had worked in Delaware as potters. In one reference to Thomas Hayhurst's former occupation, he states, "Patricia C. O'Donnell, a Friends library staff member, suggested Thomas may have been a teacher, since frequent movements such as these were a characteristic of a Quaker of that profession" (p. 71). While this is not exactly documentary evidence, Koterski does confirm that Thomas was a schoolteacher at a Friends School in Wilmington in 1830–1831. Examples of Koterski filling in the blanks in his research usually consist of his thinking of possible explanations, presenting the limited documentary evidence, and then postulating the most logical answer. Limitations of the available documentary evidence make such speculation unsatisfying.

William Hare, who not only ran Delaware's longest operating pottery (1839–1885) but also marked his wares, has a chapter of his own. "William

Hare left a distinctive mark on Delaware's pottery industry by impressing his own mark on numerous pieces of surviving stoneware" (p. 76), Koterski tells us. Using manufacturing statistics, Koterski calculates, "on the assumption then that he averaged three pounds of clay for each piece, a kiln burn would have produced nearly 4,500 pots" (p. 90). While he admits that this calculation might be a bit high, he still concludes, "Hare must have produced well over 500,000 pieces" (p. 90). For stoneware forms he believes "probably well over 95 percent was what the potter described as an air-tight preserve jar" (p. 96), and thus undecorated. Hare's decorated examples that are illustrated in this book are sparsely ornamented with cobalt blue.

Chapter 9 really sparked my interest as a political science major and economic historian. The discussion of economic depressions, foreign competition, and marketing strategies that affected pottery operations is extremely insightful. Koterski weaves a detailed story of potting history with everyday events in the potters' lives and manages to bring into focus the economic forces that determined the success or failure of a pottery.

Unfortunately, production values of this self publication are limited. The photographs of the ceramics that are depicted are small, rather dark, and lack sharp detail. Even the maps, documents, and charts appear darkened and bleak. Even though Koterski makes a valiant attempt to produce a complete story about Delaware's ceramics history, these defects will adversely affect book sales.

I can appreciate the time and effort the author spent in producing this work. His results in both genealogy and local pottery history are a valuable addition to the libraries of avid researchers and ceramic historians. Beginning with a dozen (possibly mundane by collector standards) examples of Hare-marked stoneware, Koterski has produced a book that transports the reader back in time to Delaware's elusive potters and their utilitarian wares.

George H. Lukacs, Author, *Poughkeepsie Potters and the Plague*

1. William C. Ketchum Jr., *The Pottery and Porcelain Collector's Handbook: A Guide to Early American Ceramics from Maine to California* (New York: Funk and Wagnalls, 1971), p. 105.

2. William C. Ketchum Jr., *The Knopf Collectors' Guides to American Antiques: Pottery and Porcelain* (New York: Alfred A. Knopf, 1983), p. 434.

Thomas V. Litzenburg Jr. and Ann T. Bailey. *Chinese Export Porcelain in the Reeves Center Collection at Washington and Lee University.* London: Third Millennium Publishing, 2003. 288 pp.; numerous color illus., bibliography, index. $65.00 (hardcover).

Has any other ceramic type besides Chinese export porcelain had such a far-reaching impact on the social, cultural, commercial, and technological fabric of Western culture? The lure of exotic and mysterious Chinese products, including porcelain, underpinned commercial trade from the seventeenth through the nineteenth century via various East India Companies. The rising popularity of tea drinking and notions concerning the proper use of the tea "equipage," including the abundance of ceramic vessels associated with the serving of tea, shaped many Western ideas of proper and genteel pursuits during the eighteenth century. Although largely dwarfed in scale by the other exports of the China trade, such as silks, tea, and spices, the peculiar allure of Chinese porcelain—the subject of centuries of interest on the part of western European and Middle Eastern governments and ceramics manufacturers—led to European pots with bodies, glazes, and decorations echoing to various degrees of success the remarkable Far Eastern originals.

Aimed at an audience of students and collectors alike, this catalog is meant to document and illustrate the Chinese export wares in the Reeves Center Collection at Washington and Lee University in Virginia. Litzenburg and Bailey not only succeed in this goal, they also contribute to Chinese porcelain scholarship. The authors make no attempt to offer a comprehensive history of the China trade or description of the processes of manufacture and export; that story has been told previously in a number of excellent works.

The eloquent preface by the late David Sanctuary Howard, one of the foremost scholars of Chinese porcelain and author of the monumental *Chinese Armorial Porcelain*,[1] sets the stage. A brief introduction presents a chronological overview of the export trade, creating an important economic, historical, and social context for the porcelain remnants of that trade. The visual showcase of the diverse collection, which dates from the sixteenth through nineteenth century, is made possible by Ellen M. Martin's impressive photography and provides an invaluable comparative reference. Although the catalog entries accompanying the individual object photos are limited, the text is informative, and individual vessels are cross-referenced to similar examples in other published collections.

The collection's presentation is organized into three chronological sections: late-sixteenth- and seventeenth-century, eighteenth-century, and nineteenth-century porcelain wares. Each section is profusely illustrated with superb color photographs. Section 1 on late-sixteenth- and seventeenth-century porcelains presents underglaze blue-and-white wares first, followed by *blanc de chine* pieces and polychrome-decorated jars. The text for each ceramic details the decorative embellishment, size, provenance, and date range attributed to the piece. Aside from the temporal, stylistic, and decorative information on each vessel, what makes this publication particularly informative for the student or curator of Chinese wares is that comparative examples documented within previously published works, dated shipwrecks, and the holdings of the incomparable Topkapi Saray Museum in Istanbul, Turkey,[2] are referenced. Increasing the catalog's utility is the great variety of vessel forms that are presented and discussed. In addition to typical plates and dishes, less common ceramic forms displaying the diverse array of early Chinese porcelains in the Reeves Collection include figures, baluster-shape jars, a ewer, a kendi, and ovoid jars.

The eighteenth-century porcelains in section 2, more numerous than the earlier porcelains, are presented by decorative style: underglaze blue-and-white and polychrome enamel decorated, the latter including *famille verte, famille jaune, famille noir, famille rose,* Chinese Imari, Mandarin, and *encre de chine*. After illustrating examples of these decorative styles, the authors present mostly overglaze polychrome decorated porcelains organized thematically: armorial designs; topographic and architectural views; daily life and picturesque scenes; Cornelis Pronk designs; literary and historical subjects; mythological, religious, and Masonic subjects; marine subjects; figure models; and Western forms and decoration. Important for both students learning the nuances of Chinese export wares and curators managing collections of Chinese porcelain is the stylistic breadth documented within the eighteenth-century Reeves Center Collection wares.

Highlighting section 2 is the important range of armorial holdings. Organized chronologically from the 1730s through circa 1800, the collection represents the huge breadth of artistry and beauty of armorial wares. Armorial porcelains produced for the American market are particularly significant. The initial voyage of the American vessel *Empress of China* in 1784 ushered in a period of direct trade between the new American nation and China. Eager to take their place among the multinational members of genteel Georgian society, Americans ordered a variety of armorial services, and many of these important American pieces are illustrated and discussed in the catalog. Porcelains with Western scenes painted in Chinese style, long favored by collectors of Chinese export wares, are another important category of eighteenth-century wares. Western scenes include the European trading offices at Canton (Guangzhou), known as "factories," as well as vistas copied from popular engraved prints, which prominently feature Western scenes and mythological subjects.

Section 3 details the nineteenth-century porcelains and, as with the previous sections, is arranged according to decorative style: underglaze blue and white; Fitzhugh pattern; armorial and pseudo-armorial; European decorations; American decorations; Rose Medallion, and Rose Mandarin wares. The blue-and-white nineteenth-century wares document the very common and widely used early-nineteenth-century Canton and Nanking styles, as well as the Fitzhugh pattern. The diversity of nineteenth-century overglaze porcelains, including a selection of overglaze enameled porcelains produced for the American market, is very impressive. The Reeves Center Collection of porcelains produced for the American market is nothing short of a national treasure for both the American and the Chinese people. These wares are a testament to the importance of the American trade in the overall export of Chinese porcelain during the late eighteenth and early nineteenth centuries.

The catalog concludes with a chronology of the Ming and Qing dynasties, including reign dates of the emperors, followed by a bibliography with comparative examples from other publications and collections. The contribution that this work makes to the rich variety of scholarship on Chinese wares is significant. The authors demonstrate that the Reeves Center Collection's Chinese export porcelains constitute a spectacular reference collection of this type in the upper South. The extremely high-quality full-color plates alone represent a very important contribution. In producing a copiously illustrated, well-organized, and carefully referenced compendium of the Chinese porcelains in the Reeves Center Collection, the authors provide both the novice and the seasoned scholar with an invaluable reference and guide to these extraordinary export wares.

Andrew Madsen, University of Kentucky at Lexington

1. David S. Howard, *Chinese Armorial Porcelain* (London: Faber and Faber, 1974).

2. Regina Krahl and John Ayers, *Chinese Ceramics in Topkapi Saray Museum, Istanbul: A Complete Catalogue*, 3 vols. (London: Sotheby's, 1986).

Stanley South. *An Archaeological Evolution*. New York: Springer, 2005. xi + 418 pp.; 141 bw illus., appendix, references, list of figures, index. $169.00 (hardcover).

During the summer of 1968 I had the good fortune to work for Stanley South during his last year as archaeologist for the North Carolina Department of Archives and History. It was an unforgettable experience. In four days he uncovered enough of an 1838 arsenal to force the redesign of I-95.[1] We located the site of an eighteenth-century tavern and tested the ditch and cellar of a French and Indian War fort. Between digs, Stan churned out reports at one hundred words per minute, inked drawings, and edited a newsletter. Then we started our summer's project: searching for the fine pottery of Rudolph Christ in the yard behind the potter's house in Winston-Salem. In one of the first test pits, Brad Rauschenberg came down on a drain stuffed with broken tin-glazed earthenware ("faience") bottles. The bottles, wasters from a 1790s firing, were glazed a beautiful robin's eggshell blue. Wow.[2] As soon as the sherds came out of the drain, they were washed and spread on a sheet of plywood to dry in the sun. On Friday afternoon, the clean, dry sherds rode back with us to Stan's farmhouse near Raleigh. Our weekend treat was sorting and gluing. This was archaeology in the fast lane.

The best part of the summer was Stan. He did fast leisurely, taught gracefully, filled long trips in the station wagon with intense discussion and funny stories, treated me as an equal, worked to exhaustion, but always had enough energy to enjoy a beer, dinner, and more conversation. Then on weekends, it was family, friends, and his current hobby.[3]

Stanley South was born February 2, 1928 (Groundhog Day), in Boone, North Carolina. Before he found his calling as a "mountain groundhog," his life took a few zigzags: the Navy during World War II, photography school, a degree in education, and teaching.[4] In 1950, to prepare himself to teach a science unit on the Indians of North Carolina, Stan began walking plowed fields for artifacts. Three years later, he began graduate studies in anthropology at the University of North Carolina at Chapel Hill. There he studied archaeology under Dr. Joffre Coe, a professor noted for his rigorous training in fieldwork and artifact analysis. In 1956, on completing the course work for his master's degree, Stan became archaeologist at the Town Creek Indian Mound State Historic Site, Mt. Gilead, North Carolina. Two years later, he transferred to the new State Historic Site of Brunswick Town, the archaeological site of a 1725–1776 port town on the Cape Fear River.

With his appointment as archaeologist–site manager for Brunswick Town, Stan charged into the new field of historical archaeology. In his first four years, Stan and a crew of four African-American fishermen developed a historic park, excavated nine ruins, and washed, cataloged, and analyzed tens of thousands of artifacts. Then, while continuing to dig at Brunswick Town, Fort Fisher (1865), and many smaller sites, he excavated the Moravian town of Bethabara (1753–) and a house lot in Winston-Salem (1766–). Moving to the South Carolina Institute of Archaeology and Anthropology in 1969, Stan continued to excavate sensational sites: Charles Towne Landing (1670–1680) and an underlying prehistoric ceremonial complex (ca. 1276–ca. 1387); Fort Johnson (1708–1865); Fort Moultrie (1776–1865); the Revolutionary War fortifications and siege works of the frontier town of Ninety-Six, French Charlesfort (1562); Spanish Santa Elena (1566–1587); and John Bartlam's creamware pottery at Cain Hoy (1765–1770).[5] As of this writing, Stan is still digging—starting a new season at Santa Elena.

The man from Boone loves his digging, but he equally loves the other elements of archaeology: analysis, theory building, writing, editing, and publishing. Stan had hardly begun digging at Brunswick Town before he began—with help from Ivor Noël Hume—to create detailed artifact typologies for the town. Stan began publishing Brunswick Town artifact studies in 1962, and he has continued to publish a steady stream of articles on artifacts, excavations, excavation techniques, archaeological photography, and site restoration. Beginning in 1959 he has published on ceramic types, buttons, pull tabs, flax hackles, alkaline-glazed pottery, and Native American pottery.[6]

When the Southeastern Archaeological Conference limited the participation of historical archaeologists, Stan started the Conference on Historic Site Archaeology (1960). Stan organized the meetings, typed fifteen volumes of papers, and arranged for their printing and distribution. After the conference ended, Stan started two more publication series: *Volumes in Historical Archaeology* (42 vols., 1984–2002) and *Historical Archaeology in Latin America* (16 vols., 1994–1996). Stan's organizing and editorial goals included more than information sharing. He wanted to nudge historical archaeology away from antiquarianism and into the social sciences. Toward that end, in 1967 Stan started the "Forum," a section of the conference Papers in which a particularly provocative paper was circulated for comment and criticism.

In 1971 Stan stopped nudging and began shoving. His 1960 paper (published 1962) on Brunswick Town ceramics included a stupendous chart showing the ceramic types and frequencies from eleven dwelling sites. From these and other sites, in 1971 he extrapolated a mathematical formula for establishing the mean occupation date of a site or deposit, circulated it for comment, and published it and its critiques in the "Forum" as "Evolution and Horizon as Revealed in Ceramic Analysis."[7] At the following conference, Stan fired an entire salvo—five papers on method and theory.[8]

Stan's missionary efforts reached a peak in 1977 with the publication of *Method and Theory in Historical Archaeology* and his editing of *Research Strategies in Historical Archaeology*.[9] *Method and Theory* is primarily about the distribution of artifacts in time and space, although Stan refers the reader to Ivor Noël Hume and John M. Goggin for artifact class descriptions. Since 1977 Stan's writings on archaeological theory have diminished, but his writing has never slowed. The bibliography of Stan's manuscripts, publications, and editorial projects has 222 entries.

An Archaeological Evolution is not a *Reader's Digest Condensed* version of Stan's archaeological reports; it is an autobiography. The digs are presented as detective and adventure stories.[10] The hundreds of photographs and drawings (mostly Stan's) that generously illustrate *An Archaeological Evolution* provide context and flavor to the yarn. The reader interested in the archaeological data will have to follow the references and bibliography to Stan's earlier publications.

As with any good story, the author's real challenge is more character development than plot. Despite his disclaimer to the contrary, Stan loves people. *An Archaeological Evolution* has a huge cast: family, teachers, mentors, colleagues, subjects, and crew members, and many will find their portraits among the illustrations. Stan is a gifted raconteur with a good memory. Hardly an archaeological episode passes without one or more stories: Stan being teased by his Brunswick Town crew, Ronald Reagan insisting that aliens may have left archaeological ruins on the moon, and the historical archaeology seminar in the Flowerdew Hundred toilet. (When you are famous, there is no escape.) Not all of the stories are happy. There are moving accounts of the deaths of his sister and his first wife.

Although Stan loves intellectual debate, he does not gladly suffer fools and philistines, whether Navy personnel squandering tax dollars, university presidents afraid of the word *evolution,* or National Park Service bureaucrats who insist on digging in the wrong location. His greatest, poorly concealed contempt is reserved for those who destroy our cultural heritage. These include developers who demolish historic structures, exhibit designers who abuse archaeological artifacts, and park visitors who pilfer from excavations.[11]

Stan's writing varies in quality. Some long sections read well, but others are a series of anecdotes. The book is easy to use. There is an introductory career chronology, a text organized into well-labeled chapters by subject, an excellent index, and a huge bibliography.

An Archaeological Evolution is essential reading on the development of historical archaeology. A university with a graduate program in historical archaeology should acquire several copies. At $169 per copy, however, this is an expensive book. Historical archaeologists with only a passing interest in the origins of their profession might be wise to borrow *An Archaeological Evolution* on interlibrary loan and purchase other of Stan's publications for their personal library. Those interested in Stan's groundbreaking ceramics finds should look elsewhere.[12]

Garry Wheeler Stone, Monmouth Battlefield State Park, New Jersey Division of Parks and Forestry

1. The Fayetteville Arsenal, subsequently the subject of a seven-month excavation. The surviving portions are an archaeological park.

2. The wares of the potters—principally Gottfried Aust (1755–1788) and Rudolph Christ (1766–1821)—who worked at Bethabara and Salem (now Winston-Salem) included traditional earthenwares in a wide variety of bodies and glazes; tin-glazed redware ("faience"); tortoiseshell creamware; cream-colored ware in the Queen's, Royal, and feather-edged patterns; and fine stoneware cups, saucers, teapots, bowls, and mugs—the latter with double, intertwined handles secured with floral sprigs. Aust was a cranky genius whose mass-produced milk pans, butter pots, roofing tiles, and tobacco pipes contrasted with his slip-decorated dishes, some of which were masterpieces of German art. He also turned delicate standing salts and eggcups, threw graceful substitutes for pewter tankards, and produced entire sets of yellow and redware tea services modeled after silver and porcelain. In 1773–1774 William Ellis, an unemployed English potter from John Bartlam's failed Charleston, South Carolina, "China Manufactury," worked five months at the pottery showing Aust and his apprentice, Rudolph Christ, how to make queensware and fine stonewares. Thereafter, while Aust continued wheel-throwing pots, Christ worked molding English forms. See Stanley South, *Historical Archaeology in Wachovia: Excavating Eighteenth-Century Bethabara and Moravian Pottery* (New York: Kluwer Academic/Plenum, 1999).

3. In August 1968 Stan became a vintner. A large picking of blackberries started to ferment, so Stan decided to preserve them as wine. His farmhouse has no cellar, only a crawl space. Army-surplus entrenching tools in hand, we crawled under and dug a wine cellar. Those moldy blackberries launched Stan on a multiyear trajectory through watermelon wine, tomato wine, Christmas wine. . . . Stan was never one to do anything by halves. See *Evolution,* p. 225.

4 Although it is difficult to comprehend, Stan has squeezed more into his life than archaeology. B.A. (before archaeology), he studied voice, acted, and worked as a photographer. A.A., his hobbies have included winemaking, rebuilding antique cars with his son David, creating a backyard quail habitat, and firing black-on-black pottery. As a young archaeologist at Brunswick Town, he almost was lured away by the bohemian art world, especially after discovering that his sculptures sold. Writing prose and poetry remain enduring interests.

5. Instead of John Bartlam's Cain Hoy kilns or waster dumps, Stan and colleagues located worker housing, where potters were using imperfect pottery almost as we use paper plates. In all, the archaeologists excavated 7,450 sherds from vessels made or possibly made by John Bartlam's pottery. While the archaeologists disagree about some of the small sherd groups (some may be from "seconds" imported from England), unfinished pieces and obvious wasters show that Bartlam was making a wide range of up-to-date Staffordshire-style ceramics. Bartlam's Cain Hoy operation was well staffed (potters included William Ellis, who had trained Aust and Christ at Wachovia in making English-type queensware and stoneware) and well equipped. They had plaster molds for cauliflower and pineapple tea services, pierced-edge "fruit and basket" dishes, and plates, platters, and teawares in barleycorn and dot, diaper, and basket patterns. They also had molds for a variety of floral sprigs, dies for surface impressing "partridge eye" and floral decoration, and freehand jiggers for turning vessels to standard shapes. By varying clay bodies, underglazes, and glazes, Bartlam was producing a wide variety of wares: "Carolina creamware" (darker than English), slipware, tortoiseshell ware, redware, green-glazed ware, and "china." (Bartlam's "china" was a refined but not translucent body containing a mixture of kaolin and ground lead glass, decorated under a lead glaze with cobalt blue. This ware appears to represent what the potters termed "china glaze," which contemporary usage terms "pearlware.") Painted decoration varied in quality from good to excellent. Bartlam's men also transfer-printed with glue bats. All of this remarkable production was accomplished before 1771 (when Bartlam moved to Charleston), and perhaps as early as 1768—eleven years before Wedgwood announced his "Pearl White" china. No wonder Wedgwood considered Bartlam a serious rival.

6. For a listing of Stanley South's publications and manuscripts, see *Evolution,* pp. 375–86.

7. His goals in developing a formula for finding the median date of a ceramic assemblage were modest. Inspired by J. C. Harrington and Lewis Binford's work with dating groups of tobacco pipe fragments, Stan wanted to find out if ceramic groups could be dated in a comparable way, and to see if a tool for dating ceramic assemblages would help prove the existence of ceramic "horizons" (assemblages of similar ages) throughout the Carolinas in the eighteenth century.

Stan's formula was published in *Method and Theory in Historical Archeology* (New York: Academic Press, 1977; reprint, Clinton Corners, N.Y.: Percheron Press, 2002) is straightforward. Ceramic sherds are sorted by manufacturing type, counted, and weighted by the median date of manufacture. The weighted means of all the sherd lots from the site are averaged to determine the median manufacturing date for the assemblage. The results for ceramic assemblages from four sites from Brunswick Town, North Carolina, were within one or two years of the mean dates provided by the Binford pipe stem formula, and the formula agreed equally well with site documentation and other datable artifacts from other Anglo-American sites. However, the formula did not work with German-American sites (p. 232). The formula showed that steady change in English ceramic manufactures coupled with a rapid distribution system created similar ceramic assemblages in Anglo-American supplied households of the same date. The ceramic types from these households can be used to date sites for which no documentation has been located (p. 229).

Stan has been criticized widely for providing archaeologists with a mechanical means of describing and dating ceramic assemblages without thoughtful analysis of the entire assemblage. In particular, critics have argued that fragments of ceramics or glass should be analyzed and reported not as sherds but as vessels. The critics reason that fragments of a tin-glazed teapot and those of a tin-glazed chamber pot are providing different information and should not be lumped together. Relying on the formula alone for ceramic description and analysis constitutes misuse. Stan was very explicit in defining the uses and limitations of the formula. It was only for determining the median manufacturing date of a ceramic assemblage. Studies of status and function would require analysis of vessel form and decoration. *Method and Theory*, pp. 230–31.

8. For those of us who have trouble following scientific logic, Stan provided the comic book version—delightful full-page (or fold-out) cartoons: the Dolphin Chart, or "The Hypothetico-Deductive-Inductive Scientific Cycle," the Chicken and Egg Chart, the archaeological "Data [piggy] Bank," and "Milking the Archaeological Cow." The cow has each teat neatly labeled. Since real-world bovines do not provide enough faucets, the archaeological version has five. *Evolution*, p. 282; *Conference on Historic Site Archaeology Papers* 7 (1972), pp. 131, 150, 158.

9. *Research Strategies in Historical Archaeology*, edited by Stanley South (New York: Academic Press, 1977). For the archaeologist, perhaps the most important of Stan's publications is his 1977 *Method and Theory in Historical Archaeology*, in which he demonstrates that rigorous fieldwork and analysis can lead to the recognition of patterns that help us understand "past lifeways, culture history, and culture process" (p. xiii). Ceramics—as sherd counts by type—play an important role in defining an eighteenth-century Carolina artifact pattern and a contrasting frontier artifact pattern.

10. For how to defuse live Civil War artillery shells, see pp. 168–70.

11. Stan's prize philistines were the members of the South Carolina Tricentennial Commission, who, confronted with the complexity of developing the 1670 archaeological site of Charles Towne Landing, decided to jazz it up with a fake Elizabethan village and an enormous, glass-walled exhibit pavilion. The construction of this Tricentennial Pavilion destroyed a fourteenth-century Native American ceremonial complex. Fortunately, the engineering of the twentieth-century ceremonial pavilion was faulty. It has been razed, and archaeology has resumed at Charles Towne Landing. The politics and archaeology of Charles Towne Landing and the Tricentennial Commission are described in Stanley South, *Archaeological Pathways to Historic Site Development* (New York: Kluwer Academic/Plenum Publishers, 2002).

12. The two works of most interest to the readers of this journal are Stanley South, *Historical Archaeology in Wachovia*, and South with contributions by Lisa Hudgins and Carl Steen, *John Bartlam: Staffordshire in Carolina*, South Carolina Institute of Archaeology and Anthropology, Research Series 231 (Columbia: University of South Carolina, 2004).

Historical Archaeology in Wachovia is extensively illustrated with Stan's excellent photographs (not well reproduced) and George Demmy's drawings. The text is occasionally awkward, but it is thorough, and it contains many relevant quotations from the extensive Moravian archives. This is an impressive product of three seasons of fieldwork (1964, 1966, 1968). Part 1 is about Stan's excavations of this 1753 piedmont village. In part 2, Stan describes the potters' wares (see note 2 above).

One result of Stan's discoveries in Winston-Salem was increased interest in South Carolina's early ceramic industries. Cain Hoy was an eighteenth-century brick-making center nine miles north of Charleston. In 1972, while surface collecting at Cain Hoy in the vicinity of Staffordshire potter John Bartlam's first American pottery (1765–1770), George Terry, a dean at the University of South Carolina, found fragments of unfinished creamware and "china." When Terry and Stan revisited the site in 1990 and found that it was about to be developed, they raised funds so that Stan and Carl Steen could extensively test-excavate the site in March 1992. Stan was writing the report (Stanley South, *The Search for John Bartlam at Cain Hoy: America's First Creamware Potter,* South Carolina Institute of Archaeology and Anthropology, Research Manuscript Series 219 [Columbia: University of South Carolina, 1993]) when Brad Rauschenberg of the Museum of Early Southern Decorative Arts in Winston-Salem obtained funding for Stan and James Legg to excavate more of this extremely important pottery in the autumn of 1992. For more on Cain Hoy, see South with contributions by Hudgins and Steen, *John Bartlam: Staffordshire in Carolina.* Also see Bradford L. Rauschenberg, "John Bartlam, Who Established 'new Pottworks in South Carolina' and Became the First Successful Creamware Potter in America," *Journal of Early Southern Decorative Arts* 17, no. 2 (1991): 1–66; "'A Clay White as Lime . . . of Which There is a Design Formed by some Gentlemen to Make China': The American and English Search for Cherokee Clay in South Carolina, 1745–75," ibid., pp. 67–79; "Escape from Bartlam: The History of William Ellis of Hanley," ibid., pp. 80–102; "Brick and Tile Manufacturing in the South Carolina Low Country, 1750–1800," ibid., pp. 103–13.

Staffordshire in Carolina is a two-part archaeological report with the findings of the fall 1992 excavations added to the 1993 report on the March 1992 project. It is designed to be user-friendly to both archaeologist and ceramic historian. The Bartlam and non-Bartlam ceramics are discussed in separate chapters, and all artifact illustrations are in color. Of these, forty-one plates illustrate Bartlam or possible Bartlam ceramics. This 283-page, softcover report has detailed lists of contents and illustrations, lengthy appendixes and bibliographies, and a good index.

Mark Hewitt and Nancy Sweezy. *The Potter's Eye: Art and Tradition in North Carolina Pottery*. Photography by Jason Dowdle. Chapel Hill: University of North Carolina Press for the North Carolina Museum of Art. 296 pp., 235 color illus., 1 map, index. $39.95 (hardcover).

Organized by master stoneware potter Mark Hewitt and Nancy Sweezy, former director of Jugtown (and once its head potter), *The Potter's Eye* is a voyeuristic journey into the world of North Carolina pottery from the nineteenth century to the present day. In tracing this history, Mark Hewitt and Nancy Sweezy evoke the canon of folk pottery traditions. Their premise—that North Carolina pottery "represents one of America's most revered stoneware pottery traditions"—is a telling aggrandizement of this necessary utilitarian enterprise. Perhaps it is this sense of self-righteousness that makes me leery of fully embracing the North Carolina doctrine, as I would argue that virtually every state or at least every region in this country has a pottery tradition worthy of reverence.

In a sense, then, I went to the opening of "The Potter's Eye" exhibit looking to pick a fight. Arriving at the nearly full parking lot at the North Carolina Museum of Art in Raleigh at seven o'clock in the evening, I was awed by the sheer number of attendees (at least 600)—most having paid $30 for the lavish reception and book signing. Before the evening was over, however, my contention had turned to admiration for the well-conceived and executed exhibition, its handsome accompanying catalog, and the infectious enthusiasm of the North Carolina stoneware disciples.

Hewitt and Sweezy clearly recognize and acknowledge the worthiness of other pots; both the catalog and exhibition include selected salt- and ash-glazed pots from across time and space. Indeed, the primary focus of their exhibit was to illustrate the connection that North Carolina stoneware has with other regional and world traditions, from salt-glazed examples made in New York and New England, Great Britain, and Germany, to alkaline-glazed stoneware examples made in China, Thailand, and Japan.

The book opens with several introductory essays, the most important being "New Perspectives on Old North Carolina Pots," in which Mark Hewitt discusses how the Japanese approach to pottery connoisseurship offers an alternative way to appreciate the traditionally manufactured utilitarian wares of the West.

In the next section, which represents a major part of the book (and exhibition), Hewitt puts this position into practice in his exuberant entries for the sixty-six antique pots from the regions mentioned above. He draws on the extensive vocabulary derived from the Asian language of connoisseurship, which can be poetic and at times otherworldly in its description of, for example, firing clouds that "create a vivid nebula of haloed pink, red, brown, and gray" (p. 83).

Hewitt also evokes many anthropomorphic comparisons when describing lips, shoulders, necks, and bellies. He likens a pitcher from Randolph County, North Carolina (cat. no. 26), to a "dignified dowager dressed up for a ball, gaunt but stately, bejeweled and defiant. Bedecked with a double strand of rhythmic swags on its neck and shoulder, tattoos of a bird, fish and flower on its calves, and the date of the party, 1879, right on its belly, it is going out in style" (p. 73). From the description alone, a Disney animator could easily bring this object to life on the big screen.

Certainly a modern trained decorative arts curator—constrained by formalities that inhibit any semblance of emotion—would not have written the text in this way. Only a potter could have spoken so honestly and freely. Indeed, if any of the catalog entries had been submitted to me for *Ceramics in America,* I would have stricken most of the superlatives, leaving the who, what, where, and how (while secretly agreeing with most every word I was deleting).

In the second half of the book Nancy Sweezy presents essays on the six contemporary North Carolina potters chosen for the exhibition—Kim Ellington, Mark Hewitt, Ben Owen III, Pam Owens, Vernon Owens, and David Stuempfle—each of whom has five objects included in both the show and the book. It is clear throughout Sweezy's biographical narratives that these hardworking, dedicated, and enthusiastic artists pride themselves on being informed by the past. The common experience of mastering materials, techniques, and endless repetition is the spiritual union card that all potters carry. It is this competency that non-potting scholars endeavor to capture in their descriptive language

It is interesting that while Hewitt goes to great lengths to provide background and interpretation for each antique pot represented in the exhibit, all of the contemporary pots are represented solely by photographs. This does not mean that they are mute—I want to think this was a clever device intended to allow these pots to speak for themselves. Nevertheless, I would have enjoyed the opportunity to learn, from their makers, what makes the selected pots special.

To explore my own thoughts and feelings about the show's assemblage, I allowed myself to gravitate naturally toward my pots of choice. My only conscious decision was to select one example from each period—antique and contemporary—that most convincingly spoke to me. My antique choice was a ten-gallon alkaline-glazed jar (cat. no. 63) made about 1860 by Isaac Lefevers of Lincoln County, North Carolina. The authors apparently thought a lot of this pot as well, as they devoted more text and illustrations to this one piece than to any other object in the exhibit. In choosing a contemporary example, I found that a Kim Ellington jug (cat. no. 68) caught and held my attention, which kept returning to admire the form and glaze.

With their exhibition Hewitt and Sweezy attempt to emphasize the similarities between contemporary potters and their historical predecessors. Hearing the former describe the aesthetic concerns of their profession makes me wonder what an early-nineteenth-century American potter might have thought. I find it hard to believe that utilitarian potters would have focused so much attention on individual pots. They were driven by production. A beautiful sight to them might have been at the board level or the kiln load level. Their pots, sold by the dozen, were made to store salted meats and caustic paints and solvents. Doctors' wives did not line up to have bragging rights for the choice selections from the latest kiln opening.

In defining the most basic difference between the products of the past and present, I would argue that the antique examples Hewitt and Sweezy chose embody unconscious aesthetic concerns, whereas their contemporary counterparts exude self-consciousness. This is not a bad thing, of course, but hugely important to keep in mind.

The book's photography is as important as the well-ordered text. Jason Dowdle has provided far more than an objective recording of the items. He has added his artistic interpretation in virtually every image. Photographing pots is hard work. There is tension between illumination and glare. The glazes, while beautiful, can defeat many photographers who try to eliminate the "hot spots" created by studio lighting. Dowdle has succeeded in allowing enough light to show the color and texture of the pots while incorporating the distinctive highlights into his artistic vision, and in this respect his photography contributes to the otherworldliness theme of the book. Many of his details of glazes could be mistaken for images taken by radio telescopes of celestial events.

For Hewitt, Sweezy, and Dowdle, *The Potter's Eye* is clearly a love story, a romantic exploration of the innate qualities of vernacular pottery and makers. It suggests that looking to the past and understanding the materials, tools, techniques, and history of a ceramic tradition bring us all closer to pottery bliss. With the rich language, lavish photographs, and the pots themselves, the exhibit and book are an unmitigated success. My only regret is that this lovely show did not travel outside of North Carolina. It would have done much not only to educate but to indoctrinate others about the beauty of pots. If you missed the show, buy the book. Better still, buy two and send the other to a loved one.

Robert Hunter, *Ceramics in America*

George Deike and Mickey Deike. *Feathers and Foliage: The Life and Times of the Pearlware Peafowl.* Shalimar Farm (privately by the authors), 2005. xvi + 370 pp., 358 color illus., 50 color paintings, 14 bw illus., appendixes, bibliography, index. $65.00 (hardcover).

When John and Griselda Lewis published their book *Pratt Ware*[1] in 1984, I was disappointed, thinking that their definition of Pratt ware was too limited, applying only to press-molded or slip-cast pearlware pottery decorated with high-temperature underglaze colors. When I asked them about it, their reply was that someone should write a separate book about the other polychrome underglaze-decorated wares. George and Mickey Deike have done just that.

The work focuses on the English refined white earthenware known as pearlware that was decorated with polychrome underglaze-painted peafowl in the last quarter of the eighteenth century through the middle of the nineteenth century. The Deikes' forty years of collecting and fifteen years of careful research have culminated in a beautifully illustrated book written with a humble and personal approach. The magnificence of this book, worthy of a coffee table, is the result of Bill Dillon's design and the authors' loving and painstaking production.

The book begins with a description of making pearlware that can provide much information for a beginning student of English ceramics. The first chapter deals with the origin and distribution of the materials used to make this pottery, which serves as the canvas for decoration. The clay, flint, lead, cobalt, and metal ores used were often brought in from afar to areas in England where the pottery was made. Coal, however, was always close by, since a ton of pottery requires ten tons of coal for firing.

The next chapter is about the transportation of raw materials and the finished pottery. Contemporary roads were so miserable that, in enlightened self-interest, the potters promoted the building of turnpikes and canals. There follows a chapter presenting an abbreviated history of bird-decorated ceramics resulting from the authors' search for the ancestry of these underglaze-painted examples. In the late eighteenth century, birds were painted on delftware, white salt-glazed stoneware, creamware, and porcelain. The last introductory chapter discusses many aspects of the production of pottery, especially pearlware—design, preparation of materials, potting, firing, and decorating methods.

Choosing a single but popular decoration and pattern type as a collecting and research subject has yielded a vast amount of previously unpublished details for the specialist. At the heart of the book's main section is the Aviary, chapter 5, which illustrates and describes the forty-nine families of painted birds and background settings, defined during years of collection and study. The Deikes use given names, such as Doris or Gertrude, for the styles, rather than dry descriptive labels. (Some names were given to honor particular friends or dealers.) The Deikes provide at least one portrait of each family type. Detailed descriptions of the shapes, foot rims, glaze, edge molding, and so forth—as well as factory, where known—are provided.

The remaining section of the book deals with the details of the more than 650 examples of peafowl decorated pearlware. Chapter 6 discusses the pottery shapes. Each form—plates, saucers, teapots, for example—is summarized in text tables providing details of peafowl decoration.

The next chapter deals primarily with elements of decoration other than the birds themselves, such as body color or foliage and trees. These variables are brought together to compare similarities and differences that could reveal family relationships based on recurring design elements or perhaps common factory or decorator origins.

The last chapter describes two of the many other types of polychrome underglaze painted pearlware decoration, House and Flower Basket. These patterns are discussed and illustrated to show that there is a large field of study waiting for someone with dedication and time.

Finally, all of the data are tabulated in a series of six appendixes. The first summarizes all of the 650 ceramic examples in the book. Remaining appendixes summarize, for each of the forty-nine bird families, the shapes of the pots (appendix II), the shapes of the plates (appendix III), the pose and outline features (appendix IV), the colors used (appendix V), and additional features, including foliage and decorative bands (appendix VI).

The bird family summaries are very useful for classifying a piece of pottery. It will be interesting to see whether collectors, museum curators, and auctioneers will attribute or catalog according to the Deikes' aviary. Peafowl collectors will certainly use these family names, and even if curators and auctioneers choose not to use the Deikes' unique names, they surely will no longer use simple generic descriptions for pottery with this type of decoration.

The book is lavishly illustrated with photographs by Jurgen Lorenson and paintings by Eric Werner, resulting in full-color illustrations on almost every page. On opening the book one is immediately struck with the "roll-out" view of a peafowl decorated pearlware mug on the endpapers. Justin Kerr's adaptation for *Feathers and Foliage* of this type of photographic image, which shows the decoration of an entire jug or a mug laid out on a flat surface, is the best method I have seen for turning a beautiful three-dimensional object into a beautiful two-dimensional picture. I know this technique has been used before in archaeological contexts, but to my knowledge it has never been used to this extent in a decorative arts reference.

This book shares what the authors have seen, measured, and thought about their beloved birds. In their afterword, they express the hope that readers will also find these birds to be pleasing and interesting. Given that the Deikes' boundless enthusiasm for English ceramics in general is "contagious to all who come in contact with them and their collection" (as dealer Joyce Hanes Ruskin notes in her foreword), this book should be great inspiration for dedicated collectors to explore other underglaze painted patterns.

Paul Fox, Student of English ceramics

1. John Lewis and Griselda Lewis, *Pratt Ware: English and Scottish Relief Decorated and Underglaze Coloured Earthenware, 1780–1840* (Woodbridge, Suffolk, Eng.: Antique Collectors' Club, 1984).

Barbara Wood and Robert Doares. *Old Limoges: Haviland Porcelain Design and Décor: 1845–1865*. Atglen, Pa.: Schiffer, 2005. 272 pp.; 245+ color illus.; six appendixes including trademarks, reproductions of Haviland factory drawings, and contemporary documents; bibliography, index. Color photography by the authors. $59.95 (hardcover).

In *Old Limoges* Barbara Wood and Robert Doares of Williamsburg, Virginia, present a charming personal odyssey and highly useful reference work based on heretofore unpublished primary sources. The authors visited several institutions and archives in pursuit of documentation for the first decades of American entrepreneur David Haviland's designs and production.[1] Excitement spills over in their introduction as they recount an "adventure of discovery" that has reassigned many attributions for unmarked French porcelains from "Old Paris" to David Haviland's first two decades of production at Limoges. As William G. Allman, White House Curator, states in the foreword, *Old Limoges* is "an important volume on the history of early Haviland and Limoges porcelain, in the hands of incredibly enthusiastic students of their subject."

Correlating line drawings from a recently discovered factory book of shapes (preserved at the Metropolitan Museum of Art, New York) with images in an album of painted and transfer-printed decoration (still owned by Haviland descendants) and designs in a portfolio in the archives of the modern Haviland firm, the authors have identified numerous early Haviland models, illustrating many of them with figures, vases, and teawares that they have acquired for that purpose. A fairly well-known catalog published in 1860 by the New York agency Haviland Brothers and Company facilitates attributions of table services, but "Vases, Figures, Rich Coffees, Candlesticks and an innumerable stock of Articles of Fancy" mentioned in publications associated with the Exhibition of the Industry of All Nations at the Crystal Palace, which was held in New York in 1853, have remained largely unidentified until now.

The book's seven chapters are essentially essays on a variety of topics. Chapter 1, "The Missing Haviland History," is an informative account of the Haviland family and the permutations of their business partnerships. David Haviland immigrated to France in 1841 and settled in Limoges the following year. He began exporting French porcelains made by Pillivuyt and others to Haviland Brothers and Company, New York, and soon opened design studios for the decoration of whiteware manufactured by French firms to Haviland's specifications for American taste and table customs. It was not until after 1853 that David Haviland became a licensed manufacturer of porcelain. Although the authors assert that his awareness of the superior quality and relatively low cost of French porcelain brought about the increase in demand, I suggest that status and fashion also contributed to market forces. French porcelain served the tables of many wealthy Americans at the end of the eighteenth century, and George Washington set the precedent for formal presidential dinners with his own Sèvres service.[2] French porcelain was purchased by the government for state services during the Monroe, Jackson, and Polk administrations, years before David Haviland's first trip to France.

Chapter 2 explains the relationship among the three design sources mentioned above, and chapter 3 discusses English (specifically Staffordshire) origins for several Haviland shapes. Transfer prints were incorporated into the family-held book of decoration. Among them is a view of the Fairmount Waterworks in Philadelphia, taken from W. H. Bartlett's *American Scenery* (London, 1840), that also was used on a Staffordshire table service of the same period.[3] With respect to borrowing English designs, the authors are wisely cautious about attributing pieces to Haviland based solely on near matches to the shapes in the factory documents when a common source could be the basis for similarity.[4]

The Limoges context for Haviland design is treated extensively in chapter 4 and provides a succinct review of the development of the porcelain industry in the Limousin province and the companies that supplied Haviland with whiteware during the early years. The historical context is a prelude to a discussion in chapter 5 of the early Haviland decorating studio, which began operation in 1847. David Haviland led the modernization of the French porcelain industry and trained apprentice painters to comply with standardized designs, thereby lowering costs. Images of the earliest decorating studio and the later factory for the complete process of porcelain manufacture add historical interest and clearly illustrate the development of the Haviland firm over the twenty-year period under consideration.

Chapter 6 reviews Haviland's participation in the 1853 New York Fair at the Crystal Palace and the Paris Exposition of 1855. In 1853 Haviland still decorated whiteware fabricated by other factories. His competition at the New York venue included the firm of Haughwout and Dailey of New York, who were, like Haviland, "engaged in decorating porcelain which is imported or manufactured for them."[5] Based on samples exhibited at the fair, Haughwout and Dailey were selected in 1853 by President Franklin Pierce to supply a state porcelain service comprising many shapes that had been created specifically to satisfy American table customs.[6] Was there a business relationship between Haviland Brothers and Haughwout and Dailey? Or did Haughwout and Dailey follow Haviland's lead in specifying forms to be "manufactured for them" in France?

The last chapter offers advice to collectors about searching for early, unmarked Haviland pieces, and the six appendixes include extremely useful reproductions of drawings and translations of contemporary documents for collectors and scholars (not that they are mutually exclusive!). Appendix A concerns the artists known to have worked at Haviland in the early years; appendix B translates an 1855 description of the Haviland factory; appendix C is an 1859 description of porcelain production and decoration at Alluaud and Haviland; appendix D is David Haviland's application for a decorating patent in 1858; appendix E reproduces pages from the original factory design book at the Metropolitan Museum of Art, five drawings from the archives of Haviland, S.A., and a reproduction of the Haviland Brothers and Company catalog of 1860; appendix F illustrates the few early factory and retailer marks and includes a table of later printed marks. However, one mark that I have seen occasionally—"Haviland & Cie./ Limoges"—is omitted.

The shape book at the Metropolitan begins with number 284, so we can dream that the first 283 shapes will emerge in time for the authors' proposed second volume on the earliest Haviland designs. We look forward to their continuing pursuit and the publication of more exciting discoveries.

Susan Gray Detweiler

1. Travel to France for research in the Haviland archives was supported by grants from the American Ceramic Circle and the Haviland Collectors International Educational Foundation.

2. Margaret B. Klapthor, *Official White House China* (New York: Harry N. Abrams, 1999); Susan Gray Detweiler, "French Porcelain on Federal Tables," *American Ceramic Circle Bulletin,* no. 3 (1982): 87–108.

3. A large pearlware dinner set, marked "W.R&Co" for William Ridgway of Hanley, Staffordshire, is preserved at Wyck, An Historic House and Garden, Philadelphia.

4. See p. 129. The authors indicate that an unmarked cup and saucer of Staffordshire shape at the Brooklyn Museum has been reassigned from Charles Cartlidge to Haviland based on its "consistency" with the Haviland factory designs.

5. Benjamin Silliman and C. R. Goodrich, *The World of Science, Art and Industry . . .* (New York, 1854), p. 129.

6. Klapthor, *Official White House China,* p. 77.

Compiled by Amy C. Earls

Checklist of Recent Articles and Books on Ceramics in America

For submissions or correspondence, contact Amy C. Earls, P.O. Box 121, Florence, New Jersey 08518 U.S.A. trentonpots@yahoo.com

Adams, Brian. *Bovey Tracey Potteries: A Guide and Marks*. Bovey Tracey, Devon, Eng.: House of Marbles, 2005.

Adams, Elizabeth. "Reginald George Haggar, R.I., A.R.C.A., F.R.S.A.: A Personal Memoir." In *Special Newsletter to Commemorate the Centenary of Reginald G. Haggar, R.I., A.R.C.A., F.R.S.A., 25th December 1905–2nd December 1988*, 3–7. Caernarfon, U.K.: Northern Ceramic Society, 2005.

Allington, Lindsay. "Crock Series 2003." *Ceramics in America* (2005): 252–55.

Anthony, Marilyn. "A Conversation with Gallery Owners." *Ceramics Monthly* 53 (October 2005): 114–17.

Arminjon, Catherine, and Beatrix Saule, eds. *Tables royales et festins de cour en Europe, 1661–1789: Actes du colloque international, Palais des congrès, Versailles, February 25–26, 1994*. XIIIèmes Rencontres de l'École du Louvre. Paris: La Documentation Française, 2005.

Ashton, Charles, and Richard Hunter. *Quakers, Warriors, and Capitalists: Riverview Cemetery and Trenton's Dead*. History Traced by Route 29 booklet series. Trenton: New Jersey Department of Transportation, 2005.

Atherton, David. "HMS Brazen—Liverpool 1792." *Northern Ceramic Society Newsletter*, no. 140 (2005): 33–35.

———. "Success to Middleton Market." *Northern Ceramic Society Newsletter*, no. 138 (2005): 13–18.

Austin, John C. "J. Palin Thorley (1892–1987), Potter and Designer: Part 1." *Ceramics in America* (2005): 160–201.

Bailey, Daniel N., John W. Lawrence, and Paul W. Schopp. "Rural Tenant Laborers and the Rise of the Industrial Economy: Historical Ethnography of the Heminitz Property Site (36LH267), Upper Macungie Township, Lehigh County, Pennsylvania." *Northeast Historical Archaeology* 33 (2004): 91–110.

Barber, Cindy. "Potomac Shape Profile." *White Ironstone Notes* 12, no. 3 (2006): 4–10.

Beaton, David. "Ceramic Designs in the National Archives." *Northern Ceramic Society Newsletter*, no. 141 (2006): 31–33.

Beecher, Susan. *Woodfired Pottery*. New York: Catskill Press/Mountain Arts, 2005.

Bellerjeau, Claire. "The Townsend Jug: A Remarkable Piece of Early Swinton Pearlware." *Northern Ceramic Society Journal* 22 (2005–2006): 77–88.

Bellow, Daniel. "The Long Apprenticeship." *Ceramics Monthly* 54 (March 2006): 48–50.

Beranek, Christa M. "The Social and Material Lives of the Agricultural Elite: The 18th-Century Tyngs of Dunstable, Massachusetts." *Northeast Historical Archaeology* 33 (2004): 1–12.

Berthoud, Michael, and Richard Maskell. *A Directory of British Teapots*. Bridgnorth, Shropshire, Eng.: Micawber, 2006.

Biddle, Martin. *Nonsuch Palace: The Material Culture of a Noble Restoration Household*. Oxford, Eng.: Oxbow Books, 2005.

Bikyukova, N. Y., and N. I. Kazakevich. *Masterpieces of 18th-Century Sèvres Porcelain from the Hermitage Collection*. Exhibit catalog, State Hermitage, December 8, 2005–February 26, 2006. St. Petersburg, Russia: Publishing House of the State Hermitage, 2005.

Birks, Tony. "Gabriele Koch." *Ceramics Monthly* 53 (December 2005): 40–43.

Blackie, Sebastian, ed. *Dear Mr. Leach: Some Thoughts on Ceramics*. London: A. and C. Black, 2004.

Blakey, Harold. "A Bulb Pot from Factory Z?" *Northern Ceramic Society Newsletter*, no. 139 (2005): 44–47.

Blanchett, Chris. *20th Century Decorative British Tiles: Commercial Manufacturers, A–H*. Atglen, Pa.: Schiffer, 2006.

Bolender, Charles J. *Les assiettes imprimées de Sarreguemines,*

1828–1838: La période Utzschneider. Paris: Editions SID [Société d'Information et de Diffusion], 2004.

Bonhams. *Fine Meissen, Sèvres, and Other Continental Ceramics and Glass.* Exhibit catalog, Bonhams, New Bond Street, May 18, 2005. London: Bonhams, 2005.

Bonhams and Butterfields. *Fine European and American Furniture and Decorative Arts.* Exhibit catalog, Bonhams and Butterfields, San Francisco, June 17–19, 2005. San Francisco: Bonhams and Butterfields, 2005.

Boothroyd, Noel, and David Higgins. "An Inn-Clearance Group, c. 1800, from the Royal Oak, Eccleshall, Staffordshire." *Post-Medieval Archaeology* 39, no. 1 (2005): 197–203.

Breen, Eleanor E. "Whose Trash Is It, Anyway? A Stratigraphic and Ceramic Analysis of the South Grove Midden (44FX762/17), Mount Vernon, Virginia." *Northeast Historical Archaeology* 33 (2004): 111–30.

Brody, J. J. *Mimbres Painted Pottery.* Rev. ed. Santa Fe, N.M.: School of American Research Press, 2005.

Bromberg, Francine W., and Steven J. Shepherd. "The Quaker Burying Ground in Alexandria, Virginia: A Study of Burial Practices of the Religious Society of Friends." *Historical Archaeology* 40, no. 1 (2006): 57–88.

Brooks, Alasdair. *An Archaeological Guide to British Ceramics in Australia, 1788–1901.* Sydney and Melbourne: Australasian Society for Historical Archaeology and La Trobe University Archaeology Program, 2005.

Brown, Glen R. "Dong Hee Suh: A Slice of Paradise." *Ceramics Monthly* 53 (September 2005): 60–63.

———. "Jeremy Jernegan: Surface and Form." *Ceramics Monthly* 53 (November 2005): 41–44.

———. "Margaret Bohls' Models of Tension." *Ceramics Monthly* 53 (April 2005): 32–36.

———. "Pei Xueli: Tradition and Innovation." *Ceramics Monthly* 54 (March 2006): 64–67.

Brunet, Pierre, Daniel Dufournier, and François Fichet de Clairefontaine. *Les céramiques du Pré d'Auge: 800 ans de production.* Exhibit catalog, Musée de Lisieux, France, July 3–September 30, 2004. Lisieux: Musée de Lisieux, 2004.

Bruseth, James E., and Toni S. Turner. *From a Watery Grave: The Discovery and Excavation of La Salle's Shipwreck, "La Belle."* College Station: Texas A&M University Press, 2005.

Burns, Conner. "In Pursuit of Personal Style." *Ceramics Monthly* 54 (February 2006): 48–50.

Burrow, Ian. *A Tale of Two Houses: The Lambert/Douglas House and The Rosey Hill Mansion, 1700–1850.* History Traced by Route 29 booklet series. Trenton: New Jersey Department of Transportation, 2005.

Busch, Jason T. "The Grand Salon from the Hôtel Gaillard de La Bouëxière." *Antiques* 168 (October 2005): 110–17.

Bushnell, Gordon. *The Illustrated Guide to 19th Century Coalport "Coalbrookdale" Floral Encrusted Porcelain.* Exhibit catalog, Coalport China Museum, Ironbridge, Shropshire, Eng., April 3–September 22, 2006. Privately published, 2006.

Carde, Bob. "A Marked Ridgway Sucrier." *Northern Ceramic Society Newsletter,* no. 140 (2005): 32.

Cardew, Michael. "Modern English Potters." *Interpreting Ceramics* 6 (n.d.; online journal, www.uwic.ac.uk/ICRC). Lecture, Edinburgh College of Art, November 25, 1938.

Carney, Margaret. "Hui Ka Kwong." *Ceramics Monthly* 53 (May 2005): 41–45.

Carpenter, Kim. "Musk Thistles and Sweet Gumballs: New Interpretations by Jess Benjamin." *Ceramics Monthly* 54 (April 2006): 44–45.

Carrigan, Linda, and Allan Wunsch. *Weller Pottery: The Rare, the Unusual, the Seldom Seen.* Sarasota, Fla.: Marlin Media Publishing, 2003.

Casey, Andrew. *Starting to Collect 20th Century Ceramics.* Easthampton, Mass.: Antique Collectors' Club, 2006.

Casson, Michael. "Michael Casson at Aberystwyth 1999." *Interpreting Ceramics* 6 (n.d.; online journal, www.uwic.ac.uk/ICRC). Edited transcript of a lecture given at the International Potter's Festival, Aberystwyth.

Casson, Michael, and Jack Doherty. "Mick Casson in Conversation with Jack Doherty about the Craftsmen Potters Association (CPA)." *Interpreting Ceramics* 6 (n.d.; online journal, www.uwic.ac.uk/ICRC). Transcript of a video recording by Jeffrey Jones made at Wobage Farm, Upton Bishop, Herefordshire, February 26, 2000.

Casson, Michael, and Walter Keeler. "Mick Casson and Walter Keeler Discussing a Collection of Jugs." *Interpreting Ceramics* 6 (n.d.; online journal, www.uwic.ac.uk/ICRC). Transcript of a video recording made at Wobage Farm, Upton Bishop, Herefordshire, September 5, 1995.

———. "Mick Casson and Walter Keeler Discussing a Selection of French Country Pottery." *Interpreting Ceramics* 6 (n.d.; online journal, www.uwic.ac.uk/ICRC). Transcript of a video recording made at Wobage Farm, Upton Bishop, Herefordshire, September 5, 1995.

Christie's. *British and Continental Ceramics.* Auction catalog, Christie's South Kensington, June 7, 2006. London: Christie's, 2006.

———. *British and Continental Ceramics.* Auction catalog, Christie's South Kensington, July 6, 2006. London: Christie's, 2006.

———. *Chinese Export Ceramics.* Auction catalog, Christie's New York, January 26, 2006. New York: Christie's, 2006.

———. *European Ceramics.* Auction catalog, Christie's King Street, May

22, 2006. London: Christie's, 2006.
———. *European Ceramics, Dutch Delftware, and Glass, Including the Els Wijnker Collection of English Porcelain*. Auction catalog, Christie's Amsterdam, April 6, 2005. Amsterdam: Christie's, 2005.
———. *Important English Furniture, Including Property from the Villa Maurizio, La Jolla*. Auction catalog, Christie's New York, April 7, 2006. New York: Christie's, 2006.
———. *Important European Furniture, Works of Art, Ceramics, Tapestries, and Carpets*. Auction catalog, Christie's New York, May 16, 2006. New York: Christie's, 2006.
———. *Important Mobilier et Objets d'Art, Orfevrerie, Céramiques Européenes et Art d'Asie*. Auction catalog, Christie's Paris, June 21, 2006. Paris: Christie's, 2006.
———. *The J. W. N. van Achterbergh Collection of World Ceramics*. Auction catalog, Christie's Amsterdam, November 1, 2005. Amsterdam: Christie's, 2005.
———. *19th-Century Ceramics and Glass*. Auction catalog, Christie's South Kensington, May 11, 2006. London: Christie's, 2006.
———. *Property from the Estate of Evelyn Annenberg Hall*. Auction catalog, Christie's New York, May 16, 2006. New York: Christie's, 2006.
Christman, Ernest H. "Pottery and Prophecy in the Casas Grandes Culture." *Ceramics Monthly* 53 (April 2005): 42–45.
Chuang Hsiu-ling. "Wandering between Reality and Eternity: The Work of Shih Hsuan-yu." *Ceramics Monthly* 53 (December 2005): 48–49.
Clark, Phyllis B. "Ben Owen III." *Ceramics Monthly* 53 (November 2005): 36–40.
———. "Lynn Lais." *Ceramics Monthly* 54 (April 2006): 50–53.
Clowes, Jody. *Don Reitz: Clay, Fire, Salt, and Wood*. Exhibit catalog, Elvehjem Museum of Art, University of Wisconsin-Madison, March 12–June 5, 2005, and Mint Museums, Charlotte, N.C., October 8–December 31, 2005. Madison: Elvehjem Museum of Art, University of Wisconsin-Madison, 2005.
Cluett, Robert E. *Staffordshire Pottery: Majolica, Transfer Prints, Flow Blue, Fine Bone China from Cauldon, 1858–1962*. Atglen, Pa.: Schiffer, 2005.
Coch, Carla. "Jingdezhen: Tradition and Promise." *Ceramics Monthly* 53 (September 2005): 64–65.
Cockerill, John. "The Potteries of W. Dawson and Co. at Buxtehude, near Hamburg, and Simeon Sowerbutts and Co. at Farge, near Bremen." *Northern Ceramic Society Journal* 22 (2005–2006): 47–52.
Coleman-Smith, Richard, R. Taft Kiser, and Michael J. Hughes. "Donyatt-Type Pottery in 17th- and 18th-Century Virginia and Maryland." *Post-Medieval Archaeology* 39, pt. 2 (2005): 294–310.
Cooper, Emmanuel. "Michael Casson: An Appreciation." *Interpreting Ceramics* 6 (n.d.; online journal, www.uwic.ac.uk/ICRC).
———. *The Potter's Book of Glaze Recipes*. 2nd ed. Philadelphia: University of Pennsylvania Press, 2004.
Copeland, Robert. *Wedgwood Ware*. Rev. ed. Shire Album 321. Princes Risborough, Buckinghamshire, Eng.: Shire Books, 2005.
Copeland, Robert, Catriona Maisels, and David Maisels. "A Turner Plaque? Not This Time." *Northern Ceramic Society Newsletter,* no. 139 (2005): 54–56.
Corbin, Donna. "Eighteenth-Century Vincennes and Sèvres Biscuit Figures from a Private Collection." *American Ceramic Circle Journal* 13 (2005): 17–38.
———. "Trifles from a Boston Collection." *Ceramics in America* (2005): 213–16.
Cox, Alwyn, and Angela Cox. "The Duchess of Cumberland's Rockingham Dessert Service." *Northern Ceramic Society Journal* 22 (2005–2006): 179–92.
———. "Rockingham Porcelain Figures, 1826–42." *Northern Ceramic Society Newsletter,* no. 138 (2005): 47–49.
Crawford, Gail. *Studio Ceramics in Canada, 1920–2005*. Fredericton, New Brunswick, Canada: Goose Lane Editions in association with Gardiner Museum of Ceramic Art, 2005.
Creber, Diane. *Crystalline Glazes*. Ceramics Handbooks. 2nd ed. Philadelphia: University of Pennsylvania Press, 2005.
Cruickshank, Graeme. *Scottish Pottery*. 2nd rev. ed. Shire Album 191. Princes Risborough, Buckinghamshire, Eng.: Shire Books, 2005.
Crumpton, Margaret. "Bodley: The People and Their Pots" [lecture notes]. *Northern Ceramic Society Newsletter,* no. 139 (2005): 17–20.
Currie, Christopher K. "A Zoomorphic Horticultural Urn from Gosport, Hampshire." *Post-Medieval Archaeology* 39, pt. 2 (2005): 325–27.
Curtis, Julia B., Stephen Little, and Mary Ann Rogers; ed. J. May Lee Barrett. *Trade, Taste, and Transformation: Jingdezhen Porcelain for Japan, 1620–1645*. Exhibit catalog, China Institute Gallery, New York, February 2–June 10, 2006. New York: China Institute of America, 2006.
Dawson, Charles. "Clarice Cliff: True Innovator or Clever Copier" [lecture notes]. *Northern Ceramic Society Newsletter,* no. 141 (2006): 7–12.
Dawson, David. "The Kenneth J. Barton Collection of Vernacular Pottery." *Post-Medieval Archaeology* 39, pt. 2 (2005): 322–24.
Dawson, Shirley M. "Nancy Jurs." *Ceramics Monthly* 54 (January 2006): 55–57.
de Waal, Edmund, Sebastian Kuhn, James Putnam, Jorunn Veiteberg, and Bodil Busk Laursen. *Arcanum: Mapping 18th Century European Porcelain*. Exhibit catalog, National Museum and Gallery of Wales, Cardiff, June 4–August 29, 2005. Cardiff: National Museum of Wales, 2005.
Deagan, Kathleen, and José María Cruxent. *Archaeology at La Isabela:*

America's First European Town. New Haven, Conn.: Yale University Press, 2002.
———. *Columbus's Outpost among the Tainos: Spain and America at La Isabela, 1493–1798*. New Haven, Conn.: Yale University Press, 2002.
Deike, George, and Mickey Deike. *Feathers and Foliage: The Life and Times of the Pearlware Peafowl*. Privately printed, 2005.
Denker, Ellen P. "Lenox Will Relocate Corporate Headquarters to Bristol Borough, Pennsylvania." *Trenton Potteries* [Newsletter of the Potteries of Trenton Society] 6, no. 4 (2005): 1, 3.
———. "Sister Act: The Overbeck Pottery, 1911–1955." *Antiques* 168 (July 2005): 74–81.
Denny, Walter B. *The Artistry of Ottoman Ceramics*. London: Thames and Hudson, 2005.
Desmond, Kathleen. "Joyce Jablonski: An Artist's Way of Being." *Ceramics Monthly* 53 (October 2005): 43–46.
Diemer, Jane. "Hot Water, Chocolate, and Toast Water Pitchers." *White Ironstone Notes* 11, no. 4 (2005): 4–9.
Dieringer, Bev, and Ernie Dieringer. "White Ironstone Tureens, Part 4." *White Ironstone Notes* 11, no. 4 (2005): 19–28.
Dieringer, Ernie. "Chinese Shape Profile." *White Ironstone Notes* 12, no. 2 (2005): 1, 4–13.
Dieringer, Ernie, and Bev Dieringer. "Cake Stands." *White Ironstone Notes* 12, no. 1 (2005): 4–7.
Dillon, Pamela. "Pat Antonick: Seeing beyond the Obvious." *Ceramics Monthly* 53 (April 2005): 46–49.
Dion, Jacqueline Beaudry, and Jean-Pierre Dion. "Jiggers for the Manufacture of Pottery: An Original Patent by Joseph Crossley from Trenton." *Trenton Potteries* [newsletter of the Potteries of Trenton Society] 6, no. 1 (2005): 1–4.
Doares, Robert, and Barbara Wood. *Old Limoges: Haviland Porcelain Design and Décor, 1845–1865*. Atglen, Pa.: Schiffer, 2005.
Docker, Keith. "The Mexborough Money Boxes." *Northern Ceramic Society Newsletter*, no. 140 (2005): 12–20.
Donachie, Madeleine J. *Household Ceramics at Port Royal, Jamaica, 1655–1692*. BAR [British Archaeological Reports] International Series 1195. Oxford: Archaeopress, 2003.
Earle, Joe, with Halsey North and Alice North. *Contemporary Clay: Japanese Ceramics for the New Century*. Exhibit catalog, Museum of Fine Arts, Boston, October 6, 2005–July 9, 2006. Boston: MFA Publications, 2005.
Earls, Amy C., and George L. Miller. "1830s Painted Wares from a New Orleans Importer." *Ceramics in America* (2005): 247–52.
Eliëns, Titus M., Marjoleine Groen, Sebastiaan Ostkamp, and Loet A. Schledorn. *Delftware: History of a National Product*, vol. 3, *De Porceleyne Fles*. Zwolle, Netherlands: Waanders, 2005.
Elliott, Gordon. *Oral History in the Staffordshire Ceramics Industry*. Leek, Staffordshire, Eng.: Churnet Valley Books, 2004.
Espir, Helen. *European Decoration on Oriental Porcelain*. London: Jorge Welsh Books, 2005.
Evans, Anthony. "A Danish Liaison—Minton Memorandum." *Northern Ceramic Society Newsletter*, no. 140 (2005): 44.
Falk, Cynthia G. "Celebrating Years Together: Personal Commemorations and the Colonial Revival." *Winterthur Portfolio* 40, no. 1 (2005): 1–15.
Farry, Andrew. "Regulars and 'Irregulars': British and Provincial Variability among Eighteenth-Century Military Frontiers." *Historical Archaeology* 39, no. 2 (2005): 16–32.
Faÿ-Hallé, Antoinette, ed. *De l'Immense au Minuscule: La virtuosité en céramique*. Exhibit catalog, Musée Nationale de Céramique, Sèvres. Paris: Réunion des Musées Nationaux, 2005.
Fayen, Sarah Neale. "Playful Potting: A Miniature Tin-glazed Earthenware Chair." *Ceramics in America* (2005): 209–12.
Feld, Elizabeth, Stuart P. Feld, and David B. Warren. *In Pointed Style: The Gothic Revival in America, 1800–1860*. Exhibit catalog, Hirschl and Adler Galleries, New York, April 15–June 9, 2006. New York: Hirschl and Adler Galleries, 2006.
Fisk, Rob. "With Thanks to the Newsletter: The Underglaze V Factory, Part 2." *Northern Ceramic Society Newsletter*, no. 141 (2006): 39–43.
Fithian, Charles, Claudia Leister, James Stewart, and Chris Espenshade. "Hare Pottery Research." *Ceramics in America* (2005): 239–41.
Flach, Hans Dieter. "The Ludwigsburg Porcelain Manufactory: An Undeserved Cinderella." Translated by Susan G. Detweiler. *American Ceramic Circle Journal* 13 (2005): 49–67.
———. *Malerei auf Ludwigsburger Porzellan 1759 bis um 1850*. Regensburg, Germany: Schnell and Steiner, 2005.
Forbes, Sheila, and George Haggarty. "The Ceramic History of West Pans, Part 2." *Northern Ceramic Society Journal* 22 (2005–2006): 5–22.
Fuller, Lance S. "A Danish Liaison: Stanley Pottery Co." *Northern Ceramic Society Newsletter*, no. 138 (2005): 33–36.
Galusha, Emily, and Mary Ann Nord, eds. *Clay Talks: Reflections by American Master Ceramists (Regis Masters Series, 1997–2000)*. Minneapolis: Northern Clay Center, 2004.
Gardner, Pam. "Billingsley, Brampton, Wells: Collections and Connections." *Northern Ceramic Society Newsletter*, no. 141 (2006): 24–28.
———. "The Mystery of Thomas Mason." *Northern Ceramic Society Newsletter*, no. 139 (2005): 50–53.
Gaydamak, Arcadi. *Russian Empire: Architecture, Decorative and Applied Arts, Interior Decoration, 1800–1830*. Moscow: Trefoil Press, 2005.

Gent, Tim. "Nic Collins: In Search of Space." *Ceramics Monthly* 53 (September 2005): 45–49.
Goodacre, J. Michael. *A Catalogue of the Goodacre Collection of Early Staffordshire Pottery Figures: The Early Period of Overglaze Painted Enamel Decoration, 1790–1840.* County Durham, Eng.: Bowes Museum, 2005.
Goodell, Gretchen M. "Revisiting Mount Vernon." *Antiques* 169 (January 2006): 144–51.
Goodfellow, Peter. "Pâte-sur-Pâte and Lawrence Arthur Birks: The Minton Years, 1872–1894, and the Vine Pottery, 1894–1934." *Northern Ceramic Society Newsletter,* no. 140 (2005): 26–31.
———. *The Vine Pottery, Birks, Rawlins and Co.: Influences from Minton and Spode/Copeland, Lawrence Arthur Birks, and His Sculptor/Modeller Relatives.* Easthampton, Mass.: Antique Collectors' Club, 2006.
Goss, Steven. *British Teapots and Coffee Pots.* Shire Album 446. Princes Risborough, Buckinghamshire, Eng.: Shire, 2005.
Grace, Allen. "A Tale of Two Bowls." *Northern Ceramic Society Newsletter,* no. 140 (2005): 24–25.
Gray, Jonathan, ed. *Welsh Ceramics in Context, Part 2.* Swansea, Wales: Swansea Museum, 2005.
Gregory, Ian. *Alternative Kilns.* Philadelphia: University of Pennsylvania Press, 2005.
Gundy, Barbara J., and Deborah Casselberry. "The Mansion Pottery." *Ceramics in America* (2005): 141–59.
Gurnett, Robin. "The Forgotten Woods of Brownhills, Part 1: John Wood, 1746–1797." *Northern Ceramic Society Journal* 22 (2005–2006): 28–46.
Haggarty, George. *The West Pans Pottery, East Lothian, Scotland Ceramic Resource Disk* (CD-ROM). Edinburgh: National Museums of Scotland, 2006.
Hale, Anna. "Mick Casson Interviewed by Anna Hale," no. 1: Craftsmen Potters' Association. *Interpreting Ceramics* 6 (n.d.; online journal, www.uwic.ac.uk/ICRC). Summary of an interview at Wobage Farm, Upton Bishop, Herefordshire, February 24, 2000.
———. "Mick Casson Interviewed by Anna Hale," no. 2: Biography. *Interpreting Ceramics* 6 (n.d.; online journal, www.uwic.ac.uk/ICRC). Summary of an interview at Wobage Farm, Upton Bishop, Herefordshire, February 25, 2000.
———. "Mick Casson Interviewed by Anna Hale," no. 3: Harrow Course. *Interpreting Ceramics* 6 (n.d.; online journal, www.uwic.ac.uk/ICRC). Summary of an interview at Wobage Farm, Upton Bishop, Herefordshire, possibly November 2000.
Hampson, Rodney. "The Published Work of Reginald George Haggar." In *Special Newsletter to Commemorate the Centenary of Reginald G. Haggar, R.I., A.R.C.A., F.R.S.A., 25th December 1905–2nd December 1988,* 20–33. Caernarfon, U.K.: Northern Ceramic Society, 2005.
Hanscombe, Steven. *James Giles (1718–1780), China and Glass Painter Extraordinaire.* Exhibit catalog, Stockspring Antiques, London, June 9–21, 2005. London: Stockspring Antiques, 2005.
Harvey, Ian. *Rathbone: China Manufacturers, Tunstall, Staffordshire, 1808 to 1843.* London: Harvey Publications, 2005.
Havelin, Jim L. "Danville Chadbourne." *Ceramics Monthly* 54 (January 2006): 60–63.
Haydon, Margaret. "The Journey Continues." *Ceramics Monthly* 53 (September 2005): 42–44.
Helke, Gun-Dagmar. *Johann Esaias Nilson (1721–1788): Augsburger Miniatumaler, Kupferstecher, Verleger, und Kunstacademiedirektor.* Munich: Scaneg, 2005.
Hewitt, Mark, and Nancy Sweezy. *The Potter's Eye: Art and Tradition in North Carolina Pottery.* Exhibit catalog, North Carolina Museum of Art, Raleigh, October 30, 2005– March 19, 2006. Chapel Hill: University of North Carolina Press/North Carolina Museum of Art, 2005.
Hildyard, Robin. *English Pottery, 1620–1840.* London: V & A Publications, 2005.
Hill, Steven. "An Approach to Single Firing—Further in." *Ceramics Monthly* 54 (January 2006): 50–54.
Hoffman, Sam. "Clays, Slips and Glazes for Wood Firing." *Ceramics Monthly* 53 (June/July/August 2005): 53–59.
Hoffman, William. "The Stoneware Kiln of Charles F. Decker in Washington County, Virginia." *Ceramics in America* (2005): 232–34.
Holdway, Ruth. *Harold Holdway: 20th Century Ceramic Designer.* Ashbourne, Derbyshire, Eng.: Landmark Publishing, 2006.
Holmes, Jonathan. *Les Blakebrough: Ceramics.* Sydney: Craftsman House, 2005.
Honolulu Academy of Arts. *Yakimono: 4000 Years of Japanese Ceramics.* Honolulu: Honolulu Academy of Arts, dist. by University of Washington Press, 2005.
Hopper, Robin. *Making Marks: Discovering the Ceramic Surface.* Iola, Wis.: KP Books, 2004.
Høst-Madsen, Lene. "Eighteenth-Century Copenhagen Viewed from a Refuse Dump." *Post-Medieval Archaeology* 39, pt. 2 (2005): 311–22.
Houze, Herbert G. "Samuel Colt's Porcelain Transparencies." *Antiques* 169 (April 2006): 106–15.
Howard, David Sanctuary. *The Unforgiving Minute.* Durham, Eng.: Memoir Club, 2004.
Hoysted, Marilyn. "Researching a Scottish Enterprise in Asia: The Bell Pottery of Glasgow." *Northern Ceramic Society Newsletter,* no. 140 (2005): 36–43.
Hunter, Richard W. "18th-Century Stoneware Kiln Found in Trenton." *Bulletin of the Archaeological Society of New Jersey,* no. 59 (2004): 71–73.
———. *Fish and Ships: Lamberton, the Port of Trenton.* History Traced by Route 29 booklet series. Trenton:

New Jersey Department of Transportation, 2005.

———. "Lenox Factory Buildings Demolished." *Trenton Potteries* [Newsletter of the Potteries of Trenton Society] 6, nos. 2–3 (2005): 1–9.

———. "The Pottery Decorating Shop of the Mayer Arsenal Pottery Company." *Archaeological Society of New Jersey Bulletin* 60 (2005): 77–80.

Hunter, Robert, and Marshall Goodman. "The Destruction of the Benjamin DuVal Stoneware Manufactory, Richmond, Virginia." *Ceramics in America* (2005): 37–60.

Hunter, Robert, Kurt C. Russ, and Marshall Goodman. "Stoneware of Eastern Virginia." *Antiques* 167 (April 2005): 126–33.

Hurry, Silas D. "What Is 'What' in St. Mary's City?" *Ceramics in America* (2005): 220–23.

Hurst, John G., and David Gaimster. "Werra Ware in Britain, Ireland, and North America." *Post-Medieval Archaeology* 39, pt. 2 (2005): 267–93.

Hyland, Peter. *Herculaneum Pottery: Liverpool's Forgotten Glory*. Liverpool, Eng.: Liverpool University Press, dist. in the U.S. by University of Chicago Press, 2005.

Ironbridge Gorge Museum Trust. *Caughley in Colour*. Exhibit catalog, Coalport China Museum, March 19–October 30, 2005. Coalbrookdale, Shropshire, Eng.: Coalport China Museum, Ironbridge Gorge Museum Trust, 2005.

Jack, Marlene. "A Balanced Hand." *Ceramics Monthly* 53 (December 2005): 36–39.

Jackson, Reg, Rod Burchill, David Dungworth, and Catherine Mortimer. "Excavations on the Site of Sir Abraham Elton's Glassworks, Cheese Lane, Bristol." *Post-Medieval Archaeology* 39, no. 1 (2005): 92–132.

Jacob-Hanson, Charlotte. "Fidelle Duvivier in France and the Netherlands." *Antiques* 169 (January 2006): 168–77.

Janos, Linda. "Luminosity: James Haggerty's Low-Fire Lusters." *Ceramics Monthly* 53, no. 9 (November 2005): 57–59.

Johnson, Paul. *The Vanished Landscape: A 1930s Childhood in the Potteries*. London: Weidenfeld and Nicolson, 2004.

Jørgensen, Ulla Munck. "Barbro Åberg." *Ceramics Monthly* 54, no. 3 (March 2006): 51–54.

Kaplan, Wendy, ed., with contributions by Alan Crawford, Rüdiger Joppien, Juilet Kinchin, Amy F. Ogata, Elisabet Stavenow-Hidemark, and Christian Witt-Dörring. *The Arts and Crafts Movement in Europe and America, 1880–1920: Design for the Modern World, 1880–1920*. Exhibit catalog, Milwaukee Museum of Art, May 19–September 5, 2005. New York: Thames and Hudson, 2004.

Karlson, Norman. *The Encyclopedia of American Art Tiles*. Atglen, Pa.: Schiffer, 2005.

Kelly, Henry E. *The Glasgow Pottery of John and Matthew Perston Bell: China and Earthenware Manufacturers in Glasgow*. www.bellsglasgowpottery.com.

Kentish, Trevor. "Ridgway Basalt." *Northern Ceramic Society Newsletter*, no. 141 (2006): 16–17.

Kerr, Rose, and Nigel Wood. *Ceramic Technology*. Science and Civilisation in China, vol. 5, pt. 12. Cambridge, Eng.: Cambridge University Press, 2004.

Ketz, K. Anne, Elizabeth J. Abel, and Andrew J. Schmidt. "Public Image and Private Reality: An Analysis of Differentiation in a Nineteenth-Century St. Paul Bordello." *Historical Archaeology* 39, no. 1 (2005): 74–88.

Kille, John E. "Bell Family Presentation Jar." *Ceramics in America* (2005): 235–38.

———. "Distinguishing Marks and Flowering Designs: Baltimore's Utilitarian Stoneware Industry." *Ceramics in America* (2005): 93–132.

Killock, Douglas, and Frank Meddens with Philip Armitage, Geoff Egan, David Gaimster, Chris Jarrett, Lynne Keys, Chris Phillpotts, Ken Sabel, Rachel Tyson, and Hugh Willmott. "Pottery as Plunder: A 17th-Century Maritime Site in Limehouse, London." *Post-Medieval Archaeology* 39, no. 1 (2005): 1–91.

Klingelhoffer, Eric, Tracy Collins, Sheila Lane, Margaret McCarthy, Clare McCutcheon, Sarah McCutcheon, Jo Moran, and John Tierney. "Edmund Spenser at Kilcolman Castle: The Archaeological Evidence." *Post-Medieval Archaeology* 39, no. 1 (2005): 133–54.

Koterski, James. *Early Potters and Potteries of Delaware: Historical and Commercial Perspectives, 1760–1890*. Wilmington, Del.: Cedar Tree Books, 2005.

Koverman, Jill B. "The Ceramic Works of David Drake, aka, Dave the Potter or Dave the Slave of Edgefield, South Carolina." *American Ceramic Circle Journal* 13 (2005): 83–98.

Kozakavich, Stacy C. "Doukhobor Identity and Communalism at Kirilovka Village Site." *Historical Archaeology* 40, no. 1 (2006): 119–32.

Kräftner, Johann, ed. *Baroque Luxury Porcelain: The Manufactories of Du Paquier in Vienna and of Carlo Ginori in Florence*. Exhibit catalog, Liechtenstein Museum, Vienna, November 10, 2005–January 29, 2006. Munich and New York: Prestel, 2005.

Kudryavtseva, Tamara, ed. *Circling the Square: Avant-Garde Porcelain from Revolutionary Russia*. Exhibit catalog, Somerset House, London, November 18, 2004–July 31, 2005. New York: Fontanka, dist. in the U.S. by Harry N. Abrams, 2005.

Kusakabe, Masakazu, and Marc Lancet. *Japanese Wood-Fired Ceramics*. Iola, Wis.: Krause, 2005.

Lage, Chad. *Pictorial Guide to Pottery and Porcelain Marks*. Paducah, Ky.: Collector Books, 2004.

Lange, Amanda. *Chinese Export Art at Historic Deerfield*. Exhibit catalog, Flynt Center of Early New England Life, Historic Deerfield, May 7–August 6, 2005. Deerfield,

Mass.: Historic Deerfield, Inc., 2005.

Leftwich, Rodney. *Pisgah Forrest and Nonconnah: The Potteries of Walter B. Stephen*. Bradenton, Fla.: Marlin Media, 2006.

Lenox, Stewart. "Jim Wayne's Saggar-Fired Shino Teaware." *Ceramics Monthly* 53 (April 2005): 37–41.

Levy, Martin. "Living with Antiques: In the Form of a Story: A Collection of Nineteenth-Century American and European Decorative Arts." *Antiques* 169 (March 2006): 90–99.

Lewis, John, and Griselda Lewis. *Pratt Ware: English and Scottish Relief Decorated and Underglaze Coloured Earthenware, 1780–1840*. 3rd rev. ed. Easthampton, Mass.: Antique Collectors' Club, 2006.

Livingstone, Karen, and Linda Parry, eds. *International Arts and Crafts*. Exhibit catalog, Victoria and Albert Museum, London, March 17–July 24, 2005. London: V&A Publications, 2005.

Lockett, Terence A. "English Ceramics in Australia" [lecture notes]. *Northern Ceramic Society Newsletter,* no. 139 (2005): 21–23.

———. "Reginald G. Haggar, Hon. M.Univ., R.I., A.R.C.A., F.R.S.A.: An Appreciation." In *Special Newsletter to Commemorate the Centenary of Reginald G. Haggar, R.I., A.R.C.A., F.R.S.A., 25th December 1905–2nd December 1988,* 34–45. Caernarfon, U.K.: Northern Ceramic Society, 2005.

Lomax, Barry. "Did Booths Copy Caughley?" *Northern Ceramic Society Newsletter,* no. 138 (2005): 40–42.

Longworth, Christine. "Buckley Ceramics in the Seventeenth Century: Socio-Economic Status of the Potters and Possible Design Influences." *Interpreting Ceramics* 6 (n.d.; online journal, www.uwic.ac.uk/ICRC).

Lukacs, George H. "A Pot of Butter for the Victims." *Ceramics in America* (2005): 1–7.

Mack, Charles R. *Talking with the Turners: Conversations with Southern Folk Potters*. Columbia: University of South Carolina Press, 2006.

Magid, Barbara H. "Sugar Refining Pottery from Alexandria and Baltimore." *Ceramics in America* (2005): 223–29.

Maisels, Catriona, and David Maisels. "Bridgeless Chinoiserie Pattern by Verreville." *Northern Ceramic Society Newsletter,* no. 140 (2005): 21–23.

Mallios, Seth. "Back to the Bowl: Using English Tobacco Pipebowls to Calculate Mean Site-Occupation Dates." *Historical Archaeology* 39, no. 2 (2005): 89–104.

Mansfield, Janet. *Ceramics in the Environment: An International Review*. Westerville, Ohio: American Ceramic Society, 2005.

Marshall, Frank. "Frank Willett: Potter." *Ceramics Monthly* 53 (December 2005): 53–56.

Mass, Jennifer L., Catherine R. Matsen, and Janice H. Carlson. "Materials of the Pennsylvania German Fraktur Artist." *Antiques* 168 (September 2005): 128–35.

Mau, Linda. "Jan Schachter: Potter." *Ceramics Monthly* 54 (April 2006): 40–43.

McClelland, Jim. "Helen W. Drutt English, Ceramics' Champion." *Ceramics Monthly* 53 (May 2005): 37–40.

McCuistion, John. "Homage to Palissy." *Ceramics Monthly* 53 (October 2005): 58–59.

McDonald, Eva M. "Towards a Historical Archaeology of the German-Canadians of Markham's Berczy Settlement." *Northeast Historical Archaeology* 33 (2004): 13–38.

McWilliams, Bob. "Satori Yamaoka's Kobushigama." *Ceramics Monthly* 53 (May 2005): 46–49.

Meilach, Dona Z. *Teapots: Makers and Collectors*. A Schiffer Book for Collectors. Atglen, Pa.: Schiffer, 2005.

Meyers, Allan D. "Material Expressions of Social Inequality on a Porfirian Sugar Hacienda in Yucatán, Mexico." *Historical Archaeology* 39, no. 4 (2005): 112–37.

Mickelson, Leigh T. "Reconnecting: The Expressive Utilitarian Wares of Andy Shaw." *Ceramics Monthly* 53 (October 2005): 54–57.

Mitchell, Laurence. *The Meissen Collectors' Catalogue*. Easthampton, Mass.: Antique Collectors' Club, 2005.

Mones, Nicole. *A Cup of Light*. New York: Delacorte, 2002.

Montmollin, Daniel de. *The Practice of Stoneware Glazes: Minerals, Rocks, Ashes*. 3rd ed. Translated by Brother Anthony. Vendin-le-Vieil, France: La Revue de la Céramique et du Verre, 2005.

Mueller, Shirley M. "Eighteenth-Century Chinese Export Porcelain Teapots: Fashion and Uniformity." *American Ceramic Circle Journal* 13 (2005): 5–16.

Musée des Arts Décoratifs de Bordeaux. *Marie-Antoinette à Versailles: le goût d'une reine*. Exhibit catalog, Musée des Arts Décoratifs de Bordeaux, France, October 21, 2005–January 30, 2006. Paris and Bordeaux: Somogy Édition d'Art/Musée des Arts Décoratifs de Bordeaux, 2005.

Musée National Adrien Dubouché. *Félix Bracquemond and Decorative Art: From Japanisme to Art Nouveau*. Exhibit catalog, Musée National Adrien Dubouché, Limoges, France, April 6–July 4, 2005. Paris: Réunion des Musées Nationaux, 2005.

Museum für Angewandte Kunst Frankfurt. *Faszination Keramik: The Fascination of Ceramics: Masterpieces of Modern Japanese Pottery from the Gisela Feudenberg Collection*. Exhibit catalog, Museum für Angewandte Kunst Frankfurt, April 7–August 28, 2005. Frankfurt-am-Main: Museum für Angewandte Kunst Frankfurt, 2005.

Neale, Gillian. *Miller's Encyclopedia of British Transfer-Printed Pottery Patterns, 1790–1930*. London: Mitchell Beazley, 2005.

Neff, Hector, comp. *Ceramics in Archaeology: Readings from American Antiquity, 1936–2002*. Washington, D.C.: Society for American Archaeology Press, 2005.

Nielsen, Chris. "A Natural Fixation." *Ceramics Monthly* 53 (December 2005): 59–62.

Noël Hume, Ivor. "John Bacon: Prince of Stoneware Potters?" *Ceramics in America* (2005): 20–36.

Nonestied, Mark, and Richard Veit. "The Great Wall of Terracotta: A Surprising Ceramic Discovery in South Amboy, New Jersey." *Ceramics in America* (2005): 205–8.

Norris, Scott. "Nelli Isupova." *Ceramics Monthly* 54 (February 2006): 51–53.

Owen, June M. "Gabrielle Koch, Ceramic Artist." *Northern Ceramic Society Journal* 22 (2005–2006): 137–48.

Panes, Nick. "The China Club Revisited in an Economic Context: An Idea Too Late for Its Time." *Northern Ceramic Society Journal* 22 (2005–2006): 63–76.

Partington, Matthew. "Espresso, Exoticism, and Earthenware: The London Coffee Bar Ceramics of the Picassoettes (William Newland, Margaret Hine, and Nicholas Vergette), 1952–1966." *Interpreting Ceramics* 6 (n.d.; online journal, www.uwic.ac.uk/ICRC).

Pendergast, David M. "An Unusual Caneware Admiral Nelson Jug." *Northern Ceramic Society Newsletter,* no. 138 (2005): 26–32.

Perryman, Jane. *The Mouldmaker's Handbook*. London: A. and C. Black, 2004.

———. *Naked Clay: Ceramics without Glaze*. London and Philadelphia: A. and C. Black and University of Pennsylvania Press, 2004.

Piché, Thomas, Jr., Ellen P. Denker, and Elizabeth Fowler. *Only an Artist: Adelaide Alsop Robineau, American Studio Potter*. Exhibit catalog, Everson Museum of Art, Syracuse, N.Y., March 11–May 21, 2006. Syracuse: Syracuse University Press, 2006.

Pietsch, Ulrich, ed. *Meissen for the Czars: Porcelain as a Means of Saxon-Russian Politics in the 18th Century.* Munich: Hirmer, 2005.

Pilcher, Don. "Looking at Kesl and Tilton." *Ceramics Monthly* 53 (December 2005): 44–47.

———. "Looking at Looking." *Ceramics Monthly* 54 (February 2006): 34–38.

Pine, Lynda. *Goss and Souvenir Heraldic China.* Shire Album 440. Princes Risborough, Buckinghamshire, Eng.: Shire Books, 2005.

Pogue, Dennis J., Esther C. White, and Eleanor E. Breen. "Digging for Trash and Finding Treasure at Mount Vernon." *Antiques* 168 (September 2005): 88–95.

Pomfret, Roger. "The Diorama Transport Views." *Ceramics in America* (2005): 242–47.

———. "W(***): The Case for Whitehead Re-assessed." *Northern Ceramic Society Journal* 22 (2005–2006): 111–36.

Pope, Peter E. *Fish into Wine: The Newfoundland Plantation in the Seventeenth Century*. Chapel Hill: University of North Carolina Press, 2004.

Préaud, Tamara, and Régine de Plinval de Guillebon. *Porcelaines du Musée Condé à Chantilly*. Paris: Somogy Édition d'Art, 2005.

"Quartered Rose and Memnon Shape." *White Ironstone Notes* 12, no. 4 (2006): 1, 4–11.

Rafferty, Sean, and Rob Mann, eds. *Smoking and Culture: The Archaeology of Tobacco Pipes in Eastern North America*. Knoxville: University of Tennessee Press, 2005.

Rago, David. *Zanesville and Art Pottery Sale*. Auction catalog, David Rago Auctions, Lambertville, N.J., May 21, 2006. Lambertville, N.J.: David Rago, 2006.

Raine, Polly. "A Studio in the Woods: Gerry Dinnen's Salt-Fired Pots." *Ceramics Monthly* 53 (October 2005): 47–50.

Rathbone, Belinda. "Living with Antiques: Revivalism Revived in a New York City Apartment." *Antiques* 169 (February 2006): 60–67.

Ray, Ken, and Alison Morgan. *A Guide to Dudson Sprig Moulds Used in the Production of Jasper Ware, c. 1820–1967*. Stoke-on-Trent, Staffordshire, Eng.: Dudson Museum, [2005?].

Rickard, Jonathan. *Mocha and Related Dipped Wares, 1770–1939*. Lebanon, N.H.: University Press of New England; East Nassau, N.Y.: Historic Eastfield Foundation, 2006.

Risatti, Howard. "Making Meaning: A Dialog between Nature and Culture." *Ceramics Monthly* 54 (February 2006): 39–42.

Roberts, Emlyn. "Pots and Custom." *Northern Ceramic Society Newsletter,* no. 141 (2006): 29–30.

Robison, Jim. *Large-Scale Ceramics*. Rev. ed. Westerville, Ohio: American Ceramic Society, 2005.

Rogers, Connie. "Follow-up on 'A New Version of the Dagger Border Pattern.'" *Northern Ceramic Society Newsletter,* no. 140 (2005): 45–46.

Rogers, Kory W. "Slip-Ups: Mocha Ware at the Shelburne Museum." *Antiques* 167 (June 2005): 96–103.

Rogerson, John. "Some Thoughts on the Lack of Innovation in Kiln Design in the UK Tableware Industry." *Northern Ceramic Society Journal* 22 (2005–2006): 23–27.

Rosenfield, Louise. "The Tei Tei Project." *Ceramics Monthly* 54 (February 2006): 43–47.

Russ, Kurt C., and W. Sterling Schermerhorn. "Rocketts' Red Glare: John P. Schermerhorn and the Early Richmond-Area Stoneware Industry." *Ceramics in America* (2005): 61–92.

Russo, Tom. "Fun and Games: Recent Stadiums from Douglas Baldwin." *Ceramics Monthly* 53 (October 2005): 60.

Sacks, Judith R. "Blackface Minstrel Pottery from Staffordshire: Documenting a British-American Cultural Exchange of the Nineteenth Century." *American Ceramic Circle Journal* 13 (2005): 69–82.

Savino, Kelly Averill. "Workshops: A Different Kind of Schooling." *Ceramics Monthly* 54 (April 2006): 96.

Scanlon, Des. "Exploiting the Power of Copper Plate Analysis." *Northern Ceramic Society Newsletter,* no. 141 (2006): 18–23.

———. "Separating Miles Mason Early Teaware Patterns from the WML Partnership and Thomas Wolfe." *Northern Ceramic Society Newsletter,* no. 139 (2005): 28–40.

———. "Separating Miles Mason from Thomas Wolfe: New Evidence." *Northern Ceramic Society Newsletter,* no. 141 (2006): 36–38.

———. "Wolfe-Mason-Lucock: Confirmation of Attribution?" *Northern Ceramic Society Newsletter,* no. 138 (2005): 37–39.

Schaefer, Elizabeth M., ed., with contributions by Joe K. Kindig III, Philip D. Zimmerman, Donna Ghelerter, and Michele Majer. *Wright's Ferry Mansion,* vols. 1 and 2. Seattle, Wash.: Marquand Books, dist. by Antique Collectors' Club, 2006.

Schwartz, Selma. *The Razumovsky Sèvres Service: A Porcelain Cabinet of Curiosities*. Exhibit catalog, Musée National de Céramique, Sèvres, France, November 23, 2005–March 9, 2006. Waddesdon, Buckinghamshire, Eng.: Waddesdon Manor, 2005.

Schwartz, Selma. *Sèvres: The Marton Museum Collection at Samobor.* Samobor, Croatia: Marton Museum, 2005.

Seifert, Donna J., and Joseph Balicki. "Mary Ann Hall's House." *Historical Archaeology* 39, no. 1 (2005): 59–73.

Selsor, Marcia. "La Meridiana: The International Center for Ceramic Arts in Tuscany." *Ceramics Monthly* 53 (April 2005): 50–53.

Shatz, Jayne. "The Oxidation Reverberation: How the Gas Crisis of the 1970s Influenced American Ceramics." *Ceramics Monthly* 53 (September 2005): 66–69.

Shockley, Jay, and Susan Tunick. "The Cooper Union Building and Architectural Terra Cotta." *Winterthur Portfolio* 39, no. 4 (2004): 207–27.

Skinner, Deborah. "The Haggar Collection of Ceramics." In *Special Newsletter to Commemorate the Centenary of Reginald G. Haggar, R.I., A.R.C.A., F.R.S.A., 25th December 1905–2nd December 1988,* 46–59. Caernarfon, U.K.: Northern Ceramic Society, 2005.

Skinner, Inc. *Discovery Featuring Ceramics*. Auction catalog, Skinner, Bolton, Mass., July 14–15, 2005. Bolton, Mass.: Skinner, 2005.

———. *European Furniture and Decorative Arts Featuring Fine Ceramics*. Auction catalog, Skinner, Boston, July 16, 2005. Boston: Skinner, 2005.

Slater, Greg, Jonathan Brough, Neil Mitchell, and Richard Ansell, eds. *Comprehensively Clarice Cliff: An Atlas of Over 2,000 Patterns, Shapes, and Backstamps*. London: Thames and Hudson, 2005.

Smith, Dennis W. "Working for Others or for Yourself: It's a Choice." *Ceramics Monthly* 54 (March 2006): 35–39.

Smith, Steven. *Spode and Copeland: Over Two Hundred Years of Fine China and Porcelain.* A Schiffer Book for Collectors. Atglen, Pa.: Schiffer, 2005.

Snyder, Jeffrey B. *Rookwood Pottery.* Atglen, Pa.: Schiffer, 2005.

Sotheby's. *Ceramics Special Sale.* Auction catalog, Sotheby's Hong Kong, April 10, 2006. Hong Kong: Sotheby's, 2006.

———. *Easton Neston, Northamptonshire Country House*. Auction catalog, Easton Neston, Northamptonshire, Eng., May 17–19, 2005. London: Sotheby's New Bond Street, 2005.

———. *English Country House Sale*. Auction catalog, Sotheby's New York, April 6, 2006. New York: Sotheby's, 2006.

———. *Fine Chinese Ceramics and Works of Art*. Auction catalog, Sotheby's, New Bond Street, July 13, 2005. London: Sotheby's New Bond Street, 2005.

———. *Fine French Furniture, Ceramics, and Decorative Works of Art*. Auction catalog, Sotheby's New York, May 19, 2006. New York: Sotheby's, 2006.

———. *Important English Pottery: The Harriet Carlton Goldweitz Collection*. Auction catalog, Sotheby's New York, January 20, 2006. New York: Sotheby's, 2006.

———. *Important European Ceramics and Objects of Vertu: Property from the Collection of Hanns Weinberg and the Antique Porcelain Company of New York.* Auction catalog, Sotheby's New York, May 23–24, 2005. New York: Sotheby's, 2005.

Sotheby's. *19th-Century Furniture, Ceramics, and Decorative Works of Art*. Auction catalog, Sotheby's New York, April 27, 2006. New York: Sotheby's, 2006.

South, Stanley. *An Archaeological Evolution.* Berlin: Springer, 2005.

Spavold, Janet, and Sue Brown. *Ticknall Pots and Potters*. Ashbourne, Derbyshire, Eng.: Landmark Collector's Library, 2005.

Speel, Erika. "Viennese Enamels in the Renaissance Revival Style." *Antiques* 169 (April 2006): 134–43.

Spero, Simon, and Richard Burt. *Lund's Bristol and Early Worcester Porcelain, 1750–58: The A. J. Smith Collection*. Easthampton, Mass.: Antique Collectors' Club, 2006.

Steele, Ginger, with Ellen Currans. "Everyday Dishes." *Ceramics Monthly* 54 (February 2006): 57–59.

Stern, Marc J. "Industrial Structure and Occupational Health: The American Pottery Industry, 1897–1929." *Business History Review* 77 (2003): 417–45.

Stobart, J. "Identity, Competition, and Place Promotion in the Five Towns." *Urban History* 20 (2003): 163–82.

Stradling, Diana, and William G. Liebeknecht. *"Fancy Rockingham" Pottery: The Modeller and Ceramics in Nineteenth-Century America.* Exhibit catalog, University of Richmond Museums, September 9, 2004–June 26, 2005. Richmond,

Va.: University of Richmond Museums, 2005.

Straube, Beverly A. "This Little Piggy Went to Virginia." *Ceramics in America* (2005): 217–19.

Sullivan, Faro A. "Addicted to Clay: Meira Mathison." *Ceramics Monthly* 53 (October 2005): 40–42.

Suter, Scott H. "Otto Karle: A Previously Unknown Shenandoah Valley Potter." *Ceramics in America* (2005): 229–32.

Swisher, Mariko. "A Look West from Japan." *Ceramics Monthly* 54 (March 2006): 58–61.

Talbott, Page, ed. *Benjamin Franklin: In Search of a Better World*. Exhibit catalog, National Constitution Center, Philadelphia, December 15, 2005–April 30, 2006; Missouri Historical Society, St. Louis, June 8–September 4, 2006; Houston Museum of Natural Science, October 11, 2006–January 21, 2007; Denver Museum of Nature and Science, March 2–May 28, 2007; Atlanta History Center, July 4–October 14, 2007; Musée des Arts et Métiers and Musée Carnavalet, Paris, December 4, 2007–March 30, 2008. New Haven, Conn.: Yale University Press, 2005.

———. "Franklin's Legacy: Documented Furnishings." *Antiques* 168 (December 2005): 64–73.

Tanner, Arleen, and Grahame Tanner. *Swansea's Cambrian Pottery Transferware and Other Welsh Examples*. Stow Market, Suffolk, Eng.: Polstead Press, 2005.

Taylor, Jordan. *Potter's Apprentice: Continuity and Mentorship in a Modern Tradition*. Exhibit catalog, Mahady Art Gallery, Marywood University, Scranton, Pa., October 17–November 20, 2005. Scranton, Pa.: Mahady Art Gallery, Marywood University, 2005.

Taylor, Susan. "Bullers of Hanley: A Case Study of a 19th Century Electrical Porcelain Manufacturer." *Northern Ceramic Society Journal* 22 (2005–2006): 89–109.

Teitelman, S. Robert. "Maine's Early Nineteenth-Century Barrell-Wood Family Jugs and the Remarkable Woman Who Made Them Great." *Ceramics in America* (2005): 9–19.

Tomaso, Matthew S., Richard F. Veit, Carissa A. DeRooy, and Stanley L. Walling. "Social Status and Landscape in a Nineteenth-Century Planned Industrial Alternative Community: Archaeology and Geography of Feltville, New Jersey." *Historical Archaeology* 40, no. 1 (2006): 20–36.

Tomlinson, Alan, and Janet Tomlinson. "Shards and Marks from the Ferrybridge Pottery, Yorkshire." *Northern Ceramic Society Journal* 22 (2005–2006): 149–78.

Tourtillott, Suzanne J. E., ed. *The Figure in Clay: Contemporary Sculpting Techniques by Master Artists*. Asheville, N.C.: Lark Books, 2005.

Treadway Gallery. *Decorative Arts Auction Featuring Rookwood*. Cincinnati Museum Center, Cincinnati, June 4–5, 2005.

Triggs, John R. "Social Flux at the Naval Establishment at Penetanguishene, Lake Huron, 1817–1834." *Historical Archaeology* 39, no. 2 (2005): 105–35.

van Dam, Jan Daniël. *Delffse Porceleyne/Dutch Delftware, 1620–1850*. Zwolle: Waanders; Amsterdam, Rijksmuseum, 2005.

van Lemmen, Hans. *Delftware Tiles*. Rev. ed. Shire Album 179. Princes Risborough, Buckinghamshire, Eng.: Shire Book, 2005.

Van Wormer, Stephen R., and G. Timothy Gross. "Archaeological Identification of an Idiosyncratic Lifestyle: Excavation and Analysis of the Theosophical Society Dump, San Diego, California." *Historical Archaeology* 40, no. 1 (2006): 100–118.

Veit, Richard, and Judson M. Kratzer. "Rediscovering the New Brunswick Stoneware Pottery, ca. 1862–1901." *Ceramics in America* (2005): 133–40.

Veith, Barbara. "Edward Lycett (1833–1910): An American China-Painting Pioneer." *American Ceramic Circle Journal* 13 (2005): 99–137.

Vigurs, Peter. "Introduction to the Paintings, Drawings, and Designs of R. G. Haggar." In *Special Newsletter to Commemorate the Centenary of Reginald G. Haggar, R.I., A.R.C.A., F.R.S.A., 25th December 1905–2nd December 1988*, 8–19. Caernarfon, U.K.: Northern Ceramic Society, 2005.

Vincentelli, Moira. "Gender, Pottery Technology, and Development Projects." *Interpreting Ceramics* 6 (n.d.; online journal, www.uwic.ac.uk/ICRC).

———. "Pan Casson Henry interviewed by Moira Vincentelli." *Interpreting Ceramics* 6 (n.d.; online journal, www.uwic.ac.uk/ICRC). Transcript of an interview by Moira Vincentelli, November 29, 1994; revised April 2005.

von Dassow, Sumi. "Mary Cay: A Glittering Obsession." *Ceramics Monthly* 54 (February 2006): 60–63.

Vroom, Joanita. *After Antiquity: Ceramics and Society in the Aegean from the 7th to the 20th Century A.C.: A Case Study from Boeotia, Central Greece*. Archaeological Studies 10. Leiden, Netherlands: Leiden University, 2003.

Wagner, Judith, and Richard Wagner. *Frank Beardmore: A Potter's Tale*. Self-published CD-ROM, 2005.

Walgate, Wendy. "Reflections on Accumulation." *Ceramics Monthly* 53 (September 2005): 50–53.

Wallis, Wuanda. "Joyce Nagata." *Ceramics Monthly* 54 (February 2006): 54–56.

Wargo, Ed. "Threads: Laura Peery." *Ceramics Monthly* 54 (April 2006): 46–49.

Welch, Adam. "David Packer's Poetics of Destruction." *Ceramics Monthly* 53 (October 2005): 51–53.

West, Esme, ed. *The Wallace Collection*. London: Scala, 2005.

Wilk, Christopher, ed. *Modernism: Designing a New World*. London: V & A Publications/A. and C. Black, 2006.

Wittwer, Samuel. *Die Galerie der Meissener Tiere: Die Menagerie Augusts der Starken für das Japanische Palais in Dresden*. Munich: Hirmer, 2004.

Wolf, Tom. "Healthy Forest—Happy Potters." *Ceramics Monthly* 53 (December 2005): 50–52.

Wood, Barbara, and Robert Doares. *Old Limoges: Haviland Porcelain Design and Décor, 1845–1865*. A Schiffer Book for Collectors. Atglen, Pa.: Schiffer, 2005.

Wood, Pamela. "Nottingham Stoneware" [report of meeting, Industrial Museum, Derby, February 26, 2005]. *Northern Ceramic Society Newsletter*, no. 138 (2005): 6–7.

Woodhead, Steve. *The Teapot Book*. Philadelphia: University of Pennsylvania Press, 2005.

Woolliscroft, Pam. "Flora Ceramica: Spode's Floral Patterns and Curtis's Botanical Magazine." *Northern Ceramic Society Journal* 22 (2005–2006): 53–62.

Yamin, Rebecca. "Wealthy, Free, and Female: Prostitution in Nineteenth-Century New York." *Historical Archaeology* 39, no. 1 (2005): 4–18.

Yates-Owen, Eric, and Robert Fournier. *British Studio Potters' Marks*. 2nd ed. London: A. and C. Black, 2005.

Young, Hilary. "Daniel Pattern Book Drawings Acquired by the V&A." *Northern Ceramic Society Newsletter*, no. 138 (2005): 19–25.

Zelleke, Ghenete. "Some Reflections on Changing Taste: The Transformation of a Mirror-and-Porcelain Cabinet in Prince Eugene's Winter Palace, Vienna." *American Ceramic Circle Journal* 13 (2005): 39–48.

Index

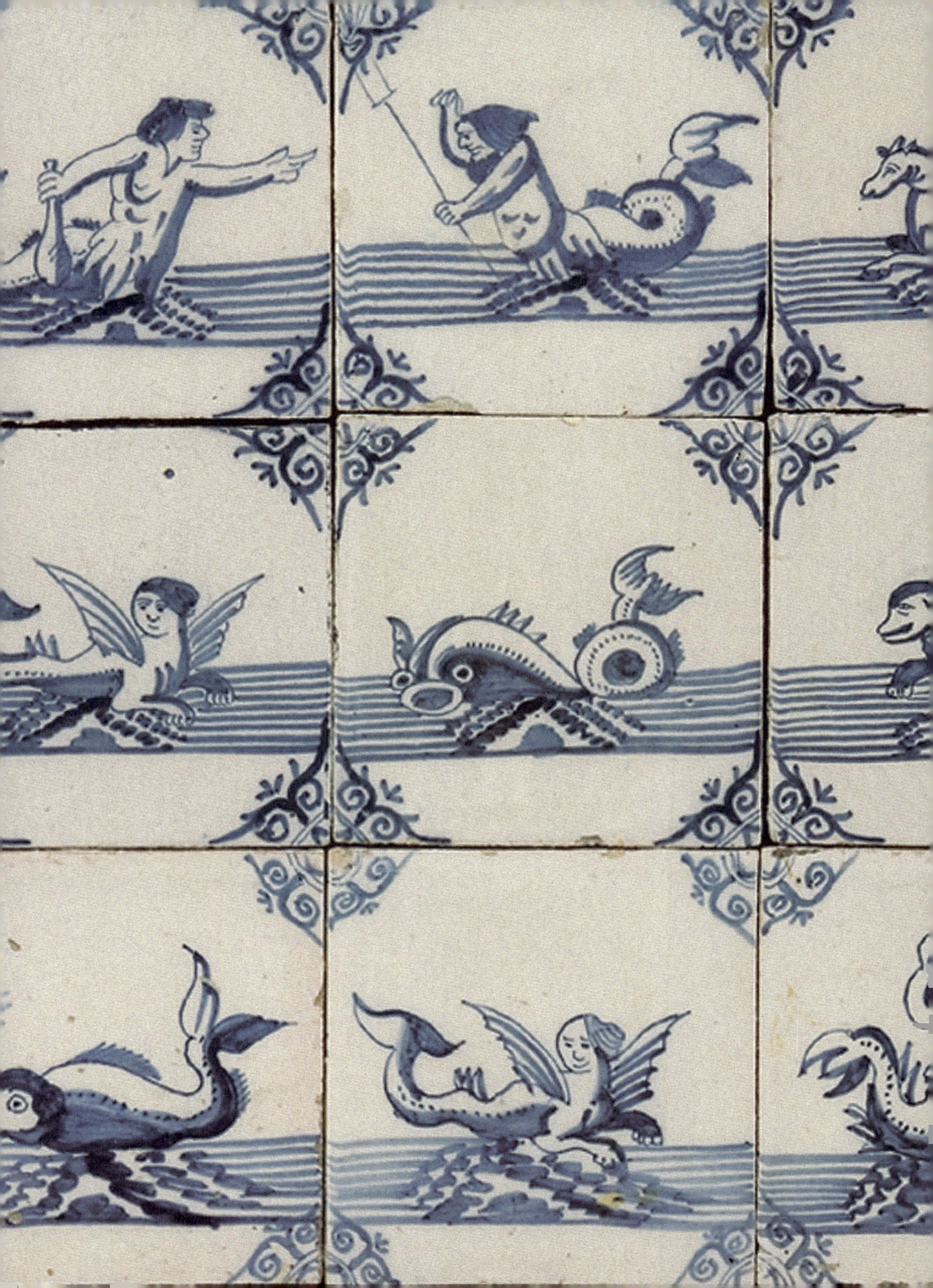